Inspirational chemistry – resources for modern curricula

Written by Vicky Wong
RSC School Teacher Fellow 2004–2005

Inspirational chemistry – resources for modern curricula

Written by Vicky Wong

Edited by Emma Kemp, Colin Osborne, Maria Pack and Kay Stephenson

Designed by Imogen Bertin

Published and distributed by Royal Society of Chemistry

Printed by Royal Society of Chemistry

For further information on other educational activities undertaken by the Royal Society of Chemistry write to:

Education Department
Royal Society of Chemistry
Burlington house
Piccadilly
London W1J 0BA

Information on other Royal Society of Chemistry activities can be found on its websites:
www.rsc.org
www.chemsoc.org/LearnNet contains resources for teachers and students from around the world.
www.chemistryteachers.org

ISBN-10: 0–85404–399–3
ISBN-13: 978–0-85404-399-6

British Library Cataloguing in Publication Data.

A catalogue for this book is available from the British Library.

Foreword

Chemistry is an exciting subject that is continually developing and changing to meet the challenges of our modern world. The Royal Society of Chemistry is pleased to be able to provide these resources to support the teaching of the new GCSE's in England , Wales and Northern Ireland starting in September 2006 and hopes they will be of use to teachers in Scotland too.

The experiments and resources range from new approaches to basic science (such as rates and rhubarb) to modern developments such as combinatorial chemistry and nanochemistry. There is a strong emphasis on sustainable development and green chemistry and it is hoped teachers will share with their students current thinking in industry and academia.

Dr Simon Campbell CBE FRSC FRS
President, Royal Society of Chemistry

RSC | Advancing the Chemical Sciences

Acknowledgements

I would like to thank:

Colin Osborne, Maria Pack, Emma Kemp, Ted Lister and John Payne from the Royal Society of Chemistry for support and encouragement throughout the project.

Tom Kempton and Paula Taylor-Moore from Didcot Girls' School for allowing me to take up this secondment opportunity.

All in the Science Department at Oxford University Department of Educational Studies, particularly Ann Childs and Jackie Coleman.

My husband, Phillip.

The following people have contributed ideas, advice, information or support and I am very grateful for their help:

Lynn Nickerson	Didcot Girls' School, Didcot, Oxfordshire
Janet Haylett	CCLRC, Rutherford Appleton Lab, Harwell, Oxfordshire
Melissa Kidd	The Soil Association
Paul Wyeth	Textile Conservation Centre, Winchester School of Art, Winchester
Sarah Howard	Textile Conservator, Hampshire County Council
Martin Carr	Department of Materials, University of Oxford
Derek Fray	Department of Materials Science and Metallurgy, University of Cambridge
John Payne	Arch Chemicals, West Yorkshire
Jeff Hardy	Department of Chemistry, University of York
Steve Robertson	CCLRC, Rutherford Appleton Lab, Harwell, Oxfordshire
Peter Borrows	CLEAPSS, School Science Service, Brunel University, Uxbridge
Shane Clark	Highcrest Community School, High Wycombe, Bucks
Rod Hebden	Hampshire Museum Service
J. P. Badyal	Department of Chemistry, University of Durham
Jason Shirley	Microscalescience.com, Cinderford, Gloucestershire
Don Sutherland	DUSC, Edinburgh

The following have helped with graphics and I am very grateful to them:

Belle and Bunty **http://www.belleandbunty.co.uk**
Boccia® info@boccia-titanium.de
CEFIC – European Chemical Industry Council **http://www.cefic.eu**
Exxon **http://www.exxonmobil.co.uk**
Guggenheim Museum, Bilbao **http://www.guggenheim-bilbao.es**
Hunterian Museum, Royal College of Surgeons **http://www.rcseng.ac.uk/museums**
Kate Jewell
Pfizer Ringaskiddy **http://www.pfizer.ie**
Shell Photographic Services, Shell International **http://www.shell.com**
Smith & Nephew **http://www.smith-nephew.com**
Techniquest, Cardiff **http://www.techniquest.org**

Vicky Wong

Contents

Introduction

This resource was produced in response to the revision of the General Certificate of Secondary Education in England, Wales and Northern Ireland for first teaching in September 2006. I hope that many of the resources will also be useful for teaching in Scotland.

The resource was produced in order to fulfil the RSC's objectives that chemistry curricula should include modern, up-to-date contexts, set where possible in everyday situations.

The writing paralleled the specification development and was accompanied therefore by consultation, to a greater or lesser degree, with awarding bodies.

Vicky Wong

How to use this resource

This resource has been designed with flexibility in mind. A whole unit could be used, but often a single lesson could be based on a selection of the material, or images or interactive activities could be used to support a lesson or part of a scheme of work.

All the material except the teacher's guide is available on the CDROM. Photocopiable, priintable and projectable materials are available in two formats:

■ Coloured pdf

■ Word document (which can be altered or edited by the teacher pror to printing.)

Each worksheet has an index number eg 2.1.1 which enables teachers to search for specific worksheets on the CDROM. These two formats allow teachers flexibility in using the resource according to their local circumstances.

CDROM instructions and system requirements

The CDROM is fully compatible with Windows NT/2000/XP and can be used on most other computer systems equipped with a CDROM drive.

In addition you will need:

■ Web browser – the content has been optimised for Internet Explorer 6 but will function correctly using most other browsers. An Internet connection is not required.

■ Java – to use the advanced search facilities Java must be installed and enabled.

■ Acrobat PDF reader – required to open PDF resource files

■ Microsoft Word – required to open Word resource files.

To use, insert the CDROM into the CDROM drive.

PC users: Your PC should run the CD-ROM automatically. If it does not, open the CDROM using My Computer and run the programe IChem.exe. You may access the resources directly from the CDROM or else install them to your PC's hard disk. Alternatively, use your web browser to open the file index.htm.

Users of other computer systems: Using your web browser, navigate to the CDROM and open the file index.htm.

The CDROM licence allows the files on the CDROM to be downloaded and to be accessible over a network. The RSC will not offer support or guidance on how best to network the files.

Note about printing the PDF version student files: if you encounter a problem where the student sheets print out slightly smaller on the page than you expected, make sure that in earlier versions of Adobe Acrobat, the option for 'Fit to page' on the print dialogue box is unchecked. In more recent versions of Adobe Acrobat this option is found by first choosing the 'Properties' button in the print dialogue box and then the 'Effects' tab. Choose 'Actual size' not 'Fit to page'.

Disclaimer

The CDROM has been thoroughly checked for errors and viruses. The RSC cannot accept liability for any damage to your computer system or data which occurs while using this CDROM or the software contained on it. If you do not agree with these conditions, you should not use the CDROM.

RSC | Advancing the Chemical Sciences

This page has been intentionally left blank.

Chapter 1
What use is chemistry?

(List of student sheets available on CDROM)

1.1   What use is chemistry?

What use is chemistry?

Index 1.1
1 sheet

This activity is based on a Sunday Times article by Sir Harry Kroto, a Nobel prize winning chemist who discovered a new allotrope of carbon – buckminsterfullerene or 'bucky balls'. The article appeared on November 28, 2004 and is reproduced overleaf as a background for teachers.

The aim is to introduce students to the scope of modern chemistry and the impact that it has on their lives, even in areas that they may not think of as related to chemistry.

An alternative exercise for more able students would be to research what was used before chemical scientists had produced a particular new product or material (*eg* silk or wool stockings before nylon, leather footballs before synthetics, grated carbolic soap before shampoo) and then to write about the difference it would make to their lives if they did not have the modern product.

Students will need:

■ Plenty of old magazines and catalogues (Argos catalogues are good as virtually everything in them would not exist without modern chemistry)

■ Large sheets of sugar paper

■ Glue and scissors.

It works well if students produce the poster in groups, but then do the written work by themselves. The activity could be set for homework.

What use is chemistry?

Some years ago I was delighted to receive an honorary degree from Exeter University recognising my contributions to chemistry – especially the discovery of a new form of carbon that has the same geometric pattern as a football and is affectionately known as the Buckyball.

With my co-workers I was awarded the Nobel prize for the discovery of this molecule, often seen as heralding the birth of nanotechnology.

Now, however, I have decided I have no alternative but to return the degree in protest at Exeter's decision, last week, to close its chemistry department, even though it has just had an intake of 107 students. I did the same thing when Hertfordshire University closed its chemistry department a few years ago. I hope that other eminent scientists who have been similarly "honoured" follow my lead.

This is yet another short-sighted slash-and-burn act of philistinism by a British university. Part of the blame, however, must be laid at the government's door for refusing to pay universities the true cost of teaching science students.

The present situation is summed up in the case of the University of Wales Swansea, whose vice-chancellor closed down chemistry in March, saying: "I don't want any chemistry undergraduates here, they're too expensive." If closures continue at the present rate there will be just six, out of about 50 where chemistry is taught, left by the end of the decade.

A university without chemistry should be stripped of the title and redesignated a liberal arts college, which is all it is.

Sir Harry Kroto

Does all this matter? Yes, because it heralds a looming disaster for Britain's economic future. If scientific research and teaching disappear from our smaller universities and become concentrated in half a dozen ponderous battleships, it will be goodbye to the sorts of laboratories where some of Britain's finest chemistry has been done: where genetic fingerprinting was invented and the experiments that paved the way to the Buckyball's discovery were carried out.

Our continued neglect of chemistry cannot fail to hamper our economic growth.

China, Japan, South Korea and India are beginning to vie with the United States as leaders in scientific research. This year about 10,000 students started psychology degrees in the UK, more than all those who began chemistry, physics and engineering degrees.

While each science student yields, on average, a 2% per year payback in tax on our educational investment, the education of psychology students results in a loss. If one adds to this the fact that the

chemistry-related industries make a £5 billion profit on a £50 billion turnover, the apparent government inaction over the looming disaster is scarcely credible.

Next spring I teach for the last time at Sussex University. I joined it in 1966 at the age of 27. It was reported earlier this year that Florida State University had offered to support my research and fund the Vega Science Trust, the science foundation I set up to make science programmes for television and the internet (www.vega.org.uk).

I did not want to leave Sussex, but in the absence of alternative support I could hardly refuse. Since the decision to leave became known a British university, as well as another American one, has offered to match Florida's offer, but it is too late.

I hope it is not too late for the vice-chancellor of Exeter to reconsider his decision and for ministers to rethink the level of support for the sciences.

In the meantime I ask all chemistry teachers to get their kids to forgo, for a week, the 20th-century contributions of chemistry to their everyday lives. How will they get by on Sunday, with no shampoo, just grated carbolic soap. Monday, no anaesthetic at the dentist. Tuesday, no food produced with inorganic fertilisers – 80% of the world would starve. Wednesday, no purified water. No adhesives on Thursday, so furniture will fall apart. Friday, no contraceptives. Saturday, no modern sports equipment.

Let's see Beckham bend a ball with the boots I had when I was a kid.

© *Harry Kroto/The Sunday Times London, 28th November 2004*

This page has been intentionally left blank.

Chapter 2
Elements, compounds, structure and reactions

(List of student sheets available on CDROM)

RSC | Advancing the Chemical Sciences

Compounds and formulae

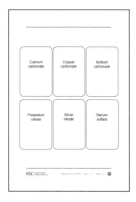

Index 2.1.1
7 sheets

Photocopy the cards, cut them up and make them into packs. You could laminate the cards to make them last longer.

Students work in pairs/small groups in one of the following ways:

■ Play 'pairs'. Lay all of the cards face down and take it in turns to turn two over. If a player selects matching name and formula cards, they keep the pair. If not, they must put the cards back. The winner is the player with the most pairs at the end.

■ Simply match the cards, starting with them all face up.

Patterns in formulae of compounds

Index 2.1.2
1 sheet

This short written activity is designed to show the power of the Periodic Table in predicting patterns. Fluorides of the elements are used to illustrate this point because fluorine reacts with most elements.

Answers to questions

1. The fluorides of the elements in Groups 1, 2, 3 and 4 have formulae in which the number of fluorine atoms matches the group number.

2. The formulae of the fluorides of Group 7 show that the number of fluorine atoms per molecule increases as you descend the group. You may want to have an explanation ready as to why such non-metals react – fluorine is the most reactive element. The increase in the number of fluorine atoms bonded to the central atom can be explained in terms of the increasing size of the central atom.

3. The fluorides of the elements in the first horizontal row have formulae in which the number of fluorine atoms per molecule is the group number or 8 minus the group number.

4. Bromine would be expected to form compounds with similar formulae to those of the fluorides.

Taboo – chemical reactions

Index 2.1.3
2 sheets

Have the cards cut up and ready to use before the lesson. You could laminate the cards to make them last longer.

Place the cards face down in a pile. Students take it in turns to take a card and try to describe the word at the top to the rest of the group. They may not use the words listed as 'Taboo', the word itself or any words derived from it. For example, to describe polymerisation they may not use the word 'polymer'.

This game can be played either as a whole class or in groups. The person who correctly identifies the word being described keeps the card and is the next to have a turn.

If the game is played in groups, the winner is the student with the most cards at the end of the game.

Heating Group 1 metals in air and in chlorine

This demonstration could follow on from work on the properties of the Group 1 metals and their reaction with water.

It is recommended that you always practise demonstrations before carrying them out in front of a class.

Equipment required

- 3 clean, dry bricks with at least 1 flat surface each
- Chlorine generator (see below) – dropping funnel, conical flask and delivery tube
- 3 gas jars with lids
- Bunsen burner
- Scalpel
- Filter paper
- Indicator paper.

Chemicals

- Lithium, sodium and potassium (**Highly flammable and corrosive**)
- Potassium manganate(VII) (**Oxidising agent, harmful**)
- Concentrated hydrochloric acid (**Corrosive**).

Chlorine

You will need to fill three gas jars with chlorine. This can be done using a chlorine generator and must be carried out in a fume cupboard, although the rest of the demonstration can be carried out on a bench if the room is well ventilated.

To make a chlorine generator, place a couple of spatulas of potassium manganate(VII) in a conical flask. Attach a delivery tube and a dropping funnel containing concentrated hydrochloric acid (see Figure 1).

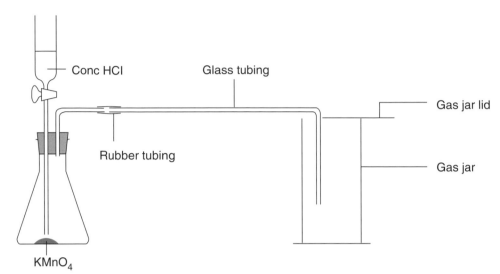

Figure 1 Chlorine generator

Check carefully that the acid is hydrochloric acid. Explosions have occurred through use of the wrong acid. Allow the acid to drip slowly onto the potassium manganate(VII) and collect the gas by downward delivery. The gas jar will appear green when enough gas has been collected. When the gas jar has been filled, seal the lid on using a little Vaseline®. Have three filled gas jars ready prior to the demonstration.

An alternative method for making chlorine is to react sodium chlorate(I) with hydrochloric acid. See CLEAPSS Hazcards 22 and 89 for more details.

Health and safety

Wear eye protection during both the demonstration and the chlorine generation.

Some chlorine will escape during the demonstration but it should be safe to carry out the experiment in a well-ventilated laboratory. The gas jars should be filled with chlorine in a fume cupboard beforehand.

Your employer's risk assessment should be consulted before carrying out this activity. It is covered by model (general) risk assessments widely adopted for use in UK schools such as those provided by CLEAPSS, SSERC, ASE and DfES. Bear in mind, however, that these may need some modification to suit local conditions.

Heating in air

Start with lithium. Cut a small cube of the metal with an edge of about 3 mm. Blot off any oil using filter paper and place the cube onto the flat surface of a brick. It is no longer recommended to remove oil by dissolving in hexane as this has caused a number of fires.

Heat the sample from above using the hottest part of a roaring Bunsen flame (just beyond the blue cone). Once the metal is on fire, remove the Bunsen flame and you should be able to observe the classic red colour of a lithium flame. (You may initially see a yellow flame, but this is the result of any remaining oil burning.)

Test the residue with damp indicator paper to show that it is an alkali.

Repeat for sodium and potassium.

The expected pattern of reactivity – *ie* increasing from lithium to potassium – may be difficult to observe as it is often hard to see the potassium burning in the absence of the Bunsen flame. This may well be because the potassium reacts faster than the other metals so an oxide coating can form almost as soon as you begin to heat it.

$$4Li(s) + O_2(g) \rightarrow 2\,Li_2O(s)$$

Sodium and potassium produce a mixture of oxides, peroxides and superoxides.

Heating in chlorine

Again, begin by cutting a small cube of metal with an edge of about 3 mm and blotting off any excess oil. Place the sample on a clean, dry brick.

Check that the mouths of the gas jars containing the chlorine are narrower than the brick to reduce the amount of escaping gas. Make sure you can see a distinct green colour in the jars – if not, there is not enough chlorine present for the demonstration to be successful.

Heat the piece of metal from above using the Bunsen burner as for the **Heating in air** demonstration. When the metal is burning, take away the Bunsen burner, invert one of the gas jars, remove the lid and immediately place the jar over the burning metal. It may be helpful to have a second pair of hands to do this. The metal will continue to burn, producing fumes of white chloride. This method avoids the production of $FeCl_3$, which can occur when the experiment is done on a combustion spoon.

Repeat for the other two metals.

Again, the trend in reactivity is harder to see than in the reaction of the metals in water.

$$2Na(s) + Cl_2(g) \rightarrow 2NaCl(s)$$

and similarly for sodium and potassium.

References

For details of how to demonstrate the reaction of Group 1 metals in water, see:

T. Lister, *Classic Chemistry Demonstrations*, London: Royal Society of Chemistry, 1995.

See also:

Safer Chemicals, Safer Reactions, Uxbridge: CLEAPSS School Science Service, 2003. This document is provided on the CLEAPSS Science Publications CDROM, which is updated annually.

The extraction of copper: a microscale version

Index 2.2.2
3 sheets

The reduction of copper with hydrogen is an interesting experiment and this microscale version is safe enough for students to perform themselves. Some exam specifications now require coursework which shows that the student has made a substance and calculated the percentage yield of the product. This reduction lends itself to that type of work. It could also be used to show the relative positions of copper and hydrogen in the reactivity series. It uses similar equipment to that needed for **Cracking hydrocarbons: a microscale version (2.5)**.

Microscale equipment suppliers

Microscale equipment can be sourced on the Internet.

Phillip Harris has a range of suitable equipment.

http://www.philipharris.co.uk (accessed November 2005).

Equipment required

Per student or pair of students:

- 1 comboplate®

- 2 cm^3 syringe

- 1 x lid number 1

- 1 x lid number 2

- 2 short lengths silicone tubing

- 1 x 10 cm piece of glass tubing of a width to match the silicone tubing

- 1 microburner filled with ethanol (**Highly flammable**)

- Copper(II) oxide (**Harmful**)

- 3 cm clean magnesium ribbon (**Highly flammable**)

- Approx 2 cm^3 2 mol dm^{-3} hydrochloric acid (**Irritant**)

- Eye protection.

If you intend to ask students to calculate their yield, they will also need access to a balance reading to 2 decimal places.

Health and safety

Eye protection should be worn at all times during this experiment.

There is a danger of the glass tube breaking when it is inserted into the silicone tubing. Warn students of this and show them how to hold the glass tube near the end that they are attaching the silicone tubing in order to minimise the risk. Technicians could perhaps do this for them.

This practical may not be appropriate for all classes as it involves putting 2 mol dm^{-3} acid into a syringe. Teachers should do a risk assessment for their class.

Answers to questions

1. As it is heated, the black copper oxide begins to turn into a shiny, orange solid. A liquid can be seen to collect near the end of the tube.

2. The products are copper and water.

3. The reactants are copper oxide and hydrogen.

4. copper oxide + hydrogen → copper + water
   $CuO(s)$ $+ H_2(g)$ $→ Cu (s)$ $+ H_2O(l)$

5. Oxygen is lost from the copper so this is a reduction reaction. (Note: The hydrogen is also oxidised – this is a redox reaction.)

6. Hydrogen is more reactive – it takes the oxygen from the copper.

7.
   a. Sodium – no.
   b. Magnesium – no.
   c. Lead – yes.

References

For a large-scale demonstration version of this experiment, see:

T. Lister, *Classic Chemistry Demonstrations*, London: Royal Society of Chemistry, 1995.

Extracting metals – words

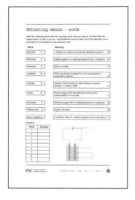

Index 2.2.3
1 sheet

Answers

Students are asked to match the words with their meaning and label the diagram.

1 = I

2 = E

3 = G

4 = B

5 = H

6 = C

7 = A

8 = D

9 = F

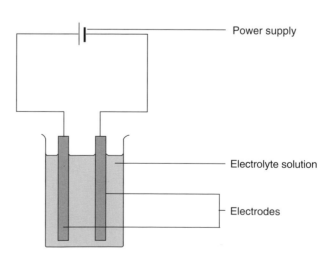

Titanium

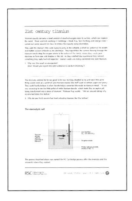

Index 2.2.4
2 sheets
Index 2.2.5
2 sheets
Index 2.2.6
3 sheets

Index 2.2.7
2 sheets
Index 2.2.8
4 sheets

This activity is aimed at students who have some knowledge of electrolysis and the extraction of metals. The reminder sheet **Extracting metals – words** could be used as an introductory activity for those who might need it.

Background information

The story of the discovery of the FFC (Fray-Farthing-Chen) Cambridge electrolysis method for extracting titanium is fascinating.

Titanium is currently extracted by the Kroll process, which was invented by William Kroll in the 1930s. This method involves carbo-chlorinating the titanium minerals rutile and ilmenite to remove oxygen, iron and other impurities and form titanium tetrachloride ($TiCl_4$) vapour. The $TiCl_4$ is reduced by treatment with magnesium metal and the magnesium chloride is removed by vacuum distillation. If sodium metal is used the process is called the Hunter Process. This step has to be carried out under a protective atmosphere of argon or another Noble gas as titanium is extremely reactive. If alloys are to be made, the titanium must be melted with the other metal(s) required. This adds to the already high cost of the process.

Kroll predicted that within 15 years his process would be replaced by an electrolytic one. So far this has not happened. Many methods have been attempted. Both molten titanium chloride and titanium oxide have been investigated as possible electrolytes and many different electrodes have been tried, but none have worked. One reason for this failure is that titanium can have a number of oxidation states and redox recycling occurs between the anode and cathode. Also, any titanium produced is in the form of fine grains which readily oxidise back to titanium oxide on contact with oxygen.

As titanium is extremely reactive, it is often contaminated with a little oxygen. This oxygen significantly weakens the metal and it was during an attempt to remove the oxygen that the electrolytic process was discovered almost by accident.

At the University of Cambridge, Prof Derek Fray, Dr George Chen and Dr Tom Farthing were trying to use electrolysis to remove the oxygen from samples of titanium. It was already known that this could be done in molten calcium chloride. A sample of titanium coated with a titanium oxide layer was used as the cathode, carbon as the anode and molten calcium chloride as the electrolyte. It had been suggested that calcium deposited on the titanium electrode reacts with oxygen in the titanium foil to form CaO, which is soluble in the $CaCl_2$, thus removing the oxygen. An alternative explanation has been put forward by the Cambridge scientists. They suggest that the oxygen reduction occurs at a more positive electrode than the calcium deposition and so direct reduction of titanium oxides to titanium metal can be achieved electrolytically, rather than via the chemical reaction with calcium.

In other words, one explanation involves the following reactions occuring at the titanium/titanium oxide cathode:

$$Ca^{2+}(l) + 2\ e^- \rightarrow Ca(l)$$

then

$$TiO_2(s) + 2\ Ca(l) \rightarrow Ti(s) + 2\ CaO(dissolved)$$

The alternative explanation involves direct electrolytic reduction of the titanium oxide:

$$TiO_2(s) + 4\ e^- \rightarrow Ti(s) + 2\ O^{2-}(dissolved)$$

Prof Fray realised that if the latter were true, then the electrolytic process might also work for a pellet of titanium oxide. When the experiment was performed, titanium was extracted from the titanium oxide, much to his excitement. The researchers realised that they had found a completely new way to extract titanium and wondered why no-one had tried it before. Other materials scientists had believed that solid titanium dioxide could not be electrolysed because it is an insulator. The Cambridge team's observations suggested that, once some oxygen is removed from titanium dioxide, it will conduct.

The process discovered by the Cambridge scientists is now called the FFC Cambridge process after the scientists themselves and the university at which they were working. It has been tried on many other metals and is showing promise as a way of extracting a variety of metals that are difficult and expensive to extract by other methods. The process has very little environmetal impact – only a very small amount of CO_2 is produced as a byproduct. It can also be used to produce alloys directly. By using a ground-up mixture of metal oxides as the cathode, the alloy of those metals is produced directly, without the need for the expensive and sometimes difficult process of melting the metals together. This may make it possible to produce alloys which have only been theoretically feasible until now.

The method is now patented and two companies have been formed to exploit the technology. The first of these companies, British Titanium plc, holds a licence solely for the extraction of titanium; Metalysis Ltd has a licence for the extraction of any other metals using the process.

References

Z. G. Chen, D. J. Fray, T. W. Farthing, *Nature*, 2000, **407**, 361–364.

S. Ashley, *Scientific American*, 2003, **289(4)**, 38–39.

S. Hill, *New Scientist*, 2001, **170(2297)**, 44–47.

K. Roberts, *Educ. Chem*, 2004, **41(3)**, Infochem 2–3.

http://www.msm.cam.ac.uk/djf/FFC_Process.htm (accessed November 2005) – webpage of Dr George Chen.

http://www.britishtitanium.co.uk (accessed November 2005) – could be understood by students.

http://www.metalysis.co.uk (accessed November 2005) – covers extraction of metals other than titanium; good section on environmental benefits and would be understood by students.

http://www.msm.cam.ac.uk/index.html (accessed November 2005) – website of the Materials Science and Metallurgy department at the University of Cambridge, where the research described above was carried out; Prof Fray's group is the 'Materials Chemistry' group.

http://www.spectore.com/process.htm (accessed November 2005) – information on the properties and uses of titanium and the traditional Kroll process.

http://aerospacescholars.jsc.nasa.gov/HAS/cirr/em/8/4.cfm (accessed November 2005) – information on the rocks found on the moon and NASA's plans to use them to produce oxygen and other resources.

Alternative activities

■ You could ask students to find out about the uses and possibly also the properties of titanium before the lesson. If necessary, they could complete the **Extracting metals – words** activity. Alternatively, begin by showing the **What's the connection?** slide. The connection is that all the items shown are made of titanium.

■ **Titanium – from discovery to Mars** – even if students do not work through the questions in this section, it is worth telling them the story.

■ Students could write a newspaper article about the discovery of the extraction method. They should think carefully about which newspaper the article is intended for and decide on a suitable headline and style of language. They could use the websites listed above to find further up-to-date information. The science should be accurate and it should be explained in a way that most readers would understand. Alternatively, students could produce an advertising leaflet for a company selling the new, cheaper titanium and explain why it is cheaper now, what its properties are and what it could be used for.

Answers

Titanium

1. Iron is extracted from its ore by heating the ore with carbon.

2. Titanium cannot be extracted in this way because it is more reactive than carbon so would not be displaced by it. In addition, titanium carbide (TiC) might form.

3. Aluminium is extracted by electrolysis.

4. Aluminium is more expensive than iron because electrolysis of its ore uses a lot of electricity, which is more expensive than the carbon used in the blast furnace to extract iron.

5. Titanium could be extracted using magnesium/calcium/sodium/potassium/any more reactive metal.

Titanium extraction

1. $TiO_2 + 2Cl_2 \rightarrow TiCl_4 + O_2$

2. Argon is used because it is a Noble gas and is extremely unreactive. It does not react with titanium so its presence does not affect the quality of the product.

3. Any other Noble gas could be used instead: helium, neon, krypton, xenon.

4. Titanium is expensive because the extraction process is slow, the chlorine and magnesium (or sodium) required are expensive, it costs a lot to heat the reactor and the process is labour intensive. Chlorine is also dangerous and difficult to handle.

5.

Atom	Number	Atomic mass	Number x atomic mass
Mg	2	24	48
Cl	4	35.5	142
Ti	1	48	48
Total			238

Table 1 Calculating the atom economy of titanium extraction

Total mass of reactants = 238 g

Mass of desired product = 48 g

$$\text{Atom economy} = \frac{\text{mass of desired product}}{\text{total mass of reactants}} \times 100\% = \frac{48}{238} = 20\%$$

The atom economy of the Kroll process is 20%.

6. The real atom economy will be even lower than the calculated value as the calculation does not take into account the atoms lost in the first stage of the process.

21st century titanium

1. The result was unexpected because titanium oxide is not a metal and would not be expected to conduct in the solid state. (It is a covalent compound – although as it is a compound of a metal and a non-metal, students may think that it would be ionic.)

2. No-one had tried extracting titanium like this before as they did not expect a covalent/non-metallic solid to be able to conduct electricity and act as an electrode.

3. $O + 2 e^- \rightarrow O^{2-}$

4. $2 O^{2-} \rightarrow O_2 + 4 e^-$

5. If a carbon anode is used, some of the oxygen produced reacts with the electrode to produce CO and CO_2.

6. This could be prevented by using an inert (unreactive) substance as the anode.

7. Mass of desired product: 48

   Total mass of reactants: 80

   $$\text{Atom economy} = \frac{48 \times 100}{80} = 60\%$$

8. If the oxygen was collected and sold the atom economy would be 100%.

9. The atom economy of the FFC Cambridge process is far better than that of the Kroll process.

10. The FFC Cambridge process is likely to be cheaper because it requires less expensive starting materials, is faster and produces less waste.

11. This is an example of Green Chemistry in the following ways:

■ Prevention is better than cure – it is better to design a process that produces no waste than to clear it up. The FFC process produces far less waste than the Kroll process.

■ Green processes should minimise waste products and put the maximum possible amount of the raw materials into the final product – the FFC Cambridge process has a much higher atom economy than the Kroll process.

■ Lower toxicity of the products – some of the byproducts of the Kroll process are extremely unpleasant whereas the FFC Cambridge process produces oxygen and a little carbon dioxide as byproducts.

Note: students will require a reference sheet of the **Twelve Principles of Green Chemistry** (Index 6.4.4) to answer this question.

Extension questions

12. Zinc chloride is not used because zinc is lower in the reactivity series than titanium. The zinc ions would be reduced and the titanium left unchanged. (Note: the reason is actually that zinc is lower in the reactivity series than oxygen, although the above is also true and is an appropriate student answer.)

13. This process is similar to the extraction of aluminium in a number of ways: both use electrolysis; oxygen is produced at the anode; if carbon is used at the anode then carbon dioxide is a byproduct; the processes are carried out at similar temperatures. They are different in that: aluminium oxide (dissolved in cryolite) is the electrolyte in aluminium extraction, whereas titanium oxide is the cathode in the titanium process and calcium chloride is used as the electrolyte.

Titanium – from discovery to Mars

1. The defence agency might use titanium for producing lighter aircraft, tanks and other vehicles. A titanium ship would be lighter, sit higher in the water and might move faster than a steel one. Students may be able to think of a range of other benefits.

2. Titanium might be useful for:

   a. Aircraft – titanium is already used in some aircraft as it is very light and strong. A whole plane made of titanium (including the engine) would use less fuel than existing aircraft because it would be lighter. However, not all parts of the engine could be made of titanium as the combustion of the fuel raises the temperature of the gases above the upper limit for the use of titanium.

   b. The motor industry – as well as being light and strong, titanium is very corrosion resistant. If the engine had titanium parts it would be much lighter and use less fuel. The steel used in car bodies is so cheap that it will be a long time before it is replaced by titanium.

   c. Engineering/building – titanium looks good and has already been used on the surface of the Guggenheim Museum building in Bilbao, Spain. This could become

RSC | Advancing the
      | Chemical Sciences

a more common practice if the price falls. In addition, titanium is very strong and corrosion resistant so engineers could use it in the design of bigger and longer-lasting bridges and skyscrapers (although it is not as stiff as steel).

3. Fuel consumption would be cut as the metal is lighter than the commonly used material (steel).

4. One answer could be to allow humans on a mission to the moon to breathe. Oxygen is possibly even more crucial as an oxidiser to burn fuel. The largest component (up to 85% by weight) of any rocket is the oxidiser and locally produced oxygen for rocket propulsion could give the greatest cost and mass saving of any non-terrestrial resource. It is therefore important for the achievement of a sustained programme to explore Mars.

Answer to **To think about:** A large proportion of the surface of Mars is made of ilmenite – a mineral that largely consists of titanium oxide. NASA wants to extract the oxygen from this ore as there is no free oxygen on the moon. For further details, see the website **http://aerospacescholars.jsc.nasa.gov/HAS/cirr/em/8/4.cfm** (accessed November 2005.)

Alloys: making an alloy

Index 2.3.1
2 sheets

In this experiment, students make an alloy (solder) from tin and lead and compare its properties to those of pure lead.

Equipment required

Per pair or group of students:

- About 2 g lead

- About 2 g tin

- Crucible

- Pipe clay triangle

- Bunsen, tripod and heatproof mat

- Spatula

- Carbon powder – 1 spatula per student

- Tongs

- 2 sand trays or sturdy metal lids

- Sand

- Access to a balance

- Eye protection.

Health and safety

The most likely incident in this experiment is a student burning themselves so warn them that the equipment will be hot.

Pouring molten metal can be hazardous if you are not sure how to use tongs correctly – it would be worth demonstrating how to use them safely. Some tongs in schools do not grip well. Every pair must be checked before the start of the experiment.

Eye protection should be worn.

Lead is a toxic metal. If it is heated for too long or too high above its melting point it could start to give off fumes. Ensure that the laboratory is well ventilated, warn students against breathing in the fumes given off by their sample during the experiment and tell them to heat the metals no longer than is necessary to get them to melt.

RSC | Advancing the
      | Chemical Sciences

Results

Hardness testing should show clearly that the alloy is harder than the pure lead. The alloy can be used to scratch the lead convincingly. The lead does not leave a mark on the alloy.

(Students may need to be reminded how to do this simple test – just try to scratch one metal with the other.)

The density of the alloy should be less than that of the lead, but this test is fairly subjective.

The lead melts first, followed by the tin, whilst the alloy has the highest melting point. This demonstrates that the alloy has very different properties from its constituent metals.

Extension

This experiment can stand alone as a demonstration of how the properties of a metal can be changed by alloying. Alternatively, you could follow it up with more able students by asking them to explain the results of the hardness testing in terms of the structure of the metals.

A good answer would include a reference to the layer structure of metals and describe how alloying can prevent the layers from sliding over each other, making it more difficult to change the shape of the metal. This makes the alloy harder than the pure metal as it is more difficult to change the shape of the alloy (which is necessary to scratch it).

Alloys: modelling an alloy

Index 2.3.2
4 sheets

This experiment enables students to experience how alloying can be used to change the properties of a metal. Plasticine is mixed with varying amounts of sand and the ductility of the sample is measured in a simple test. The plasticine is used to represent the main metal in an alloy, *eg* iron, and the sand represents an added substance, *eg* carbon in steel.

The practical is suitable for students of all abilities and can remain at the level of observation for the less able or act as a springboard to the explanation of the properties of alloys for the more able.

Equipment required

Per pair or group:
- 4 x 35 g samples of plasticine – one with no sand added, one with 2 g, one with 4 g and one with 6 g sand

- Magnifying glass.

Extension activity only:
- 35 g sample of plasticine with either 3 g or 5 g sand added (you need to know how much).

Samples

The samples can be prepared by a technician or by students so practical details are given on a separate sheet. Samples can be used several times so it is worth colour coding them for ease of identification (*eg* all samples containing 2 g sand are made using blue plasticine). For the experiment to give good results it is very important that the sand is mixed thoroughly and evenly with the plasticine. If samples are to be stored, wrap them in cling film or place them in plastic bags to prevent the plasticine from drying out.

Sand from a builders' merchants is the best as the particle sizes tend to be fairly uniform. Sand from a fire bucket usually has a wide range of particle sizes and is often dirty so is not recommended for use in this experiment.

If students make their own samples, remove the one containing 6 g sand from them until later in the experiment.

If students carry out the extension activity, they will require a further 35 g plasticine lump with either 3 g or 5 g sand added for them in advance.

Notes

If the same plasticine samples are used repeatedly they will always snap in the same place. To solve this problem simply re-mould the plasticine for a couple of minutes until all the sand is evenly distributed again and the sample is warm.

The plasticine and sand could be further investigated and other factors such as temperature tested for their effect on the properties of the samples.

Students often get the terms brittle, malleable or ductile and strong muddled up. It is worth ensuring at some stage during the lesson that they are happy with the use of these terms. A material that can be stretched or drawn into wires is ductile (malleable means that it can be moulded into shape when cold). If it does not stretch but snaps, then it is brittle. A material can be strong but brittle – and indeed many are. The opposite of ductile is brittle – not weak.

The sheet **Alloys of iron – steels** can be used after the **Alloys: modelling an alloy** practical activity. It may help students to understand the structure of metals and how they can be changed by alloying if you demonstrate this to them using polystyrene balls in a tray or a similar model.

Answers

Alloys: modelling an alloy

1. Ductile means that the material can be drawn into wires.

2. The iron is represented by the plasticine and the carbon by the sand.

3. The fracture surfaces are different sizes and there is a pattern. As more sand is added the fracture surface gets larger.

4. Before the plasticine broke it was stretched into a thinner shape. (This is called 'necking').

5. Before the plasticine broke, the force required to pull it apart at a steady rate reduced.

6. The fracture surfaces are rough with several small peaks. They contain more sand than the surface where the plasticine has simply been snapped.

7. As more sand is added to the plasticine it becomes less ductile and more brittle.

8. and **9.** The lump with 6 g sand in it follows the trend of the other samples – it is less ductile, thins out less and has a larger fracture surface than the others. The difference between the sample with 4 g sand and that with 6 g is less marked than that between 0 g and 2 g or between 2 g and 4 g sand. (The results of this part of the experiment may vary depending on the type of sand used – there may be very little difference between the 4 g and 6 g samples.)

Modelling an alloy – making the model mixtures (extension activity)

You might expect students to decide to form a similar sized cylinder of plasticine and to perform the same test as for the other samples. By comparing the size of the fracture surface with their results from the previous experiment they should be able to deduce a possible range of values for the amount of sand in the sample.

Alloys of iron – steels

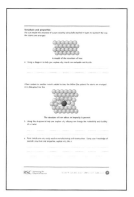

Answers

1. The properties listed could include any of the following:

■ Conduct electricity

■ Conduct heat

■ Ductile

■ Malleable

■ Shiny

■ Sonorous

■ Strong

■ Hard

■ Dense

2. Mild steels might be expected to be more ductile (and malleable) than high carbon steels. (Properties like conduction and density would probably be a little different.)

3.

   a. Scalpel and other surgical instruments – high carbon steel as it is less malleable and stronger than mild steel.
   b. Paper clip – mild steel as it is easy to bend.
   c. Hammer – high carbon steel, otherwise it might change shape when used to hit something.

4. The layers of the metal can slide over each other, allowing it to change shape easily:

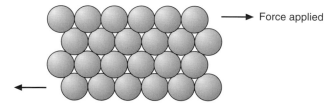

Figure 1 The arrangement of atoms in a metal

The layers can slip over each other to form the pattern shown below:

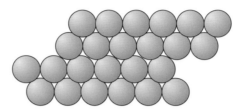

Figure 2 Atoms in a metal can slide over one another

5. Alloying can change the malleability and ductility of a metal by preventing the layers from sliding over each other. (This is called 'pinning' the position of the atoms.)

6. Pure metals are not often used in manufacturing and construction as they are soft and readily deformed because the layers of metal atoms can easily slide over each other. In contrast, alloys are harder and more difficult to deform because other atoms are present in the structure and these prevent the movement of the layers.

Acknowledgements and references

The practical work in this resource is based on an idea of Dr Martin Carr from the Department of Materials at Oxford University.

The following websites contain further information on iron, steel and alloying and should be easily understood by most students:

http://www.schoolscience.co.uk/content/4/chemistry/steel/index.html (accessed November 2005)

http://learningzone.coruseducation.com/schoolscience/KS5specialiststeels/ index.html (accessed November 2005).

Electrolysis of molten zinc chloride

Index 2.4.1
1 sheet

This demonstration shows that an ionic salt conducts electricity when molten but not when solid. Lead(II) bromide used to be used for this demonstration but this is no longer recommended because of the toxicity of both the salt and the decomposition products. Also, lead bromide decomposes into its elements to some extent upon heating, without the need for electricity.

Timing

This demonstration can be done in a 50 minute lesson. Some quite long periods of waiting are required during the experiment and it would be a good idea to have other activities planned for students during these times.

Equipment

- Bunsen burner
- Tripod
- Heatproof mat
- Pipe clay triangle
- Crucible
- 2 graphite electrodes supported in an electrode holder or bung
- Leads
- Low voltage (0–12 V) power pack
- Ammeter and/or bulb
- Clamp and stand
- Zinc chloride
- Metal spatula
- Indicator paper and/or starch iodide paper

- Tongs

- Distilled water

- Plastic beaker

- Filter paper and funnel

- Circuit tester (optional)

- Fume cupboard if possible, or a very well-ventilated laboratory.

- Eye protection.

Procedure

Set up a heatproof mat, tripod, Bunsen burner and pipe clay triangle. Put the crucible on the pipeclay triangle, ensuring that it is stable and in no danger of falling through.

Set up the electric circuit with the power pack, ammeter and/or bulb and electrodes in series. Complete the circuit at the electrodes with a key or the metal spatula to satisfy yourself and the students that the circuit works. If it does not work and the electrodes are mounted in a bung, check that they are not broken as this is often the cause of failure of the circuit.

Clamp the electrodes just above the crucible so that they almost touch the bottom but do not touch each other. Fill the crucible to within about 5 mm of the top with powdered zinc chloride. When the solid melts it will decrease in volume as air escapes. It is important that the level of the liquid does not drop below the bottom of the electrodes. Make sure the leads are well out of the way of the Bunsen burner flame. Using long electrodes can help with this.

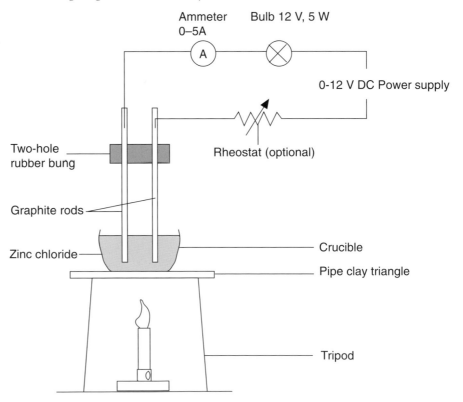

Figure 1 Apparatus for the demonstration

Show that the solid zinc chloride does not conduct electricity.

Begin to heat the crucible with a low to medium Bunsen burner flame. Watch the leads, and the bung if you are using one, to ensure you are not over-heating them. The zinc chloride should take about three or four minutes to melt. It may be tempting to use a roaring Bunsen burner flame to speed the melting up but if you do so a crust may form over the top of the zinc chloride which will prevent students from seeing what is going on. Care should also be taken that the liquid does not boil (see below).

As the salt melts, the bulb will light up and/or the ammeter will give a reading. Turn the Bunsen burner down a bit at this point. The electric current will have a heating effect and this on its own may be enough to keep the zinc chloride molten (as in the industrial electrolysis of aluminium oxide). The boiling point of zinc chloride is about 750 °C, which can easily be reached through a combination of heat from the Bunsen burner and the heating effect of the electric current. If the zinc chloride does begin to boil then it could boil over the sides of the crucible. Boiling also produces fumes of zinc chloride in the air, which rapidly solidify to form a fine powder. The bubbles of zinc chloride fumes could be confused with those of the chlorine gas being formed.

Bubbles of gas form at the positive electrode and this product smells of bleach and swimming pools. To confirm that the gas is chlorine, hold a piece of moist indicator paper close to the bubbles – the paper turns red and the edges may start to bleach. A more convincing test is to use moist starch iodide paper, which turns black in the presence of chlorine.

It is also possible to see crystals of zinc forming on the negative electrode. These can form a bridge across the electrodes, effectively shorting the circuit.

Electrolyse the molten salt for about 15 minutes with the current adjusted to about 0.5 A. Check the current every few minutes to ensure it remains more or less constant as there is a tendency for it to increase slowly. During this period there is little for students to do so they could carry out another activity, such as writing up the experiment. If you have access to a webcam or video camera and a data projector these could be set up to allow students to see what is going on inside the crucible as the experiment progresses. If not, allow the students to view the experiment in groups of two or three. They should observe which electrode the bubbles are forming at, smell the bleachy gas being produced with great care and look for crystals of zinc around the negative electrode.

You could set the apparatus up on a wooden board to show students and then move it into the fume cupboard for the electrolysis. If you do this, check beforehand that the stands etc fit in the fume cupboard and move the apparatus before you start the electrolysis.

It is not recommended that you try to remove the Bunsen burner and cool the salt while still electrolysing it to show that the salt only conducts when molten. The heating effect of the electric current will keep the salt molten for several minutes and when it does cool, a crust will form that is very difficult to melt again. Instead, ensure that you pointed out at the start of the experiment that the solid does not conduct.

After 15 minutes, turn off the power pack and Bunsen burner and remove the electrodes from the crucible. If this is not done while the salt is still molten, the electrodes will stick. Leave the crucible to cool for about 10 minutes. You may be able to see zinc crystals on the electrode and on the surface of the mixture in the crucible. You could stop at this point, but to convince students that a metal really has been made you could separate the zinc from the remaining zinc chloride as described below.

When the crucible is cool to the touch, put it into a beaker of distilled water. If the water is at all basic (like most hard tap water), the zinc ions will flocculate, forming large particles that are very hard to remove from the zinc metal so do use distilled water if you possibly can. The zinc chloride will dissolve in the water (this may take some time) and can be decanted off. Swirl the beaker so that the zinc metal concentrates in the centre then decant off most of the liquid. Filter the remainder and show students the shiny pieces of metal left on the paper.

Dry the metal carefully between further sheets of filter paper and test it with a circuit tester to prove that you have a metallic product. Given that the starting material was zinc chloride and you have made chlorine during the electrolysis, most students will have little difficulty in accepting that the metal is zinc.

Health and safety

The electrolysis can be carried out in a well-ventilated open laboratory as the amount of chlorine generated in 15 minutes is small enough (less than 60 cm^3) to be acceptable in an averaged sized laboratory. However, chlorine can cause asthma attacks and you may prefer to do the experiment in a fume cupboard to reduce the smell.

Your employer's risk assessment should be consulted before carrying out this activity. This activity is covered by model (general) risk assessments widely adopted for use in UK schools, such as those provided by CLEAPSS, SSERC, ASE and DfES. Bear in mind, however, that these assessments may need some modification to suit local conditions.

It is recommended that you always rehearse demonstrations before carrying them out in front of a class.

Answers

1. Zn^{2+} and Cl^-.

2. The salt does not conduct when it is solid because the charged particles (ions) from which it is made are held in place in the ionic lattice and cannot move. Therefore, these particles cannot carry the charge.

3. Once the salt is molten, the ions are no longer held in place and are free to move. As they are charged, they can carry the current.

4. Moist indicator paper can be used to test for chlorine gas. The paper turns red and is then bleached if chlorine is present. Alternatively, moist starch iodide paper turns black in the presence of chlorine.

5. Chlorine is made at the positive electrode. Chloride ions are negative and so are attracted to the positive electrode, where they give up an electron to become neutral.

6. $2\ Cl^- \rightarrow Cl_2 + 2\ e^-$

7. Zinc metal is made at the negative electrode. Zinc ions are positive and so are attracted to the negative electrode, where they receive electrons to become neutral.

8. $Zn^{2+} + 2\ e^- \rightarrow Zn$

References

A description of the electrolysis of lead(II) bromide can be found in:

T. Lister, *Classic Chemistry Demonstrations*, London: Royal Society of Chemistry, 1995.

A microscale version of the experiment is described in:

Safer Chemicals, Safer Reactions, Uxbridge: CLEAPSS School Science Service, 2003. This document is provided on the CLEAPSS Science Publications CDROM, updated annually.

A colourful electrolysis

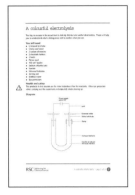  

Index 2.4.2
3 sheets

This experiment is an interesting introduction to the electrolysis of brine. It is probably not best used as the first example of electrolysis that students encounter – without any prior experience they are likely to struggle to explain for themselves what is going on.

The experiment could be followed by work on the electrolysis of salt solution in industry – perhaps using the video and question sheets provided on the RSC's *Alchemy?* CDROM.

Prior knowledge required

■ The process of electrolysis and relevant terminology, *eg* electrode

■ Structure and bonding in ionic compounds

■ Writing balanced half equations

■ Acids, alkalis and the colours of Universal Indicator

■ How to test for chlorine gas

■ For Q15 only: understanding of the term 'atom economy'.

Equipment required

Per pair or group of students:

■ U-shaped test-tube

■ Clamp and stand

■ 2 graphite electrodes

■ 2 electrode holders or other suitable means of securing the electrodes (not bungs)

■ 2 leads

■ Power pack

■ 100 cm^3 beaker

■ Sodium chloride (salt) – two spatulas per group

■ Spatula

■ Universal Indicator – approx 0.5 cm^3 per group

■ Stirring rod

■ Distilled water – if this is a problem then tap water could be used, but it may affect the colours produced, especially in areas with hard water

■ Eye protection.

Health and safety

The products of this experiment (hydrogen, chlorine and sodium hydroxide solution) are all more hazardous than the reactants. Ensure the current is turned off as soon as a trace of chlorine is detected. Chlorine can be a problem for very asthmatic pupils. If the directions in the students' notes are followed then very little chlorine is produced.

Sodium hydroxide is corrosive.

Ensure students wear eye protection, especially when they are clearing up after the experiment.

Answers to questions

1. When the power supply is switched on gas bubbles are formed at both electrodes. At the negative electrode the indicator turns purple. At the positive electrode the indicator initially turns red and is then bleached colourless. A cautious sniff of the product at the positive electrode reveals that it has a bleachy smell. The indicator in the U-tube between the electrodes remains green.

2. Sodium chloride has a giant ionic structure.

3. When sodium chloride dissolves, the ions separate and spread or diffuse through the water.

4. H^+, OH^-, Na^+, Cl^-

5. Chlorine is made – it turns the indicator red and then bleaches it. It also smells of bleach.

6. $2\ Cl^-(aq) \rightarrow Cl_2(g) + 2\ e^-$

7. H^+, OH^-, ✘, Na^+

8. H^+, Na^+

9. Hydrogen gas.

10. $2\ H^+(aq) + 2\ e^- \rightarrow H_2\ (g)$

11. ✘, OH^-, Na^+, ✘

12. Na^+, OH^-. These ions form sodium hydroxide.

13. The indicator goes purple because sodium hydroxide is produced and it is an alkali.

14. The products are hydrogen, chlorine and sodium hydroxide.

15. 100% – all the reactants are converted into products and there are no byproducts.

16. Uses of the products:

Sodium hydroxide – cleaning products, making other chemicals, making fibres such as nylon, making paper.

Chlorine – making bleach, making other chemicals (such as PVC), making solvents, water purification.

Hydrogen – making margarine, making other chemicals.

Cracking hydrocarbons: a microscale version

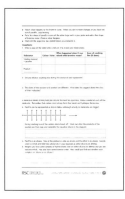

Index 2.5
4 sheets

Paraffin can be cracked to form ethene and a shorter alkane as a class practical. However, the safety concerns of many teachers and equipment shortages in some schools have meant that this key and interesting experiment is often performed as a demonstration. While there is much value in demonstrations as part of students' experience of chemistry, many prefer to do as much as possible hands-on. Microscale experiments offer a potential solution to the problem.

This experiment uses a far smaller quantity of chemicals than the traditional set-up and avoids the problem of suck-back of cold water into a hot tube. You may not wish to buy the microscale equipment for just this one experiment, but if you already have the comboplates® in your school or are considering purchasing some then this is a great way to use them. **The extraction of copper: a microscale version** in this resource also uses comboplates®.

The RSC sent the book Microscale Chemistry by John Skinner to all schools in 1996, along with a small trial kit. This book is available online at **http://www.chemsoc.org/networks/learnnet/microscale.htm** .

Acknowledgements

This experiment was developed by Jason Shirley of Microscalescience.com and is presented here with his permission.

Microscale equipment suppliers

Microscale equipment can be sourced on the Internet.

Phillip Harris has a range of suitable equipment.

http://www.philipharris.co.uk (accessed November 2005).

Equipment required

Per pair or small group of students:

■ 2 comboplates®

■ 1 x 10 cm^3 syringe

RSC | Advancing the
      | Chemical Sciences

- 1 x 10–15 cm straight glass tube, approx 2–3 mm bore – must be narrow enough to allow the silicone tubing to expand over it

- 2 x L-shaped pieces of glass tubing – 5–6 cm in height, 2–3 cm long L piece and 2–3 mm bore

- 4 x 1–2 cm length silicone tubing – 3 mm bore, 1 mm wall

- 1 plastic pipette

- 1 microburner filled with ethanol

- 2 x lid number 2 for the comboplate® (with one long and one short port)

- Mineral wool

- Aluminium oxide – microspatula per group

- Liquid paraffin – about 0.5 cm^3 per group

- Bromine water (less than 1%) – about 3 cm^3 per group (**Harmful and Irritant**)

- Eye protection.

A molecular model kit such as a Molymod® or Orbit kit would be very helpful for students when answering the questions that follow the experiment.

Health and safety

- Eye protection should be worn at all times during this experiment, including during set-up and dismantling.

- Ethanol is highly flammable. The burner must be held upright to prevent spillage of any fuel and should be presented to students already filled.

- Care should be taken in handling glass tubes to ensure that they do not break and cut hands. The glass tube should be held near the end to which the silicone tubing is being attached to minimise the risk of breakage.

- Bromine water is toxic and corrosive at or above a concentration of 1%. A dilute solution shows the results well so a more concentrated one is not necessary. The dilute solution is harmful and an irritant.

- Allow all equipment to cool before dismantling.

Diagram

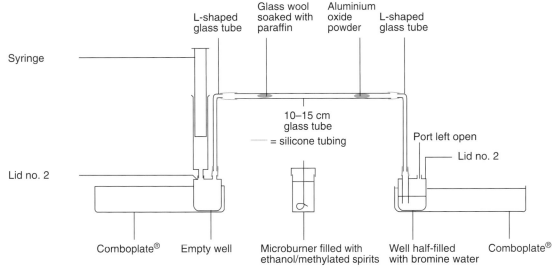

L-shaped glass tube
Glass wool soaked with paraffin
Aluminium oxide powder
L-shaped glass tube
Syringe
10–15 cm glass tube
= silicone tubing
Port left open
Lid no. 2
Lid no. 2
Comboplate®
Empty well
Microburner filled with ethanol/methylated spirits
Well half-filled with bromine water
Comboplate®

Note: attach the lids to wells F3 or F4 on the comboplates.

Figure 1 Apparatus for the experiment

Notes

- The equipment could be set up for the students if they are likely to find putting all the pieces together very difficult.

- As the hole and therefore the drop size of a pipette can vary, it is a good idea to trial this practical before the students do it to ensure that sufficient ethene is generated to discolour the bromine water. If the bromine water is dilute (0.1%) then this is less of a problem.

- The catalyst should be heated strongly initially. Once it is hot, the paraffin can be heated: the flame should remain under the catalyst and the occasional flick of the burner used to heat the paraffin – it is possible to see when the paraffin is boiling.

- Continual gentle depression of the syringe will ensure that no suck-back takes place and the gases flow through the system into the bromine water. When the syringe is empty, simply remove it, pull the plunger back, re-attach and continue depressing it.

- The disadvantage of this experiment is that it is not possible to collect enough product to fill a container and show that it is a gas at room temperature. However, at the end of the reaction any remaining ethene can be lit as it comes out of the open port hole on the lid over the bromine. This should be done with care, but an impressive 10–15 cm flame can be achieved. You could compare the behaviour of paraffin by trying to set fire to it. Put a few drops in a watch glass and attempt to light it using a splint NOT the microburner. It does not light easily. An extra column can be added to the table in the students' notes to allow for observations relating to this additional aspect of the experiment.

Timing

This experiment is far quicker than the traditional version. How long it takes a class will depend on how familiar they are with the microscale equipment, but students should be able to complete it easily within 30 minutes. This leaves time for the model-making and theory work in the same lesson.

Answers

1.

Substance	Colour	State	What happened when it was mixed with bromine water?	Ease of catching fire (if done)
Starting material – paraffin	Colourless	Liquid	No reaction	Does not catch fire easily, although it will make a lighted splint burn with a larger flame
Product – ethene/alkene	Colourless	Gas	Reacts and turns the bromine water from orange/ yellow to colourless	Burns easily and quickly

Table 1 Expected experiment results

2. Students may observe a second, liquid product towards the end of the tube. This is often mistaken for unreacted paraffin but is usually the smaller alkane product. It is sometimes yellowish in colour because it contains impurities.

3. The starting material is a liquid and the product a gas, which suggests that the reactant has larger molecules than the product.

4.

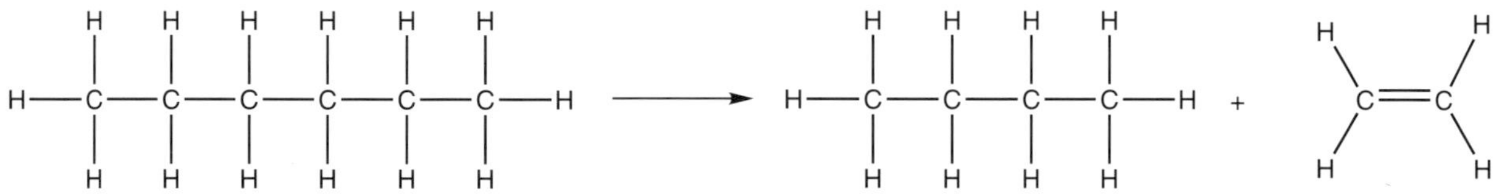

Figure 2 Equation for the cracking reaction

5. Alkane   →   Alkane   +   Alkene

6. Add a few drops of bromine water to a little of each sample. If there is no reaction it is an alkane, if the bromine water turns colourless it is an alkene.

7.

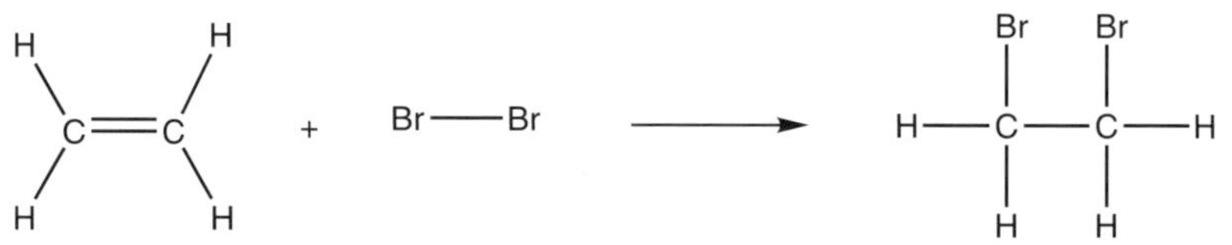

Figure 3 Reaction of bromine with an alkene

8. Alkanes do not react with bromine because they do not have a double bond. The double bond of an alkene can open up and allow other atoms to bond to the carbon atoms. As the carbon atoms in an alkane are each already bonded to four atoms, there is no room for any other atoms to attach.

Nail varnish removal

Index 2.6.1
2 sheets
Index 2.6.2
2 sheets

Aims

■ To revise the terms solvent, solute and solution.

■ To appreciate that not everything will dissolve in water and that different solvents work well for different solutes.

For the more able:

■ To introduce the idea that water dissolves things which have a charge or are polar but not things which are uncharged or non-polar.

■ To provide an opportunity to discuss the concepts of the reliability and limits of evidence.

Notes on the activity

This activity is straightforward. It will take no more than 15 minutes (including time for students to fill in the table) once the nail varnish is dry.

The nail varnish will take a while to dry so ask students to paint it on a white tile early on in the lesson. If behaviour is a problem, you might wish to consider using a fast-drying nail varnish. Students can copy the table ready for completion and answer the introductory questions while the nail varnish dries.

Hydrochloric acid is included in the table because students often think acids will dissolve anything.

The practical activity itself does not take very long – its purpose is to provide an opportunity for discussing why things dissolve and the limits of the evidence available.

The extension sheet entitled **Dissolving** asks students to explain why some substances dissolve and others do not. This exercise is aimed at more able students.

Equipment required

■ Nail varnish – a dark colour will work best (try not to use old, dried up nail varnish and have several bottles available to reduce waiting times)

■ White tile – if these are in short supply, glass beakers also work well but do not use anything made of plastic

■ Ethanol (**Highly flammable**) – ensure all solvent bottles have the correct hazard labels on them

- Propanone (acetone) (**Highly flammable and irritant**)

- Ethyl ethanoate (**Highly flammable and irritant**)

- Hydrochloric acid 1 mol dm^{-3} (or less concentrated) (**Irritant**)

- Cotton wool pads – you can cut them into quarters to make them last

- Access to water

- Tongs

- Eye protection.

You may like to provide students with bottles of commerical nail varnish remover for comparison, especially if you can find one with the solvents it contains listed on the packaging.

Answers – nail varnish removal

1. The solvents are the ethanol, propanone etc.

   The solute is the nail varnish.

   The solution is the nail varnish dissolved in the solvent (on the cotton wool).

2. This is not really a fair test – the amount of solvent, the thickness of the nail varnish, how hard you rub in your attempt to remove it and which nail varnish you use could all affect the result. The best remover could be ethyl ethanoate or propanone – it is hard to tell. (This last part of the answer may vary depending on which nail varnish you are using but these are usually the best two solvents.)

3. To make the test fair and therefore the evidence more reliable, you would need to find a way of making the nail varnish the same thickness for each test and use a controlled amount of solvent. Students may think of various ways of solving the problem.

4. No, you cannot tell which solvent would be the best nail varnish remover. The experiment does not take into account how the solvents might affect the nails/skin of the person using the nail varnish remover.

5. The further information needed could include the effect of the solvents on skin/nails, likelihood of allergic reaction etc.

6. Testing on humans/animals – you could use this question to start an ethical debate if you wish.

7. This nail varnish avoids solvents that may harm young children's skin. An alternative answer is that children should not be using anything containing flammable solvents without close supervision.

8. Nail varnish does not dissolve in water because its particles are not attracted to the water particles. Water is polar (it has an uneven distribution of charge in its molecules) and dissolves other charged or polar substances. Nail varnish dissolves in uncharged solvents better as it is itself uncharged. Students are unlikely to come up with this on their own, but it is good to get them thinking and talking about some of their ideas before giving them the explantation or the **Dissolving** extension sheet. Answers to the questions on the extension sheet are provided below. A further explanation of dissolving is provided in the RSC Particles in Motion CD ROM.

Answers to the extension sheet – Dissolving

1.

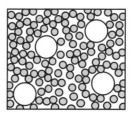

Figure 1 A dissolved solid (solid particles unshaded)

2. The solid particles have to separate/break apart, *ie* the bonds between the particles have to break.

3. No, the nail varnish did not dissolve in the water.

4. No, the nail varnish does not have charged particles.

5. No, ethyl ethanoate and propanone do not have charged particles.

    Note that they do have small charges on their particles – but these charges are not as great as those on water and ethanol particles, which can form hydrogen bonds. With the most able students, you could discuss the fact that the size of the charges on the solid and solvent particles is important, as is the strength of the bonds between the particles in the solid. Most students at this level do not need to appreciate the subtleties of this, but the gifted may find it interesting.

6. Propanone and ethyl ethanoate have uncharged particles so they are better than water (which is polar) at dissolving the uncharged nail varnish particles.

This page has been intentionally left blank.

RSC | Advancing the
Chemical Sciences

Carbon monoxide

Index 2.7.1
4 sheets
Index 2.7.2
1 sheet

Carbon monoxide kills around 50 people in the UK each year, with a further 150 requiring hospital treatment. These are preventable deaths. Educating students about the causes of CO poisoning and how to prevent it may help to reduce the number of incidents. The way the poison works is an interesting piece of blood biochemistry which is easily understood by most students and adds an extra dimension to the activity. There are two available worksheets:

■ **Carbon monoxide – the silent killer** is a comprehension exercise

■ **Carbon monoxide alert** is a research exercise that requires students to make a poster or leaflet to present the information they have found.

Prior knowledge required

None for the poster making.

For the comprehension exercise:

■ Balancing equations

■ Reversible reactions.

Answers to questions

1. CO is colourless, odourless and tasteless. We cannot sense it without special equipment.

2. $CH_4 + 2\,O_2 \rightarrow CO_2 + 2\,H_2O$
    $CH_4 + 1^1/_2\,O_2 \rightarrow CO + 2\,H_2O$ or
    $2\,CH_4 + 3\,O_2 \rightarrow 2\,CO + 4\,H_2O$

3. The first equation uses two molecules of oxygen for every molecule of methane. The second uses only 1.5 molecules of oxygen for every molecule of methane. When less oxygen is present, carbon monoxide is formed.

4. Answers could include any gas appliances (*eg* gas cooker, fire, central heating boiler), wood- or paraffin-burning appliances and cigarette smoke. Electric appliances do not produce CO.

5. Carbon monoxide prevents oxygen getting to the cells in your body. This makes you feel tired.

6. $\rightleftharpoons$ means the reaction is reversible.

RSC | Advancing the
      | Chemical Sciences

7. The reaction represented by Equation 4 is not reversible. This means that once carbon monoxide has bonded to your haemoglobin it does not come off easily and so prevents the haemoglobin from carrying oxygen as it should. If there is no CO present, haemoglobin binds oxygen and releases it again easily when it reaches cells that need it.

8. Carbon monoxide changes the haemoglobin and prevents it from giving up any of the oxygen it is carrying to the body cells. In other worlds, it stops the reaction represented by Equation 3 from being reversible.

9. Unborn babies' haemoglobin binds even more carbon monoxide than the mother's. This means that the baby suffers more at low CO levels than its mother. (It can also alleviate the mother's symptoms as the baby removes CO from her blood.)

Chapter 3
Large molecules

(List of student sheets available on CDROM)

3.1 Polymers

3.2 Emulsifiers

3.3 Textile conservation

3.4 Epoxy glues and the ATLAS project

Polymers in everyday things

Index 3.1.1
5 sheets
Index 3.1.2
8 sheets
Index 3.1.3
6 sheets

Background information for teachers

Polymers are a part of everyday life and examples can be found almost anywhere. Many people think of polymers simply as plastics used for packaging, in household objects and for making fibres, but this is just the tip of the iceberg.

Areas in which polymers are important include:

1. Kitchen applications and food

2. Medical products for wound care, dentistry and in contact lenses

3. Sportswear and sporting materials

4. Protective equipment for work and leisure activities

5. Home and personal care products.

Further information on some of the uses of polymers in these areas is given below.

Polymers are produced by addition reactions or condensation reactions.

1.   Polymers and food

Polymers are used very widely in the production, distribution, packaging and preparation of food. some examples of such uses are listed below.

Farming:
■ Sheeting to protect crops

■ Encapsulation of seeds (gels and nutrients)

■ Protective clothing for farm workers.

Distribution:
■ Packaging in an inert atmosphere

■ Vacuum packing

■ Insulated packaging.

Retail:
■ Carrier bags (now biodegradable)

■ A variety of packaging types

■ Display units.

In the kitchen:

■ Storage (sealable containers, cling film, vacuum packing machines)

■ Food preparation (plastic cutting boards, microwave-safe transparent containers, flexible utensils, cook-in-the-bag techniques).

In food

■ Swelling of starch – perfect chips, roast potatoes, risotto.

■ Denaturation of protein and connective tissue – cooking meat at low temperature.

■ Thickening of soup using starch, gelatine (a heteropolymer of amino acids) or insulin (a non-digestible polysaccharide).

2.   Polymers in medical products

Contact lenses

The material used in contact lenses was originally made by bulk free radical polymerisation, which was carried out very slowly to minimise stress. The polymer rods were then cut into buttons, which were shaped on a lathe to give the correct optical shape. Nowadays, cast moulding with UV initiation is the preferred technique.

To extend the wearing time of contact lenses, researchers had to look at a long list of required properties:

■ Mechanical properties

■ Resistance to dehydration

■ Fluid transport

■ Recovery characteristics

■ Transparency

■ Wetability

■ Resistance to lipid, protein and environmental debris

■ Oxygen transmissibility (to let oxygen reach the cornea).

The polymers used in contact lenses are silicones and high water content materials.

Wound care

There are several types of polymer that give physical and biological protection (*ie* act as tissue sealants):

■ Fibrin glue, which is naturally occurring and is formed by mixing fibrinogen and thrombin

■ A cross-linked protein formed by mixing a natural protein (albumin) with a synthetic cross-linker, *eg* PEG(SS)$_2$ (PEG=polyethylene glycol)

■ Cyanoacrylates – exact properties depend on the alkyl chain.

These materials are easily polymerised and form strong bonds to the tissue.

Transparent adhesive dressings consist of a polyurethane film plus a pressure-sensitive adhesive. The adhesive is normally an acrylate.

Dental polymers

The silver/mercury amalgam used for fillings in the past has been replaced by polymeric materials because of concerns about the poisonous nature of mercury vapour and because larger amounts of tooth have to be removed to provide a key for the amalgam.

Tooth enamel is hydroxyapatite. Dentine is 40% protein and 60% hydroxyapatite. Any material used to fill a tooth must be resistant to moisture, extremes of heat and cold, abrasion, mechanical stress, bacterial microflora and shrinkage stresses and must have an acceptable appearance. A resin composite that can be shaped and cured in situ is the most commonly used material.

The procedure involves first preparing the tooth by etching it with acid to remove debris. An adhesive is then applied and the solvent evaporated. Light is used to cure the adhesive then the filling paste is added and light cured. Finally, the filling is polished.

As well as the properties listed above, the mixture needs to be biocompatible, suitable for light curing but not too light sensitive, stable and easy to use. It must have a low heat of polymerisation and a long shelf life.

The resins used are cross-linking methacrylate-based thermosets. However, they are not strong enough for use in fillings on their own so glass and other fillers are used to produce a composite.

An adhesive is used to bond the resin to the tooth, *eg* PENTA, which contains a phosphate group that chelates with the calcium ions in the tooth.

3.   Polymers in sport

Sporting equipment often consists of many different types of polymer and can usefully be broken down into clothing and footwear, protective equipment, and games or event equipment.

Trainers (athletic footwear)

Almost all parts of a modern trainer rely on polymers, from the upper part of the shoe to the sole.

A trainer upper can contain polymers in the laces, foam padding and non-woven liner. Nylon™ is often used in the fabric and synthetic leather is also a polymer.

The soles consist of an outer sole, a mid-sole, and a stability bar. The outer sole is rubber. Coloured soles are made of synthetic rubber, whilst black ones are the natural material. The mid-sole is designed to be energy absorbing. It consists of a foam made of open cell polyurethane with physical cross-linking that gives a porous flexible solid.

Protective equipment

Protective equipment for sports needs to contain both hard plates to spread the load of an impact (these are often polycarbonates or aramids, which are aromatic polyamides) and padding for fit and comfort (foams).

Examples are helmets worn in cricket, baseball, American football and cycling.

Polymers in games or event equipment

Polymers have been used to improve performance in events such as the pole vault and in games like golf.

The pole vault

The rules for the pole vault state that the pole may be made of any material or combination of materials and may be any length or diameter. Over the years the material used has progressed from a rigid ash or hickory pole to bamboo to aluminium or steel to flexible fibreglass.

The pole is a substantial help to the vaulter as some simple physics shows. The kinetic energy of the vaulter if fully transferred to potential energy at the top of the vault can be expressed as

$$^1/_2 \, mv^2 = mgh$$

So, if the vaulter is running at 10 m s^{-1}, the maximum height she/he can reach is 5.1 m. However, Sergei Bubka's world record is 6.14 m!

There has to be something in the technique (run-up, plant, swing, rock-back and clearing the bar) or in the pole that allows the extra height to be gained. In fact, when the pole is bent it stores elastic strain energy and this allows the extra height to be gained.

Golf balls

More than 1000 million golf balls are produced annually and the evolution of balls and clubs that have allowed improvements in performance. Originally golf balls were wooden or consisted of a feather ball in a leather case. In 1845 smooth balls made from 'gutta percha', a trans-1, 4-polyisoprene polymer, were introduced. In 1902 balls with a rubber thread core and a gutta percha cover were first made. Dimples were introduced to improve the aerodynamic qualities of the ball.

The development of Ziegler-Natta catalysts allowed a polymer of cis-1,4-polybutadiene to be produced and used in golf balls. The latest balls have polybutadiene cores and covers made of materials such as Surlyn®.

Carbon fibre in sport

Many natural materials are composites and have anisotropic properties. Such composites have enhanced properties that cannot be achieved with the individual component materials, such as an enhanced stiffness to weight or strength to weight ratio. Nowadays, carbon fibres can be bonded with epoxy resins to make synthetic composites that are used in fly-fishing rods, snowboards, high performance cycles (frame weight 1 kg), golf shafts, tennis racquets, windsurfing masts and safety cells for drivers of Formula 1 racing cars.

Carbon fibres are made by pyrolysing polymeric materials. Atactic polyacrylonitrite is dissolved in a solvent, extruded to give multifilaments and thermally stabilised at 200–300 °C to give a ladder polymer. This polymer is then heated to 1000 °C in an inert atmosphere, which removes the non-carbon elements and some nitrogen. In the polymer the intramolecular C≡N groups try to get as far apart as possible whilst the intermolecular cyanide groups interact.

The ladder polymer is a conjugated nitrile which, on heating eliminates N_2, H_2 and HCN as well as some aromatic fragments.

4.   Polymers in protective equipment for work and leisure activities

Aramids

Many items of protective equipment are made from meta and para aramids. The meta compounds have the trade names Nomex® or Teijinconex® and the para compounds Kevlar® or Twaron®.

The aramids are stable because the bond dissociation energies of C-C and C-N bonds in aromatic systems are 20–30 % higher than those in aliphatic systems. The materials are polar, have high rigidity, crystalline and form large amounts of char when exposed to a flame.

Kevlar® consists of perfectly orientated chains in an extended configuration. It has a low density and high performance at low weight.

Useful properties of para-aramids	Useful properties of meta-aramids
Thermal stability	Thermal stability
Chemical stability	Chemical stability
Flame resistance	Flame resistance
Toughness	Dielectric properties
Damage tolerance	Intumescences (porous charring)
Dimensional stability	
Low creep	

Table 1 Comparison of aramid properties

Products could be formed as yarns, staple (a short fibre), floc (formed by the aggregation of a number of fine suspended particles), fabric, paper, pulp and composites.

The materials have a myriad of applications:

■ Soft and hard ballistic protection

■ Protective apparel

■ Fibre optics, ropes and cables

■ Composites

■ Friction products, gaskets, hot gas filtration

■ Electrical insulation, electronics

■ Tyres.

For example, Kevlar® is used in:

■ Bullet proof vests and helmets, although it would not stop a knife

■ Store rooms for tornado protection – these are tested by firing pieces of wood at the Kevlar® surface.

Polymers in mountaineering

Mountaineering and hill walking would be much less pleasant and safe without modern polymers. Ropes are made from nylon and boots from high friction rubber. Smooth solid rubber shoes are used for climbing. These work because as the shoe moves across the rock the sole is alternately compressed and allowed to expand so that it grips the surface.

Waterproof, breathable coats are made from nylon laminated with a polymer membrane. It is important that a coat keeps out the rain but it is much more comfortable if sweat can escape so an impermeable coat is not the best option.

Surface tension causes water to form beads. If the surface of the coat is rough and is comprised of peaks and troughs of a suitable size, these water droplets are too big to get into the troughs so they run off the coat. Many modern waterproof coats are covered in a layer of expanded PTFE membranes, which are very hydrophobic. Behind this layer is a hydrophilic porous barrier layer made of polyurethane. The material formed by these layers is known as GORE-TEX®.

A new material called eVENT® is now also available. In this material, the polyurethane of GORE-TEX® is replaced by a perfluoroalkyl acrylic copolymer, which also coats the insides of the pores in PTFE membrane.

Polymers in everyday things

This set of activities illustrates some uses of polymers. Much of the material will be particularly useful for teachers of GCSE Applied Science specifications, in which the applications of materials feature prominently.

A large amount of chemistry background is given here for teachers' information only. It is not intended that students would need to know some of the details given.

The Polymers in everyday things activity set consists of three student worksheets:

■ **Polymers in everyday things – contact lenses**

■ **Polymers in everyday things – dentistry**

■ **Polymers in everyday things – mountaineering**.

Each worksheet provides background information and a series of questions that guide students to relate the properties of materials to their uses in particular contexts. Students also compare the advantages and disadvantages of some synthetic materials with those of naturally occurring ones.

Notes on using the activities

The format of the activities is quite flexible and they can be used as a teaching aid when a teacher is present or as material for part of a cover lesson when a specialist teacher is not available. Students can use knowledge from previous lessons, reference books and internet searches (if web access is available) to help them answer the questions. To get the most out of **Polymers in everyday things – mountaineering**, internet access is advisable.

Each activity requires students to read the information provided on the worksheet and answer some associated questions. A number of suggested questions have been provided which teachers may use or adapt to suit the needs of their students. Once this written exercise has been completed, it is advisable to organise a follow-up class or group discussion about the information on the worksheet and students' answers to the questions. This will help ensure students have fully understood the material and will allow them to express their opinions about what they had read.

You may also wish to set students the task of further investigating some aspect of the topic using reference books and/or the internet. The information they gather could be used to produce a poster or presentation, for example.

If students are to use the internet at any stage of these activities, be aware of the following points:

■ Contact lenses – entering 'polymers in contact lenses' into a search engine such as Google produces a selection of websites with relevant information; **http://www.contactlenscouncil.org/scon-history.htm** shows a time line of the history of contact lenses; **http://www.contactlenses.org/whatare.htm** provides useful information about gas-permeable lenses (both sites accessed January 2006).

■ Dentistry – a search for 'dental polymers' with the Google search engine gives websites that are a bit difficult for the target group; the search term 'teeth fillings' gives hits of general interest, whereas 'materials in tooth fillings' gives websites more specifically relevant to the topic discussed here. You may wish to check the search results given by whichever search engine you favour before asking students to use the internet.

■ Mountaineering – internet searches should be used with caution: the top hits produced by a search for 'clothing', for example, are usually commercial sites with little science interest; the search term 'moderate protection' is likely to give some inappropriate sites on more exotic clothing. The worksheet directs students to two websites to help them answer particular questions. If access to the internet is difficult, you may wish to provide alternative sources of information or to print out selected material from the suggested websites before the lesson.

Answers to suggested questions

Polymers in everyday things – contact lenses

1. Cosmetic/cultural reasons (*eg* appearance) / convenience (*eg* sports people).

2. Preference (*eg* may not want to have to put contact lenses onto eye) / some contact lenses need to be kept moist/dry out and this could cause inconvenience.

3. Transparent / can refract light / rigid / can be made to the correct shape.

4. Glass could shatter/break.

5. Artificial polymers had not been discovered.

6. If something is transparent it allows light rays to pass through it so that objects behind it can be clearly seen.

7a. So that it does not tend to fall off the eye / comfort.

7b. So that any chemicals on the surface of the eye do not react with the material in the lens (which may eventually degrade it).

7c. Cost reasons / so that it is affordable to those who need them.

8. Methylmethacrylate

9. PMMA was not very comfortable for users of contact lenses made from PMMA. There were problems such as oxygen not passing through to the cornea and the need to use wetting solutions.
These are not problems when it comes to using PMMA for aquariums where the durability of the material and its clarity are most useful properties.

10. Because polyacrylamide is hydrophilic / accept 'much of the lens is water'.

11. After cross-linking the polymer absorbs water.

12. Comfort / there is no need to use a wetting solution (as the water is already there in the lens).

13. Hard (lenses)

14. Fluorine

15. Silicon

16.

	Advantages	Disadvantages
Soft lenses	Soft and flexible Material is hydrophilic No need for wetting Cheaper than other types Can use some types for just one day and discard	Fragile Clarity of vision may be affected
Hard lenses (PMMA)	Good clarity of vision Very durable	Not very comfortable Need to use a wetting solution on eyes Takes some time to get used to them Do not allow oxygen to get to the surface of the eye.
Rigid gas-permeable lenses	More comfortable Oxygen can pass through to eye surface (cornea) Quickly get used to wearing them Good rigidity/clearer vision Good for astigmatism or bifocal needs	High cost Some inflexibility

17. Almost any answers are possible here. If the answers are to focus on polymers then any of the suggestions made in this teachers' sheet are applicable. The main idea is to emphasise that scientists continue to try to find materials to improve the quality of people's lives.

Polymers in everyday things – dentistry

1. Calcium, phosphorus, oxygen, hydrogen.

2. 22

3. Calcium

4. Carbon

5. pH 8 is only very slightly alkaline but pH 0.5 is very acidic.

6. It is still possible to chew but the gum does not now put sugar around your teeth which would cause plaque.

7. Eat less of the foods that causes plaque to form, brush teeth regularly.

8. Mercury, silver, copper and zinc.

9. Cosmetic reasons (to avoid seeing the grey filling on the back teeth) / back teeth do more chewing and the mercury amalgam filling is strong and wear resistant.

RSC | Advancing the
      | Chemical Sciences

10. The best answers will refer to environmental problems associated with disposal.

11. Objectivity/the need to have other scientists (who are not connected with the original work) check the results / to avoid anyone just putting out there own thoughts without thorough scrutiny.

12. If dentists remove fillings then there is mercury vapour about, although this is at an acceptable level for the patient (who just breathes it as it is removed), the dentist may have to do this several times a day.

13. Some method of extracting vapour / good ventilation in room.

14. Mark as a level of response answer. Science discovered the use for mercury amalgam in fillings for teeth. It has helped millions of people. As possible issues about its safety arise, scientists can carry out research to pin down whether there is a substantial danger or not. They can decide levels of acceptable safety. This can only be done by detailed research where results can be duplicated and supported by others. Meanwhile scientists carry out more research to find materials with better properties still and which carry even less risk.

15. Cosmetic reasons

16. Composite materials exploit the advantages of each material from which it is composed without suffering the weaknesses of these materials.

17. It is a harmful substance, avoid swallowing, breathing in or skin contact.

18. $C_4H_6O_2$

19.

20. The resin would set with background light before the composite is added.

21. Resin has shrunk and allowed a small gap to form, this has been exploited by bacteria allowing dental decay (caries) to form, when the caries reaches the dentin the person will feel pain and know there is a problem.

22. Best advice is to use a white composite filling. Explain that it is less durable than mercury amalgam but that it will not suffer too much mechanical stress. Its appearance will be much better in such a prominent place in the mouth. Also it is only a small filling and so it is OK to use composite. Its colour can be matched to the tooth it is in. It does not contain mercury. However it may cost more than mercury amalgam and could wear out faster. [These are some of the key ideas; there could be others.]

23. The best advice is to probably have a mercury amalgam filling. The teeth at the back of the mouth suffer very large mechanical stresses as they chew food. Mercury amalgam is very durable unlike composite fillings. The silvery colour of the amalgam should not be a cosmetic problem as they teeth are at the back of the mouth. It is a large filling and composite material is not so suitable. Although they contain mercury, safety guidelines show that the level of risk is acceptable. [These are some of the key ideas; there could be others.]

24. Phosphate

25. Attractive in appearance

26. Release of fluoride ions will help other teeth in child's mouth / although filling won't last a long time, neither will the child's milk teeth / it gives a good seal to rest of tooth.

27. The glass ionomer filling gives a good seal to the existing tooth, the composite can the give a good colour match to the filling at the surface.

28.

Advantages of glass ionomer fillings compared to composite fillings	Disadvantages of glass ionomer fillings compared to composite fillings
Interacts with the enamel and dentin of existing tooth forming an excellent (chemical and biological) seal with the tooth	Poorer colour match to existing tooth
The fluoride ions that are part of the filling are slowly released and since these are next to the teeth they react with the tooth enamel further strengthening the teeth	They are weaker under normal chewing forces.

Polymers in everyday things – mountaineering

1.

animal skins	fibres from plants	wool	silk
10 000	8000	6000	3000

number of years ago

2.

Examples of plants used to make clothing	Examples of animals used to make clothing
Hemp	Sheep (for wool)
Nettles	Cows (for hide)
Flax	Pigs (for hide)
Cotton	Silk worms

3. a. Colder climates in far north or far south of the hemispheres.
   b. Hotter climates (tropical or equatorial).
   c. It hadn't been discovered.

RSC | Advancing the
      | Chemical Sciences

4.

	Polyester	Nylon
Year when first used	1942	1935
What it is used for now	Clothing and handles	Used to make bearings, blow mouldings, and clothing fabric.
Useful properties	It has excellent dimensional stability, high dielectric strength, and good toughness. It has moderate chemical resistance, low resistance to strong acids and bases, is notch sensitive, and is not recommended for outdoor use or in hot water. Copyright © 2006 eFunda	It has high lubricity and moderate strength. It is tough, inexpensive, and has poor dimensional stability due to water absorption (hygroscopic nature). Copyright © 2006 eFunda

This excellent site would allow teachers to ask many more questions of this nature. Time lines would also be possible.

5. Thermosetting: a plastic that can be moulded into shape during manufacture but which sets permanently rigid on further heating (due to cross links forming which cannot be reversed).
Thermoplastic: a plastic that can be repeatedly softened on heating and then will harden on cooling (no cross links form).

6. More continuous movement created more sweat (which then condensed).

7. Water vapour particles are much smaller.

8. Monomer

9. Alkenes

10.

$$\left[\begin{array}{cc} F & F \\ | & | \\ -C-C- \\ | & | \\ F & F \end{array} \right]_n$$

11. It must be stretched.

12. It could crack if repeatedly bent in the same place / fires could cause it to melt.

13. PTFE has holes in it that are big enough to allow water vapour to pass through, but small enough to prevent liquid water passing through / the PTFE itself is quite fragile and so it is joined to layers of other stronger, more durable fabrics.

14. It helps the Gore-Tex® resist contamination by body oils. If this were not done, the oils could prevent the Gore-Tex® fabric from breathing properly.

15. From the home page click on 'what is it'. In the article that follows click on 'dry venting' TM and 'dry system' TM.

Monomer – Polymer card game

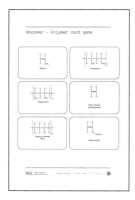

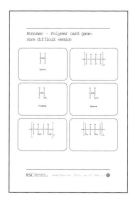

Index 3.1.4
3 sheets

How to set up

Make enough sets of the cards to give one set for every two or three students in the class. Laminating the cards makes them last longer.

What to do

Divide the students into pairs or small groups and give each group a set of cards. Students lay all the cards face down on the table and take it in turns to pick up two cards. If a player picks up a 'pair' – *ie* a polymer card and the card showing the monomer that can be used to make that polymer – then they keep the cards. Otherwise, they put them down and the next player has a turn. Picture cards have also been included to encourage students to make links between monomers, polymers and objects made from polymers.

How to use this activity

This activity could be used as a plenary after a teaching session on monomers and polymers.

Time

5–10 minutes, depending on the ability of the students.

IUPAC names

Styrene – phenylethene

Vinyl acetate – ethoxyethene

Teflon – polytetrafluoroethene

Monomer – Polymer card game: more difficult version

A more difficult version of the above card game is given where the polymers are not named to encourage the students to look carefully at the polymer structures. Picture cards of the polymers have also been included; the students can link monomers, polymers and objects made from polymers.

RSC | Advancing the
      | Chemical Sciences

Changing the properties of polymers and plastics

Index 3.1.5
7 sheets

The properties of polymers and plastics can be changed in a number of ways. One way is of course to produce a different polymer or plastic with a different chemical structure, but there are also other possibilities. This series of activities allows students to explore some of them:

■ The use of plasticisers – see **Making a plastic from potato starch** or the activity in this section on plasticised and unplasticised PVC (polyvinyl chloride)

■ Cross-linking polymer chains – see **Making slime** or the activity on rubber

■ The effect of changing the length of the polymer chains

■ The effect of the presence of branched chains

Four of the activities (PVC, rubber, chain length and branched chains) could be set up as a circus and carried out by the whole class in turn. Alternatively, groups of students could each do one of the four experiments and then report back to the class. **Making a plastic from potato starch** and **Making slime** are likely to be very popular activities and are probably best done as class experiments. Details of these activities are given separately.

Equipment required

Set 1: Plasticised and unplasticised PVC
■ Plasticised PVC – squeezy toy and/or PVC cling film

■ Unplasticised PVC – piece of uPVC pipe or guttering.

Set 2: Cross-linking (or vulcanising) rubber
■ Rubber – microscope slide with a thick coating of Copydex glue (set up at least the night before)

■ Vulcanised rubber – thick rubber band.

The glue can be peeled off the microscope slide by students when they are ready to use it. One sample will not last long enough for all groups to use it so have several in reserve.

Set 3:   Changing the length of polymer chains

■ Candle wax or candle

■ Polyethene (*eg* carrier bag)

■ Sample of medium chain length hydrocarbon (*eg* kerosene) in a sealed tube.

Set 4:   Branching chains in polyethene

■ High density polyethene – a 'rustly' supermarket carrier bag

■ Low density polyethene – a 'quiet' department store carrier bag.

All the above items should be labelled.

Additional items:

■ Assorted worksheets and helpsheets as chosen by the teacher for each group

■ Some groups may need glue and scissors.

Differentiating the activity

Students should describe the properties of the samples in each set and try to explain the difference they observe. The activity can be left very open-ended and students given no help, or the support documents listed below can be used:

■ List of possible properties for students to choose from (**List of properties**)

■ Table to complete (**Table**)

■ Explanations for use in completing the table (**Explanations**)

■ Cut and stick diagrams (**Diagrams**)

■ Information sheet (**Information sheet**).

Each item in the list is provided as a separate document so teachers can choose how much help to give to each class or group of students.

The most able students could be given the samples and perhaps the diagrams. They can discuss in groups why there are differences between the polymers in each group of substances. If they get stuck or wish to check their answers when they have thought of an explanation they could be given the information sheet.

The least able students could be given the properties list and the table to begin with, then the diagrams and explanations when they have decided on the properties. Giving them only one step at a time will help to prevent them from becoming overwhelmed.

Possible answers to questions

The depth at which these questions can be answered could vary widely depending on the abilities of the students concerned.

Example answers

Changes to polymer	Properties and differences in properties	Diagram of the molecules	Why the plastics have different properties
Plasticisers Plasticised PVC	Very flexible; soft.		The chains with small plasticiser molecules between them slip and slide over each other more easily. This makes the plastic more flexible. Without plasticiser molecules the plastic is tougher and more rigid because the chains line up in rows and hold on to each other more tightly.
Unplasticised PVC	Rigid; hard.		
Cross-linking Rubber (Copydex)	Very elastic and stretchy; does not always go back to original shape; soft.		The substance made of molecules that are cross-linked is harder and less flexible than the one without cross-links. The links stop the molecules moving over each other so the structure is more rigid.
Cross-linked rubber (elastic band)	Elastic – but not as elastic as the Copydex; harder than the Copydex.		
Length of chains Kerosene	Viscous liquid		Longer molecules can get tangled up in each other. They also stick together better. This means that substances made of longer molecules are harder and have higher melting points.
Candle wax	Soft solid		
Polyethene	Stretchy/elastic solid		
Branching chains High density polyethene	Rustles when moved; stretchy/elastic		The plastic which has no side chains on its molecules has a higher density because more molecules can pack into the same amount of space. The molecules can slide over each other more easily if there is nothing in the way to stop them. When there are side chains or branches in the way, the polymer chains cannot slide over each other so easily because there are more interactions between the chains.
Low density polyethene	Not as stretchy as the high density polyethene		

Table of example answers

Polythene bags

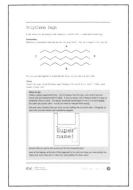 

Index 3.1.6
2 sheets

This is a quick 10 minute activity which could be used at the end of a lesson on polymer structure or as a review at the beginning of the next lesson. Students think about the orientation of polythene molecules and the implication for the properties of plastic bags. The activity could also be set as a practical homework exercise, although students do find it helpful to discuss their ideas.

Equipment required

For each group of students:

■ At least $^1/_4$ of a plastic bag. Supermarket carrier bags are the best.

■ Scissors.

Running the activity

Make sure students have thought about the structure of polymer chains or recap the relevant ideas if this is being used as a 'starter' activity. Less able students may find a class dissusion prior to the activity helpful. Pulling the plastic to bits only takes a few seconds.

Answers

Students should draw diagrams of what they have observed.

1. If the chains are pulled over each other the plastic is more likely to stretch.

   Note: It is important to make sure students understand that the molecules move over each other but do not themselves stretch. The chains do unravel first but the idea that the molecules stretch when the plastic stretches is a common misconception and students who believe this will not understand the topic.

2. The horizontal piece of plastic.

3. Stretching occurs from C to D.

4. Carrier bags are all likely to be produced with the polymer molecules aligned horizontally, otherwise if you put heavy shopping in them they would quickly stretch and break. If they are produced in this way, they are much stronger in the vertical direction as the molecules are held together by intermolecular forces. These ideas could be tested by trying a larger sample of plastic bags from a number of different stores.

RSC | Advancing the
Chemical Sciences

Making a plastic
from potato starch

Index 3.1.7
7 sheets

In this activity students make a plastic from potato starch and investigate the effect adding a 'plasticiser' has on the properties of the polymer they have made. Students can begin either with potatoes or with commercially bought potato starch. The practical is straightforward; the main hazard is the possibility of the mixture boiling dry.

Timing

Extracting the starch takes about 15–20 minutes and making the plastic about 20 minutes.

Using the activity

This can be used simply as a practical to enhance the teaching of a polymers or plastics topic, it can be used as an introduction to further work on biopolymers and bioplastics and/or it can be used as an example of the effects of plasticisers. A number of student sheets are provided so you can choose to do the whole activity or just selected parts of it.

Extracting starch from potatoes

For each group of students you will need:

■ 100 g potatoes

■ Grater

■ Tea strainer

■ Distilled water

■ 2 x 400 cm^3 beaker

■ Pestle and mortar.

Making the plastic film

For each group of students you will need:

■ 250 cm^3 beaker

- Large watch glass
- Bunsen burner and heat proof mat
- Tripod and gauze
- Stirring rod
- Potato starch
- Propan-1,2,3-triol (glycerol)
- Hydrochloric acid 0.1 mol dm^{-3} (**Minimal hazard**)
- Sodium hydroxide 0.1 mol dm^{-3} (**Irritant**)
- Food colouring
- Petri dish or white tile
- Universal Indicator paper
- Eye protection
- Pipettes
- Access to a balance
- 25 cm^3 measuring cylinder
- 10 cm^3 measuring cylinder.

Health and Safety

Wear eye protection.

Propan-1,2,3-triol (glycerol) has no hazard classification but may be harmful if ingested in quantity.

Notes

1. If students have extracted their own potato starch they will need to use about 4 g of it for the next part of the experiment as it is a wet slurry rather than a dry powder. They should add about 22 cm^3 water. If they do not have enough extract then a bit of bought potato starch added to the mix will be fine.

2. If access to a balance is difficult then a heaped spatula of starch can be used rather than 2.5 g.

3. If access to 10 cm^3 measuring cylinders is difficult, 4 pipette squirts of hydrochloric acid and 3 squirts of propan-1,2,3-triol are suitable amounts.

4. If you have a drying cabinet, the plastic film should dry in about 90 minutes at 100 °C.

5. Warn students not to let the mixture boil dry because this can cause it to 'pop' and it shows a tendency to jump out of the beaker. For this reason, students should wear eye protection at all stages of the practical.

6. Food colouring – while use of this is optional, it does enhance the product and the colour makes the plastic film look more like plastic. Only one drop is needed or the film is too dark.

7. If too much water is used, the polymer does not solidify and remains a liquid.

Advancing the
Chemical Sciences

Explanation

Starch is made of long chains of glucose molecules joined together. Strictly, it contains two polymers: amylose, which is straight chained, and amylopectin, which is branched. When starch is dried from an aqueous solution it forms a film as a result of hydrogen bonding between the chains. However, the amylopectin inhibits the formation of the film. The addition of hydrochloric acid breaks the amylopectin down, allowing a more satisfactory film formation. This is the product formed in the student activity without the addition of propan-1,2,3-triol. The straight chains of the starch (amylose) can line up together and make a good film. However, it is brittle because the chains are so good at lining up – areas of the film can become crystalline, which causes the brittleness.

The addition of propan-1,2,3-triol has an effect because of its hydroscopic (water attracting) properties. Water bound to the propan-1,2,3-triol gets in amongst the starch chains and inhibits the formation of crystalline areas, preventing brittleness and resulting in more 'plastic' properties. In the notes for students, reference to water has been omitted to allow them to concentrate on the effect of the propan-1,2,3-triol itself.

Answers – Making a plastic from potato starch

1. Without propan-1,2,3-triol: a brittle, transparent film is produced. It has probably cracked in the drying process.

   With propan-1,2,3-triol: a plastic, transparent film is produced. It feels slimy and can be bent and manipulated without breaking.

2. The propan-1,2,3-triol has the effect of changing the properties of the product – the propan-1,2,3-triol makes the film flexible and plastic instead of brittle.

3. Students should draw a diagram showing wavy lines all lined up to represent the brittle film. To represent the plastic film, they should draw small molecules in between the polymer chains, which should not be lined up in this case:

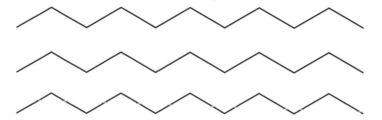

Figure 1 Polymer chains lined up – the product is very brittle

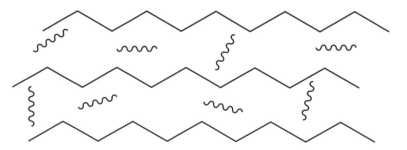

**Figure 2 Polymer chains have small molecules between them,
which prevents them from lining up**

Answers – Using plastics from potato starch

1. Plastic from potato starch will probably be biodegradable because all the starting materials are biodegradable.

2. The plastic could be mixed with some compost, left in a warm place for a few weeks and then examined. If it shows signs of decomposing, it is biodegradable. Another similar answer would be acceptable.

3. The energy needed to extract the starch would come from electricity, which is mostly made from petrochemicals (coal, oil and natural gas.)

4. Most plastics are made from petrochemicals/oil.

5. A bioplastic is a plastic made from a plant or other living thing. Biodegradable means that it is broken down by living things.

6. Plants are renewable raw materials and will not run out; the plastics produced are usually biodegradable; the products can be marketed as 'greener' plastics.

7. Oil is still used up as a lot of energy is needed to produce the bioplastics. Lots of fertilisers are used to grow the crops and these can cause pollution if they get into the water supply.

8. As the plastic is soluble it cannot be used for packing anything that is wet, contains water or is likely to get wet because the packaging will just fall apart. Possible answers could include: not used for packaging drinks, fresh food, plants. Could be used for: packing dry goods, dried food, protective packaging.

9. Mark by impression. Look for a good use of the scientific information and clear communication of the facts.

Investigating cross-linking – making slime

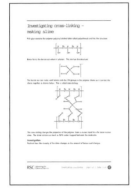

Index 3.1.8
2 sheets

This experiment is great fun and always popular with students. It can be made as difficult or as straightforward as you wish, depending on the level and ability of the class. For the least able, making the slime and observing the changes in its properties may be enough. For the more able, detailed measurements and observations are possible.

Equipment required

Per student, pair or group:

■ PVA solution approx 100 cm^3 per group – see notes below

■ Borax solution – 0.8 g borax in 20 cm^3 water will allow students to make about four batches of slime; scale up according to the number of groups (**Minimal hazard**)

■ Food colouring or a water soluble dye such as fluorescein (optional) – can be added when mixing the solutions

■ Approx 4 x 100 cm^3 beakers

■ 1 x 25 cm^3 measuring cylinder

■ 1 x 10 cm^3 measuring cylinder

■ Stirring rod.

Notes

PVA solution can be made from PVA glue (wood glue or white paper glue). The composition of the glue may vary depending on its source so check that the mixture works and adjust the glue/water mix accordingly. Two parts glue to one part water generally works well.

Alternatively, it is possible to buy solid PVA. To make the PVA solution using this starting material, add about 4 g PVA to 100 cm^3 water at 90 °C. Stir with a magnetic stirrer until the PVA dissolves. Cool and add enough water to make the volume up to 100 cm^3 (to replace any water lost by evaporation).

SEP (Science Enhancement Programme) slime kits, which include PVA, borax and dye, are available through Middlesex University Teaching Resources. See **http://www.mutr.co.uk** (accessed November 2005) for more information.

Health and safety

Borax and PVA both represent a minimal hazard. As borax is a weak alkali, it can cause skin irritation in those with eczema, sensitive skin or cuts. Disposable gloves should be made available for these students. Most others will enjoy handling the slime as it is very tactile.

Borax is toxic if large quantities are ingested so students should wash their hands at the end of this activity and certainly before any food or drink is consumed.

The best slime

Typically the 'best' slime can be made by using a 5:1 PVA to borax mixing ratio, but this can vary, especially if diluted wood or paper glue is used.

Possible alternative investigations

Students could investigate:

■ The amount of stretch achieved in a given time or the time taken to stretch a given amount of slime (both can be converted to a stretch rate of mm s^{-1})

■ Slime with the greatest bounce

■ Slime that can stretch the most without breaking.

Instead of plotting viscosity against the amount of borax used, students could investigate one of the following factors:

■ Temperature – the slime can be left in a water bath until it has reached the desired temperature

■ pH – the pH of the slime mixture affects the degree of cross-linking and can be varied by adding borax/borate buffers.

RSC | Advancing the Chemical Sciences

Cross-linking polymers – alginate worms

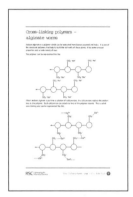  

Index 3.1.9
3 sheets

This practical provides a fun look at cross-linking and a chance to explore some of the many and varied uses of sodium alginate.

Sodium alginate is a polysaccharide of repeating monosaccharide units, each containing a carboxylate ion:

$$CO_2^-\ Na^+ \qquad CO_2^-\ Na^+$$

$$CO_2^-\ Na^+ \qquad CO_2^-\ Na^+$$

$$CO_2^-\ Na^+ \qquad CO_2^-\ Na^+$$

$$CO_2^-\ Na^+ \qquad CO_2^-\ Na^+$$

Figure 1 Structure of sodium alginate

The naturally occuring form of this polymer is alginic acid (the protonated form), which can be extracted from brown seaweed and kelp. It usually has a molar mass of around 240 000.

In this experiment, the sodium ions of sodium alginate are replaced with calcium ions. Since each calcium ion can bond to two carboxylate groups, the ions can cross-link the polymer chains, which results in the formation of an insoluble, gel-like substance.

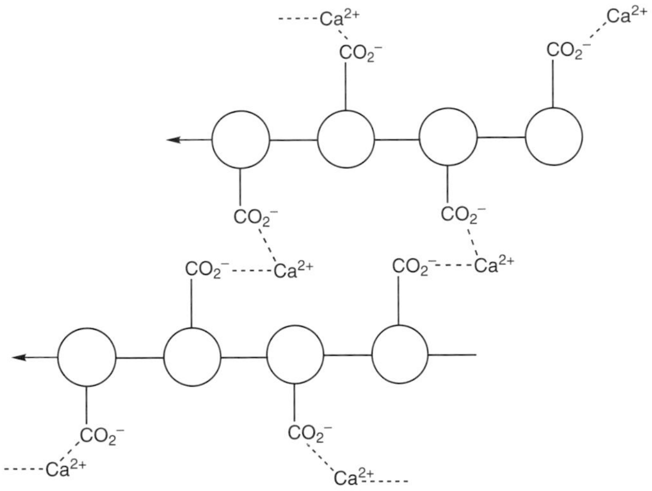

Figure 2 The cross-linked polymer

If the cross-linked alginate is placed in a solution of sodium ions, these replace the calcium ions and the gel-like worms fall apart again.

Equipment required

Per pair or group of students:

- 5 cm^3 sodium alginate suspension – see notes

- Dropping pipette

- 2 x 150 cm^3 beakers

- 100 cm^3 saturated sodium chloride solution – made in distilled water

- 100 cm^3 calcium chloride solution – 1 g calcium chloride per 100 cm^3 distilled water

- Labels for the beakers or pens to label the glass

- Eye protection.

Health and safety

Solid calcium chloride is an irritant. A solution of the suggested concentration is not an irritant.

Wear eye protection.

Notes

To make the alginate suspension, add 2 g sodium alginate to 100 cm^3 distilled water. Do not use tap water, especially in a hard water area. It is best to make the suspension at least a day in advance and allow it to stand overnight so that it becomes homogeneous. For this experiment, the results are clearer if a few drops of food colouring are added. This can either be done when making the suspension or students could do it themselves and perhaps be offered a choice of colours.

Gaviscon® can be used as an alternative to sodium alginate. This will give coloured and opaque worms as a result of the presence of other ingredients in the antacid, which can be quite fun. Any leftover Gaviscon® cannot be used medicinally.

It is also possible to cross-link alginate with other ions. Both nickel(II) chloride and copper(II) chloride can be used. Nickel gives pale green and copper pale blue worms. These worms do not break down in the presence of sodium chloride and so are not suitable for this experiment but could be used to create a colourful demonstration. If you carry out such a demonstration, be aware that nickel salts are sensitisers. Avoid skin contact and wash your hands after use.

If the 'worms' are not removed from the calcium chloride straight away they can take a long time to break down in the sodium chloride. If left in the calcium chloride, the worms feel very different – they are much harder. The Gaviscon® worms break down faster in the sodium chloride solution than the sodium alginate ones.

It is also possible to try to make worms in sodium chloride solution. The Gaviscon® will form rather pathetic worms which disintegrate in time. The sodium alginate does not form worms at all if squirted straight into sodium chloride.

Disposal

The solutions can be filtered or strained and the worms put in the rubbish bin. The easiest way to do this is with a tea strainer.

Answers to questions

1. The alginate or Gaviscon® is a viscous liquid.

2. a. Sodium ion: Na^+.
   b. Calcium ion: Ca^{2+}.

3. Calcium has two positive charges and can therefore attract/bond to two of the negatively charged ions on the alginate. Sodium has only one positive charge and can form only one bond with the polymer.

4. Students should be able to predict a thickening of the polymer.

5. The polymer thickens and becomes a solid. It should agree with the prediction.

6. When the worms are placed in sodium chloride solution they become much softer, get fatter and eventually fall apart.

7. The polymer is cross-linked when it is put into the calcium chloride solution. As the polymer chains are linked they are unable to move independently of each other and the material becomes a lot thicker and solidifies. When the worms are put into the sodium chloride solution, the sodium ions displace the calcium ions and the cross-links begin to fall apart. Eventually, when most of the calcium ions have been displaced, the worms turn back into a liquid.

8. Sodium chloride solution could be used to rinse the wound.

9. Foods listed could include: some fruit drinks and snack bars, various sauces, cheese spread and many other processed foods, stuffed olives. Alginate is almost always used as a gelling agent, a thickener or a stabiliser.

Polylactic acid

Index 3.1.10
2 sheets
Index 3.1.11
3 sheets

⚠ **This experiment is aimed only at more able and more sensible students as an introduction to other studies on polylactic acid. It is not recommended for students who would find it difficult to be calm and concentrated while heating a liquid to a high temperature for at least 10 minutes.**

The student sheet **Using polylactic acid** covers some of the environmental issues surrounding the industrial production of this plastic and its disposal.

Background information

Lactic acid (2-hydroxypropanoic acid) can be obtained by fermenting glucose or maltose or can be extracted from milk. It is used as the starting point for the production of polylactic acid, also known as poly (2-hydroxypropanoic acid), which can also be made from petrochemicals.

Polylactic acid is a condensation polymer – a molecule of water is produced for every link made in the chain. This is not specified in the students' notes, but you might wish to draw their attention to it. When the water is removed, an oligomer made of 10–30 lactic acid units is formed. An oligomer is essentially a short length of polymer. The production of true polylactic acid involves catalytic depolymerisation of the oligomer to a lactide intermediate followed by polymerisation of the lactide to high molecular weight polylactic acid. In the experiment described here only the oligomer is made. The product therefore has different properties from the polylactic acid used in packaging. It may be useful to discuss with students how the different properties are related to the chain length of the polymer molecules.

Figure 1 Lactic acid or 2-hydroxypropanoic acid

During polymerisation, the OH on the acid group (CO_2H) group of the monomer reacts with the OH group on the second carbon atom of another monomer. This is an esterification reaction and involves the loss of water. The resulting polymer is shown in Figure 2.

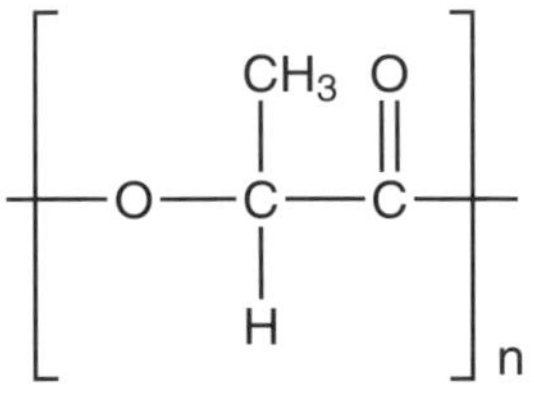

Figure 2 Polylactic acid or poly (2-hydroxypropanoic acid)

Equipment required

For each group of students:

- Test-tube

- Test-tube holders

- Bunsen burner and heat proof mat

- Anti-bumping granules

- Lactic acid (2-hydroxypropanoic acid) (**Irritant**)

- Hydrochloric acid 2 mol dm^{-3} (**Irritant**)

- Petri dish or white tile

- Eye protection.

Health and safety

The boiling point of lactic acid is 122 °C so when students pour it out it will be very hot. Warn students to take care.

Notes

Lactic acid is very viscous so do not attempt to get students to use measuring cylinders. A $^1/_5$ full test-tube works fine and measuring the lactic acid in this way saves making several measuring cylinders very sticky.

The product is very sticky, so discourage students from touching it. They should use a glass rod or a wooden splint instead of their fingers.

Answers – making polylactic acid

1. The molecules will have got bigger because during polymerisation, the small molecules join together in chains.

2. The lactic acid is quite viscous, but the product is a solid (or a very viscous liquid). The product does not run off a petri dish or fall off if you turn it upside down. Note: You can also test for water solubility – the starting material is water soluble but the product is not.

3. As the molecules get larger, the intermolecular forces holding them together also get larger. This causes the material to become stickier and more viscous. (Larger molecules also tend to move around less so materials with large molecules are often solids.)

Answers – using polylactic acid

1. Most plastics are made from petrochemicals/oil.

2. The raw material for polylactic acid is corn.

3. Energy is required for harvesting the corn, transporting it and processing it. All this energy comes from fossil fuels.

4. Production of polylactic acid releases more carbon dioxide into the environment than production of polyethene because a greater amount of fossil fuels is burnt to release energy.

5. Polylactic acid is made from biodegradable material (corn), whereas polyethene is made from fossil fuels, which will release carbon dioxide if the plastic is burnt as waste. (Note: polylactic acid releases carbon dioxide and methane when it degrades.) Also can trap carbon dioxide during photosynthesis.

6. The procedure for producing polylactic acid uses a lot of fossil fuel energy, which is expensive. It is also a complex procedure and involves a lot of steps. For example, the bacteria used to ferment the sugar in the raw material must then be separated from the lactic acid produced.

7. Production of polylactic acid uses less oil.

8. Mark by impression. Students should produce a reasoned argument to support their opinion.

9. Most British rubbish is dumped in landfill sites (where only some of it will biodegrade over hundreds of years) or is incinerated (which can produce toxic fumes). Students could be directed to **http://www.foe.co.uk/resource/factsheets/plastics.pdf**) (accessed November 2005) for more information on the disposal of plastics.

10. Mark by impression. Students might be expected to suggest disposable products because of the biodegradable nature of polylactic acid and a marketing campaign that emphasises the 'green' aspect of the plastic.

This page has been intentionally left blank.

Emulsifiers

Index 3.2.1
2 sheets
Index 3.2.2
2 sheets

In this activity, students are given the role of a scientist in a food technology laboratory. They are asked to test a range of kitchen substances to find out which will act as an emulsifier. The activity is intended as a follow-up to work on emulsifiers and emulsions and students will need to be familiar with these terms.

Two versions of the student sheet are provided – one includes experimental instructions (**Emulsifiers (1)**) and the other asks students to plan the experiment themselves (**Emulsifiers (2)**).

Equipment required

- Boiling tubes and bungs
- Pipettes
- Spatulas
- Cooking oil – corn oil is particularly good as it is dark in colour, which makes it easier to see
- Access to water
- Washing up liquid
- Sugar
- Flour
- Mustard powder – Colman's is good and lasts far longer than ordinary mustard so can be used from year to year
- Salt
- Egg white
- Egg yolk
- Other test substances can be used if prefered.

Health and safety

Raw egg can be a cause of salmonella. Use eggs marked with the lion symbol. Use a disposable pipette to transfer the egg to boiling tubes and avoid handling it.

Warn students against tasting anything, *eg* sugar.

RSC | Advancing the Chemical Sciences

Notes

The boiling tubes must be very clean – in particular, ensure they are not contaminated with detergent as this would result in all the test substances appearing to act as emulsifiers. Although the use of boiling tubes rather than test-tubes means that larger quantities of chemicals are consumed, the results are easier to see on this scale and it is much easier to clean up afterwards.

If the eggs are fresh it is fairly easy to separate the yolk from the white. Ensure no yolk contaminates the white; contamination the other way round is less of a problem.

This experiment could easily be done in a kitchen and adapted so that students initially make a salad dressing using oil and vinegar rather than oil and water. The test mixtures could be tasted as well as observed. If you do this, the mixture containing raw egg should not be tasted.

An emulsifier is a substance that stabilises an emulsion (a mixture of one liquid dispersed in another). Washing up liquid, egg yolk and mustard are emulsifiers. The other substances listed above are not. Students may observe colloidal mixtures in the other tubes, but these are not oil and water emulsions and it should be easy to identify two separate layers.

Answers

1. Washing up liquid, egg yolk and mustard are emulsifiers.

2. Emulsifiers are used in foods to hold a mixture of oil and water together so that it does not separate into unappealing layers.

3. The best emulsifier is usually the washing up liquid.

4. No. Further work/changes to the experiment could include: ensure the mixture is shaken with the same force for the same amount of time; use the same quantities of oil, water and emulsifier each time; leave each mixture for the same amount of time before making observations or observe them at regular intervals over a set period of time; repeat the experiment.

5. The best emulsifier for a salad dressing would be mustard. Although washing up liquid is a better emulsifier it would not taste good and egg yolk could contain harmful bacteria (although this is the emulsifier used in mayonnaise).

6. The answers to this could vary widely.

7. Lecithin is commonly used, as are various E-number additives.

8. Foods that contain both oil and water need to be stabilised and therefore contain emulsifiers. Many 'diet' and 'low fat' foods contain emulsifiers because the oil and fat usually found in these foods is often replaced by water so the mixture needs to be stabilised. Low fat margarine is one possible example.

Textile conservation

Index 3.3.1
3 sheets
Index 3.3.2
5 sheets
Index 3.3.3
4 sheets

A textile is any filament, fibre, or yarn that can be made into fabric or cloth but the word 'textile' also refers to the fabric or cloth itself. Textiles can be used to make a wide range of products – probably the most obvious is clothing, but others include car interiors, yacht sails, furnishings and the wings of early aircraft to name just a few.

Textile conservation can be an interesting context within which to teach organic chemistry because the topic covers both fibres known from ancient times, such as cotton and linen, and modern materials like PVC and rubber.

Conservators are used to dealing with the traditional materials and have various tried and tested methods for the conservation of such materials at their disposal. Modern materials often present more of a challenge because the way they decay is poorly understood. Research is currently underway to find the best methods for preserving what may become important cultural artefacts for future generations.

In this resource

- **Suggested starter activities:** A selection of introductory activities (see below).

- **Textile conservation – introduction:** A short introduction to some of the issues in textile conservation

- **Textile conservation – the structure of cotton and linen:** Organic chemistry in the context of textiles. Two student sheets are provided – a question sheet and a diagram sheet.

- **Case study – the Victory sail:** Infrared spectroscopy in the context of conservation. The material includes information on recent research carried out at the Textile Conservation Centre in Winchester. Some prior knowledge of general and organic chemistry is required for this activity and students have the opportunity to apply what they learnt from the exercise on the structure of cotton and linen.

Suggested starter activities

There are a number of ways in which this topic could be introduced and it is recommended that you do at least one of the following starter activities before attempting the other work provided in this resource. Several of these starters could be combined to form the basis of a satisfying unit of work for less able students on 'materials and their properties'.

- Show a series of images of costumes/textiles from different time periods and ask students to discuss their reactions to the pictures. You could include 1960s 'flower

RSC | Advancing the
      | Chemical Sciences

power' style items, 70s brown flares, 80s shoulder pads, pictures of assorted current celebrities (perhaps with their faces blanked) etc. Try and include the 'yuck' factor and the 'wow' factor.

■ Provide students with a set of garments or other textiles (charity shops are good sources of these) and ask them to decide what each item is made of and where the material in question came from. For example, a blouse might be made of polyester with cotton thread and have plastic buttons. Polyester and the plastic used for the buttons are made from petrochemicals; cotton comes from a plant. Clothing labels can be helpful but include a few items without labels to stimulate discussion. Students could be asked to think about the following questions: How could you find out what the fabric is made of? How would museum conservators do so?

■ Cover two display boards with matching samples of various materials in a range of colours. You could include materials such as silk, delicate cotton, more robust cotton, velvet, PVC, polyester. Put one board somewhere where students can touch the samples and encourage them to do so. After a couple of weeks, compare the untouched samples with those that have been handled.

■ Collect samples of a variety of fabrics in different colours and cut each in half. Hang one half in a sunny window and leave the other half in the dark. In the summer months a change can be observed within a couple of weeks – for best results leave the samples for a couple of months. Note which colours have faded the most. This experiment can also be done with sugar paper. Instead of using two pieces of each colour, attach a square of thick card to the centre of each piece of sugar paper. The change of colour can be seen where the paper was left uncovered but no change occurs in the square that was protected from the light by the thick card.

Answers

Textile conservation – introduction

1. A damp atmosphere would make the textile heavier.

2. Delicate fibres may break if the weight they are supporting increases.

3. As you heat an object it expands; as it cools, it contracts.

4. If the textile has dust between its fibres and the temperature changes, the dust acts as an abrasive and rubs against the fibres as they expand and contract and move over each other. This can cause the fibres to break. Without the movement caused by a temperature change, dust particles can do less damage. A temperature change in the absence of dust only causes the fibres to rub against each other so the damage is less severe.

5. Textiles in museums are often displayed in glass cases because this keeps out dust and prevents people from touching the items. The glass also helps to limit any temperature changes that might take place within the case. The light level is kept low to minimise the amount of damage done to the dyes in the textiles and to prevent the colours from fading.

Textile conservation – the structure of cotton and linen

1. (A copy of the cellulose structure is supplied as part of the student sheets).

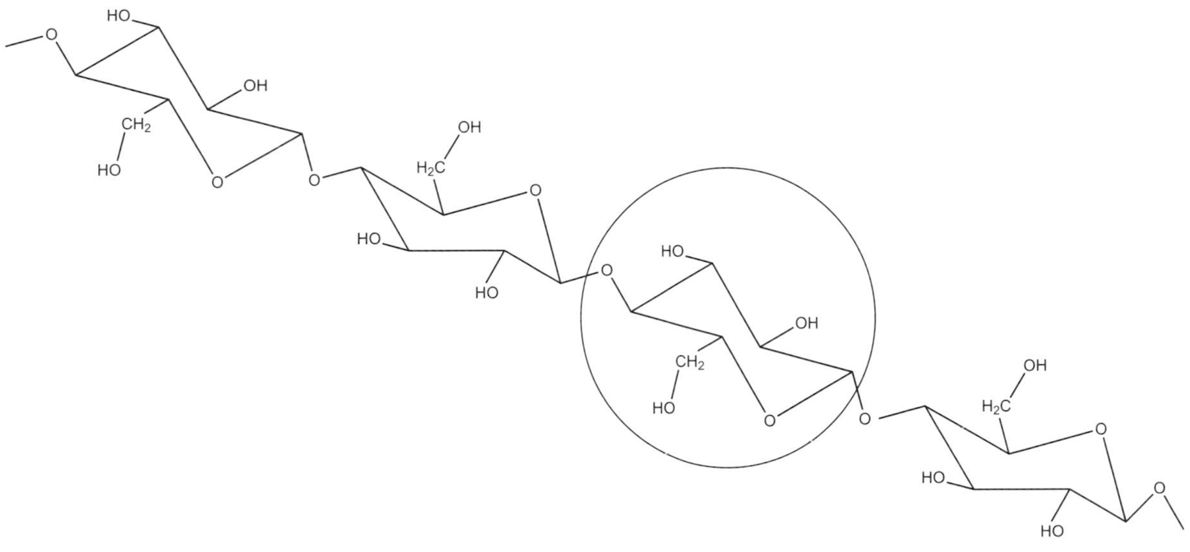

Figure 1 The circle marks one cellulose monomer

2. Starch is unable to form fibres because it has branched chains. The branches prevent the chains from lining up in an ordered way so they cannot form a long ordered crystal (*ie* a fibre).

3. If the water was removed from linen it would become brittle and would be more likely to break.

4.

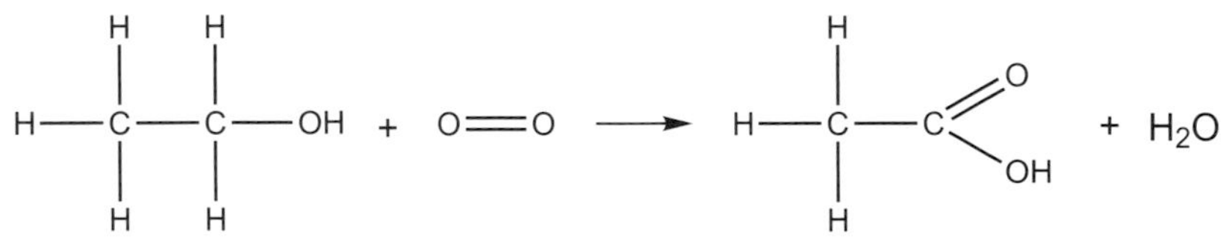

Figure 2 Reaction of ethanol with oxygen

5. This is an oxidation reaction. Oxygen is added to the ethanol and hydrogen is removed.

6. The organic molecule is ethanoic acid.

7. It contains a carboxylic acid group.

8. Solutions containing hydrogen ions are acids.

9. The presence of H^+ ions will increase the rate of this reaction.

10. Catalysts take part in the reaction but are not used up during it. This means they can be re-used so they can take part in several reactions. Since the H^+ ions are not used up when they catalyse the polymer breakdown, they can do more damage than if they reacted with the textile (which would use them up).

11. This reaction will make the textile far more fragile and weak because it breaks down the long chains which give the material its strength.

12. An alkali would react with the H^+ ions and remove them.

13. Conservators use anionic or neutral detergents because the fabric is slightly negatively charged as a result of the damage caused by oxygen. Cationic detergents would be attracted to the fibres and would not be washed away. This would change the chemical make-up of the textile. Anionic and neutral detergents are not attracted to the fibres so they can be washed away easily.

14. This is only a temporary solution because the bonds formed by the cations are not very strong. Also, the fabric will continue to be oxidised if it remains exposed to air and there will be no calcium or magnesium ions available to react with the newly formed negatively charged groups.

15.
$$H^+ + NH_3 \rightleftharpoons NH_4^+$$

16. This is only a temporary solution because the reaction is reversible. Given time, the NH_4^+ ion will turn back into ammonia and hydrogen ions.

17.

$$2 \ R-C \overset{O}{\underset{O^-}{\diagup\!\!\!\backslash}} \ H^+ \ + Ca(OH)_2 \longrightarrow 2\left[R-C \overset{O}{\underset{O^-}{\diagup\!\!\!\backslash}} \right] Ca^{2+} + 2H_2O$$

Figure 3 Reaction of a damaged fibre with calcium hydroxide

Case study – the Victory sail

1. It was important to vacuum the sail to get rid of dust (and fungi spores). Dust acts as an abrasive and damages the fibres, especially when the sail is moved.

2.

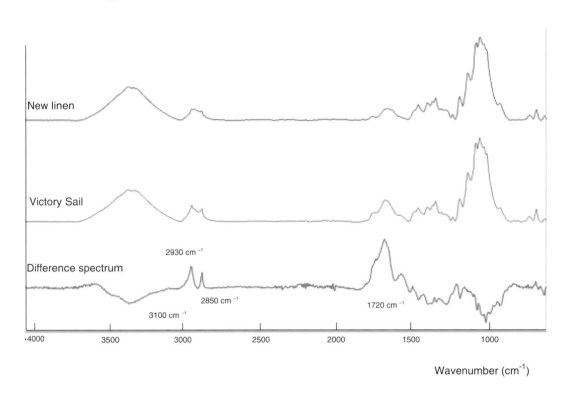

New linen

Victory Sail

Difference spectrum

2930 cm⁻¹

3100 cm⁻¹

2850 cm⁻¹

1720 cm⁻¹

,4000   3500   3000   2500   2000   1500   1000

Wavenumber (cm⁻¹)

Figure 4 Linen spectra and difference spectrum

3. The band at 1720 cm⁻¹ in the difference spectrum supports the idea that the linen
   has been oxidised because it shows that the oxidation product (the old sail)
   contains C=O groups but the new linen does not. If the formula shown in **Textile
   conservation – the structure of cotton and linen** is correct, then the original linen
   contained alcohol groups but no carbonyl (C=O) groups. The carbonyl groups must
   have been produced by oxidation.

4. Yes, this band confirms that the reaction has occurred. If the number of carbonyl
   groups has increased as a result of oxidation of alcohol groups to carbonyls, then
   the number of alcohol groups present should have decreased. The spectrum shows
   that this is the case, which further supports the idea that the linen is being oxidised.

5. The more damage that has occurred (at least in terms of oxidative damage), the
   more carbonyl (C=O) groups there will be in the linen – *ie* the peak at 1720 cm⁻¹
   in the difference spectrum will become more pronounced as the damage increases.
   At the same time, the number of alcohol groups present will decrease – *ie* the band
   at 3100 cm⁻¹ will become even more negative. If these bands are only shallow, the
   linen being tested is very similar to new linen and not much damage has occurred.
   If the bands are very marked then a lot of damage has occurred and the old linen is
   very different from new linen. The damaged old linen will be more brittle than
   new linen and extra care will be needed when handling it.

This page has been intentionally left blank.

Epoxy glues and the ATLAS project: the biggest experiment ever?

Index 3.4.1
2 sheets
Index 3.4.2
1 sheet
Index 3.4.3
4 sheets

ATLAS is a particle physics experiment that will explore the fundamental nature of matter and the basic forces that shape our universe. The ATLAS detector will search for new discoveries by colliding extraordinarily high-energy protons head on with each other. ATLAS is the largest collaborative effort ever attempted in the physical sciences. Over 1800 scientists from more than 150 universities and laboratories in 34 countries (including the UK) are participating.

The protons will be accelerated in the Large Hadron Collider, an underground accelerator ring 27 kilometres in circumference at the CERN Laboratory in Geneva, Switzerland. The particle beams will be steered to collide in the middle of the ATLAS detector. It is hoped that the debris of the collisions will reveal fundamental particle processes and may help in the discovery of new particles. The energy density in these high-energy collisions is similar to the particle collision energy of the early universe less than a billionth of a second after the Big Bang.

The amount of information generated by this experiment will be huge. There will be four experiments and each will generate about one petabyte (10^{15}) of data (one thousand million million bytes, equivalent to a billion copies of the Sunday Times newspaper) every year. This information will need to be sent to researchers in various institutions around the world. Current information technology cannot handle this quantity of data and so new computing systems will be needed. The world wide web was originally developed to handle the data produced in an earlier experiment at CERN and a new generation of the internet is likely to result from the ATLAS project. Much of the development of the new technology is taking place in the UK – for further details see the PPARC (Particle Physics and Astronomy Research Council) website at the address given below.

The idea of colliding particles to find out more about matter is not new. Ernest Rutherford discovered that the majority of the mass of an atom is in a tiny region in the centre (which we now call the nucleus) by bombarding gold leaf with alpha particles. This is described in the RSC publication *Chemists in a Social and Historical Context* by Dorothy Warren. Many other discoveries have since been made by studying collisions.

The ATLAS project will seek to accelerate protons to greater speeds than ever before in the hope that the exceptionally high energy of their collisions will lead to new discoveries and either confirm or help to develop current theories about the make-up of matter.

Aim

The aim of this activity is to give students some idea of the scale of the projects undertaken in science and to make them aware of how scientists from dozens of different countries can work collaboratively to enhance our understanding of the universe.

The activity focusses on one aspect of the ATLAS accelerator/detector – that of the superconducting magnets and, more specifically, the glue which is used to hold them together.

Students will develop their own glue to solve a specific problem and then look at how the issue was tackled by the research team at CCLRC (Council for the Central Laboratory of the Research Councils) in Didcot, Oxfordshire.

Background knowledge required

Developing glue is accessible to the majority of students; the rest of the activity may not be. This activity could be carried out as an enhancement activity or as a science club activity.

Students will need to have some understanding of the structure of the atom. The video (see Resources available below) goes way beyond what students will need at this stage. However, with a little background knowledge and interest they will be able to understand enough to grasp the point of the ATLAS experiment.

Students will need some knowledge of organic chemistry, acids and bases.

Resources available

- **http://atlas.ch/** (accessed November 2005) – The ATLAS website is an excellent source of information with diagrams, graphics and well presented information. Interested students could be directed to it for further details. In particular, the video is superb: **http://atlas.ch/movie/index.html** (accessed November 2005). It is 18 minutes long and very well worth watching as an introduction to the experiment as it gives the background and the scale of the project. Note that there are two English language versions – the USA server version is different from the Swiss server one. The USA version is far more accessible and the question sheet provided in this resource is based on this version of the movie. The ATLAS website provides details of how to obtain a CD copy of the movie if you prefer not to work online. Some suggested questions to accompany the video are provided with the student sheets (see below).

- **http://www.pparc.ac.uk/frontiers/archive/feature.asp?id=18F3&style=feature** (accessed November 2005) – Information from PPARC on the UK scientists involved in ATLAS.

- **http://www.pparc.ac.uk/frontiers/archive/feature.asp?id=8F2&style=feature** (accessed November 2005) – Information from PPARC on the computing power required for ATLAS.

- **http://www.pslc.ws/macrog/eposyn.htm** (accessed November 2005) – Information on epoxy resins and glues.

In this resource

■ Questions for use with the video provided on the ATLAS website – **The biggest experiment ever? – video question sheet**.

■ Glue development worksheet – **Developing glue**

■ Information sheet for students (includes questions) – **The biggest experiment ever?**

Developing glue – equipment needed

For making the glue:

■ Milk – full fat, semi-skimmmed and skimmed – 100 cm^3 of each per group

■ Vinegar

■ Bases, *eg* sodium hydrogencarbonate, magnesium carbonate, calcium carbonate, milk of magnesia

■ Spatulas

■ Stirring rods

■ Measuring cylinders (100 cm^3 and 25 cm^3)

■ Beakers (100 cm^3 and 250 cm^3)

■ Bunsen burner, mat, tripod, gauze

■ Filter funnel and paper

■ Conical flask (100 cm^3 or above)

■ Universal Indicator paper (some students may request it).

■ Lolly sticks

For testing the glue:

■ Weights (the ones on a hook are ideal) to be added aprox. 100 g at a time

■ Sand tray.

Health and safety

Eye protection should be worn whilst making the glue.

To avoid the danger of weights landing on students' feet, make sure a sand tray is placed underneath the sticks as they are tested.

If the sticks snap then there is a risk of splinters. Students should wear eye protection during testing.

Check students' proposals for testing before they are carried out – pay particular attention to their plans for safety.

Notes on the method

Factors that could be varied in the glue making include: the type of milk – ordinary full fat, semi-skimmed, skimmed milk; the base (*eg* sodium hydrogencarbonate, magnesium carbonate, calcium carbonate) – carbonates are good as the mixture bubbles when they are added so it is easy to tell when the glue has been neutralised without using an indicator; the pH of the glue (*ie* vary the amount of base added, from not enough for neutralisation to excess); the acid used to curdle the milk.

Students can make one 'batch' with one type of milk and then divide the curd and water mixture into portions and add a different base to each portion. A large number of different glues can then be made without too much effort.

Testing the glue can be awkward as it is so strong that if paper is used, the paper will almost certainly break before the glue does, which would not allow relative strengths of glue to be tested. One alternative is to use two lolly sticks overlapped by about 2 cm and glued down at the overlap. Once the glue is dry, balance the sticks over the gap between two tables separated by a suitable distance. Place a sand tray underneath and hang weights on the lower of the two sticks until the glue breaks and the sticks come apart. Quite a significant amount of force is required for this – if you are short of weights, make the overlap between the sticks smaller to reduce the force needed. As the results vary, each glue should be tested at least twice.

Table 1 gives some sample results. Two variables were tested – type of milk and base used.

| Type of milk | Mass required to break the glue (g) | |
	Magnesium carbonate	Sodium hydrogencarbonate
Full fat	2500	1200
	3000	2100
Semi-skimmed	2900	2100
	3000	2400
Skimmed	3100	2400
	3000 – stick broke	2300

Table 1 Sample glue test results

The results in the table show that skimmed milk tends to be the best type for this purpose, and magnesium carbonate the best base, although there is some variability within these results.

The glue consists of molecules of the protein casein which are precipitated from the milk by the addition of acid. Polymerisation of these protein molecules forms the glue. The fat in the milk can get in the way of these polymer chains – lubricating them like oil on a bicycle chain – and prevent them from sticking together as effectively as in the absence of fat.

Answers

Video Question sheet

1. Copernicus and Gallileo questioned the geocentric model. They lived in the 16th century.

2. Satellites, mobile phones and broadcasting rely on Newton's equations.

3. The current model of matter is the Standard Model.

4. The four forces in nature are gravity, electrostatic force, the weak and the strong forces.

5. ATLAS will try to find out if the Higgs particle exists, what it is like, what causes particles to have mass, why different particles have different masses, how mass was obtained from energy, why there are only 12 basic particles and why there are three families of particles. It will also try to find evidence of supersymmetry.

6. 34 countries are involved in building ATLAS.

7. High speed protons will be made to collide.

8. The accelerator will contain supermagnets made of superconductors.

9. These will be the biggest supermagnets ever built.

10. The ATLAS accelerator/detector is 45 m x 22 m (as big as a five storey building).

11. The equipment is precise to 1/100 mm (in some cases it can be even more precise than this).

The biggest experiment ever?

Note: The full structures of the molecules used to make the epoxy glues described in the student sheet are given at the end of this section.

1. The aim of the ATLAS experiment is to find out what protons are made of and to discover new particles.

2. During the experiment, protons will be accelerated to very high speeds and made to collide. (Detectors then look at what is produced.)

3. The accelerator is to speed up the protons.

4. The epoxy resin will be used to hold the coil of the superconductor together and prevent individual loops of the coil from touching each other.

5. The glue will need to be an insulator and must not break when exposed to very low temperatures.

6. The glue will need low viscosity so that it can be put in between the coils of the electromagnet without difficulty. It will need to be a good insulator to prevent short circuiting, which would make the temperature of the magnet rise. It will need a long working time so that the coils can be made before the glue dries. It will need to be flexible rather than brittle so that it does not snap when it undergoes extreme temperature changes.

7. The brackets and the number five mean that the group of atoms within the brackets is repeated five times in a chain.

8. Resin B has hydrogen atoms attached to the carbon atom in the middle (between the two carbon chains), whereas resin A has methyl (or CH_3) groups in this position.

9. The resins all contain a group of three atoms in a ring – two carbon atoms and an oxygen atom:

Figure 1 The epoxy group

10. Resin C and hardener 2 both contain long chains in their molecules. These long chains are what gives the glue its flexibility.

11. The two hardeners share an $-NH_2$ group.

RSC | Advancing the Chemical Sciences

The full structures of the molecules used for the epoxy glues are:

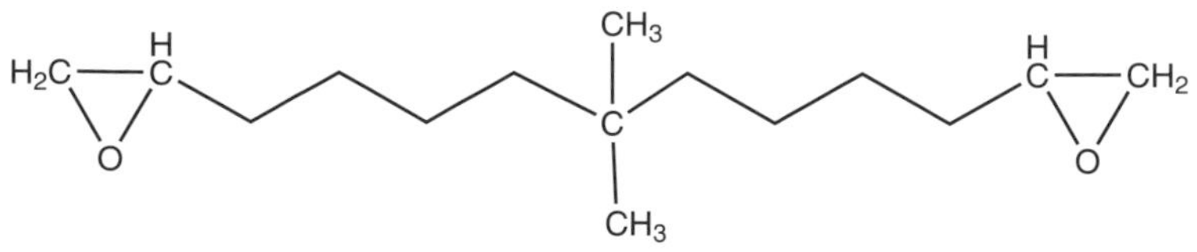

Figure 2 Resin A

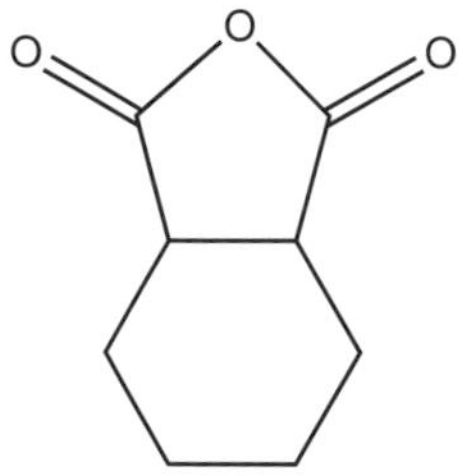

Figure 3 Hardener 1

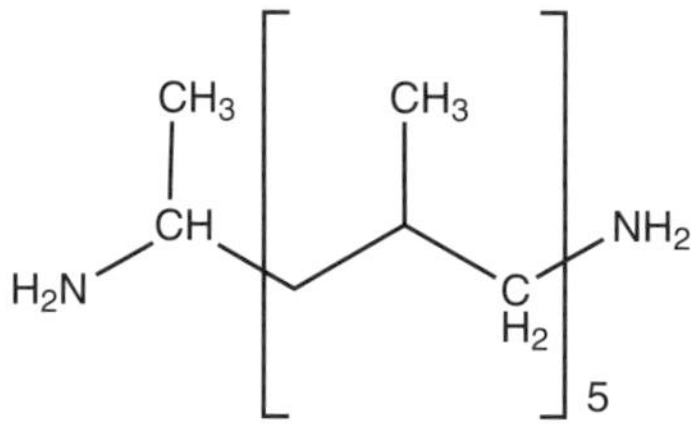

Figure 4 Hardener 2

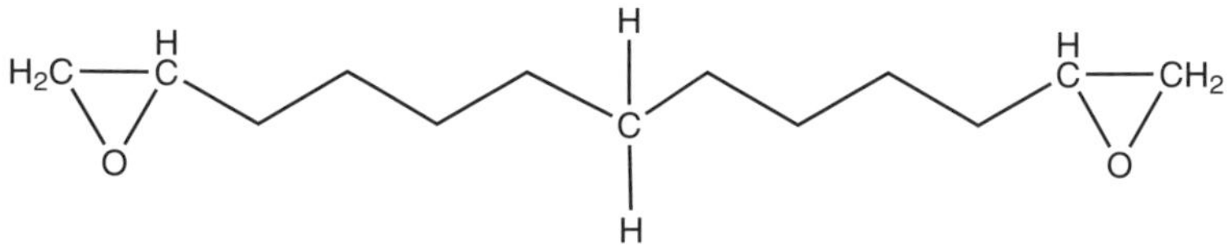

Figure 5 Resin B

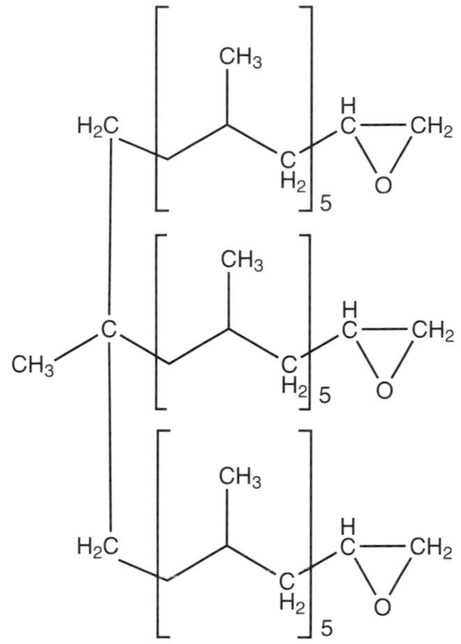

Figure 6 Resin C

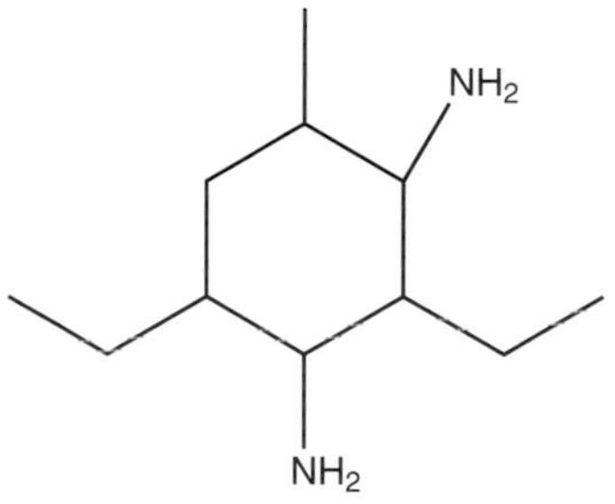

Figure 7 Hardener 3

The structures used in the student sheets are simplified representations only and not the full correct structures. The pictures given enable the students to answer all the questions and understand the key ideas in this topic.

This page has been intentionally left blank.

Chapter 4
Modern applications

(List of student sheets available on CDROM)

RSC | Advancing the Chemical Sciences

Cooking potatoes

Index 4.1.1
2 sheets

The aim of this practical activity is for students to observe and appreciate the changes that take place when a potato is cooked. The vast majority will be aware that the potato changes – this activity will help show what those changes are.

The activity takes at least 30–40 minutes.

Prior knowledge required

Students will need to know the structure of a plant cell, or should be reminded of it at the start of the lesson. A diagram of a plant cell is shown below:

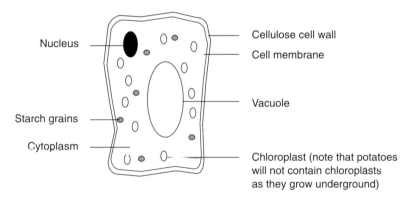

Figure 1 A plant cell

Equipment required

For each student or group of students:

■ Microscope

■ At least 3 microscope slides and cover slips

■ Sharp scalpel – warn students to take care with this

■ 250 cm^3 beaker

■ Bunsen burner, tripod, gauze and mat

■ 0.01 mol dm^{-3} iodine solution – either in a dropping bottle or with a plastic dropper pipette

■ Timer

- Knife

- Potato (one average sized potato will be enough for 4 or 5 groups)

- Tweezers or tongs

- Labels or pens to label microscope slides.

Notes

For students who struggle with practical work (or if you wish to avoid handing out sharp scalpels) you could have the potato already cut into pieces and very thin slices. These slices will not keep and will dry out very quickly so this cannot be done much in advance of the lesson. You may wish to amend the worksheet.

Using a kettle to boil water will help cut down the time required.

Students can look at one sample under the microscope while another is boiling if they are able to do this and keep track of the time.

It may help students if you set up a digital microscope and link it to a television or computer screen (particularly if you can project the images). Students with particularly good slides could show these to the class. You can then discuss what changes take place during cooking. This is better done alongside a class practical rather than instead of it.

Answers to questions

1 and 2. The main changes that take place in the cells of a potato when it is cooked are:

- The cell membrane ruptures

- The membrane around the vacuole breaks

- The membrane around the starch grain breaks and the starch grain swells up, although it initially remains intact

- The cell wall breaks down and the contents, including the starch, begin to disperse.

The first two can be difficult to see under the microscope. The cell membrane is very thin and is usually in close contact with the cell wall. The latter two points show up very well. The starch grains swell noticably and iodine staining makes them clearly visible. You can also see that the starch initially remains in the cells. As time progresses, the iodine-stained starch begins to spread into the gaps between cells and the grains no longer have a distinct spherical shape. You can see some breakage of the cell walls.

3. As the potato is cooked its texture becomes softer and 'squishier'.

4. The reason for the change in texture is the rupturing of the cell walls. When they are intact they hold the potato in a rigid shape. They are strong and hard to break just by gently pressing on the potato. As the walls break down they no longer have a strong rigid structure. Breaks in the structure mean that the potato begins to collapse, which gives the softer, 'squishier' texture.

Baking powder

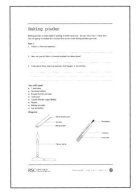  

Index 4.1.2
4 sheets

This activity introduces students to the action of baking powder, its chemistry and some of the ways in which it can be used. The study can be extended to cover all carbonates. Ideally, the latter two stages of this activity should not be done in a laboratory but in a clean evironment where students can be allowed to eat the product of their experiment (always very popular). This usually means that Part 1 has to be done on a separate day.

Equipment required

Part 1

■ Test-tubes

■ Test-tube holders

■ Bunsen burners and mats

■ Limewater

■ Cobalt chloride paper (**Toxic**)

■ Pipettes

■ Baking poweder (or $NaHCO_3$ labelled as 'baking powder').

■ Eye protection.

Part 2

■ Sponge cake (see note below).

Part 3

■ Sugar

■ Golden syrup

■ Baking powder

■ Baking paper or aluminium foil

■ Saucepan

■ Wooden spoon

■ Cooker or Bunsen burner, tripod, mat and gauze.

Important note: For best results, baking powder must be bicarbonate of soda **only**. (Some baking powders contain other ingredients too.) In some brands this is 'baking soda'.

Health and safety

■ Wear eye protection.

■ Cobalt chloride is a carcinogen if inhaled and a skin sensitiser. Minimise handling, *eg* use forceps and wash hands well after use. Refer to Hazcard 25 (CLEAPSS 2000) for more information.

Part 1

This is a simple experiment to show what happens when a hydrogencarbonate is heated. To engage student interest, you could withhold the name of the 'chemical' until they have completed the practical work. Students may find it difficult to believe that something as 'normal' as baking powder is a chemical so you may wish to have a labelled laboratory container of sodium hydrogencarbonate ready for them to test as well.

Students may not have come across cobalt chloride paper as a test for water before. If they have not, demonstrate the test at the start of the lesson so they know what they are looking for. If all your paper has gone pink, either put it in a dessicator for a few days or heat some gently in a boiling tube. It will turn blue again and be ready to use. Bring it to the lesson in a dessicator.

When the baking powder is heated the gas produced can be passed through a delivery tube into limewater in a test-tube. However, there is a danger of suck-back with this method so it may be better for students to use a pipette to collect the gas. As long as they do not squirt cold liquid into the hot test-tube, this is an easy and safe method. For best results, ensure students use only about 1 cm^3 limewater.

The limewater should go milky and students should recall that this is a test for CO_2. The cobalt chloride paper should turn pink, which indicates the presence of water. This can be harder to see but some water usually condenses on the inside of the boiling tube and the cobalt chloride paper changes colour if placed over this.

Students should be familiar with word equations but this may be the first symbol equation they have met so they could need some guidance.

Part 2

Show students some cake. Look at the recipe and discuss why baking powder might be included in the list of ingredients. More able students will be able to do the questions without the need for the visual prop of a cake in front of them but many students will find it helpful – especially if they get to eat it afterwards.

You might wish to prepare a cake without baking powder for comparison (it can be kept in a freezer and re-used over several years). Less able students could then discuss the reasons for the differences between the two cakes.

Part 3

This part of the activity can be done in the laboratory but will be most effective if students have the opportunity to eat the proceeds of their hard work. If they are to do so, the experiment needs to be done in a food technology room or a kitchen.

RSC | Advancing the
      | Chemical Sciences

Answers

1. A chemical reaction is a change that you cannot easily reverse where new substances are made.

2. You can tell a chemical reaction has taken place if a colour change occurs, a gas is produced, the temperature changes.

3. Answers could include making toast, boiling an egg, cooking meat. Do not accept physical changes such as making ice, melting chocolate.

4. Condensation forms on the inside of the tube and the cobalt chloride paper turns from blue to pink.

5. These observations indicate that water is present.

6. The limewater turns milky.

7. Carbon dioxide is being made.

8. $2\ NaHCO_3 \qquad\rightarrow\qquad Na_2CO_3 \qquad + CO_2 \qquad + H_2O$

9. Sodium hydrogencarbonate → sodium carbonate + carbon dioxide +water

10. The baking powder decomposed during cooking and released carbon dioxide gas.

11. This makes the cake rise by putting bubbles of gas in it. Baking powder is included in the recipe to make the cake light and fluffy.

12. The baking powder makes the syrup mixture puff up and turn into a honeycomb with lots of little bubbles in it

13. This happens because the baking powder breaks down to form carbon dioxide. Carbon dioxide is a gas and it causes the bubbles in the mixture.

14. $2\ NaHCO_3 \rightarrow Na_2CO_3 + CO_2 + H_2O$

   Sodium hydrogencarbonate → sodium carbonate + carbon dioxide + water

Rates and rhubarb

Rhubarb contains oxalic acid, which has the formula $C_2H_2O_4$:

Figure 1 Structural formula of oxalic acid

Oxalic acid reacts with acidified potassium manganate(VII) and is oxidised to carbon dioxide and water:

$$2\ MnO_4^- + 5\ C_2H_2O_4 + 6\ H_3O^+ \rightarrow 2\ Mn^{2+} + 10\ CO_2 + 14\ H_2O$$

The potassium manganate(VII) decolourises, which provides a convenient and easy to measure end point for the reaction.

This experiment is probably most suited to less able students who do not need to be given the details of the reaction or to try to relate the rate back to the equation.

The reaction is autocatalysed (catalysed by a product of the reaction) by the Mn^{2+} ions. This could lead to some confusion if students analyse the results too closely.

More able students could be expected to plan their own experiment.

It is also possible to use this reaction to look at how a change in temperature affects the rate of reaction.

Timing

The practical work can easily be completed in a one hour session.

Equipment required

■ Rhubarb – fresh if possible (frozen also works if the pieces are long enough and should be fine whatever the size for the concentration experiment)

■ 100 cm^3 beakers (at least 2 per pair of students)

■ 50 cm^3 measuring cylinders

■ Dilute acidified potassium manganate(VII) solution (**Irritant**) – see note below

- Timer (1 per pair)

- White tile (1 per pair)

- Knives (4–6 per class should be fine)

- 250 cm³ beaker

- Bunsen burners, heatproof mats, tripods and gauzes

- Filter funnels and filter paper or tea strainers

- Eye protection.

Potassium manganate(VII) solution

Put 2 or 3 crystals of potassium manganate(VII) into a beaker with about 500 cm³ distilled water and stir until the crystals dissolve. Add about 500 cm³ 2 mol dm⁻³ sulfuric acid (**Corrosive**) and stir to mix. The solution should be a light purple colour. If necessary, dilute further with a little more water. The exact concentration is not critical.

Health and safety

The sulfuric acid used for making the potassium manganate(VII) solution is 2 mol dm⁻³ and is corrosive. Once it is diluted with potassium manganate(VII) it is approximately 1 mol dm⁻³ and the resulting solution is an irritant. Wear eye protection when making and using the solution.

Warn students not to consume anything in the laboratory – they may be tempted to taste the rhubarb.

If you use home-grown rhubarb, ensure that the leaves are removed before it is given to students as they contain far more oxalic acid than the stalk and are toxic.

Answers

Surface area
1. This question could either be answered simply by saying that the surface area increases or students could measure the surface area of the rhubarb using graph paper or another method.

2. This could be answered simply by saying that the rate of reaction increases as the surface area increases. However, if students have calculated the total surface area of each sample then they could look for a mathematical relationship, *eg* as the surface area doubles, the rate of reaction doubles – or whatever is appropriate for their results.

3. There are a number of possibilities here. The stick of rhubarb may not be the same width along its length, so the length of the rhubarb pieces may vary. Measuring length is not an accurate way of ensuring that you have the same quantity of rhubarb. The outside of the rhubarb is red and the inside white – they may have different amounts of oxalic acid near them. It can be hard to tell exactly when the potassium manganate(VII) has decolourised. Students' plans for improving the experiment will vary depending on what problems they have identified.

Concentration
1. As the number of drops increases, the concentration of rhubarb in the reaction mixture increases.

2. This is not a fair test because the volume of rhubarb juice is not the same each time.

3. It is probably not a big problem in this experiment because the volume of rhubarb juice is always very small compared to the overall volume of liquid.

References and acknowledgements

This experiment is based on an idea of Don Sutherland at DUSC (Development to Update School Chemistry) in Scotland.

More information on rhubarb and the acids it contains may be found at: **http://www.rhubarbinfo.com/rhubarb-poison.html** (accessed December 2005)

Details of the autocatalysis of the reaction carried out in this experiment, together with a possible demonstration and video clip are available from the Journal of Chemical Education:

http://jchemed.chem.wisc.edu/JCESoft/CCA/CCA3/MAIN/AUTOCAT/PAGE1.HTM (accessed December 2005)

The importance of structure: chocolate

Index 4.1.4
3 sheets
Index 4.1.5
2 sheets

This activity would be a good introduction to the topic of structure and bonding. It emphasises how important it is to know about the structure of a substance and demonstrates that structure can make a big difference to the properties of a material. The concept of polymorphism is introduced and can be built on later.

The activity is fun to do, interesting and the taste tests are always popular. It is adapted from material on the Seeing Science website of the CCLRC (Council for the Central Laboratory of the Research Councils). The activities on this website are aimed at 11–14s. For more information, see **http://www.seeingscience.cclrc.ac.uk/** (accessed December 2005).

Prior knowledge required

Students need to know about changes of state and be able to interpret graphs showing changes of state.

Background information

Since its discovery in the 1500s when the Spanish sailed to the New World, chocolate has been an increasingly popular confectionery material in Europe. The first bar of chocolate was produced by Fry and Sons in 1847 and milk chocolate was produced for the first time by Henry Nestlé about 30 years later.

Chocolate is made from naturally occurring ingredients. It is a mixture of many chemical compounds, of which about 400 have been identified. Taste, texture, gloss, 'snap' and other properties can be varied according to how the mixture is processed and chocolate manufacture is a very complex, multi-step process. Successful chocolate making also requires an understanding of how the consumer perceives the product. Tastes in the UK and the USA are very different and European chocolate is different from both the UK and the USA versions. The taste is only partly dependent on the recipe used; the manufacturing process is also key.

Chocolate is made from cocoa beans, which are the seeds of the *Theobroma cacoa* tree. (Theobroma means food of the gods and is nothing to do with bromine.) The beans form in pods. These pods are harvested and the beans are extracted, fermented and roasted before they are shipped to chocolate manufacturers.

The process of making a chocolate bar begins with mixing and grinding. The ingredients are mixed together and ground until the particles are the correct size. The particle size is critical to the 'mouth feel' of the product and is typically about 0.02 mm. The next stage is known as 'conching' and involves the removal of volatile compounds and adjustment of the moisture content and viscosity. This gives the end product its desired flavour. The mixture is melted, sheared (stirred) and cooled in a complex process known as tempering. The temperature and shearing have to be very carefully controlled or the chocolate ends up brittle, crumbly and with the wrong taste. This part of the process and its effects are modelled in the experiment the students carry out.

The complexity of chocolate is the result of one of its key ingredients: cocoa butter. Cocoa butter is a triglyceride produced by the reaction of glycerol (propan-1, 2, 3-triol) with various fatty acids to form a molecule with the shape of the capital letter E. Cocoa butter fats are polymorphic (they can take on a number of different crystal forms). Each different form has its own characteristic melting point and this affects how the product feels in the mouth. The form favoured by the chocolate industry is Form V which, with a melting point of 33.8 °C, 'melts in the mouth' and is the one generally favoured by consumers.

Assessing exactly which forms are present and the precise conditions under which they are produced is an area of on-going investigation. Scientists at the CCLRC Daresbury Laboratory have used X-rays to investigate the structure of chocolate and a video clip of their experiments is available (see next page).

References and further information

http://www.cclrc.ac.uk/activity/ACTIVITY=SRDAnnualReport9697;SECTION=469 (accessed December 2005) – a report on research into chocolate structure at CCLRC.

P. Fryer and K. Pinschower, *Materials Research Society Bulletin*, 2000, **25**, 25–29 – 'The Materials Science of Chocolate,' an excellent (although long) article on the manufacture of chocolate, with reference to its structure.

G. Tannenbaum, *J.Chem. Ed*, 2004, **81**, 1131–1135 – 'Chocolate: A Marvellous Natural Product of Chemistry.'

http://www.cadbury.co.uk (accessed December 2005).

Equipment required

Chocolate to eat:

■ At least two squares of milk chocolate per student. Half needs to be melted and re-hardened first. Take a whole chocolate bar (the ones that are fully wrapped in one sealed wrapper are best). Put it somewhere warm (such as on a radiator) and allow it to melt. Once it has melted, put it in a refrigerator (not the one where the chemicals are stored) to harden quickly. Once it has set, remove it and allow it to return to room temperature prior to the lesson. The remaining chocolate should be of the same make and type but simply stored at room temperature.

Practical work – for each pair or group of students:

■ 1 square of milk chocolate

■ 1 square of the same type of chocolate which has been pre-melted and re-hardened (see above)

■ 2 boiling tubes

- 250 cm^3 beaker or access to a hot water bath
- Kettle (for boiling water)
- Thermometer
- Timer.

Health and safety

Students should not eat in a laboratory.

Check whether any students are diabetic or have other disorders that preclude them eating chocolate.

Boiling water can cause scalding. Warn students to take care.

Suggested lesson plan

- The introduction must be carried out away from the laboratory because students need to eat the chocolate samples. Give each student two pieces of milk chocolate, one from an ordinary chocolate bar and one from a bar of the same kind which has previously been melted and quickly re-hardened. Students should note the differences that they can see, try snapping the pieces and note what happens then eat the chocolate and notice any differences in taste and texture. This should not be a blind trial – they should know which piece is which. Once they have eaten the chocolate they can carry out the rest of the activity in a laboratory. They should be warned not to eat any more of the chocolate during the lesson.

- Discuss the differences between the two pieces of chocolate. Emphasise to students that the chemical composition of the chocolate that was melted is unchanged because it was heated and cooled within the wrapper. Nothing has been added or removed but the chocolate has clearly changed. The way the atoms are arranged and the structure of the components have changed. Ask students how chemists find out what a substance is and what its structure is like. Explain that they will see how scientists at CCLRC in Harwell, Oxfordshire found out about the structure of chocolate, then they will plan and carry out an experiment for themselves.

- Students work through the information on the student sheet and attempt the first few questions.

- Show the video clip from **http://www.seeingscience.cclrc.ac.uk/Activity/Food;SECTION=5237** (accessed December 2005). This clip is also available on a CCLRC CDROM (details are on the website).

- Students plan and carry out the practical work. A set of clue cards is included in this resource to assist students who struggle with the planning aspect of the activity. Photocopy and cut out the clue cards before the lesson. Students should take one at a time and think about how the information helps them to plan an experiment to find out whether chocolate samples contain Form V cocoa butter. A few blank cards have been provided should you wish to create your own clues.

  They should aim to use as few clues as they can when they design their experiment.

Answers

1. Chocolate manufacturers are keen to ensure that their product contains mainly Form V cocoa butter as this is the one that tastes best and has the best texture. It also has a melting point close to body temperature so that the product 'melts in the mouth.'

2. The difference between Form V and the others is the way the molecules are arranged in the structure. They all have the same chemical formula.

3. To test the two samples of chocolate for Form V it will be necessary to measure their melting points. If either melts at around 33–34 °C then it contains Form V. Place each sample in a boiling tube and put it in hot water (50 °C is a good temperature to use). Measure the temperature every 15–30 seconds. The sample should be stirred between temperature readings. Continue until the sample is completely molten. Alternatively, data logging equipment could be used. Plot a graph of temperature against time. If the graph has a plateau at around 34 °C then the sample contains Form V.

Standard chocolate		Pre-melted chocolate	
Time (s)	Temperature (°C)	Time (s)	Temperature (°C)
0	23	0	23
30	28	30	33
60	34	60	37
90	35	90	40
120	39	120	42
150	43	150	46
180	46	180	48
210	48	210	50

Table 1 Sample results

Results will vary slightly depending on the type of chocolate used, the temperature of the water, the quantity of chocolate used and how well the mixture is stirred.

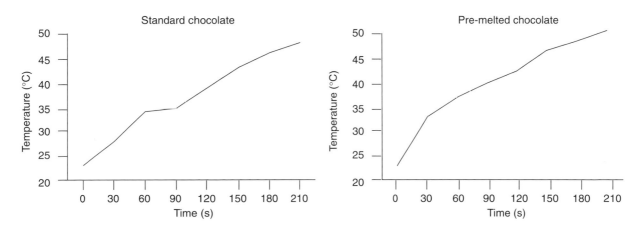

Figure 1 Graphs of sample results

RSC | Advancing the Chemical Sciences

4. The standard chocolate contains Form V chocolate. On the graph there is a plateau or levelling out at 33–34 °C, which is the melting point for Form V. This is caused by the change of state occuring at this temperature.

5. The other sample melts almost as soon as it is put into the hot water. It is hard to tell from the data which form it contains, but as it is not V or VI it must be one of the others. (It is probably a mixture of Forms III and IV).

6. You could remove Forms I–IV by heating the chocolate above 27.3 °C but keeping the temperature below the melting point of Form V (33.8 °C). The molten and solid components could then be separated.

7. Knowing about structures is important to chocolate manufacturers because they want to sell lots of chocolate. Customers only buy chocolate they know they will like so the product needs to be consistently good. As the structure of the cocoa butter has so much impact on the flavour and texture of the product, it is important that manufactures know how to get the required structure.

Extension question

The pre-melted chocolate has a lower melting point than the other sample. If the chocolate is liquid for longer while it is in the mouth then it is more likely that the volatile flavour compounds will be released and will reach the mouth and nose. If more flavour compound molecules are released, a greater sensation of flavour is achieved. You may wish to discuss why this form of chocolate is not used instead. The answer is that it does not rate as well in terms of texture, 'mouth feel' and 'snap.'

Making new medicines – combinatorial chemistry

  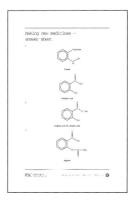

Index 4.2.1
1 sheet
Index 4.2.2
7 sheets

Making medicines and drugs is a topic of interest to many students as virtually all will have taken some kind of medicine at some point in their lives. The pharmaceutical industry is also a major contributor to the UK economy and employs a large number of chemists. Pharmaceutical companies carry out nearly a quarter of all industrial research and development in the UK. However, medicinal chemistry rarely features in pre-16 chemistry courses because the organic chemistry involved in a lot of the processes is too complex.

This activity looks at some cutting edge industrial chemistry – the development of combinatorial chemistry. It does not require students to have an understanding of organic chemistry beyond the simple addition reactions that feature in many pre-16 courses. Students will be guided to:

■ Notice differences between the structures of given molecules

■ Predict the products of a reaction when given the reactants and products of a similar reaction

■ Calculate the number of possible products of a combinatorial synthesis when given appropriate information.

While this activity can stand on its own, it is probably best used alongside the resource **Making medicines** on the RSC's CDROM *Alchemy?* This resource includes a five minute video clip which will be accessible to the majority of 14–16 year olds. The question sheets provided on the CDROM are more suited to post-16s, but alternative questions are suggested here. The video clip will help students understand the written material provided here. It would fit well after the section on aspirin. If the question sheet is used, the film should be shown at least twice. Alternatively, students could be given individual access to the clip and allowed to pause it when required.

This activity could be introduced using the history of medicine timeline available on the Association of the British Pharmaceutical Industry's website: **http://www.abpischools.org.uk/resources04/history/index.asp** (accessed November 2005).

It may be worth explaining to students the use of the terms 'drug' and 'medicine.' A drug is a substance that affects how the body works – either for better or for worse. A medicine improves health. A medicine contains beneficial drugs as the active

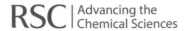

ingredients (or a precursor which forms the active ingredient once inside the body) as well as other substances that make it easy or convenient to take.

Make sure students appreciate that chemists are required to design the synthetic methods and routes used in combinatorial chemistry. The involvement of a robot does not mean that chemists are not required. The syntheses must be designed to be general methods which can be used to make several members of a class of compounds. Once a particular molecule has been selected for further work, an alternative synthesis may be developed that produces the specific compound of interest more efficiently.

Once combinatorial synthesis is complete, mass spectrometry and NMR (nuclear magnetic resonance) spectroscopy are used to confirm the identity of the compounds that have been made, although this is often only attempted for the substances which have shown evidence of the required activity.

References and further information

T. Lister, *Cutting Edge Chemistry*, London: Royal Society of Chemistry, 2000.

Alchemy? CDROM from the Royal Society of Chemistry, sections entitled Making medicines and Combinatorial chemistry. For further information, see **http://www.chemsoc.org/networks/learnnet/alchemy.htm**.

Chemistry Now – Combinatorial Chemistry leaflet from the Royal Society of Chemistry which can be downloaded free from **http://www.chemsoc.org/networks/learnnet/chemnow_combi.htm**

Answers

Video questions

1. A pharmaceutical company makes around 10 000 new chemicals in the process of developing each new medicine.

2. The three main reasons for failure are that the compound is not effective, produces side effects and/or is toxic.

3. About 1 g of a compound is generally made by traditional methods.

4. A few micrograms are required for biological activity testing.

5. The types of chemical used as a starting point are:
   – A chemical found in the body
   – An existing drug
   – A chemical that chemists think might react with an enzyme (they decide by looking at the structure of the compound).

6. An acid (usually trifluoroethanoic acid) is used to detach new compounds from resin beads.

7. Up to 100 kg can be made at a pilot plant.

8. It takes around 10 years to develop a drug.

Making new medicines

1.

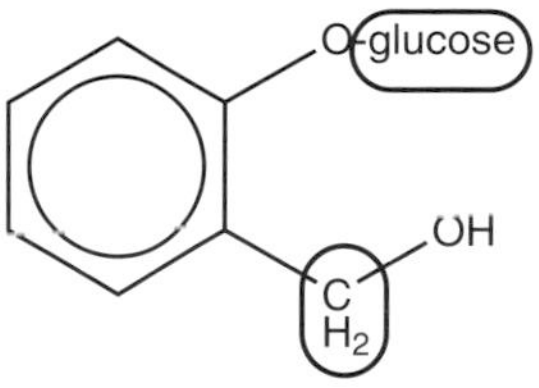

O
glucose
OH
C
H_2

Salicin

O
C
OH
This group is different
OH
No CH_2 group

Salicylic acid

2.

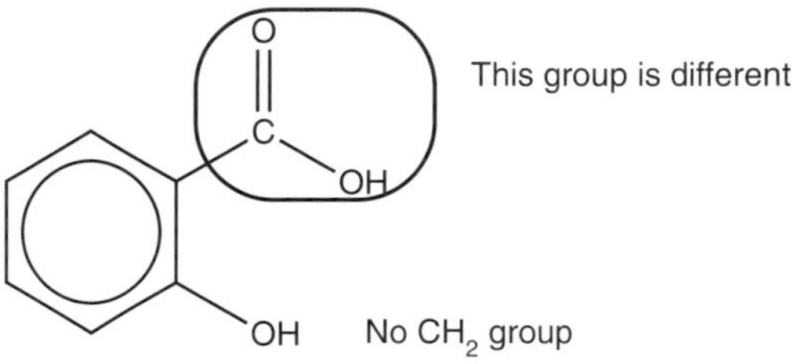

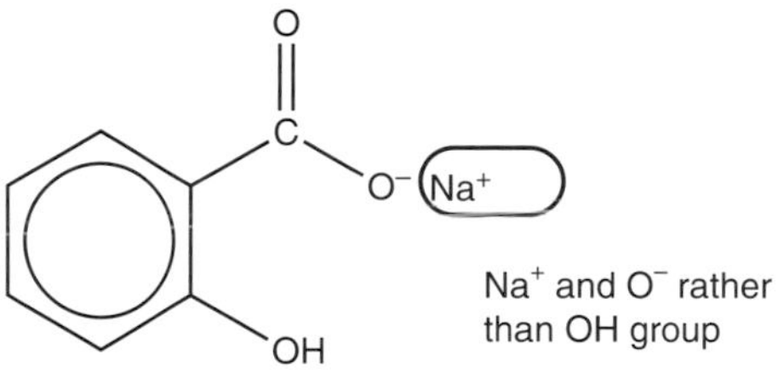

O
C
O^-
Na^+
Na^+ and O^- rather than OH group
OH

Sodium salt of salicylic acid

3.

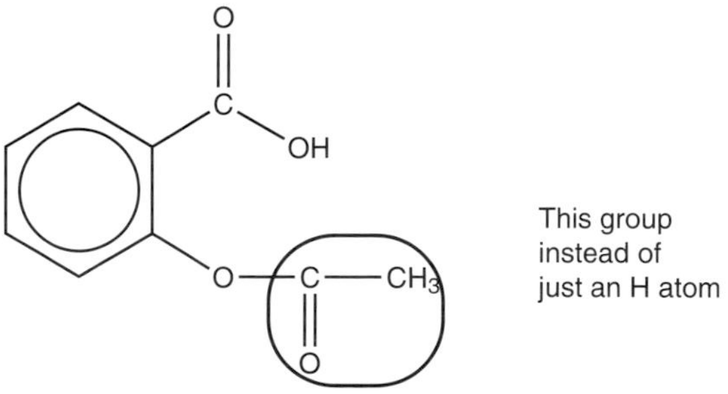

O
C
OH
This group instead of just an H atom
O
C
CH_3
O

Aspirin

4. 12 products can be made.

5. The product from bromine and ethene would be:

Br Br
H C C H
H H

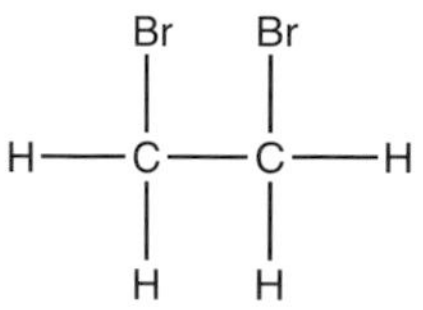

RSC | Advancing the
      | Chemical Sciences

The product from chlorine and butene would be:

The product from iodine and cyclohexene would be:

6. 96 compounds would be made in a 12 x 8 synthesis.

7. Reaction vessel 3 will contain the following at stage 2:
   Bead + ●■
   Bead + ▲■
   Bead + ■■

8. $3^4 = 3 \times 3 \times 3 \times 3 = 81$.

9. $4^3 = 4 \times 4 \times 4 = 64$.

A composite material: concrete

  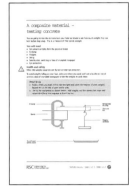

In this activity students make concrete with various additives then investigate how changing the additives can change the properties of the resulting composite. Concrete is a mix of cement, water, sand and gravel. In the first lesson students prepare samples of concrete and leave them to set. In the second lesson they test their samples for strength.

The written activity **Composite materials** could be used alongside the practical work (**A composite material – making concrete** and **A composite material – testing concrete**).

Equipment required

Quantities will depend on how many samples each group or pair of students is to make.

■ Cement (**Corrosive**)

■ Sand

■ Gravel

■ Teaspoons (at least one each for cement, sand and gravel per class)

■ Paper clips or wire of similar thickness

■ Plaster of Paris (**gets hot when mixed with water**)

■ Wooden sticks or splints

■ Other possible additives that could be tried include talc, flour, clay

■ Plastic cups or yoghurt pots

■ Disposable stirrers (lolly sticks or plastic spoons would be suitable)

■ Newspaper

■ Moulds – see note below

■ Large rubbish bag

■ Eye protection

■ Disposable gloves.

For testing the concrete:

■ Set samples from the previous lesson

RSC | Advancing the Chemical Sciences

- G-clamp
- Weights – the type that fit onto a hook are best
- String – needs to be strong or it will break before the cement samples
- Sand bucket or sand tray
- Eye protection.

Health and safety

Session 1 – making the cement: Cement is corrosive. Avoid raising dust. Wear eye protection. Avoid skin contact. Those with cuts, eczema or sensitive skin should wear gloves. Students should wash their hands at the end of the experiment.

Session 2 – testing the cement: Wear eye protection. Care should be taken that weights do not fall onto students' feet. A tray of sand or a box containing tightly scrunched up newspaper should be placed where the weights will fall.

Note on moulds

There are a number of possibilites for the moulds. Students can make their own from card but this is fiddly and time consuming. Ice cube trays can be used but it is sometimes hard to extract the samples and the resulting cubes of cement are short so they can be difficult to test. The best moulds are the trays that new test-tubes come in, cut up into individual pieces. These can be re-used several times and give a good shape for testing. Similarly, boiling tube trays can be used. However, about twice as much cement is required to fill these and far more force is needed to break the resulting samples.

The instructions on the student sheets are based on the quantities required for a mould made from a test-tube tray.

Avoiding blocked sinks

Ensure that students are aware that none of the materials used or produced in this experiment should be washed down the sink. Any spare cement mix can be tipped onto newspaper on their benches and thrown away. Plastic pots and disposable stirrers do not need washing up and can simply be re-used for each sample required or discarded and a new pot used each time.

To keep students away from sinks, water bottles or beakers of water could be provided for use when making the mixtures.

Possible alternatives

Students could investigate how changing the size of the pieces of gravel or the proportion of it in the mix changes the properties of the resulting concrete.

The compression strength of the samples could be tested instead of, or as well as, their tensile strength. G-clamp a piece of the sample to the table very lightly so it is just held in place. Count how many quarter turns of the clamp it takes until the sample breaks. This test could be carried out on samples that have been tested for tensile strength or on separate samples.

Further information on composites

http://www.science.org.au/nova/059/059key.htm (accessed December 2005) – an excellent Australian site with a good overview of composites; includes links to some interesting activities.

http://www.newscientist.com (accessed December 2005) – a search for 'composite materials' gives several articles, many of which are interesting and relevant. A subscription to the magazine or the site is required to view several, but not all of these.

Answers

Composite materials

1. A composite is a material made by combining two or more other materials. It is possible to tell the component materials apart as they do not dissolve or blend into each other.

2. Collagen alone would not be much use in the skeleton because it is very flexible so it would not provide the support required by the body.

3. Composites are particularly important in nature as structural materials. Trees could not stand up and therefore could not grow so tall without wood. We could not stand up and move about without bone.

4.
   a. Mud is the matrix and straw the reinforcement.
   b. Cement is the matrix and gravel (and sand) the reinforcement.

5. Answers will vary. Glass and carbon composites can be used in applications ranging from construction to sports equipment. Shiny helium balloons are made from a composite of aluminium foil and polyester sheeting. There are many composites in use today in a wide variety of ways.

6. Composites are important in the A380 because they help to keep it light. This is very important in getting such a big plane off the ground.

7. Using composites will mean that cars and aircraft can be made lighter. If vehicles are lighter they require less fuel to move them from place to place. Less fuel means lower emissions.

Testing concrete

1. Answers will vary. Mark by impression.

2. The answer to this should be no. Just one sample was used for each test. The amount of water used to make the various samples may have varied and this could affect the strength. There may have been a crack in the sample already. Various other things could prevent one particular sample from being representative of all samples made to a set of specifications.

3. In many ways it is not. The exact position of the G-clamp on the sample and where the weights hang will affect exactly what force is exerted. The weight of the samples varies.

4. Answers will depend on students' results. Gravel of the right size will increase the strength but if the pieces are too big the strength may be reduced. Paper clips will almost certainly make the concrete stronger. Adding steel makes steel-reinforced concrete, which is widely used as a building material.

RSC | Advancing the
      | Chemical Sciences

Investigating a natural composite – chicken bone

Index 4.3.4
3 sheets

This experiment takes three sessions with about 20 minutes of work in each session. It involves taking apart a natural composite, chicken bone, and looking at the consequences of the removal of each of the components. Students can do this as a practical exercise. Alternatively, they could be given prepared samples to examine.

The experiment helps to show that composites are not only manufactured materials but also exist in nature. Bone is generally about 30 % collagen and 70 % hydroxyapatite (calcium phosphate) dry mass. The water content varies.

Prior knowledge required

■ This activity assumes that students know what a composite is.

■ Students should be able to calculate density from mass and volume but may need reminding how to do it.

■ Students should be able to calculate volume – the easiest way is for them to submerge the bone in a measuring cylinder of water and measure the volume of water displaced.

Equipment required

Per pair or group of students:
■ 2 cleaned chicken bones – see note below

■ Beaker – large enough to hold the chicken bone

■ 1 mol dm^{-3} hydrochloric acid – sufficient to cover the bone in the beaker

■ Measuring cylinder – 25 cm^3 will probably suffice; it needs to hold the chicken bone

■ Crucible or other heat proof dish for heating the bone

■ Eye protection.

Per class:
■ Access to a balance (accurate to two decimal places)

■ Oven or other way of warming the bones overnight at 60 °C

■ Furnace or hot oven to remove the collagen.

An alternative to heating the bone in an oven or furnace is to roast it in a Bunsen burner flame. This should be done in a fume cupboard and eye protection worn. The bone may spit and crack so care should be taken. This step could possibly be done by students.

Note on chicken bones

The bones should be raw and clean. They are probably best cleaned by the technician prior to use. A pan scourer usually works well – as much meat and gristle as possible should be removed. Avoid using a scalpel or other sharp instrument. Alternatively, simmering (not boiling) in water with a small amount of sodium carbonate loosens the flesh and it is then easy to remove, *eg* with an old toothbrush. The bones should be rinsed after this procedure.

Health and safety

Chicken bone is a potential source of salmonella and other diseases. Ensure that students are aware how to handle the bones safely. They should wash their hands after handling bone and before eating or putting anything else into their mouths. If this is a concern then the bones can be sterilised after de-fleshing, *eg* by leaving them in domestic bleach overnight. They should be rinsed prior to use.

Wear eye protection when using 1 mol dm^{-3} hydrochloric acid.

The bone ash produced in the final stage of this activity may contain calcium oxide. Avoid contact with skin and eyes. Eye protection should be worn and gloves made available for students with sensitive skin.

Possible homework activity

It might help students if they find out about the structure of bone between sessions 2 and 3.

Answers

1. Calcium phosphate is hard and brittle. Collagen is soft and flexible.

2. The bone is less dense than the components of the bone.

3. Students may come up with various answers. By the end of the activity they should realise that the low density of the bone is a result of its structure – it has a number of holes or cavities within it. At this stage the aim of the question is to set them thinking.

4. Calcium phosphate was removed by the acid.

5. This has made the bone far floppier and softer. It is now more flexible and bendy.

6. Bones like this would not support the body. They would not allow movement because they would bend when the muscles pulled on them rather than moving at the joints as they should.

7. Answers will vary.

8. Density = mass/volume. Answers will vary.

9. Collagen has been removed.

10. Mass of collagen = mass of bone after water removed – mass of bone after collagen removed. Answers will vary.

11. The bone is far more brittle and crumbly. It has far less strength than before.

12. The bones would break far too easily.

13. The bone contains cavities within its structure, as well as spaces for blood vessels.

Note: An image search on the internet will produce a number of interesting pictures of bone. The website **http://invsee.asu.edu/Invsee/invsee.htm** (accessed December 2005) hosts a gallery of images, including a number of electron microscopy pictures of bone (some of which are chicken bone).

14. Bones containing cavities are lighter than solid bones. Therefore less energy is needed to move the animal about. The cavities also allow vital nutrients and oxygen to be delivered to the living cells within the bone.

15.
    a.   Bones from different parts of the skeleton are likely to have different densities. Long tubular bones have a central shaft filled with bone marrow; short bones do not. Flat bones (*eg* in the skull) consist of two layers of compact bone and vertebrae are different again.
    b.   Bones from different animals are likely to have different densities. For example, birds have lighter bones with more hollows in them than many other animals. This allows them to take off.

Hydrogels – smart materials

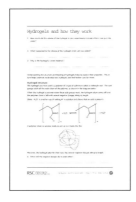

Index 4.4.1
3 sheets
Index 4.4.2
4 sheets
Index 4.4.3
2 sheets

In this series of activities students investigate hydrogels, a type of polymeric smart material. Hydrogels are found in many commonly available products, including disposable nappies, cheap hair gel and plant water storage crystals. The practical work (**Experiments with a smart material – hydrogels**) – is fun to do and the results are clear and easy to see (in two cases, they are both sudden and dramatic). This is followed by written work (**Hydrogels and how they work**). Information is provided on the structure of hydrogels and students consider how this structure relates to the properties they have observed. Finally, a possible future use for very small particles of hydrogel (microgels) in drug delivery systems is introduced (**Drug delivery and smart materials**).

This material could be used to enhance teaching of ionic and covalent bonding or equilibrium. Alternatively, hydrogels could be considered as interesting polymers and as an example of smart materials and nanotechnology.

Prior knowledge required

Students will need to have some knowledge and understanding of:

■ Ionic and covalent bonding

■ Reversible reactions

■ Acids and bases

■ Rates of reaction and particle size – for **Drug Delivery and smart materials** only.

Hydrogels – background information

Hydrogels are polymers that can retain many times their own weight in water. They are often polymers of carboxylic acids. The acid groups ionise in water, leaving the polymer with several negative charges along its length. This has two effects. Firstly, the negative charges repel each other and the polymer is forced to expand. Secondly, polar water molecules are attracted to the negative charges. This increases the viscosity of the resulting mixture because the polymer chain now takes up more space and resists the flow of solvent molecules around it.

RSC | Advancing the
      | Chemical Sciences

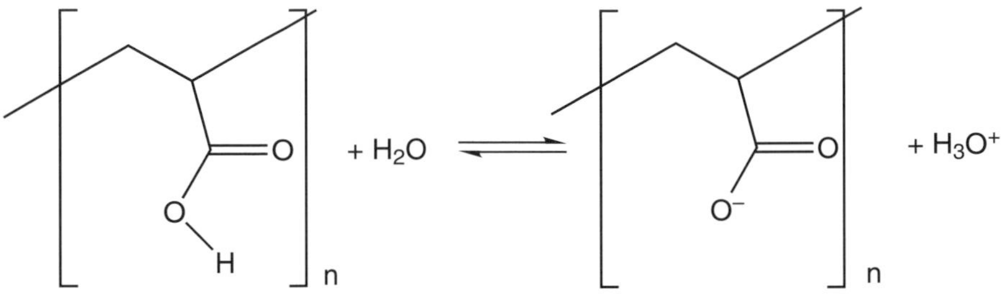

Figure 1 Carboxylic acid groups on the polymer ionise in water

The polymer is in equilibrium with the water around it, but the equilibrium can be disturbed in a number of ways. If the the ionic concentration of the solution is increased, for example by adding salt, the positive ions attach themselves to the negative sites on the polymer, effectively neutralising the charges. This causes the polymer to collapse in on itself again. Adding alkali removes the acid ions and causes the position of equilibrium to move to the right; adding acid has the opposite effect.

There are a large number of hydrogels and they expand and contract at different pH values, temperatures and ionic concentrations. By using a mixture of monomers to create the polymer these characteristics can be fine-tuned.

The commonly available hydrogels that are suggested for use in this practical activity are sensitive to salt concentration but do not show much change across the pH range which can be investigated readily in the classroom. They lend themselves very well to a range of investigative practical work. For example, their volume in different amounts of water or in different salt concentrations could be measured.

Equipment required

Do not be put off by the long list of requirements given below; many items are needed for all the experiments. The equipment is listed separately for each experiment so that just one or two parts of the activity can be prepared if preferred.

Plant water storage crystals

This experiment is referred to in the worksheet **Drug Delivery and smart materials**. If this worksheet is to be used, then tea must be included in the experiment. If you do not intend to use the worksheet, the water crystals can be coloured with a few drops of food colouring (for wonderful, lurid colours) or not at all (but they look great when coloured).

For each pair or group:

Part 1
■ 1 teaspoon water crystals – available from garden centres and sold under various names, *eg* Phostrogen Swellgel

■ Large (at least 1 dm^3) beaker or plastic tub – ice cream or similar tubs are fine

■ 500 cm^3 strong tea – use 1 tea bag per 500 cm^3, pour on boiling water and leave to brew overnight (this tea will stain some containers).

Part 2
■ Sieve (plastic ones are fine) or large funnel lined either with paper towels or with filter paper – groups will be able to share these

■ 2 x 250 cm^3 beakers

- 200 cm^3 very concentrated or saturated sodium chloride (table salt) solution

- 200 cm^3 distilled water

- Dessert spoon or similar – plastic disposable spoons are fine and could be re-used

- White paper (to place under beakers to make it easier to see the results)

- 2 x stirring rods

- Sieve or tea strainer – if a funnel was used earlier, tea strainers are needed now; the same sieves could be used throughout

- 2 petri dishes – lids not required.

Hair gel

For each pair or group:

- Approx 1 large teaspoon hair gel – the cheaper and nastier the better

- Salt

- Petri dish or lid

- Teaspoon or similar – an ordinary spatula is a bit small.

Disposable nappies

For each pair or group:

- A disposable nappy – the ultra-absorbent type

- Scissors

- A large ice cream tub or similar container for collecting the inside of the nappy – this is safer than using newspaper or similar; if tubs are in short supply, large zip-lock bags could be used (students put the nappy in the bag, zip it up and manipulate it until all the hydrogel has been extracted)

- Approx 500 cm^3 distilled water – tap water can be used but the results are not as spectacular

- Salt

- Dessert spoon or similar measure

- Stirring rod

- Large beaker or plastic tub – at least 600 cm^3

- Eye protection

- Plastic gloves for those with sensitive skin.

Note: As an alternative to using nappies and extracting the hydrogel, sodium polyacrylate can be ordered from Sigma-Aldrich.

Hydrogel and sugar

Students are asked to predict the outcome of this experiment towards the end of the worksheet **Hydrogels and how they work**, and then to test their prediction. The experiment should be carried out in the same way as part 2 of **Plant water storage crystals** but using sugar instead of salt solution.

- Remaining hydrated plant water storage crystals

RSC | Advancing the
       | Chemical Sciences

■ Sugar or sugar solution in distilled water

■ Distilled water – it is important that distilled water is used, both in the sugar
  solution and as the 'plain' water in this experiment

■ 2 x 250 cm^3 beakers

■ Tea strainer or sieve.

Timing

It will take over an hour to do all the practical work at once. If lessons are shorter than
that then part 1 of **Plant water storage crystals** can be done in a separate lesson before
the rest of the experimental work. The crystals will keep for a few days in the tea
solution, although the tea may stain some types of container. The remaining practical
work should fit into an hour if students are organised. They should set up the parts of
the experiments that need to be left then do the rest of the work while they wait.

The time required for the written work will depend on the ability of the group. The
Hydrogels and sugar experiment is best done part-way through the written work. The
hydrated crystals can be kept for a few days if covered with water.

Answers

Experiments with a smart material – hydrogels
Students should make detailed notes on their experiments. They should record changes
in volume, colour and any other observations they make. Some expected observations
are described below.

Plant water storage crystals
The crystals swell up from about 5 cm^3 to about 500 – 600 cm^3. They take on the
colour of the tea, which shows that the tea has also been absorbed.

The hydrated crystals swell up more when added to distilled water than in salt solution.
In distilled water, the tea remains absorbed in the crystals and the water does not
change colour. In salt water, the crystals begin to shrink and the water changes colour
as the tea is released.

It is possible to measure the approximate size of individual pieces of the hydrogel and
to show that the pieces have swollen or shrunk.

Hair gel
The hair gel shrinks very quickly when salt is added. After a couple of minutes only
liquid is left in the petri dish.

Disposable nappies
About 10 cm^3 hydrogel can be extracted from the nappy core (the exact quantity
depends on the make and size of the nappy used). The hydrogel absorbs water and
swells up far faster than the plant water storage crystals. It will absorb about 500 cm^3
distilled water. A very viscous mixture results. On adding salt, the viscosity is
immediately reduced and the mixture becomes easier to stir. The hydrogel releases the
water and settles on the bottom of the beaker.

Hydrogel and sugar
For the best results, be sure to use distilled water for this experiment.

The hydrated crystals in the sugar solution have the same volume as those in the distilled water. If they are left to stand, the tea is not released for about 15 minutes. (After this time, the water in the hydrated crystals is in equilibrium with the water in the beaker and some tea may begin to be observed.)

Hydrogels and how they work

1. The volume should increase from about 1–2 cm^3 to about 500 cm^3, *ie* by about 500 times. Exactly how much the hydrogel expands depends on the make of water crystals and the tap water used. Students' results with the nappies may differ from those with the plant water crystals. This is partly due to the difference in the hydrogels but is also a result of the use of tap water in one experiment and distilled water in the other.

2. In the presence of salt the volume of the hydrogel decreases dramatically and the water is released again.

3. The hydrogel is a smart material because it changes shape when a change occurs in its environment – in this case, a change in the concentration of ions.

4. The negative charges will repel each other.

5. This will force the polymer chain to open up and take up more space to allow the negative charges to move apart.

6. The water molecules (the positive ends) will be attracted to the negative charges on the polymer molecules.

7. $NaCl(s) \rightarrow Na^+(aq) + Cl^-(aq)$

8. The Na^+ ions will be attracted to the polymer chain.

9. This will neutralise the charges on the polymer chain.

10. This will cause the molecule to collapse back to its original shape.

11. The symbol means the reaction is reversible.

12.
    a. If you add acid the equilibrium moves to the left in order to remove H_3O^+ ions (*ie* acid) from the solution.
    b. If you add alkali it reacts with the acid on the right hand side of the equation and removes it. The equilibrium will move to the right, forming more acid.

13. Yes, the hydrogel would be expected to show smart behaviour in response to changes in pH.

14. The shape of the hydrogel does not change when it is placed in sugar solution. Sugar is a covalent molecule and will not be attracted to the hydrogel, nor will it react with any of the molecules in the polymer/water equilibrium. Hydrated hydrogel crystals in a solution of sugar in distilled water should be the same size and take up the same volume as hydrated hydrogel crystals in distilled water.

15. The nappy hydrogel will probably absorb about 500 cm^3 distilled water.

16. The nappy will absorb less urine than distilled water. Urine contains dissolved salts, which will reduce the amount of water the hydrogel can absorb.

17. Manufacturers might put hydrogels in hair gel to increase its volume and make the product look good. The main ingredient of many hair gels is water.

18. BARRICADE® gel contains a large quantity of water. The water removes the heat from the fire before it can get to the coated item. Without heat, the fire triangle is broken and the fire cannot burn.

For further information on BARRICADE® gel, see **http://www.barricadegel.com** (accessed December 2005).

Drug delivery and smart materials

It may be a good idea to allow students to discuss questions 1 and 2 in groups then talk them through the answers and the following paragraph. They can then complete the written work on their own.

1. Methods currently used to get drugs into the body include swallowing, injection or drip, inhalation (*eg* asthma drugs), patches placed on the skin (certain bandages and nicotine patches), slow release implants (used for some contraceptives such as Norplant), topical creams placed on the skin. Where the target site is external, such as the skin or eyes, it is relatively easy to deliver the drug selectively to the target. If the drug is required internally it is harder to make it site-specific.

2. Most pills enter the digestive system, pass into the blood stream and then travel through the whole body.

3. The tea is the drug and the hydrogel the carrier.

4. When the hydrogel is soaked in tea, both water and tea enter it. The gel changes colour so you can see that it has absorbed the tea as well as the water. The water is absorbed because the hydrogel has many negative charges on its molecules. Water molecules are polar (they have small charges on them) so they are attracted to the negative charges on the hydrogel polymer.

5. In salt solution, the tea is released from the hydrogel; in distilled water, it is not released. The salt causes the hydrogel to change shape and release some of the water it has absorbed. As it does so, the tea is also released. In distilled water, the hydrogel does not release any of its bound water and so the tea also remains bound.

6. If the particles are very small the overall surface area is very large. This will increase the rate at which the drug is released when the conditions are right.

Superconductors

Index 4.5.1
3 sheets
Index 4.5.2
1 sheet

This activity enables students to look at the topic of structure and bonding of metals and giant ionic compounds in the context of superconductors. It introduces some complex formulae and shows that they work in the same way as the simple formulae students are used to. The material also offers the opportunity for a discussion of the concept of 'blue skies' research.

Notes

The activity could be introduced by showing students a video clip illustrating the levitation properties of superconductors. There are some good, short clips on **http://www.fys.uio.no/super/levitation/** (accessed December 2005).

A list of uses of superconductors is given in the student material. You may prefer to ask students to research this for themselves. There is a very good section on the website **http://www.superconductors.org** (accessed December 2005), which would be an excellent starting point for their research. The site has details of several uses and potential uses of superconductors and provides links to other related web pages. Students could research one use and produce a poster on it. The posters could then be shared and discussed with the whole class. Ask students to think about the advantages and disadvantages of superconductors for each application.

Some students may be interested in how superconductors work. The above superconductors website has some good, clear explanations of this, particularly at **http://www.superconductors.org/oxtheory.htm** (accessed December 2005).

Answers to questions

1. A conductor is a substance that conducts heat or electricity. Metals usually conduct when solid.

2. A superconductor is a substance that conducts electricity without any resistance.

3. Advantage – no loss of energy.
   Disadvantage – high cost of cooling with liquid helium.

4. From 1911 to 1962 – 51 years.

5. Giant ionic compounds do not conduct when solid as they have no free electrons or other charged particles that can move to carry the current. As they do not ordinarily conduct, it is surprising to find them behaving as superconductors.

6. Nitrogen is far more common. Also, since nitrogen has a higher boiling point than helium, less pressure is needed to liquefy it, which means less energy is required to produce liquid nitrogen.

7. Yttrium, 1 atom; barium, 2; copper, 3; oxygen, 7. This compound is called yttrium barium copper oxide (but is usually referred to as YBCO).

8. The compound could be made to superconduct by cooling in liquid nitrogen, which is a lot cheaper than liquid helium.

9. The material contains mercury, thallium, barium, calcium, copper, oxygen.

10. Mercury, 1 atom; barium, 2; calcium, 2; copper, 3; oxygen, 8.

11. If the planes (or sheets) of atoms in a metallic structure slide over each other, the atoms end up in an identical environment to the one they have just left and they are still held in place by the 'sea' of delocalised or free electrons. If planes of ions slide over each other in an ionic structure, a displacement of one ion across leads to an arrangement in which ions of like charge are touching. These ions repel each other and the structure shatters.

Blue skies research

These questions are designed to stimulate discussion, rather than requiring full written answers. Examples of discoveries that have been made by chance or when researchers were looking for something else include the glue in Post-it® notes (scientists were searching for a strong glue), polythene and Teflon®.

The discovery of the hole in the ozone layer is an example of something that was missed by those with the most data. The credit for the discovery went to Molina and Rowland in the mid-80s but NASA had relevant data right back to the mid-1970s. However, NASA scientists missed the discovery because their data had not been properly analysed. There are more details on this story on the NASA website **http://www.nas.nasa.gov/About/Education/Ozone/history.html** (accessed December 2005).

Chapter 5
Nanotechnology

(List of student sheets available on CDROM)

Nanotechnology size and scale

Index 5.1
1 sheet

Further activites on nanochemistry size and scale can be found in the RSC publication *Contemporary chemistry for schools and colleges*, London: Royal Society of Chemistry, 2004.

Name	Symbol	Number in standard form	Image
Terametre	Tm	10^{12} m	Bigger than the diameter of the solar system, less than the distance to the closest star
Gigametre	Gm	10^{9} m	The sun is 1.5 Gm across.
Megametre	Mm	10^{6} m	Earth
Kilometre	km	10^{3} m	Angel Falls, Venezuela – 980 metres
Hectometre	hm	10^{2} m	Football pitch
Decametre	dm	10 m	Orca
Metre	m	1 m	Royal Python snake
Centimetre	cm	10^{-2} m	Width of a fingernail
Millimetre	mm	10^{-3} m	Mite
Micrometre	μm	10^{-6} m	Bacterium
Nanometre	nm	10^{-9} m	Buckyball – approximately 1 nm
Picometre	pm	10^{-12} m	Atom diameters range from 30–600 pm

Nanotechnology and smelly socks

Index 5.2.1
1 sheet
Index 5.2.2
7 sheets

This activity is based on the science behind the consumer product 'Purista', a treatment for textiles which helps to kill the bacteria that grow on clothing and cause it to smell.

The activity requires background knowledge of a number of areas of science:

■ Bacteria

■ Acids/bases and neutralisation

■ Reversible reactions

■ Ionic bonding.

Further information about the product is available on the company's website **http://www.purista.co.uk** (accessed November 2005) but there is very little scientific information there.

'Purista' is the marketing name of the bacteriocide PHMB or Poly(hexamethylene biguanide hydrochloride). It has long been used in swimming pools and contact lens solutions but has now been developed by Arch Chemicals to provide near-permanent antibacterial protection for clothing.

The polymer contains positive groups that appear regularly along its chain:

$$\left[-CH_2-CH_2-CH_2 \underset{\underset{H}{N}}{\overset{\overset{NH}{\|}}{C}} \underset{\underset{H}{N}}{\overset{\overset{\overset{+}{NH_2}\quad \overset{-}{Cl}}{\|}}{C}} \underset{\underset{H}{N}}{} CH_2-CH_2-CH_2 - \right]_n$$ n – where n is about 16

Figure 1 PHMB or Poly(hexamethylene biguanide hydrochloride)

These positive groups can bind to cellulose fibres, such as cotton or viscose. The fibres contain carboxylic acid groups formed by oxidation of alcohol groups during processing of the raw material:

RSC | Advancing the
      | Chemical Sciences

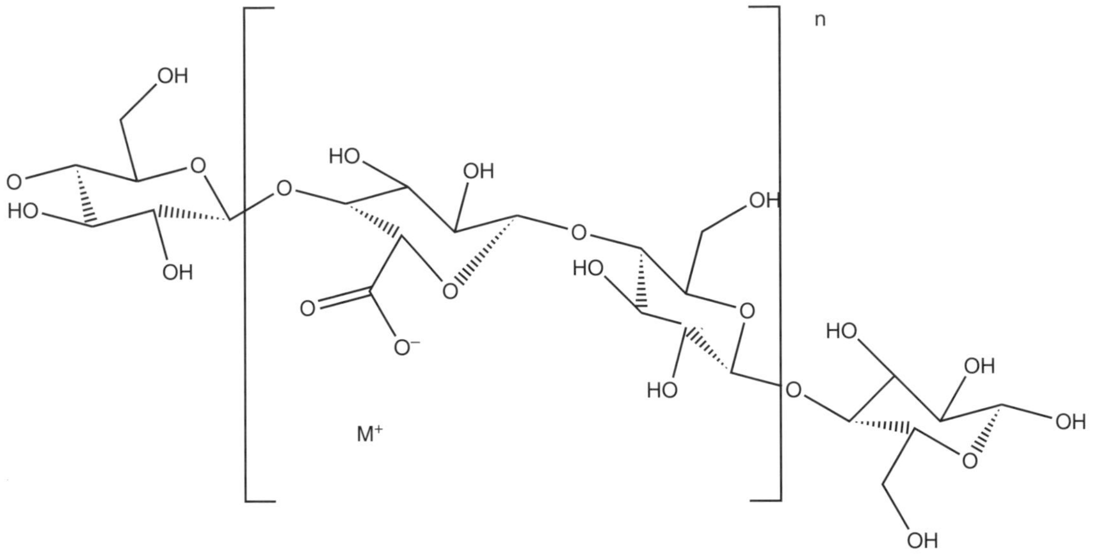

Figure 2 A cotton fibre after processing

In solutions above about pH 6, the carboxylic acid groups dissociate to give carboxylate ions, which are negatively charged. These carboxylate ions bond to the positive charges on the PHMB molecule and hold it in place. Several bonds form between each PHMB molecule and cotton fibre, which join together a little like Velcro®. This holds the PHMB in place even during washing.

No bacteria have yet been found that are resistant to PHMB, even though it has been used in swimming pools for many years.

Suggested lesson plan

■ To introduce the lesson you could show the slide **Odd one out?** provided in this resource. The bacteria is the odd one out because the rest of the items contain PHMB, sometimes known as 'Purista®', which kills bacteria.

■ The Belle and Bunty website (**http://www.belleandbunty.co.uk**, accessed November 2005) sometimes has a video of their latest show, which you could use as an introduction.

■ Students work through the information and questions. For greater impact, you could put the pictures of bacteria up on a screen using a data projector. Some sensitivity may be required if you have students in the class who have a body odour problem.

■ The material could lead to a discussion of how science is presented to the general public. Why is such a scientific product marketed without any mention of how it works? Can students think of other products where this is also the case?

Answers to questions

1. Bacteria on clothing and on your skin have warmth, food, water and shelter. These conditions allow them to thrive.

2. Washing is now largely done at 40 °C, which is not a high enough temperature to kill bacteria. Thus the bacteria can survive from one wash cycle to the next, their numbers building all the time. (They are also stuck to the clothing in a biofilm which is not broken down at 40 °C.)

3. Although the bacteria are being killed, sweat, oils and dirt still accumulate on the clothing and must be removed.

4. Delicate clothing needs to be washed at low temperatures that do not kill bacteria so Purista can help get rid of the bacteria instead. Socks are very good places for bacteria to grow because they are worn inside shoes where the bacteria are kept warm and moist and the colony can grow rapidly. Purista helps solve this problem.

5. All acids contain the H^+ ion.

6. $HCl + NaOH \rightarrow H_2O + NaCl$

7. $H^+ + OH^- \rightleftharpoons H_2O$

8. The symbol means reversible reaction/reaction at equilibrium.

9. You could ensure that there are several negative charges on the fabric by removing the H^+ ions, which would cause the equilibrium to move to the right. You could do this by adding alkali to the solution. The alkali would react with the H^+ ions and remove them.

10. The cotton fibres need several negative charges so that the PHMB can attach to the fibres in several places and be held in place more effectively.

11. An ionic bond will form between the cotton and the PHMB.

12. There are few negative charges on the cotton at low pH so it is unable to bind to the positive PHMB. Above pH 6, there are very few H^+ ions to stick to the cotton so the PHMB can stick instead. Adding an alkali removes H^+ ions, which moves the equilibrium to the right so that more negative charges are formed, providing more places for the PHMB to bind.

13 and 14. In questions 13 and 14 students are asked for their opinion.

RSC | Advancing the Chemical Sciences

Nanotechnology and smelly socks – introduction

Which is the odd one out?

This is available as a slide on the CDROM
– Index 5.2.1

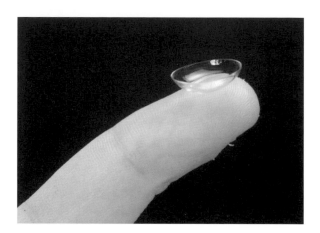

Picture reproduced by kind permission of Belle & Bunty (http://www.belleandbunty.co.uk)

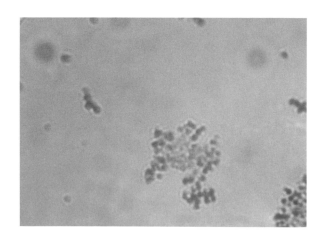

Nanosilver in medicine

Index 5.3
2 sheets

This resource would fit into work on rates of reactions and particle size or nanotechnology. It aims to show how scientists are using nanotechnology to improve something as simple as a wound dressing.

Further information on these dressings can be found on the following websites:

- ■ **http://www.nucryst.com** (accessed November 2005) – in the section on SYLCRYST® medical coatings

- ■ **http://www.smalltimes.com/document_display.cfm? section_id=46&document_id=8027** (accessed November 2005)

- ■ **http://www.smith-nephew.com/investors/portfolio/wound-Acticoat.html** (accessed November 2005).

Answers

1. The properties of silver that make it useful for jewellery are: it is unreactive, shiny and does not dissolve in water.

2. These properties would make silver less useful for medicine because if it does not dissolve it will be hard to get it into the body and if it is unreactive it will not do anything even if it does get to the source of the problem.

3. The SILCRYST® nanocrystals on the Acticoat® dressings are much smaller than the normal crystals and their surface is much rougher. This means they have a much larger surface area than the normal crystals.

4. This will make the silver more reactive as it is the atoms at the surface that are able to react or dissolve. As there are far more atoms at the surface of the nano silver it is much more reactive than normal silver.

5. The bacteria will multiply very rapidly if they are not killed. The conditions in a wound are close to optimum for bacteria and they could double in number every 20 minutes, which would quickly lead to an infection. The infection could then spread elsewhere in the body.

6. The nanoparticles are more effective against the bacteria as they dissolve far more quickly than larger particles. Nanosilver can therefore get to work faster than normal silver.

7. If dressings are improved, this may mean that fewer people get infections that could become serious and make them very ill or lead to an amputation. Improving simple treatments could help reduce the need for more difficult ones.

This page has been intentionally left blank.

The surfaces of substances

Index 5.4.1
3 sheets

The aim of this activity is to introduce students to the idea that the surface of a substance may behave differently from the bulk material. This is achieved by observing two familiar experiments closely. It is worth carrying out both experiments to get the point across.

Practical activity 1 – equipment

- Hydrochloric acid – 2 mol dm^{-3} works well

- Marble chips – at least 7 mm in diameter

- Calcium carbonate powder – label as 'crushed marble'

- Small beakers or boiling tubes

- Measuring cylinders

- Timers.

Practical activity 2 – equipment

- Oil – corn oil works well as it is slightly darker than sunflower oil

- Detergent – good quality washing-up liquid

- Blue food colouring – optional, but it does make the microscopy easier.

Answers

1. The marble powder will react faster. This is because the acid can only react with the atoms/particles on the surface of the marble as particles from the two substances have to collide in order to react. The powdered marble has more atoms/particles on the surface overall and so more reactions can take place at the same time.

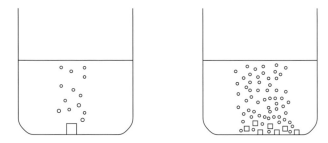

Figure 1 If the surface area is increased, the reaction becomes faster

2. It is probable that students will be unable to time the reaction: the powder reacts too quickly and the large lump too slowly. This does not matter as qualitative observation is enough to get the point across.

3. The results should back up the prediction, otherwise an explanation similar to that given in the answer to question 1 should be included here.

4. Students should be able to observe that gas is only given off from the surface of the large lump of marble – they should therefore be able to deduce that only the surface reacts with the acid.

5. As the pieces get smaller, the reaction should get faster as more atoms/particles are at the surface and available to react with the acid.

6. The water and oil are touching each other only at the surface between them.

7. Without detergent, the oil and water do not mix at all to start with – they form two completely separate layers. When the boiling tube is shaken, they mix but quickly separate back out into two layers.

When the detergent is added the oil and water start to mix at the boundary between the layers. Small balls of oil sink through the water, but most of the oil remains separate from the water.

When the detergent mixture is shaken, the liquids mix and remain mixed.

8.

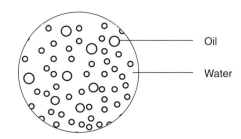

Figure 2 A dispersion of oil in water

9. The oil and water are touching at the surfaces only, but there are more surfaces and the total surface area is much bigger.

10. The amount of oil touching the water has increased.

11. Foam has formed at the surface where water touches air.

12. Particles at the surface do not behave the same as the ones in the middle.

Changing the surface

Index 5.4.2
1 sheet
Index 5.4.3
1 sheet
Index 5.4.4
3 sheets

This activity is based on the chemistry of functionalising the surface of a textile to make it water and stain resistant. The key idea is that a short length of $-(CF_2)n-$ polymer can be added to the surface of clothing to change its characteristics.

The polymer is polytetrafluoroethene (also known as a perfluoroalkyl) – the same polymer as that used to make Teflon® coatings. Teflon® and these textile coatings work in the same way. The tightly-bound, non-bonding electron pairs surrounding each fluorine atom are not easily polarised. This prevents the atoms both from hydrogen bonding with water, and from forming dispersion interactions with nonpolar liquids such as oils.

The technology for these coatings was developed by Prof J P Badyal at the University of Durham during the 1990s. The work was done in collaboration with the Ministry of Defence. Staff at the Ministry were interested in producing suits for armed forces personnel that would repel toxic chemicals such as mustard gas (a dispersion that consists of very tiny particles of liquid).

The mechanism of reaction is more complex than that of normal addition polymerisation. The details of the mechanism are still not fully understood and there are a number of different possibilities for how it might occur. This is not covered in the student material.

Prior knowledge required by students

Students will need to have studied addition polymerisation or at least be aware of what polymers are. They should also know something about surfaces – at a minimum, they should be aware that the surface behaves differently from the bulk of a substance. They could perhaps have done the practical activity **The surfaces of substances**.

Suggested lesson plan

■ Begin by showing students the slide entitled **What's the connection?** This could be displayed as students are coming into the room. It shows pictures of four seemingly disparate items. The connection between these items is that the properties of each are governed not by what the bulk of the item is made of, but by its surface.

■ Demonstrate super-repellant surface properties, either by dropping water onto a super-repellant surface (such as waterproof trousers for hikers), by showing the slide **Water-repellant surfaces,** or by showing the video clip **An electrifying way to stay dry** available at **http://www.research-tv.co.uk/stories/science/plasma/** (accessed November 2005).

Alternatively, students could try dropping water onto a variety of surfaces themselves and looking at the shape of the drop formed on the surface. They could try cotton and other fabrics, Teflon® sheets, glass slides and waterproof trousers. (Note that the waterproof trousers on the market are generally not made using the technology featured in this activity and often do not have such convincing water-repellant properties as the fabrics shown on the slide and in the video clip.)

■ Students do the first few questions on the student sheet.

■ Show the video clip from **http://www.research-tv.co.uk/stories/science/plasma/**

■ If possible do a demonstration with a plasma ball: As the electricity passes through the gas in the ball it excites the gas. You can see this happening because light is produced. By putting your finger on the outside of the ball you can localise the glow on your finger. If a textile is used instead of your finger on the glass surface, then it is possible to coat the textile with the excited gas, which forms a polymer coating.

■ Students complete the rest of the student sheet.

■ You could conclude with a discussion of surfaces that may be in use in the future (see question 9 on the student sheet).

Answers

1. There are a number of possibilities here: aprons for chefs/cooks; babies' changing mats; tents; chairs and other surfaces in restaurants and bars etc. Get students to use their imaginations.

2.

$$\begin{array}{c} F \qquad\qquad F \\ \diagdown \qquad\qquad \diagup \\ C = C \\ \diagup \qquad\qquad \diagdown \\ F \qquad\qquad F \end{array}$$

Figure 1 Monomer used to make PTFE

3. High temperature, pressure and a catalyst are usually used to make a polymer from a monomer in addition polymerisation.

4. Using a large number of solvents is expensive, especially if they are not recycled. Solvents escaping into the atmosphere can cause a number of environmental problems (*eg* chlorinated solvents destroying the ozone layer). A company may be fined if they are caught releasing a large quantity of pollutants into the environment and local residents may well complain. In addition, many solvents are flammable and can be a major hazard if a high concentration is present in a factory. Some solvents are hazardous to human health.

5. If the fabric were coated with a thick layer when only a thin one was needed it would cost a lot more to coat the fabric than necessary. A thick coating may well change the 'feel' of the fabric and make it less pleasant to wear.

6. The process has minimal environmental impact because so few resources are used. Since only a small amount of electricity is required, less CO_2 must be released during electricity generation than for a process that uses more electricity. Very little waste chemical material is produced, which reduces the need for waste disposal (often a cause of environmental damage).

7. If you can coat a wide range of materials then the process will find a wide variety of applications. Also, you do not need to put such careful controls in place to restrict which material is put through the process.

8. A shirt made of normal waterproof material is unlikely to be able to breathe and will feel very 'sweaty' and unpleasant next to the skin. An ordinary shirt with the polymer coating described will feel like an ordinary shirt and be far more pleasant to wear.

9. An open question aimed at encouraging students to think about some potential applications of fuctionalised surfaces. They may consider fragranced surfaces, other types of repellancy, clothing that could sense if you were sick, surfaces that kill bacteria or almost anything else. You could set them the challenge of searching the internet to find out if their idea is already being researched somewhere (depending on what they have thought of).

References

J.P.S. Badyal, *Chemistry in Britain*, 2001, 37, 45.

S.R Coulson, J.P.S. Badyal et al, *Langmuir*, 2000, 16, 6287.

http://www.research-tv.co.uk/stories/science/plasma/ (accessed November 2005)

What's the connection?

This is available as a projectable image on the CDROM – index 5.4.2

Wear of car brake pads

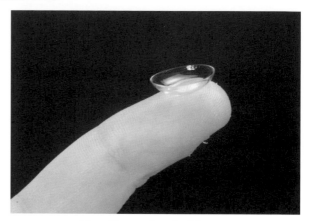

Cleanliness of optical lenses

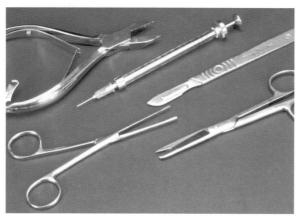

Resistance of medical devices to bacteria

Speed of computer hard disks

Water-repellant surfaces

Image of a surface treated for superrepellancy kindly supplied by Professor JP Badyal, University of Durham see http://www.dur.ac.uk/chemistry/Staff/jpsb/jpsb.htm

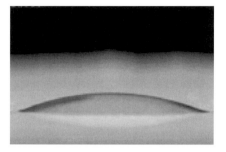

Before surface treatment

The same surface after surface treatment

RSC | Advancing the
Chemical Sciences

This page has been intentionally left blank.

Chapter 6
Plastics

(List of student sheets available on CDROM)

6.1   Plastics

6.1.1   What use is plastic?
6.1.2   Degradable plastics – information sheet
6.1.3   Degadable plastics

6.2   Packaging

6.2.1   Active and intelligent packaging
6.2.2   Disposable cups and the environment – information cards
6.2.3   Disposable cups and the environment
6.2.4   Managing waste and rubbish

6.3   Nappy choice and the environment

6.3.1   Nappy choice – information cards
6.3.2   Nappy choice and the environment

6.4   Dry cleaning and green chemistry

6.4.1   Dry cleaning and Green Chemistry – part 1
6.4.2   Dry cleaning and Green Chemistry – part 2
6.4.3   Dry cleaning and Green Chemistry – part 3
6.4.4   The twelve principles of Green Chemistry

6.5   Feed the world: artificial nitrogen fertilisers

6.5.1   Feed the world – briefing sheets

6.6   Sustainable development

6.6.1   Green Chemistry, atom economy and sustainable development
6.6.2   Making oil from waste – which is the odd one out? (Powerpoint slide)

Plastics

Index 6.1.1
1 sheet
Index 6.1.2
2 sheets
Index 6.1.3
3 sheets

What use is plastic?

This activity is intended as an introduction to work on plastics and polymers. It aims to get students thinking about the advantages as well as the disadvantages of using plastics. It could be set as homework or done as a group activity in class.

Answers

Item now made of plastic	What it used to be made from	Advantages of using plastic	Disadvantages of using plastic
Bucket	Metal	Lighter, does not rust, more colourful, cheaper	
Milk bottle	Glass	Lighter, cheaper	Cannot be re-used
Rope	Plant-based materials such as cotton or hemp	Lighter, can be coloured, does not corrode	If it is dropped as litter, *eg* in the sea, then does not break down and can be hazardous to wildlife
Disposable cups	Paper, but often not used	Lighter, cheaper	Cannot be broken down (degraded) if dropped as litter
Chairs	Wood, metal	Lighter, greater variety of colours, cheaper	
Water pipes	Metal (lead or copper) in homes; concrete for mains water pipes	Lighter, non-toxic, less likely to crack	
Nappies	Cloth	Waterproof, cheaper	Cannot be re-used or recycled, not biodegradable
There are various possibilities that could be included here			

Table 1 Sample answers

There are other possible answers that could be included in Table 1. The remaining questions can be used as part of a class discussion. Some points you may wish to consider are:

The main advantage of plastics is that they are light. This reduces transport costs and means that less petrol/oil is used to move them around than heavier materials. Many students do not appreciate this. The use of plastic parts in cars has greatly reduced their weight and contributed to the improved effeciency of modern cars.

The main disadvantage of plastics is that they are often not biodegradable so they remain in the environment when thrown away. This is very noticable in the the case of rope. Rope made from plastics (often polypropene) is often washed up on beaches after having been in the ocean for some time. It can kill sealife.

However, the longevity of plastics was initially seen by chemists as a real advantage. For example, if a plastic is used for underground pipe work the fact that it lasts a long time means that it needs less frequent replacement and less water is wasted as a result of cracks and leaks than might be the case with some other materials.

It is when we litter or are wasteful that the disadvantages of plastics become most noticable. It might seem easy to say that we should not use plastic packaging at all, but without it far more food would spoil and become inedible before it could be used. This would lead to greater fertiliser and pesticide use in an attempt to increase production to compensate for food wastage.

This is a brief introduction to some of the complex issues raised in the activities provided within this resource that relate to polymers, plastics and the environment.

Degradable plastics

The aims of this activity are to:

■ Consider what makes a plastic biodegradable

■ Learn the difference between biodegradable and photodegradable plastics

■ Consider the social and environmental consequences of using either degradable or non-degradable plastics.

The activity is aimed at able 14–16 year olds and would fit into a unit on plastics, polymers or the environment. Two student sheets are required:

■ **Degradable plastics** – worksheet including questions and a little information

■ **Degradable plastics – information sheet** – more detailed background information; students will need this sheet if they are to tackle all the questions on the Degradable plastics worksheet.

Further information

http://www.co-op.co.uk/ – type 'consumer issues' into the search facility then click on 'Consumer issues index page.'

http://www.foe.co.uk/resource/factsheets/plastics.pdf – this link goes directly to a pdf file. Alternatively, go to http://www.foe.co.uk and type 'plastics' into the search facility.

http://www.degradable.net/ – this company sells degradable bags and chemicals that can be added to plastic bags to make them degradable.

http://www.guardian.co.uk/supermarkets/story/0,12784,1274047,00.html – a newspaper story on degradable bags. Copies of the articles could be given to students and used to introduce the activity.

(All sites accessed December 2005.)

RSC | Advancing the Chemical Sciences

When using websites, students should be encouraged to consider:

- Who is paying for the website?

- What is the sponsors'/owners' point of view on the topic being researched?

- Why are they likely to hold that view?

- Is the information likely to be biased? If so, in which direction would you expect the bias to be?

Students should be encouraged not simply to take the information at face value.

Answers

1. Biodegradable means the material can be broken down by the action of living organisms.

   Photodegradable means it can be broken down by the action of light.

2. This question could be used to highlight the fact that there is no accepted international (or national) definition of biodegradable. The following answers are generally true.

   Polypropene: Not biodegradable – it is like polythene but with a group of atoms substituted for one of the hydrogen atoms.

   PVC: Not biodegradable – it is like polythene but with a chlorine atom substituted for one of the hydrogen atoms.

   Teflon: Not biodegradable – it has four fluorine atoms substitued for the four hydrogen atoms of polythene.

   Polyethene: Will biodegrade if its molecular weight is low enough.

   Polylactic acid: Biodegradable – it contains $-CH_2-CO-$ groups in its chain. Bacteria can easily remove this group, which will cause the chain to fall to pieces.

3.

Advantages	Disadvantages
Very little plastic is actually recycled so using degradable plastic reduces waste build-up.	Most plastic goes into landfill and even biodegradable plastic cannot decompose there.
Even in landfill, ordinary plastic does not decompose.	If photodegradable plastics are mixed with ordinary plastic during recycling the resulting plastic mix is sometimes useless because it degrades.
Recycling can take more energy than making the plastics in the first place.	Some degradable plastics do not degrade completely and little bits of plastic which do not break down remain.
Recycled plastic cannot be used for any product that will come in contact with food so it makes sense for food packaging to be made of degradable plastic.	

Table 1 Some advantages and disadvantages of degradable plastics

4. If photodegradable plastics are mixed with other plastics during recycling, the chemical additive used in these photodegradable plastics could begin to degrade the plastic mixture produced. Whatever item was made from the recycled plastic would degrade too. This is not appropriate for all applications. For example, a garden chair made from such recycled plastic could collapse after a year or two of use. This could lead to an even bigger disaster if it caused consumers to think recycled plastic is worthless.

5.

   a. There are fewer than 19 ethene molecules in a chain with a molecular weight of less than 500.
   b. Commonly used polythene consists of chains of 10s of 1000s of ethene molecule units.
   c. No, polythene bags are not biodegradable.

6. Mark letter/essay/leaflet/poster according to the science students have used to explain their arguments. They should distinguish between bio- and photodegradable plastics and consider the arguments for and against the use of each.

RSC | Advancing the
Chemical Sciences

This page has been intentionally left blank.

Active and intelligent packaging

Index 6.2.1
8 sheets

Introduction

This activity requires students to use their existing knowledge of redox reactions, word equations, respiration and micro-organisms, as well as knowledge of polymer structure and function, and to apply this knowledge in a new context – that of packaging materials. The student sheet guides them through some of the research and development chemists have carried out in recent years in the area of food packing. They may need access to texts to assist them in recalling what they have learnt previously.

The activity provides an example of a positive contribution science is making to people's everyday lives, even though they are largely unaware of it.

Suggestions for running the activity

Students could work on their own or in small discussion groups. They could either all write down the answers to the questions or have one scribe and then work together to produce a leaflet or poster to explain the new packaging to customers.

At the end of the activity, students are asked for their opinions and feelings on the issues raised. Question 27 on the student sheet could be set as a follow-up homework task.

Timing

1 lesson + possible homework.

Answers

1. Reasons could include: keeps liquids in one place, prevents spillages, protects food, standardises the amount in a pack, allows marketing/branding of products, can put list of ingredients on pack, helps stop the food going off, hygiene, convenience, improves how food looks.

2. Possible aims for packaging include: pre-weighed packs, stop food going off/increase shelf-life, increase speed at check-outs. Students might give a range of other answers.

3. Widgets are expensive so they are only used in relatively expensive products and where there is really no alternative.

RSC | Advancing the Chemical Sciences

4. The reasons people like fresh food include: taste, texture, contains more vitamins and minerals, they enjoy cooking it.

5. Fresh food goes off. Since you cannot keep it for a long time, you either have to go shopping more often or accept a certain amount of wastage.

6. Moulds and bacteria grow on the food and cause it to go off. (Another cause is oxygen reacting with the food, but students will probably not give this answer based on their previous knowledge of science.)

7. Warm, damp conditions make food go off quickly.

8. Glucose + oxygen → carbon dioxide + water
    $C_6H_{12}O_6 + 6O_2 \rightarrow 6CO_2 + 6H_2O$

9. Wet swimming kit left in a plastic bag for a few days starts to smell.

10. Micro-organisms, especially moulds, grow on the kit because there is plenty of water for them. They release the gases that you smell.

11. Fresh fruit and vegetables contain living cells which are respiring. They produce water which cannot escape from the bag. The water encourages micro-organisms (especially moulds) to grow.

12. Moulds need oxygen to survive.

13. An oxygen scavenger is something that removes oxygen.

14. If you remove oxygen from the pack, moulds are less likely to grow because they need oxygen to respire.

15. Oxidise in this context means to react with oxygen.

16. Iron + oxygen → iron oxide

17. They might think their food could get contaminated with a non-edible/poisonous substance.

18. If consumers cannot see the oxygen scavenger, they will not know it is there. Consumers are used to seeing food wrapped in clear plastic and will not notice a difference.

19. Retailers might like the packaging because it will help make the food last longer so less will go past its 'sell-by' date before it is sold. This will reduce the retailer's waste and save money.

20. The new packaging might be more expensive than conventional packing.

Questions 21–24 require knowledge of polymers, their structure and function. If this topic has not been covered, you may wish to direct students to leave these questions out or remove them from the worksheet.

21. Polymer: a long chain molecule made of lots of small molecules (monomers) joined together.
    Side chain: a group of atoms sticking off/attached to the main polymer chain.

22. As the polymer warms up, the molecules move more and vibrate faster.

23. A polymer with long side chains needs more energy to move apart from the other polymer molecules. This is because there are stronger intermolecular forces holding the molecules with long side chains together so it will take more energy to pull them apart.

24. Long side chains are likely to lead to a higher melting point as there are stronger attractions between the polymer chains. The side-chained polymers move more as the material warms up, leading to the opening of pores in the polymer film.

25. Supermarkets might want to use the labels to give consumers confidence in what they are buying.

26. Various answers are possible here – students might answer 'yes' because the labels would allow consumers to see if what they are buying is fresh or not; they might answer 'no' because unscrupulous supermarkets could just change the labels.

27. Paragraph about students' own reaction to the information. Mark by impression.

References and further information

http://www.foodscience.afisc.csiro.au/actpac.htm – good factsheet from the Australian government.

http://www.newscientist.com – a search for 'active packaging' gives some interesting results (the full content of some articles can only be accessed by subscribers).

http://www.chemsoc.org/chembytes/ezine/2003/birkett_oct03.htm – an interesting article but probably beyond most 14–16 year olds; gives more detailed chemical explanations about how the packaging works.

http://www.dupont.com/packaging/structures/index.html – the Dupont website has information about various types of packaging, including food packaging.

(All sites accessed December 2005.)

Disposable cups and the environment

Index 6.2.2
3 sheets
Index 6.2.3
4 sheets

Our society is becoming more conscious of the need to look after our environment and conserve its natural resources. We try to make choices that are 'environmentally friendly' and are sometimes prepared to pay more for products that seem to be better for the environment. One choice that could affect the environment is the choice of material to be used for making disposable cups.

In this activity, students put themselves in the role of a manager in a take-away hot drinks company. They have to choose which of two materials will be best for making the cups for their drinks – polystyrene or paper. Most students will assume at the start that paper cups would be better for the environment because paper is made of a natural, renewable material. However, on examination of the evidence this appears not to be the case.

Suggested lesson plan

Starter

You could begin this activity by displaying headlines or articles from newspapers that express concern about the use of plastics. Alternatively, you could discuss with students which material they feel is more 'environmentally friendly' – paper or polystyrene. Ask them what evidence they have to support their views. Discuss why people who are environmentally conscious might use disposable cups. Reasons could include: they are more hygienic; they are more convenient and appropriate for use in certain situations.

Main activity

Give each student a worksheet and a table to complete and each pair or group a thoroughly shuffled pack of information cards. The cards could be laminated to protect them and make them easier to reuse. Alternatively, they could be copied onto cardboard.

Students read the information on the cards and use it to complete the table. This allows them to analyse the environmental impact of each of the two options for making disposable cups. They then complete the questions on the worksheet.

Plenary

Discuss students' findings and highlight in particular those facts which surprised them. Discuss why the general perception is that paper would be the more environmentally friendly material to use. Ask students if there are any aspects of the debate about which they feel they do not have enough information to make an informed decision.

Follow up
Students make a leaflet to share their findings with the general public. This could be set for homework or done in a subsequent lesson.

Answers

1. Consumers might want a disposable cup because they do not want to stay in the shop where they have purchased their drink – they may wish to take the drink on a train or to work for instance. They may also prefer disposable cups because they perceive them to be more hygienic.

2. Students should answer based on their own views.

Item	Paper cup	Polystyrene cup
Making the cups (per cup)		
Mass of wood and bark needed (g)	33	0
Mass of petroleum needed (g)	4.1	3.2
Mass of other chemicals needed (g)	1.8	0.05
Tick the material whose manufacture uses the most		
Steam	✓ (12 times as much)	
Electricity	✓ (36 times as much)	
Cooling water	✓ (twice as much)	
Tick the material whose manufacture produces the most		
Waste water	✓ (580 times as much)	
Water pollution	✓ (10 times as much)	
Metal salts		✓
Waste gases	✓ (per cup)	
Using the cups		
Mass of 1 cup (g)	10.1	1.5
Cost of 1 cup (pence)	5	2
After use		
Can the cup be reused?	✗	✓
Can the material be recycled?	✗	✓
Can it be burnt?	✓	✓
How much energy will you get from 1 kg if you burn it?	20 MJ	40 MJ
What mass of material would go in a landfill from 1 cup?	10.1 g	1.5 g
Is it biodegradable?	✓	✗

Table 1 Data table for the cups

RSC | Advancing the
Chemical Sciences

3. The paper cups will have higher transport costs as they are heavier and more petrol/diesel will therefore be needed to get them to the shops.

4. The main advantage of paper cups is that they are biodegradable.

5. The main advantages of polystyrene cups are that they use fewer raw materials and their manufacturing process consumes less energy and produces less waste than that of paper cups. Polystyrene cups are also lighter than paper ones, which leads to lower transport costs.

6.

■ Anaerobic decomposition is decomposition without using oxygen (breaking down)

■ Biodegrade – to be broken down by living organisms

■ Greenhouse gas – a gas that contributes to global warming.

7. This will depend on the answers given earlier. It is likely that the main advantage of using paper (the fact that it is biodegradable) now appears to be less of an advantage than previously thought.

8. The polystyrene cup is probably better for the environment when all the factors are taken into consideration. It is not only the disposal of the cup that needs to be considered, but the materials used to make and transport it too.

9. The answer to this is up to the student. They may choose polystyrene as it probably is more 'environmentally friendly' or paper because of the persistent belief among consumers that it is better for the environment.

You could mark the leaflet for:

■ Scientific accuracy

■ Persuasiveness of the arguments

■ Presentation.

References and further information

M. Hocking, *Science*, 1991, **251**, 504.

J. Emsley, *New Scientist*, 1991, **1791** (19 October), 132.

Managing waste and rubbish

Index 6.2.4
3 sheets

This activity would follow on well from **Disposable cups and the environment** or **Making polylactic acid**. If students have already done one or both of these other activities, they will have been introduced to the idea that landfill is a problem. They should also already be aware that waste does not rot in a landfill site so as long as we dump everything in a hole in the ground it does not matter much what the waste is made of, the effect remains the same.

Managing waste and rubbish introduces composting as a way of managing rubbish by turning it into a useful product – not just garden waste but the majority of biodegradable waste.

The activity focuses on two projects, one in Dorset in the UK and the other in the German town of Kassel.

Full details of the Kassel project can be found at **http://www.modellprojekt-kassel.de/eng/downloads/BUW_Kassel_Orbit_Text_2003.pdf** (accessed December 2005). More able students should be able to read and understand this report but it is a very long document (10 pages).

If access to the internet or to computers for a whole class is difficult then the web pages mentioned in the student worksheet could be printed out and photocopied prior to the lesson. It is not necessary for students to do an internet search.

Answers to questions

1. Most rubbish in Britain ends up in landfill sites. There not much happens to it and it can remain unchanged for decades.

2. Answers could include the points listed in Table 1. There are other possible answers too – this list is not exhaustive.

Advantages	Disadvantages
Simple and quick. No sorting required as all waste goes in together.	Takes up a large amount of space and is ugly.
	The waste does not go away and does not decompose much – it just sits there using up space.
	It is wasteful – the Earth only has limited resources and they are all ending up buried in landfill sites.
	Toxins can leach out of the rubbish into water supplies.
	The government is taxing landfill now and fining local authorities who use it too much.
	Gases such as methane can build up in the rubbish.

Table 1 Advantages and disadvantages of landfill

3. Answers could include the points in Table 2.

Advantages	Disadvantages
Reduces the amount of waste going into landfill.	Not all waste can be composted.
You get a useful product out at the end.	If it is not done properly the compost heap can smell.
Can also reduce the amount of peat bought by people for their gardens. Peat is a limited natural resource.	Not all homes have a garden with enough space for a compost heap (this problem and the point above could be solved by having a municipal compost heap).

Table 2 Advantages and disadvantages of composting

4. This question asks for students opinions so answers will vary.

Few people compost their waste because:

■ It takes effort and it is easier to throw things in the bin and forget about them

■ People do not know about the advantages of composting or the disadvantages of landfill

■ There are no (or limited) municipal composting facilities for domestic rubbish in the UK

■ People still do not know which wastes are biodegradable.

5. Yes, you can use the compost to help plants grow.

6. The ordinary plastics in the compostable waste would not break down and would contaminate the final compost. The biodegradable plastics would mix in with the ordinary plastics being recycled without causing major problems.

The task of matching the statements about the experiment is intended to make students think about how an experiment is carried out. They may initially think the Kassel project is not really like a science experiment because it is about people's behaviour and attitudes. This exercise should help them to realise that it is not so very different from an experiment conducted in the laboratory.

The scientists carried out their experiment only in the town of Kassel and not across the whole country.	Using a small sample gives you enough data to work with and is less expensive than trying to do the experiment on a very large scale.
They told people how to tell biodegradable and ordinary plastics apart.	This made it possible to collect biodegradable polymers with other biodegradable waste.
They tested the compost produced using the biodegradable polymers to see if plants would grow in it.	If plants were poisoned by the compost or could not grow in it then it would be useless and there would be little point in making it.
They analysed the waste people put into the different waste bins.	They wanted to see if the waste was being separated correctly.
They chose a town that already had a separate 'compostable waste' bin.	They were trying to find out if people could separate the plastics – not if they could separate all compostable waste.
They put a clear logo on the disposable plastics.	This made it easy to tell the plastics apart so people could do it even if they were in a hurry.

Table 3 Statements about the experiment aligned with the correct explanations

You may wish to discuss the questions in the **To think about** section with students before they write their report.

The report could be set for homework or used as another follow-up activity. Students should focus on possible ways for the county council to reduce landfill but wherever possible should also highlight the difficulties that could arise.

RSC | Advancing the
      Chemical Sciences

Nappy choice
and the environment

Index 6.3.1
3 sheets
Index 6.3.2
2 sheets

In this activity students put themselves in the position of a nursery/day care centre manager with the responsibility of choosing the type of nappy to be used on the children in their care.

The activity is based on the idea of 'life cycle assessment,' which involves assessing the environmental impact of the production, use and disposal of a product. Students consider these three parts of the life cycle of a nappy and consider which has the least impact on the environment. The work-based context helps interest and engage them.

Students analyse information provided on a set of cards, find out about costs, come to a decision about which type of nappy they would use and then communicate what they have found out to their customers (the parents). They could search for additional information on the internet. A selection of websites devoted to this subject is given below.

For pricing information, students could look at supermarket websites such as: **http://www.tesco.co.uk** and **http://www.sainsburys.co.uk** – for disposable nappies **http://www.mothercare.co.uk** or **http://www.boots.com** – for reusable nappy start-up costs.

They could also check a nappy laundering service: **http://www.nappytales.freeserve.co.uk**

The Women's Environmental Network has some facts and figures relating to nappy use: **http://www.wen.org.uk/nappies/facts.htm** – includes a good section on cost comparisons. Although the costs considered are per baby and relate more directly to family than to nursery use, the information is clear and students may find it helpful.

Most county council websites have a section devoted to waste management and many of these mention nappies. Check your own for local information or look at Oxfordshire County Council's website:

http://www.oxfordshire.gov.uk – search the site for 'Real nappies' or follow the links to the section on how to reduce your household waste, which includes the page on nappies. The nappies page provides several links to local and national sources of information.

There is a large number of American websites that offer more general information on nappies and the various options available. These sites may confuse students and if you use them you will need to explain that the American word for nappy is 'diaper'. If you type 'disposable diaper' or 'real diaper' into a search engine, many sites will be listed. Several have a political agenda and are very biased so warn students to think about this if you do ask them to search for futher information on the internet. Many of the sites base their arguments on the idea that all chemicals are nasty and bad for the baby whilst conveniently ignoring all the chemicals that are used to produce a cotton nappy.

There are various websites offering information on disposable nappies. Pampers is one example of a well known brand that students may be familiar with:

http://www.uk.pampers.com – only limited information on what the nappies are made of is provided.

(All websites accessed December 2005)

Running the activity

Give each student a worksheet and a table to complete and each pair or group a thoroughly shuffled pack of information cards. The cards could be laminated to protect them and make them easier to reuse. Alternatively, they could be copied onto cardboard.

Much of the information on the cards is based on 1000 nappy changes. This is because it is hard to make direct comparisons between the two types of nappy without defining what is being compared in this way. Students may need to have this explained to them. In the case of disposables, '1000 nappy changes' refers to 1000 nappies; for reusables it refers to 1000 changes using a smaller number of nappies.

Students could work in groups to complete the table. If you have internet access then they can look up current prices. If internet access is difficult or time is limited you could provide them with the following information:

- On average, a disposable nappy costs 17.9p

- Babies require an average of six changes a day

- Over two years (average length of time a child is in nappies) the total cost of the disposable nappies required is about £700 based on the information above

- The total cost of 20 reusable nappies and waterproof pants is at least £60

- When washing costs are factored in, the cost of using reusables for two years is about £185

- The average price of a nappy laundering service is £8.50 per week.

These figures are based on 2004 prices and information taken from the Women's Environmental Network website.

Some basic questions are provided that could be used to help students make sense of what they have written in the table. The more able probably will not need them.

The leaflet can be marked on the quality of the science it contains, as well as on presentation and persuasiveness.

Answers

Factor	Disposable nappies	Reusable cotton nappies
What the nappy is made of	Plastic backing, hydrogels, paper	Cotton nappy with wool or plastic outer (Note: the outer has not been included in the figures below)
Making the nappies Mass of plastics used (kg)	15	0
Mass of paper used (kg)	108	0
Mass of cotton used (kg)	0	2.3
Energy used (kWh)	338	57
Water used which will be polluted afterwards (dm^3)	10500	1100
Are pesticides used?	no	yes - lots
Total raw materials used (kg)	123	2.25
Using the nappies How many times is the same nappy used?	1	167 (as 6 nappies used)
Energy used in laundry (kWh)	none	141
Mass of detergents used (kg)	0	32
Water used which will be polluted afterwards (dm^3)	1500	6200
Disposing of the nappies Mass of waste (kg)	221	30
Where solid body waste (faeces) ends up	In landfill sites	In the waste water supply with other toilet waste

Table 1 Data table on nappies – all data per 1000 nappy changes

1. Reusable nappies use less raw materials.

2. The reusables cause more water pollution – from pesticide use and from the laundry.

3. The reusables produce less solid waste.

4. This is a fairly subjective question – students are likely to answer that the reusable nappies cause less harm. This is probably true but the answer is not entirely clear-cut.

5. Many parents use disposables because they are so convenient. Also, parents are often unaware of the environmental problems such nappies cause.

Acknowledgements and references

This activity is based on the document:

C. Lehrburger, J. Mullen and C. V. Jones, *Diapers: Environmental Impacts and Lifecycle Analysis*, 1991.

This document is a report to:

National Association of Diaper Services
2017 Walnut Street
Philadephia
Pennsylvania 19103

The report was kindly supplied by the National Association of Diaper Services (NADS) for use in the development of this activity. NADS is not responsible for any errors or omissions in the activity.

Subsequent to this activity being written, the UK Environment Agency commissioned an independent study to establish the true environmental impacts of disposable and reusable nappies. They concluded that there is little to choose between them. A summary of the findings (as well as the full life cycle assessment) is available at **http://www.environment-agency.gov.uk/yourenv/857406/1072214/** (accessed June 2006).

Dry cleaning and Green Chemistry

Index 6.4.1
3 sheets
Index 6.4.2
2 sheets
Index 6.4.3
1 sheet
Index 6.4.4
1 sheet

The purpose of this activity is to look at the effect that applying the principles of Green Chemistry (see **The twelve principles of Green Chemistry**) can have on the environmental impact of a well known process: dry cleaning.

Students first examine what happens in 'ordinary' cleaning and consider a simple explanation of how a detergent works. They then look at dry cleaning and the impact the solvents used in this process can have on the environment. Finally, the 'greening' of this process through the use of liquid carbon dioxide as a solvent is discussed.

This material is aimed at 14–16 year olds. A more advanced version is available at **http://www.chemsoc.org/networks/learnnet/green/co2/index.htm**.

Background information – surface tension

Water molecules hold on to each other tightly and create a surface tension. In order to use water to clean grease from an item, the surface tension has to be reduced to allow the water to wet the thing you are trying to clean. Surface tension is the force that makes a blob of water stay together and not spread out. It allows pond skaters and other insects to walk across water and also enables a pin to float.

It you look closely at drops of water you can see that they try to form spheres. Gravity stretches out drops that cling to an eye dropper or a tap. However, when the drops fall, they become spherical.

The shape of a water drop is a result of surface tension. Water is composed of molecules each consisting of two hydrogen atoms and one oxygen atom. These molecules are attracted to each other. In the middle of a drop of water, each molecule is surrounded by other molecules on all sides so it is pulled equally in all directions by these attractive forces. On the surface, however, the molecules only experience attractive forces in certain directions: across the surface and inward. This causes the water to try to form a shape with the smallest possible surface area – a sphere. Gravity causes water drops resting on a surface, to flatten out as shown in Figure 1.

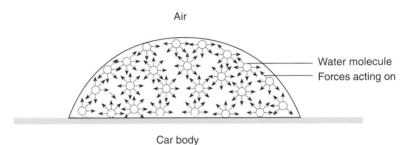

Air

Water molecule
Forces acting on

Car body

Figure 1 A water drop on a surface

The molecules at the surface of a body of water behave like an elastic membrane. You can easily see the elastic membrane effect if you float a needle on the surface of a glass of water. Lower the needle gently onto the water surface with a pair of tweezers, let go and examine the water near the needle. You will see that the water surface is depressed slightly as though it were a thin sheet of rubber.

The addition of a surfactant like liquid soap or washing up liquid to water reduces its surface tension. Water molecules do not bond as strongly with soap molecules as they do with each other. Therefore, the force that enables the molecules to behave like an elastic membrane is weaker if soap is present. If you put a drop of liquid soap in the glass of water with the needle the surface tension is greatly reduced and the needle quickly sinks.

Part 1 Practical work

Equipment required – normal cleaning

- Test-tubes

- Oil – in plastic dropper bottles or with pipettes

- Washing up liquid – in plastic dropper bottles or with pipettes

- Access to water

- Bungs.

Emphasise to students the need to mix the oil and water without shaking so hard that a lot of foam is formed.

Equipment required – how detergents work

- A few clean 2p coins

- Plastic dropping pipettes

- Paper towels

- Washing up liquid

- Solid soap

- Cooking oil

- Paraffin wax

- Sugar

- A soft lead pencil (the 'lead' is actually graphite).

Emphasise to students the need to make sure the 2p coin is cleaned properly after each test. If there is any trace of washing up liquid or soap on the coin when another substance is being tested the results could be inconclusive.

Answers – Part 1

1. Water alone will not remove the oil because oil and water do not mix.

2. Water and washing up liquid may remove the oil because the washing up liquid helps the water and oil to mix. This allows the oil to be washed away in the water.

3. The water forms a curved or bulging shape on the top of the coin as shown in Figure 2.

Figure 2 Water on top of the coin

4. The washing-up liquid and soap made the number of drops that would fit onto the coin decrease.

5. The oils made the number of drops of water that would fit on the coin increase.

6. Adding a detergent to the water enables the oil and water to mix together. The oil will be held in the water by the detergent molecules, which act as a bridge between the oil and water molecules. This allows the oil and grease to be washed away in the water and the dishes become clean.

Part 2 Dry cleaning

This section of the work emphasises the environmental impact of the usual dry cleaning process at the moment to help students understand why it might need to be changed. Students will need a copy of **The twelve principles of Green Chemistry**.

Before students begin this section, carry out two demonstrations. First, demonstrate how dry cleaning works. Put some oil into a test-tube and show that it can be dissolved in organic solvents. Explain that the process does not need a detergent because the oil or grease dissolves in this kind of solvent without it.

Next, demonstrate how carbon dioxide can be made to form a liquid. Students should already know about carbon dioxide gas and they may have seen dry ice (solid carbon dioxide) but they will probably be unfamiliar with the idea that it can form a liquid too. Either do the demonstration detailed below or show the video clip of this demonstration found at **http://www.nottingham.ac.uk/~pczctg/Video_Clips_Menu.htm** (accessed January 2006).

Equipment required for the carbon dioxide demonstration

- Eye protection
- Tongs or gloves
- One disposable plastic pipette
- One zip-sealing plastic sandwich bag
- A pair of pliers
- A few small pieces of dry ice.

Health and safety

Wear eye protection.

Wear gloves and handle the dry ice with care – prolonged contact with the skin can cause frostbite.

What to do

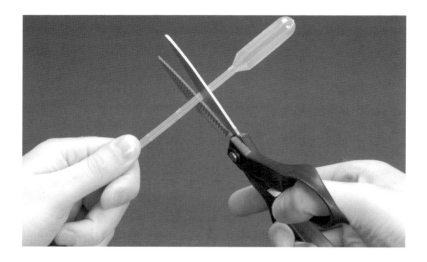

Figure 3 Preparing a pipette for the demonstration

1. Cut the tip end off a plastic pipette as shown in Figure 3. Put 4–5 small pieces of dry ice into the pipette. Slip the pipette into a zip-sealing sandwich bag, allowing only the tip end to protrude. Partially zip the bag closed. Use a pair of pliers to clamp the open end of the pipette so that it is sealed. It may be helpful to fold the tip end once or twice to help shut the opening and ensure that no gas escapes.

2. Once the carbon dioxide has liquefied, loosen your grip on the pliers to release the pressure.

3. Reseal the pipette with the pliers and repeat the process of building up and releasing pressure to observe several changes of state.

Answers – Part 2

1. PERC is the solvent, oil or grease the solute and the resulting mixture of PERC and grease is the solution.

2. PERC can cause problems to humans and animals when it is breathed in. If too much is inhaled it can cause dizziness and confusion. If it gets into drinking water and animals and humans drink it then the liver and kidneys can be damaged. It will stay in fatty tissue in the body. If it is released into the air and there are other polluting chemicals present then PERC can contribute to photochemical smog.

3. Chemists are keen to find an alternative to PERC because it can cause so many problems. It can be a particular problem to people who work in the dry cleaning industry and are exposed to PERC regularly.

4. If carbon dioxide is released into the environment it will become a gas and gets blown away.

5. This is better for the environment because it does not cause liver and kidney damage to animals and does not play a part in causing photochemical smog.

6. This is worse for the environment because it will contribute to the rising levels of CO_2 in the atmosphere, which may be causing global warming. However, it should be emphasised that much of the CO_2 used for dry cleaning is the byproduct of other industrial processes; if it were not used in dry cleaning, it would be released into the environment anyway. The quantities on an industrial scale are relatively small.

7. The answer is the students' own response to what they have read. Most will probably conclude that CO_2 is less harmful than PERC. They may decide they need further information on some aspect of the subject, *eg* whether using CO_2 will require detergents and what effect those detergents might have on the environment.

Answers – Part 3

1. You do not need to use a detergent with PERC because the PERC dissolves grease without one. A detergent is not required to make the grease stay in the solvent as it would be with water.

2. After the oil and water were mixed, they separated again.

3. When a detergent was added, the oil stayed in the water and the substances remained mixed.

4. When carbon dioxide and oils are mixed they separate again if they are left to stand because they do not dissolve each other.

5. This would not help in cleaning clothes because the grease would not mix with the carbon dioxide but would stay on the clothes.

6. This process uses a solvent that is renewable, relatively inexpensive and can be recycled. (It also uses less energy overall but students are unlikely to realise this unless they do extra research on the process.)

The twelve principles of Green Chemistry

1. **Prevention**
   Try not to make waste, then you do not have to clean it up.

2. **Atom economy**
   The final product should aim to contain all the atoms used in the process.

3. **Less hazardous chemical synthesis**
   Wherever it is possible, production methods should be designed to make substances that are less toxic to people or the environment.

4. **Designing safer chemicals**
   Chemical products should be designed to do their job with minimum harm to people or the environment.

5. **Safer solvents**
   When making materials try not to use solvents or other unnecessary chemicals. If they are needed then they should not be harmful to the environment in any way.

6. **Design for energy efficiency**
   The energy needed to carry out a reaction should be minimized to reduce environmental and economic impact. If possible, processes should be carried out at ambient temperatures and pressures.

7. **Use of renewable feedstocks**
   A raw material should be renewable wherever possible.

8. **Reduce derivatives**
   Try not to have too many steps in the reaction because this means more reagents are needed and more waste is made.

9. **Catalysis**
   Reactions that are catalysed are more efficient than uncatalysed reactions.

10. **Design for degradation**
    When chemical products are finished with, they should break down into substances that are not toxic and do not stay in the environment.

11. **Real-time analysis for pollution prevention**
    Methods need to be developed so that harmful products are detected before they are made.

12. **Inherently safer chemistry for accident prevention**
    Substances used in a chemical process should be chosen to minimise the risk of chemical accidents, including explosions and fire.

Feed the world: artificial nitrogen fertilisers

Index 6.5.1
12 sheets

The problems caused by artificial fertilisers have had a fair amount of media coverage in recent years. Should we still use them? Should we all 'go organic'? What is best for the environment? What sustainable solutions are there to the problems associated with growing food?

This activity allows students to consider the benefits and drawbacks of using artificial fertilisers and encourages them to do so from a scientific, rather than an emotive point of view. Students will probably bring their own existing ideas and beliefs to the activity and they should be encouraged to think about whether the scientific data supports their views.

Overview

Students take on the roles of a number of experts. Some information is provided for each role and students can do their own research to find out more about the likely views of the expert they are playing.

One of three routes can be taken through the activity:

1. If video cameras are available, students could make a TV programme. This could either be done in the form of a TV debate in the style of programmes like 'Question time', or 'Richard and Judy,' or it could be a mini documentary in the style of 'Panorama,' 'Horizon' or another similar programme. Students would need to be in reasonably large groups and the activity would need careful management.

2. Alternatively, the whole class could prepare for a radio show on the subject. The final show will require a presenter with a reasonable grasp of the arguments both for and against the use of artificial fertilisers, three experts on each side of the debate and a group of listeners. Divide the class into seven groups to prepare for the show and give each group the information sheet for one of the possible roles. Students should then discuss the information and the views their character might hold. Each group must nominate a spokesperson to take the role of the presenter or one of the experts in the final show. The remainder of the class becomes the listeners, who can phone the radio station during the show to state their opinions. You could use a microphone to aid management of the discussion – students are only 'on air' when speaking into the microphone and must remain silent otherwise. It is important that a strong student plays the role of the presenter, who should hold the experts in line, take calls and facilitate discussion. If your presenter group is

struggling with preparing for this role, you could discuss some of the arguments they might hear from the other students with them before the show starts.

Student sheets required:

Feed the world – general briefing sheet (TV)
Feed the world – general briefing sheet (radio)
Expert briefing – presenter
Expert briefing – farmer
Expert briefing – spokesperson for the charity 'Food for All'
Expert briefing – spokesperson for the charity 'Action on Habitat Destruction'
Expert briefing – 'Organic Food Producers and Consumers Association'
Expert briefing – 'Green Earth' environmental charity
Expert briefing – 'Water Quality Campaign'

The teacher takes the role of the producer of the show, supporting the presenter in managing the discussion and taking calls.

3. Another alternative is to ask students to do some research then hold a class debate about the issues involved.

Experts

Neutral
Presenter

For artificial fertiliser use
Farmer

Spokesperson from the charity 'Food for All'

Spokesperson from 'Action on Habitat Destruction'

Against artificial fertiliser use
Spokesperson from the 'Organic Food Producers and Consumers Association'

Spokesperson from 'Green Earth' environmental charity

Spokesperson from the 'Water Quality Campaign' group

Each student will need a copy of the **General briefing** (either the TV or the radio version as appropriate) and a copy of the **Expert briefing** for the group to which they have been assigned. The **General briefing** includes some suggested websites for further research.

Note

Remind students that this debate/TV or radio programme is about fertilisers and not about the use of pesticides, antibiotics or genetic modification. If they do their own research to add to the information supplied in the notes they are given, then they should make sure they stick to the main point of the debate, *ie* 'Should we use artificial fertilisers?'

Follow-up

After the debate, each student could write a newspaper or magazine article about the issues raised. They should include arguments from both sides of the debate, as well as their own opinions.

A more able group could be asked to research and present ideas for sustainable solutions to the problem of feeding 10 billion people. Possible starting points could include the websites listed on the General briefing sheet and perhaps also the following site: **http://www2.essex.ac.uk/ces/default.htm#top** (accessed December 2005) – students will need a high level of reading and comprehension skills to use this site but it contains a lot of useful information.

Less useful agriculture

If students are up in arms at the environmental damage agriculture is causing, you may wish to point out that tobacco is also produced on farms. These farms cause habitat destruction and pollution from fertilisers and the final product does not even feed anyone!

References

G.J. Leigh, T*he World's Greatest Fix: A history of nitrogen in agriculture*, Oxford: Oxford University Press, 2004.

N. Borlaug, *Feeding a world of 10 billion people*, Alabama, USA: IFDC, 2003 (available on the internet at **http://www.ifdc.org**).

V. Smil, *Scientific American*, 1997, **227**, 76.

Green Chemistry, atom economy and sustainable development

Index 6.6.1
5 sheets

The competing needs of development and environmental protection are often mentioned in the media and many students have opinions on the subject. Chemistry is seen as a 'polluter,' which partly accounts for its poor image among students and the general public. This activity introduces the concept of atom economy in the context of sustainable development and Green Chemistry. It aims to show that development is necessary but can be achieved in a way which limits environmental damage.

Prior knowledge required

Students need to know/be able to:

■ Calculate relative molecular mass (RMM or M_r)

■ Know what a reversible reaction is and how the yield of a reaction might be affected by its reversible nature

■ Calculate percentage yield – a section on this is included in the resource but it would be better if students had already covered it so that they do not get it mixed up with atom economy.

Further information

Further information on sustainable development is available on various websites, including:

http://www.uyseg.org/sustain-ed/Index.htm – this website of the Chemical Industry Education Centre is a good introduction to why development is necessary and how it can be made more sustainable. Students can calculate their personal sustainability and also how much carbon dioxide they produce in a year. Examples of chemical industries that are going greener are provided, along with links to several other sites. This could provide a useful basis for some project-type work on the topic.

http://www.uyseg.org/greener_industry/index.htm – the Chemical Industry Association's Greener Industry website. There is an interesting section on greener cars, as well as information on other areas of chemical industry.

http://www.makepovertyhistory.org/schools/resources.shtml – this page from Make Poverty History has a list of resources on the topic of development that are available free to schools, mainly from various charities working in the developing world. These are not specifically chemistry-based resources (and most, if not all, do not mention chemistry at all) but they may provide good background material on the need for development.

http://www.chemsoc.org/networks/learnnet/green/index.htm – this is a resource from the Royal Society of Chemistry and is mainly aimed at post-16s, but the site contains some interesting information and further details about atom economy and Green Chemistry.

(All sites accessed December 2005.)

Answers

1. Various answers are possible, including: not all pollution may be cleared up; energy will be wasted producing unwanted products; it is better not to produce hazardous materials at all then there is no danger of them being spilled or falling into the wrong hands.

2. Using a catalyst (or more efficient catalyst) may mean that the process can be carried out at a lower temperature, which saves energy. (It may also increase the yield of one product over that of another, unwanted one.)

3. Percentage yield = 21/28 x 100 = 75%

4. Maximum mass of SO_2 that could be produced = 16 g

    Percentage yield = 14/16 x 100 = 87.5%

5. Atom economy = 100%

6. Atom economy = 224/356 x 100 = 63%

7. **a)** Displacement = 48/128 x 100 = 37.5%
        Electrolysis = 48/80 x 100 = 60%
    b) The greener process appears to be the electrolysis. Before finally deciding it might be important to know how much energy is used in each process, what (if any) solvents are used and what the yield is for each reaction.
    c) 100 %

8. **a)** Atom economy = 28/142 = 20%
    b) Atom economy = 142/142 = 100%
    c) If only one product can be sold then the rest is wasted; if both can be sold then there is no waste product and the atom economy is greater.

9. **a)** Atom economy = 100%
    b) The reversible reaction symbol ($\rightleftharpoons$) suggests that the reaction is unlikely to have a very high yield.

10. Answers may include: more efficient use of resources; less waste; less pollution; less energy used. These things are important for sustainable development because they all help to ensure that we leave the planet healthy for future generations and do not use up all the available resources.

Making oil from waste

Index 6.6.2
1 sheet
1 Powerpoint
slide

Much scrap organic matter can be turned into useful oil by a process which is similar to that by which oil was formed in Earth's crust but takes only a fraction of the time. Production of oil from organic waste is already underway in the USA and there are plans to build a facility in Europe before too long.

Further information about the American plant is available on the following websites:

http://www.res-energy.com

http://www.changingworldtech.com

http://www.guardian.co.uk/life/feature/story/0,13026,960689,00.html

(All sites accessed December 2005.)

The initial feedstocks come from a turkey processing plant and use the unwanted bits of turkey that cannot be turned into food. The plant is turning something that was once a problem waste material into a renewable energy resource. About 80 % of the energy from the waste is present in the oil produced – the rest is lost as heat or used in the conversion process. The company hopes eventually to reduce the dependence of the USA on imported oil.

This activity focuses on understanding how oil was formed in the Earth's crust and how scientists put their understanding of that process to good use to develop a new process of their own.

Suggested lesson plan

Begin by showing the slide **Which is the odd one out?** or handing out photocopies of the pictures and asking which is the odd one out. The answer is the oil rig – the other three are connected because oil can be made from poultry waste in a processing plant.

Explain that scientists have developed a technology to turn waste products such as tyres, plastics, paper, animal waste and agricultural waste into oil and that they did so by gaining an understanding of how oil is made naturally, then using that understanding to develop a faster process.

Organise a class debate on this topic. There are a number of issues that could be debated, including the potential political, environmental and social, (For example, should vegetarians use oil made from turkey waste, and would they be able to avoid it if they wanted to?), impacts of the process if it were to become widely used.

RSC | Advancing the Chemical Sciences

More able students could explore the res-energy site listed above and produce an information leaflet or newspaper article about the process. The most valuable information is in the 'Press Kit' at **http://www.res-energy.com/press/presskit.asp** (accessed December 2005) – the 'Cornerstone Technology' pdf file is particularly useful. Directing students to the press kit will make it seem more as if they are really behaving like a journalist or science correspondent. Make sure students compare the new process with what happens in the Earth's crust. They should think about what type of newspaper or magazine they are writing for and who their audience will be.

You could then show them the article in the Guardian (see link above) to complete the lesson.

Which is the odd one out?

This page has been intentionally left blank.

Chapter 7
Analysis

(List of student sheets available on CDROM)

RSC | Advancing the Chemical Sciences

Reactions of positive ions with sodium hydroxide

Index 7.1
3 sheets
Index 7.2
2 sheets

This activity is a microscale version of a common test-tube practical. The main advantages of the microscale version are the tiny quantities of chemicals consumed and the lack of test-tubes to wash up.

Two student sheets are provided:

■ **Reactions of positive ions with sodium hydroxide**

■ **Reactions of transition metal ions with sodium hydroxide** – a shorter version of the positive ion activity using only transition metal ions.

The student sheet should be laminated or placed inside a plastic document wallet. It can then be wiped clean over the sink and the residues washed away with plenty of water.

Equipment required

Bottles of those chemicals from the list below that are appropriate to the student sheet you have chosen to use will be required for each group of students. If enough bottles are supplied, students will not need to wander around looking for the reagents they need. Dropper bottles are best. The concentrations are not crucial, and all are minimal hazard at the concentrations specified but the sodium hydroxide should be **below** 0.5 mol dm^{-3} to minimise the hazard. Exactly which salt is used is also not critical.

■ Laminated worksheets, or photocopied worksheets and plastic wallets

■ Red litmus paper

■ Sodium hydroxide < 0.5 mol dm^{-3} (**Irritant**)

■ Iron(II) sulfate 0.2 mol dm^{-3}

■ Iron(III) nitrate 0.2 mol dm^{-3}

■ Copper(II) sulfate 0.2 mol dm^{-3}

■ Aluminium sulfate 0.2 mol dm^{-3}

■ Calcium chloride 0.2 mol dm^{-3}

- Magnesium chloride 0.2 mol dm^{-3}
- Ammonium chloride 0.2 mol dm^{-3}

Health and safety

Eye protection should be worn.

Answers

Teachers should select the answers appropriate to the student sheet chosen from those given below.

1.

Positive ion solution	Sodium hydroxide
Iron(II), Fe^{2+}	Grey-green solid formed
Iron(III), Fe^{3+}	Orange solid formed
Copper(II), Cu^{2+}	Light blue solid formed
Aluminium, Al^{3+}	White solid formed. Dissolves in excess sodium hydroxide.
Calcium, Ca^{2+}	White solid formed
Magnesium, Mg^{2+}	White solid formed
Ammonium, NH$_4^+$	A gas is evolved that turns damp red litmus paper blue.

Table 1 Sample observations

2. A solid formed by reacting two solutions is a precipitate.

3. The solids made are: iron(II) hydroxide, iron(III) hydroxide, copper(II) hydroxide, aluminium hydroxide, calcium hydroxide and magnesium hydroxide.

4. Students could give any three of the following equations:

$$Fe^{3+} + 3\ OH^- \rightarrow Fe(OH)_3$$

$$Cu^{2+} + 2\ OH^- \rightarrow Cu(OH)_2$$

$$Al^{3+} + 3\ OH^- \rightarrow Al(OH)_3$$

$$Ca^{2+} + 2\ OH^- \rightarrow Ca(OH)_2$$

$$Mg^{2+} + 2\ OH^- \rightarrow Mg(OH)_2$$

5. The gas formed by the reaction between the ammonium and hydroxide ions is ammonia.

6. $NH_4^+ + OH^- \rightarrow NH_3 + H_2O$

This page has been intentionally left blank.

Testing for negative ions

This activity is in two parts – in the first, students make observations while carrying out the tests for various negative ions. In the second, they use their observations to help them identify the negative ions present in a number of unknown solutions. To make the second part of the exercise more challenging, tests for positive ions could be introduced and students could be asked to identify both the positive and negative ions present in a solution.

Equipment required

The exact concentrations of the test solutions are not important. Use approximately 0.1–0.5 mol dm^{-3} for salt solutions and 0.5–1.0 mol dm^{-3} for acid solutions, except for nitric acid, which is corrosive at such concentrations (use 0.4 mol dm^{-3} instead).

- Test-tubes

- Dropping pipettes (these can be used just for the carbon dioxide testing or also for dispensing solutions; if the latter, far more pipettes will be required)

- Nitric acid 0.4 mol dm^{-3} (**Irritant**)

- Silver nitrate solution 0.1 mol dm^{-3}

- Barium chloride solution 0.1 mol dm^{-3} (**Harmful**)

- Hydrochloric acid

- Aluminium powder (**Highly flammable**)

- Sodium hydroxide solution less than 0.5 mol dm^{-3} (**Irritant**)

- Limewater

- Red litmus paper

- Ammonia solution 0.4 mol dm^{-3}.

For the initial observations

- Sodium or potassium chloride solution

- Sodium or potassium bromide solution

- Sodium or potassium iodide solution

- Sulfate solution, *eg* sodium sulfate

- Carbonate solution, *eg* potassium carbonate

- Nitrate solution, *eg* potassium nitrate.

For testing unknowns

The number of unknowns required depends on the time available. It is a good idea to use at least four solutions to ensure students are challenged. Label the solutions A, B, C etc and make sure you know which is which.

Health and safety

Wear eye protection.

Barium chloride solid is toxic; the 0.1 mol dm^{-3} solution is harmful. Wash your hands after use and warn students to do the same.

Ammonia solution is an irritant when concentrated but not at the concentrations used by students in this activity. However, it can give off ammonia vapour, which can irritate the eyes and lungs. Keep the lid on the bottle when not in use.

Nitric acid is an irritant.

Silver nitrate solution can stain skin and clothes.

Further problem solving ideas

There are several suggestions in C. Wood, *Creative Problem Solving in Chemistry*, London: Royal Society of Chemistry, 1993.

Tests for negative ions – expected observations

Negative ion	Test	Observations
Cl^- chloride	Add a few drops of dilute nitric acid followed by a few drops of silver nitrate solution. Let the mixture stand for a few minutes and then add some ammonia solution.	A white precipitate forms which discolours on standing. The precipitate is soluble in ammonia solution.
Br^- bromide	"	A cream precipitate forms which discolours a little on standing. The precipitate is slightly soluble in ammonia solution.
I^- iodide	"	A yellow precipitate forms which does not discolour on standing. The precipitate is insoluble in ammonia solution.
SO_4^{2-} sulfate	Add a few drops of barium chloride solution and then a few drops of hydrochloric acid.	A white precipitate forms.
CO_3^{2-} carbonate	Put a small amount of limewater into a test-tube (no more than 1 cm^3). Put your sample in a separate test-tube and add a few drops of hydrochloric acid. Using a pipette, collect the gas given off and bubble it through the limewater. (Note: you can also do this test on a solid sample.)	Bubbles of gas form. The gas turns the limewater milky, which shows that it is carbon dioxide.
NO_3^- nitrate	Add a few drops of sodium hydroxide solution and a little aluminium powder. Warm the solution in a Bunsen flame and test any gas given off using red litmus paper.	A gas is given off which turns the litmus blue. This shows that the gas is ammonia.

Equations

$$NaCl(aq) + AgNO_3(aq) \rightarrow NaNO_3(aq) + AgCl(s)$$

$$Cl^-(aq) + Ag^+(aq) \rightarrow AgCl(s) \text{ (and similarly for } Br^- \text{ and } I^-)$$

$$Na_2SO_4(aq) + BaCl_2(aq) \rightarrow 2\ NaCl(aq) + BaSO_4(s)$$

$$SO_4^{2-}(aq) + Ba^{2+}(aq) \rightarrow BaSO_4(s)$$

$$2\ HCl(aq) + Na_2CO_3(aq) \rightarrow 2\ NaCl(aq) + H_2O(l) + CO_2(g)$$

$$CO_3^{2-}(aq) + 2H^+(aq) \rightarrow CO_2(g) + H_2O(l)$$

For completeness, the reaction with the nitrate ion is shown below. It is unlikely that students will be able to construct this for themselves and the student sheet does not ask them to do so.

$$8Al(s) + 3NO_3^-(aq) + 5OH^-(aq) + 18H_2O(l) \rightarrow 8[Al(OH)_4]^-(aq) + 3NH_3(g)$$

Cold light

Index 7.4
2 sheets

This practical on cold light provides an interesting way to emphasise the importance of making detailed observations. It allows students to practice following a flow chart and is a good lead-in to teaching spectroscopy. It could also be followed up with individual or group projects on the applications of cold light – these could include examples of bioluminescence or research on how a TV or flat screen works for instance. If teaching a biology unit, the example of chlorophyll from spinach may well be worth using to show a little more clearly how chlorophyll works.

Equipment required

A darkened room or a box for each group of students to create their own dark area.

For each group of students:

■ White light – from a lamp or torch

■ UV lamp (these can be purchased at reasonable prices from many internet sites – they are often sold as security devices for checking whether bank notes are genuine)

■ 2 or 3 spinach leaves

■ Knife (blunt ones such as ordinary table knives are fine)

■ Tile or plate

■ Approx 20 cm^3 ethanol

■ Beaker (100 or 250 cm^3)

■ Boiling tube with bung

■ Glass rod

■ Wrapped clear boiled sweet – if you can get Lifesavers® from the USA in 'oil of wintergreen' flavour they work best, but mints such as Fox's Glacier Mints or similar are fine

■ Clear plastic bag (eg sandwich bag)

■ Pliers

■ White paper

■ Washing machine powder

■ Fluorescent pens

■ Brown paper (back of an old envelope is fine)

■ Glow-in-the-dark sheet (available at a reasonable price from **http://www.mutr.co.uk** (accessed November 2005))

■ Tonic water (the bottle does not need to be opened so the same one will last for a long time)

■ Samples of gyspum, calcite and/or fluorite minerals

■ Some resealable envelopes – optional (they give a flash of light when opened in a darkened room but not all work so it is worth checking before giving them to students)

■ Eye protection.

Health and safety

Warn students not to look at the UV lights. They should always be pointing away from the eyes and students should shine the light only on the item that they are observing.

Ethanol is flammable – it should not be used near flames.

Pliers can crush fingers if they are not used correctly.

Warn students not to eat in the laboratory.

Sample results and observations

Item	White light on	White light off	UV light on	UV light off	Luminescent	Type of luminescence
boiled sweet	nothing	gives out a flash of light as it is crushed	nothing	nothing	yes	triboluminescence
fluorescent pens	glow a little	nothing	glow brightly – especially against the brown paper	nothing	yes	fluorescence
Glow-in-the-dark sheet	glows a little	glows brightly	glows a little more	glows extremely brightly and for far longer than with white light	yes	phosphorescence
tonic water	nothing	nothing	glows a dull greeny blue	nothing	yes	fluorescence
mineral samples	nothing	nothing	glows – colour depends on which mineral is used	nothing	yes	fluorescence
spinach solution	looks green	nothing	glows red	nothing	yes	fluorescence
white paper	nothing	nothing	glows	nothing	yes	fluorescence
washing powder	nothing	nothing	glows a bluey colour	nothing	yes	fluorescence due to optical brighteners in the powder

If you wish, you can also add chemiluminescence to the above. Experiment details are in the following RSC publications.

T. Lister, *Classic Chemistry Demonstrations*, London: Royal Society of Chemistry 1995.

How it works

The way in which the light is produced varies according to which type of luminescence occurs. Fluorescence and phosphorescence are most closely related to the way in which Infra-red (IR) and UV-VIS spectroscopy work.

In both fluorescence and phosphorescence, light is absorbed at one wavelength and released at another. When the substance is exposed to certain wavelengths of light this radiation is absorbed and causes electrons in the atoms to be excited to a higher energy level. As these electrons fall back to a lower energy level (which may or may not be the original one) they release the 'extra' energy in the form of light. The diagram below helps to explain this.

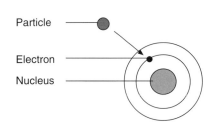
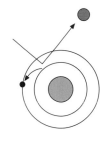
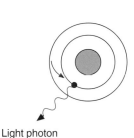

Light photon

1. A particle collides with the atom

2. This causes an electron to jump to a higher energy level.

3. As the electron falls back to its original energy level, light energy is released in the form of a photon

Figure 1 How atoms emit light

In the case of chemiluminescence (and bioluminescence, which is just a form of chemiluminescence) the light is the energy that is released during the course of a reaction. Some reactions produce light rather than heat or sound.

Triboluminescence is light emitted when mechanical forces act on the structure of a substance.

Spectroscopy

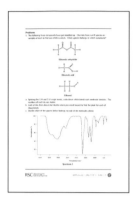
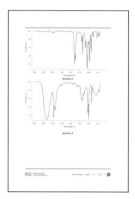

Index 7.5
5 sheets

Three activities that could be used to teach spectroscopy are presented within this resource.

A good introduction to spectroscopy is to do the **Cold light** practical or to demonstrate that many objects absorb UV light and then emit light of a different wavelength/frequency. Explain that the same thing happens with infrared radation (and many other types of electromagnetic radiation) except that we cannot see it. Instead, a detector is used to read which wavelengths of infrared are being absorbed (or transmitted). Every substance has its own unique infrared (IR) spectrum. A spectrum obtained during an experiment can be matched against a data bank of sample spectra to identify a substance. It is also possible to interpret certain individual peaks on the spectrum using a data book/sheet.

When IR radiation interacts with matter and is absorbed, it causes the bonds in the substance to vibrate. They can either bend or stretch – this is the basis of the activity Chemical aerobics, which is another good introduction to the topic of spectroscopy.

Two problem-solving activities relating to IR spectroscopy are given in the student sheet entitled Spectroscopy.

For further examples of spectra and for video clips showing how an IR spectrometer is operated, see the RSC CDROM *Spectroscopy for schools and colleges* or the website **http://www.chemsoc.org/networks/learnnet/spectra/index2.htm**.

Chemical aerobics

When infrared radiation interacts with a substance, certain wavelengths cause particular bonds in that substance to vibrate. Only certain wavelengths will do this and each different type of bend or stretch of a bond will require a different wavelength of radiation.

Get the class to stand up and pretend that they are each a water molecule. Their heads are the oxygen atoms, their hands the hydrogen atoms and their arms the bonds. If you want to make this really clear, you can make large labels for the heads and hands. Ask the students to try to work out which bends or stretches are possible.

They should be able to work out that there are three possible kinds of vibration, although they almost certainly will not get the names for them. The three vibrations are:

■ symmetric stretch (both hands moving back and forth away from the body together and both arms bending together)

- asymmetic stretch (one hand moves forward as the other moves back towards the body, or one arm bending as the other straightens)

- bending (hands moving towards each other in front of the body and back again with arms staying straight).

Symmetric stretch   Asymmetric stretch   Bending

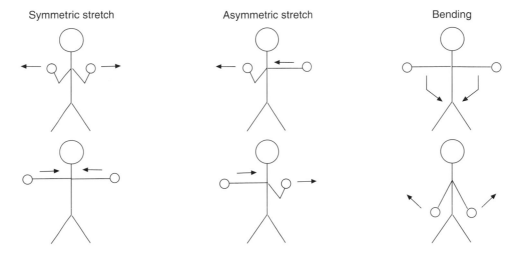

If you wish to make this exercise more fun, you could have some music clips prepared on a tape or CD. Assign a different singer/group/song to each movement. When the students hear the singer they have to make the correct move – if there is a singer with no movement assigned then they stay still. You can be as cheesy or up-to-date as you like but the singers do need to be well known by the vast majority of the class and the clips need to be short.

Solutions to spectroscopy problems

1. Spectrum 1 is ethanoic acid; 2 is ethanoic anhydride and 3 is ethanol. Labelled spectra are given below.

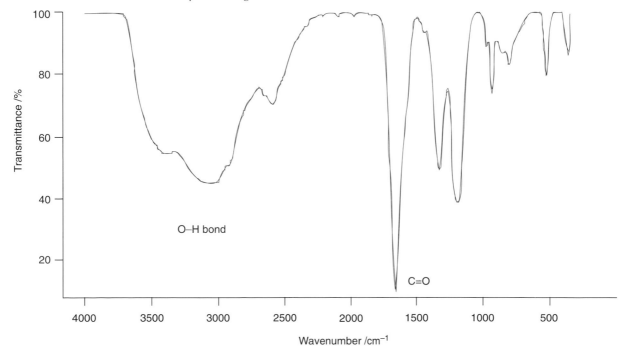

Figure 1 Spectrum 1 – ethanoic acid

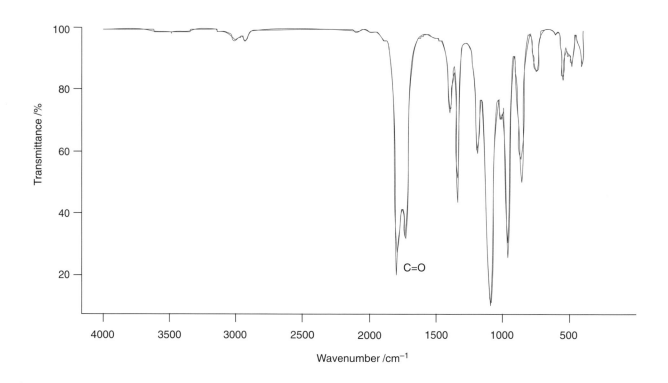

Figure 2 Spectrum 2 – ethanoic anhydride

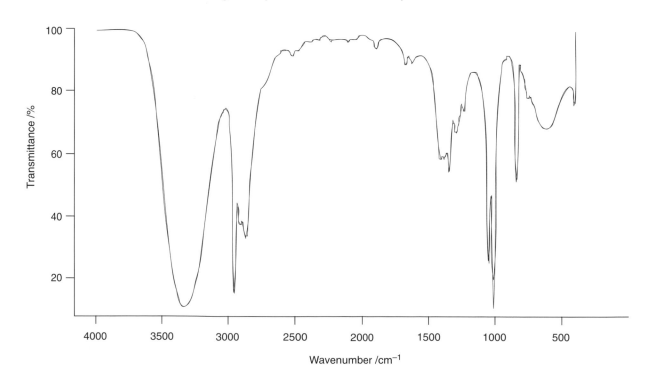

Figure 3 Spectrum 3 – ethanol

2. Smell Fresh deodorant contains chloroethane and Jaguar contains butane. Labelled spectra are given below:

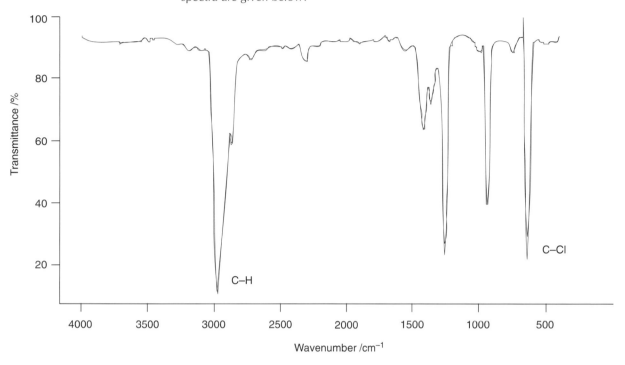

Figure 4 Smell Fresh deodorant – chloroethane

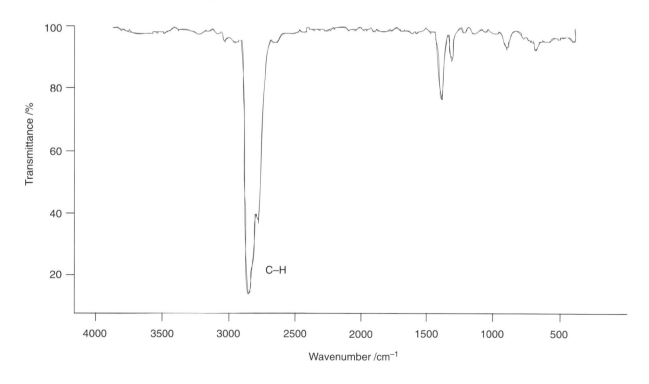

Figure 5 Jaguar deodorant – butane

3. Yes, they have obtained the product they wanted. The annotated spectrum is shown below. There is no O-H stretch at around 3200-3400 cm^{-1} but a peak corresponding to a C=O stretch can be seen.

Note: This reaction is reversible and in reality it would not go to completion.

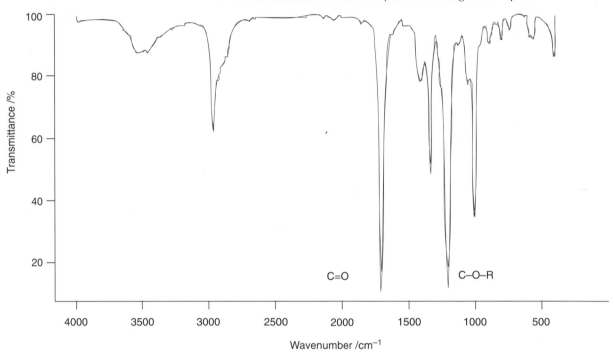

Figure 6 Spectrum of ethyl ethanoate

LEARNING TO TEACH IN THE SECONDARY SCHOOL

Learning to Teach in the Secondary School 5th edition offers a comprehensive, in-depth and practical introduction to the knowledge and skills you need to become an effective teacher. With a focus on evidence-informed and reflective practice, this book contains many examples of how to apply theory to practice and how to analyse practice to maximise pupil learning. It gives you a framework of underpinning theory and evidence to enable you to reflect on your practice and become a teacher who makes a difference to the lives of the young people in our society.

Written by experienced practitioners, this book provides practical help for many of the situations and potential challenges you are faced with in school. Tasks in each unit engage with issues in your school experience, provide opportunities for you to reflect on your own learning and performance and provide you with an analytical toolkit. This 5th edition is updated to reflect recent developments in professional knowledge and practice. Masters level tasks and annotated further readings respond to the requirements for teachers to engage in M level work.

Drawing on effective practice established from research and evidence, each unit covers a key concept or skill, including personalised learning, and the relationship between the brain and learning, which are new to this edition, plus:

- planning lessons and schemes of work
- managing behaviour for learning
- assessment
- ways pupils learn
- differentiation, progression and pupil grouping
- inclusion and special educational needs
- using ICT in teaching and learning
- getting your first teaching post
- guidance on writing.

This core text is supported by the Learning to Teach Subjects in the Secondary School series by the same editors and is an essential purchase for every aspiring secondary school teacher.

Susan Capel is Professor and Head of the School of Sport and Education at Brunel University, UK.
Marilyn Leask is Professor of Education at Brunel University, UK.
Tony Turner was Senior Lecturer in Education at the Institute of Education, University of London, UK.

LEARNING TO TEACH SUBJECTS IN THE SECONDARY SCHOOL SERIES

Series Editors: Susan Capel, Marilyn Leask and Tony Turner

Designed for all students learning to teach in secondary schools, and particularly those on school-based initial teacher training courses, the books in this series complement *Learning to Teach in the Secondary School* and its companion, *Starting to Teach in the Secondary School*. Each book in the series applies underpinning theory and addresses practical issues to support students in school and in the training institution in learning how to teach a particular subject.

LEARNING TO TEACH IN THE SECONDARY SCHOOL

A companion to school experience

5th Edition

Edited by

Susan Capel, Marilyn Leask and Tony Turner

Routledge
Taylor & Francis Group

LONDON AND NEW YORK

Fifth edition published 2009
by Routledge
2 Park Square, Milton Park, Abingdon, Oxon OX14 4RN

Simultaneously published in the USA and Canada
by Routledge
711 Third Ave, New York, NY 10017

First published 1995 by RoutledgeFalmer
Second edition published 1999 by RoutledgeFalmer
Third edition published 2001 by RoutledgeFalmer
Fourth edition published 2005 by Routledge

Routledge is an imprint of the Taylor & Francis Group, an informa business

© 1995, 1999, 2001, 2005, 2009 Susan Capel, Marilyn Leask and Tony Turner
for editorial material and selection. Individual chapters the contributors

Typeset in Times New Roman by
Florence Production Ltd, Stoodleigh, Devon

British Library Cataloguing in Publication Data
A catalogue record for this book is available from the British Library

Library of Congress Cataloging in Publication Data
Learning to teach in the secondary school: a companion to school
    experience/edited by Susan Capel, Marilyn Leask, and Tony Turner. – 5th ed.
        p. cm. – (Learning to teach subjects in the secondary school series)
        Includes bibliographical references and index.
    1. High school teaching – Handbooks, manuals, etc. 2. Classroom
    management – Handbooks, manuals, etc. I. Capel, Susan Anne, 1953–.
    II. Leask, Marilyn, 1950–. III. Turner, Tony, 1935–.
LB1737.A3L43 2009
373.1102 – dc22   2008050122

ISBN10: 0–415–49532–6 (hbk)
ISBN10: 0–415–47872–3 (pbk)

ISBN13: 978–0–415–49532–5 (hbk)
ISBN13: 978–0–415–47872–4 (pbk)

CONTENTS

CONTENTS ■ ■ ■ ■

ILLUSTRATIONS

FIGURES

TABLES

TASKS

CONTRIBUTORS

Françoise Allen is a primary modern foreign languages school improvement adviser in a local authority. Previously she was a lecturer in education at Brunel University, where she taught on the PGCE secondary and primary programmes.

Michael Allen is Lecturer in Science Education at Brunel University.

Steve Bartlett is Professor of Education at the University of Wolverhampton.

Rob Batho is School Standards Adviser with the Department for Children, Schools and Families, having previously been an English teacher in secondary schools, Head of Education at University of Chichester and a senior adviser with the Secondary National Strategy.

Richard Bennett is Senior Lecturer in Education at the University of Chester. Previously he was a headteacher, an Ofsted inspector and an advisory teacher.

Diana Burton is Professor of Education and Pro-Vice-Chancellor for Research and Student Experience at Liverpool John Moores University.

Graham Butt is Reader in Geography Education, Director of Academic Planning and Deputy Head of the School of Education, University of Birmingham.

Susan Capel is Professor and Head of School of Sport and Education at Brunel University.

Jon Davison has been Professor of Teacher Education in four universities, including the Institute of Education, University of London, where he was also Dean.

Philip Garner is Professor of Education at the University of Northampton.

Misia Gervis is Senior Lecturer in Sport Psychology and Coaching at Brunel University.

Andrew Green is Senior Lecturer in Education and Course Leader for the Secondary English PGCE and MA Education at Brunel University.

Terry Haydn is Reader in Education at the School of Education and Lifelong Learning at the University of East Anglia.

Graham Haydon is Reader in Philosophy of Education at the Institute of Education, University of London.

Susan Heightman is an independent education management professional, previously a school senior manager and teacher educator.

Ruth Heilbronn is Lecturer in Education at the Institute of Education, University of London.

Judy Ireson is Professor of Psychology in Education and Dean of the Faculty of Children and Health, Institute of Education, University of London.

Margaret Jepson is Senior Lecturer in the Faculty of Education, Community and Leisure at Liverpool John Moores University.

Julia Lawrence is Principal Lecturer and Subject Group Leader for Physical Education Teacher Education at Leeds Metropolitan University.

Marilyn Leask is Professor of Education at Brunel University.

Hilary Lowe is Assistant Dean at the Westminster Institute of Education, Oxford Brookes University.

John McCormick taught science before moving to Liverpool John Moores University to lead the PGCE programme. He is currently Secondary Partnership Manager for Initial Teacher Education.

John Moss is Dean of Education at Canterbury Christ Church University.

Colette Murphy is Senior Lecturer in Education at Queen's University Belfast and former Head of Pre-service Education.

Andrew Noyes is Associate Professor in Education at the University of Nottingham.

Nick Peacey is Coordinator of the Special Educational Needs Joint Initiative for Training (SENJIT), Institute of Education, University of London.

Janet Pritchard is Professor and Head of College of Education and Lifelong Learning at Bangor University.

Jonathan Sharples is currently Manager of Partnerships at the Institute for Effective Education, at University of York. He previously worked at the Institute for the Future of the Mind, at University of Oxford, where he explored how insights from brain-science research can help inform decision-making in learning and education.

Alexis Taylor is Lecturer in Education and Programme Leader for the Doctorate of Education at Brunel University.

Allen Thurston is Senior Lecturer at The Stirling Institute of Education, University of Stirling.

Rob Toplis is Lecturer in Secondary Science Education and Course Leader for the Secondary Postgraduate Certificate in the School of Sport and Education at Brunel University.

Keith Topping is Professor of Educational and Social Research in the School of Education at the University of Dundee.

Tony Turner was, before retirement, Senior Lecturer in Education at the Institute of Education, University of London.

Barbara Walsh is Centre Leader for Sport, Dance and Outdoor Education in the Faculty of Education, Community and Leisure at John Moores University.

Mike Watts is Professor and Subject Leader for Education at Brunel University.

Carrie Winstanley is Principal Lecturer at Roehampton University and a National Teaching Fellow.

Bernadette Youens is Lecturer in Science Education and Director of of Initial Teacher Education, at the School of Education, University of Nottingham.

Paula Zwozdiak-Myers is Lecturer in Education at Brunel University.

INTRODUCTION

Susan Capel, Marilyn Leask and Tony Turner

The book introduces the professional knowledge and skills required by teachers, including general principles of effective teaching. The book is backed up by subject specific and practical texts in the *Learning to Teach in the Secondary School* series by the same editors and by *Readings for Learning to Teach in the Secondary School: A Companion to M Level Study*. The text *Starting to Teach in the Secondary School* by the same editors is designed to support you in the transition from student teacher to newly qualified teacher. The *Reader* and *Starting to Teach in the Secondary School* texts provide extension reading around key areas of professional knowledge underpinning teaching.

This fifth edition of *Learning to Teach in the Secondary School* has been updated to take account of the move in England and Wales to require teachers to have Masters level (M level) qualifications. Student teachers now on initial teacher education courses usually gain credits towards Masters level accreditation. This edition of *Learning to Teach* includes new sections, for example on personalised learning and neuroscience as well tasks which can support work being undertaken for Masters level accreditation.

The website which accompanies the text www.routledge.com/textbooks/9780415478724 includes further information and PowerPoint summaries of each unit. It also contains additional materials on the *Every Child Matters Agenda* in England and on *Safeguarding Children*, as well as a unit 'Using research and evidence to inform your teaching' from the text *Starting to Teach*. This text is written for teachers in their early years of teaching and provides advice for you in undertaking the kind of action research project which could lead to M level accreditation.

Teaching is a complex activity and is both an art and a science. In this book we show that there are certain essential elements of teaching that you can master through practice that help you become an effective teacher. However, there is no one correct way of teaching, no one specific set of skills, techniques and procedures that you must master and apply mechanically. This is, in part, because your pupils are all different and each day brings a new context in which they operate. Every teacher is an individual and brings something of their own unique personality to the job and their interactions with pupils. We hope that this book helps you to develop skills, techniques and procedures appropriate for your individual personality and style and provides you with an entry to ways of understanding what you do and see that you can bring together into an effective whole when a lesson goes well.

An effective teacher is one who can integrate theory with practice, use evidence to underpin their professional judgement and one who can use structured reflection to improve practice.

We also hope that the text provides the stimulus for you to want to continue to learn and develop throughout your career as a teacher.

The content of this book, some of the tasks, further readings and the *Reader* together with the extra materials on the website are intended to support M level work within your initial teacher education (ITE) course.

DEVELOPING YOUR PHILOSOPHY OF TEACHING

On your ITE course much of your time is spent in school. You can expect your ITE course to provide not merely *training* but also to introduce you to wider educational issues. What we mean by this is that ITE is not an apprenticeship but a step on the journey of personal development in which your teaching skills develop alongside an emerging understanding of the teaching and learning process and the education system in which it operates. This is a journey of discovery that begins on the first day of your course and may stop only when you retire. Teachers are expected to undertake further professional development throughout their career. Thus, we use the term initial teacher *education* rather than initial teacher training throughout this book.

The school-based element of your course provides the opportunity to appreciate at first hand the complex, exciting and contradictory events of classroom interactions without the immediacy of having to teach all the time. It should allow you time, both in the classroom and the wider school to make sense of experiences that demand explanations. Providing such explanations requires you to have a theory of teaching and learning.

By means of an organised ITE course that provides for structured observation, practical experience and reflective activity suitably interwoven with theoretical inputs and evidence, you begin to develop your own theory of teaching and learning which is embedded in practice. Theoretical inputs and evidence to underpin practice can come from a range of sources including tutors and teachers, lectures and print- and web-based resources. Theory also arises from practice, the better to inform and develop practice.

Everyone who teaches has a theory of how to teach effectively and of how pupils learn. The theory may be implicit in what the teacher does and teachers may not be able to tell you what their theory is. For example, a teacher who is a disciplinarian is likely to have a different theory about the conditions for learning than a teacher who is liberal in their teaching style. Likewise, some teachers may feel that they do not have a philosophy of education. What these teachers are really saying is that they have not examined their views, or cannot articulate them. What is your philosophy? For example, do you consider that your job is to transfer the knowledge of your subject to pupils? Or are you there to lead them through its main features? Are you 'filling empty vessels' or are you the guide on a 'voyage of discovery'? On the other hand, perhaps you are the potter, shaping and moulding pupils.

It is recognised that an ITE course only enables you to start developing your own personal understanding of the teaching and learning process. There are a number of different theories about teaching and learning. You need to be aware of what these are, reflect on them and consider how they help you to explain more fully what you are trying to do and why. Through the process of theorising about what you are doing, reflecting on a range of other theories as well as your own, and drawing on the evidence base you understand your practice better and develop into a reflective practitioner, that is, a teacher who makes conscious decisions about teaching strategies to employ and who modifies their practice in the light of experiences.

An articulated, conscious philosophy of teaching emerges only if a particular set of habits is developed, in particular, the habit of reviewing your own teaching from time to time. It is these habits that need to be developed from the start of your ITE course. This is what many authors mean when they refer to 'the reflective practitioner'. This is why we (as well as your course tutors) ask you to evaluate your own teaching, to keep a diary of your evaluations (reflective practice), a folder of your lesson plans and other material to develop a professional development portfolio (PDP) to record your development and carry that forward from your ITE course to your first post. If you are learning to teach in England you are required to compile a self-evaluation tool which includes reflection and evidence.

HOW TO USE THIS BOOK

Structure of the book

The book is laid out so that elements of appropriate background information and theory along with evidence from research and practice introduce each topic. These are interwoven with tasks designed to help you identify key features of the topic.

A number of different inquiry methods are used to generate data, e.g. reflecting on reading and observation or on an activity you are asked to carry out, asking questions, gathering data, discussing with a tutor or another student teacher. Some of the tasks involve you in activities that impinge on other people, for example, observing a teacher in the classroom, or asking for information. If a task requires you to do this, *you must first of all seek permission of the person concerned.* Remember that you are a guest in school(s); you cannot walk into any teacher's classroom to observe. In addition, some information may be personal or sensitive and you need to consider issues of confidentiality and professional behaviour in your inquiries and reporting.

In this book, we use the symbol **M** to denote tasks which can be designed to meet the requirements of M level work but it is up to your tutors to design assignment titles which meet the requirements of the higher education institution with which you are registered. Most student teachers are on programmes which provide accreditation towards a Masters degree which can be completed through further research and study focused on the workplace in your early years of teaching. M level and Doctorate in Education programmes are designed to support your further professional development through research, reflection and wider reading.

An appendix on writing and reflection is included to help you with the written assignments on your ITE course. A glossary of terms is also included to help you interpret the jargon of education.

We call school children *pupils* to avoid confusion with *students*, by which we mean people in further and higher education. We refer to those learning to teach as *student teachers*. The important staff in your life are those in school and higher education institution; we have called all these people *tutors*. Your institution will have its own way of referring to staff.

Meeting the requirements of your course

The range and type of requirements (standards) you are expected to meet during your ITE course have been derived from those for all student teachers in the country in which you are learning to teach. The units in this book are designed to help you work towards meeting these requirements. Your tutors in school and in your institution help you meet the requirements for your course. At appropriate points in the text you should relate the work directly to the specific requirements for your ITE course.

Reflective practice and your professional development portfolio (PDP)

As you read through the book, undertake other readings, complete the tasks and undertake other activities as part of your course, we suggest you keep a professional development portfolio (PDP). You may want to keep a diary of reflective practice to record your reactions to, and reflections on, events, both good and bad, as a way of letting off steam! It enables you to analyse strengths and areas for development, hopes for the future, and elements of your emerging personal philosophy of teaching and learning.

Your PDP holds a selective record of your development as a teacher, your strengths as well as areas for further development, and is something that you continue to develop throughout your teaching career. It is likely that your institution has a set format for a PDP. If not, you should develop your own. You can use any format and include any evidence you think appropriate. However, to be truly beneficial, it should contain evidence beyond the minimum required for your course. This further evidence could include for example work of value to you, a response to significant events, extracts from your diary of reflective practice, good lesson plans, evaluations of lessons, teaching reports, observations on you made by teachers, outcomes of tasks undertaken, assessed and non-assessed course work.

At the end of your course you can use your PDP to evaluate your learning and achievements. It is also used as the basis for completing applications for your first post; and to take to interview. It can form the basis of a personal statement describing aspects of your development as a teacher during your course. Your PDP could include teaching reports written by teachers, tutors and yourself. It can also help provide the basis of your continuing professional development (CPD) as it enables you to identify aspects of your work in need of development and thus targets for induction and CPD in your first post, first through your self-evaluation tool if you are learning to teach in England, then as part of the appraisal process you will be involved with as a teacher.

Ways you might like to use this book

With much (or all) of your course being delivered in school, you may have limited access to a library, to other student teachers with whom to discuss problems and issues at the end of the school day, and, in some instances, limited access to a tutor to whom you can refer. There are likely to be times when you are faced with a problem in school which has not been addressed up to that point within your course and you need some help immediately, for example before facing a class the next day or next week. This book is designed to help you address some of the issues or difficulties you are faced with during your ITE course, by providing supporting knowledge interspersed with a range of tasks to enable you to link theory with practice.

The book can be used in a number of ways. You should use it alongside your course handbook, which outlines specific course requirements, agreed ways of working, roles and responsibilities. It is designed more for you to dip in and out of, to look up a specific problem or issue that you want to consider, rather than for you to read from cover to cover (although you may want to use it in both ways of course). You can use it on your own as it provides background information and supporting theory along with evidence from research and practice about a range of issues you are likely to face during your ITE course. Reflecting on an issue faced in school with greater understanding of evidence of what others have written and said about it, alongside undertaking some of the associated tasks, may help you to identify some potential solutions. The book can also be used for collaborative work with other student teachers or your tutors. The tasks are an integral part of the book and you can complete most individually. Most tasks do, however, benefit from wider discussion, which we encourage you to do whenever possible. However, some tasks can be carried out only with other student teachers and/or with the support of a tutor. You should select those tasks that are appropriate to your circumstances.

This book will not suffice alone; we have attempted to provide you with guidance to further reading by two methods: first, by references to print and web-based material in the text, the details of which appear in the references; second, by further readings and relevant websites at the end of each unit.

There is much educational material on the Internet. In England 'TeacherNet', the QCA and the TDA Teacher Training Resource Bank are important resources, as are the subject association websites. Keep a record of useful websites.

If you see each unit as potentially an open door leading to whole new worlds of thought about how societies can best educate their children, then you have achieved one of our goals: that is, to provide you with a guide book on your journey of discovery about teaching and learning. Remember, teaching is about the contribution you make to your pupils, to their development and their learning and to the well-being of society through the education of our young people.

Finally, we hope that you find the book useful, and of support in school. If you like it, tell others; if not, tell us.

Susan Capel, Marilyn Leask, Tony Turner
January 2009

1 BECOMING A TEACHER

Through the units in this chapter, the complexity and breadth of the teacher's role and the nature of teaching are explored. You are posed questions about your values and attitudes because these influence the type of teacher you become. Society is constantly changing and so the demands society places on teachers change. Consequently as your career progresses you will find you need new skills and knowledge about teaching and learning (pedagogy), so you can expect to continue to learn through your career. Professional development is therefore a lifelong process which is aided by regular reflection on practice and continuing education. Evidence about effective practice is becoming increasingly easy to access to support your development. In the UK, you can find a wealth of material on government-supported websites to support you as a teacher. Each unit lists relevant websites.

Each unit in this chapter examines different facets of the work of student and experienced teachers. Unit 1.1 covers wider aspects of the teacher's role, including academic and pastoral roles, and we consider the necessity for regular curriculum review as society changes.

In Unit 1.2 we discuss the expectations which the tutors responsible for your training have of you. The meaning of professionalism is discussed and the idea that you will have your own philosophy of teaching is introduced. Phases which mark your development as a teacher are identified. We suggest that as your own confidence and competence in managing the classroom grow, you can expect the focus of your work to move from your self-image and the mechanics of managing a lesson to the learning taking place generally and, as you become more experienced, to the learning for the individual pupil.

Unit 1.3 provides advice for managing time, both inside and outside the classroom, for preventing stress and for managing stress that you cannot prevent. Managing competing demands on your time gives you time to enjoy your work and have time for leisure. Unit 1.4 introduces you to ways you are expected to use ICT to support teaching and learning in your classroom as well as your own professional development.

To become a teacher you need to supplement your *subject content knowledge* with *professional pedagogic knowledge* (about teaching and learning) and to develop your *professional judgement*, for example about managing situations which arise with pupils. Ways of developing your professional knowledge and judgement provide themes running throughout the book.

You may come to recognise your situation in the following poem called 'Late'.

You're late, said miss.
The bell has gone,
dinner numbers done
and work begun.

What have you got to say for yourself?
Well, it's like this, miss
Me mum was sick,
me dad fell down the stairs,
the wheel fell off my bike
and then we lost our Billy's snake
behind the kitchen chairs. Earache
struck down me grampy, me gran
took quite a funny turn.
Then on the way I met this man
whose dog attacked me shin –
look, miss you can see the blood
it doesn't look too good,
does it?
Yes, yes sit down –
and next time say you're sorry
for disturbing all the class.
Now get on with your story
fast!
Please miss, I've got nothing to write about.

(Judith Nicholls in Batchford (1992) Assemblies for the 1990s)

Readings for Learning to Teach in the Secondary School: A Companion to M Level Study
brings together essential readings to support you in your critical engagement with key issues raised in this textbook.

UNIT 1.1

WHAT DO TEACHERS DO?

Andrew Green and Marilyn Leask

Now, what I want is, Facts. Teach these boys and girls nothing but Facts. Facts alone are wanted in life. Plant nothing else, and root out everything else. You can only form the minds of reasoning animals upon Facts: nothing else will ever be of any service to them. This is the principle on which I bring up my own children, and this is the principle on which I bring up these children. Stick to Facts, sir!

(Thomas Gradgrind – Dickens, *Hard Times*)

Tell me, I will forget. Show me, I may remember. Involve me, and I will understand.

(old Chinese proverb)

INTRODUCTION

Everyone has their own view of what teachers do, partially formed by their own experiences of school – often idealised by the passage of time – and the well-formed literary and media-inspired images they have imbibed over the years. The media, television, cinemas and literature all contain various representations of teachers. Mr Chips, Miss Jean Brodie, John Keating, Thomas Gradgrind, the teachers of Grange Hill and Waterloo Road are the stuff of urban myth. What English teacher has not dreamt of inspiring the passion and devotion to literature that Keating inspires in *Dead Poets' Society*? The fact is that nobody entering the teaching profession does so as a blank canvas; we have all experienced education ourselves and this has shaped our particular views of what teachers are and do. These views are central to your development as a teacher; however, it is important to recognise that not all of these views are either valid or useful in the context of the current school system and the demands it places on teachers.

Views of what teachers do are socially and culturally constructed. Views vary from one culture, one era and one group in society to another. The two epigraphs at the beginning of the unit illustrate this. Thomas Gradgrind's view of education contrasts radically with the old Chinese philosophy of teaching. The existence of a plurality of views of what teachers do is increasingly evident in the multicultural society in which we now live.

Many people have a vested interest in what teachers do, the most obvious being teachers, pupils, parents and carers. Views of what teachers do may vary considerably, depending on who you ask and when you ask. Other groups exert influence (sometimes considerable) over what teachers do, for example, politicians; local authorities (LAs)/councils; teachers' unions; professional subject associations; educational researchers. Each of these groups holds valid and important views about education and the teacher's role within it.

What a parent may expect from you as a teacher, what their child may expect and what you as a teacher believe you should provide may differ significantly. On the other hand, for example, union advice on an issue may be in opposition to legislation.

What teachers do, therefore is complex. How to manage this without compromising either the needs of the individual pupil, the requirements of parents and carers or your own professional integrity is the focus of this unit.

OBJECTIVES

At the end of this unit you should be able to:

■ describe various aspects of a teacher's role and responsibilities, including academic and pastoral roles and administration
■ account for the multi-faceted nature of the knowledge required for effective teaching
■ explain the rights and responsibilities of teachers and learners within classrooms
■ explain how teachers can proactively manage the learning environment.

Check the requirements of your course to see which relate to this unit.

TEACHERS AS INDIVIDUALS

Teaching is a deeply personal profession. There are as many ways of being an effective teacher as there are effective teachers. Your course of Initial Teacher Education (ITE), therefore, should provide you with opportunities both to explore individually the kind of teacher you wish to be, and also to understand the context in which you are working and the demands this places upon you as a teacher (see Hayes, 2004; Moore, 2004; Wragg, 2004).

But let us start where education should start, with the pupils. Above all, pupils respond to individuals. Think back to your own schooldays and the teachers you had. What do you remember about them? What did they do? Who are the teachers you most liked, and why? Which teachers did you least like, and why? Almost certainly the issues you identify are to do with personality: e.g. enthusiasm; intelligence; humour; disinterest; eccentricity; conformity; efficiency; incompetence. Similarly, one of the first things *your* pupils pick up on is you as a person; how you present yourself as an individual and as a practitioner. Their parents and carers look at you as a person, but are also interested in a different set of issues: e.g. are you likely to form supportive relationships with their child. Your head of department and senior staff apply a third set of criteria: e.g. what skills and interests you might have that could be of benefit to the department or wider school curriculum. Your university tutor and school mentor have yet another: e.g. how you might be supported to do your job well.

All of these perspectives about you as a teacher and what you do are important. It is particularly important, however, to familiarise yourself with the model of teaching written into the requirements for qualifying to teach, relevant to your course, a copy of which you should have received from your course leader. These requirements should be used to provide a basis for a programme which is coherent, intellectually stimulating and professionally challenging. The components of your course are planned to provide a secure framework for you to develop your professional knowledge and skills by providing opportunities to make links between subject expertise, principles, theory and developing experience to meet the course requirements. Task 1.1.1 focuses on these requirements.

Task 1.1.1 FOCUSING ON STANDARDS REQUIRED OF NEWLY QUALIFIED TEACHERS (NQTS)

To understand what is expected of you as a NQT you need to be familiar with the standards you are required to reach by the end of your course. These can be found in your course handbook and other documentation provided by your institution. Look at them now. What do they suggest your role as a teacher encompasses? What do they mean to you at the beginning of your ITE course? Ask your tutor to explain what achieving the standards might look like at the end of your course. Is there anything in these standards that surprises you?

Which aspects of the standards do you feel most prepared to meet? Which do you believe you need more help to meet? How do you see yourself developing your capability over the course of your career as a teacher? Check, and carefully log (in your Professional Development Portfolio (PDP)), your progress against these standards regularly so that you are aware early of any areas where you may require additional input and support. You may find it useful to discuss these areas with an experienced teacher, thinking about how you can record your achievements to provide information for your first employer and a basis for your continuing professional development (CPD).

YOUR ROLE AS A TEACHER

The teacher's job is first and foremost to ensure that pupils learn. To a large extent, *what* (i.e. the lesson content) pupils should learn in maintained (state) schools is determined through legislation and the requirements are set out in various national curriculum documents. On the other hand, *how* you teach so that the pupils learn effectively (i.e. the methods and materials used) is left to the professional judgement of the individual teacher, department and school.

In your first days in school, it is likely that you will spend time observing a number of experienced teachers. You are unlikely to see two teachers the same. You may see teaching styles with which you feel at home, while others do not seem as appropriate to your own developing practice. There is no single, correct way to teach. Provided effective teaching and learning takes place, a whole range of approaches from didactic (formal, heavy on content) to experiential (learning by doing) is appropriate – often in the same lesson. Unit 5.3 provides more details about teaching styles and Unit 4.1 ways to group pupils for learning. Increasingly, as student teachers you have the opportunity to use video to analyse the teaching styles of others and to have your own lessons videoed for analysis to improve practice.

Your role as a teacher falls into two distinct categories. You have responsibility for both the academic and the pastoral development of your pupils. Table 1.1.1 lists the main activities in each of these areas that you are expected to undertake.

In addition, teachers have a role to play in supporting the school ethos by reinforcing school rules and routines, e.g. on behaviour, in dress and in encouraging pupils to develop self-discipline so that the school can function effectively and pupils can make the most of opportunities available to them.

From the beginning of your school experiences, it is worth developing efficient ways of dealing with administration to save time (see Unit 1.3). Developing your word-processing and spreadsheet skills is essential in helping you prepare teaching materials and recording and monitoring progress and keeping up to date. Some teachers keep their mark books electronically using spreadsheets and many schools have Management Information Systems which are used to monitor pupil performance and assessment. For advice in your subject area, see the subject-specific and practical texts in this *Learning to Teach in the Secondary School* series (see p. ii of this text).

THE WORK IN THE CLASSROOM – THE TIP OF THE ICEBERG

On the surface, teaching may appear to be a relatively simple process – the view that the teacher stands in front of the class and talks and the pupils learn appears to be all too prevalent. (Ask friends and family what they think a teacher does.) The reality is somewhat different.

Classroom teaching is only the most visible part of the job of the teacher. The contents of this book are designed to introduce you to what we see as the invisible foundation of the teacher's work: *professional knowledge* (see Table 1.1.2) about teaching and learning and *professional judgement* about the routines, skills and strategies which support effective teaching. An effective teacher draws on these factors in planning each and every lesson; and the learning for a particular class is planned ahead so that there is *continuity and progression* in pupils' learning. Each lesson is planned as part of a sequence of learning experiences (see Unit 2.2).

The following analogy may help you understand what underpins your work in the classroom. Think of a lesson as being like an iceberg. The work in the classroom represents the tip of an iceberg (20–30 per cent). Supporting this tip, but hidden in the base (70–80 per cent), are the elements of the teacher's professional expertise (see Figure 1.1.1). These elements include:

■ *planning* of a sequence of lessons to ensure learning progresses and planning for a specific lesson
■ *evaluation* of previous lessons
■ *planning and preparation* for the lesson
■ *established routines and procedures* which ensure that the work of the class proceeds as planned
■ personality, including the teacher's ability to capture and hold the interest of the class, to establish their authority
■ *professional knowledge* such as subject content knowledge; pedagogic knowledge about effective teaching and learning; knowledge of learners; knowledge about the educational context in which you work – local and national (see Table 1.1.2).
■ *professional judgement* built up over time through reflection on experience.

Throughout your course, you should expect to develop confidence and new levels of competence in all the areas in Figure 1.1.1.

■ **Table 1.1.1** Some of the activities which teachers undertake in their academic and pastoral roles

The academic role	The pastoral role and spiritual and moral welfare
This encompasses a variety of activities including: • subject teaching • lesson preparation • setting and marking of homework • monitoring pupil progress • assessing pupil progress in a variety of ways, including marking tests and exams • writing reports • recording achievement • working as part of a subject team • curriculum development and planning • undertaking visits, field courses • reporting to parents • planning and implementing school policies • extra-curricular activities • being an examiner for public examination boards, e.g. GSCE and GCE A level boards • keeping up to date (often through work with the subject association).	These roles vary from school to school. They often include: • getting to know the pupils as individuals • helping pupils with problems • being responsible for a form/tutor group • registering the class, following up absence • monitoring sanctions and rewards given to form members • reinforcing school rules and routines, e.g. on behaviour • writing reports, ensuring records of achievement and/or profiles are up to date • working as part of a pastoral team • teaching Personal, Social and Health Education (PSHE) and citizenship • house/year group activities (plays/sports) • liaising with parents • ensuring school information is conveyed to parents via pupils • giving careers and subject guidance • extra-curricular activities, e.g. educational trips • taking part in a daily act of worship required by legislation • liaising with primary schools.

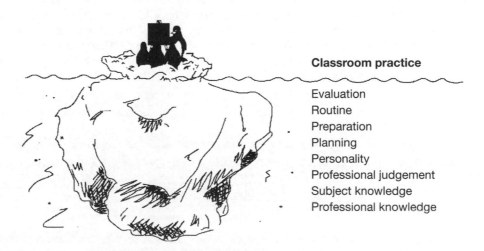

Classroom practice

Evaluation
Routine
Preparation
Planning
Personality
Professional judgement
Subject knowledge
Professional knowledge

■ **Figure 1.1.1** The work in the classroom – the tip of the iceberg
With kind permission of Simon Beer

PROFESSIONAL KNOWLEDGE FOR TEACHING

This section gives an overview of the forms of knowledge you need for teaching.

Teaching requires you to transform the knowledge you possess into suitable tasks which lead to learning (sometimes called pedagogic knowledge). Knowing a lot about your subject does not automatically make you an effective teacher. Your professional knowledge is built from a number of different components.

The forms of knowledge teachers need has been described in different ways. Shulman (1986; 1987) identifies seven knowledge bases which form what he regards as the minimum knowledge for teaching. These are summarised in Table 1.1.2. This is a starting point for thinking about the forms of professional knowledge you may need to acquire.

(Subject) Content knowledge

This is a declared body of knowledge about your subject; the concepts and skills pupils are expected to acquire. You amass this knowledge from a variety of sources; your education at home, at school, at university, as well as through personal study and reading; together, these influence the amount and organisation of knowledge you have. Although your awareness of and engagement with all of these different sources of content knowledge clearly varies, your content knowledge is likely to be the area of the greatest confidence for you as you begin your teaching career. You should actively

---

■ **Table 1.1.2** Forms of professional knowledge for teaching

---

1   *Content knowledge*: the content that is to be taught. Schwab (1964) identifies two components of content knowledge:
    - substantive: knowing what are the important concepts and skills in the subject
    - syntactic: knowing how the concepts and skills are structured and organised within the subject.

2   *General pedagogic knowledge*: the broad principles and strategies of classroom management and organisation that apply irrespective of the subject.

3   *Curriculum knowledge*: the materials and programmes that serve as 'tools of the trade' for teachers.

4   *Pedagogical content knowledge*: the knowledge of what makes for effective teaching and deep learning that is the basis for the selection, organisation and presentation of the content teachers want their pupils to acquire; i.e. the integration of content and pedagogy for teaching the subject; that which makes the content instructional. Grossman (1990) breaks pedagogical content knowledge into four components:
    - knowledge and beliefs about the *purposes* of teaching a subject at different grade levels
    - knowledge of pupils' understanding, *conceptions and misconceptions* of subject matter
    - knowledge of *curriculum* materials available for teaching a subject and knowledge of horizontal and vertical curricula for the subject
    - knowledge of *instructional strategies* and representations for teaching particular topics.

5   *Knowledge of learners and their characteristics*: both knowledge of learners of a particular age range (empirical or social knowledge) and cognitive knowledge of learners, comprising knowledge of child development and knowledge of a particular group of learners.

6   *Knowledge of educational contexts*: including a specific school, catchment area and the wider community.

7   *Knowledge of educational ends (aims), purposes, values and philosophical and historical influences*: both short and long term goals of education and of a subject.

---

seek to extend the range of your content knowledge. This process supports your confidence for teaching and engages you with your subject on a personal level. A word of caution, however. You may see this body of knowledge as the key measure of your likely effectiveness as a teacher, but it is the way you transform that knowledge into effective teaching that is most important. Task 1.1.2 asks you to audit your content knowledge.

Task 1.1.2 **AUDITING YOUR SUBJECT CONTENT KNOWLEDGE**

Analyse a copy of the curriculum for your subject, identifying the areas you can cope with now, those for which you require some additional work, and those for which totally new learning is required. Set yourself targets for developing your knowledge in the areas for development you identify. Discuss these areas for development with your tutor. Plan a course of action for this development. Keep a record of your progress, filing it in your PDP.

General pedagogic knowledge

This is the body of knowledge and understanding you require for the effective transformation of your content knowledge into meaningful learning activities for pupils. This knowledge includes the broad principles and strategies that are designed to guide class instruction, organisation and management (e.g. settling a class, managing the learning environment for effective learning, managing resources and other equipment, gaining and sustaining the attention and interest of the class, encouraging the disaffected, supporting the less able and extending the most able). It also requires you to adapt your content knowledge, planning for the immediate demands of teaching. You also need to consider carefully how you should introduce pupils to processes within your subject. How, for example, should pupils engage with writing reports or essays in your subject? Or what processes should they go through when researching information? By broadening your general pedagogic knowledge, your classroom becomes a more varied and stimulating place both for yourself and your pupils. For further details see the subject-specific text books in the *Learning to Teach* series.

Curriculum knowledge

This is the full range of programmes designed for the teaching of particular subjects and topics at a given level, the variety of instructional materials available in relation to those programmes and the set of characteristics that serve as both the indications and contraindications for the use of particular curriculum or programme materials in particular circumstances. It includes knowledge of your subject national curriculum and the curriculum in your placement school, the public examinations they serve and the requirements of that examination.

Pedagogical content knowledge

This has been described as that special amalgam between content and pedagogy, uniquely the province of the teacher; it goes beyond knowledge of content per se to the dimension of content knowledge for teaching. 'The most regularly taught topics in one's subject area, the most useful forms of representation of those ideas, the most powerful analogies, illustrations, examples, explanations, and demonstrations – in a word, the ways of representing and formulating the subject that makes it comprehensible to others' (Shulman, 1986: 9). It includes how you build assessment into your planning so that feedback enhances your understanding of pupil learning and enables you to plan the next lesson. In due course your pedagogic content knowledge should include the historical development of your subject, how it came to be as it is. This dimension enhances your sense of what your subject is about and why it is studied. Table 1.1.2 gives a description of several aspects of pedagogical content knowledge.

Knowledge of learners and their characteristics

There are different kinds of knowledge of the learner. These include empirical or social knowledge of learners, i.e. what children of a particular age range are like, how they behave in classrooms and school, their interests and preoccupations, their social nature, how contextual factors such as weather or exciting events can have an affect on their work and behaviour and the nature of the pupil–teacher relationship; and cognitive knowledge of learners, which consists of two elements: knowledge of child development which informs practice; and knowledge of a particular group of learners, the kind of knowledge that grows from regular contact with these learners, of what they can and cannot know, do or understand.

Knowledge of educational contexts

Knowledge of educational contexts refers to all settings where learning takes place: schools, classrooms, nursery settings, universities and colleges, and the broader educational context of the community and society. This knowledge ranges from the workings of the group, classroom, school governance and financing, to the character of communities and cultures. It includes the range of teaching contexts which affect development and classroom performance. These include the type and size of school, the catchment area, the class size, the extent and quality of support for teachers, the amount of feedback teachers receive on their performance, the quality of relationships in the school, and the expectations and attitudes of the headteacher, as well as school policies, the curriculum and assessment processes, monitoring and reporting, safety, school rules and expectations of pupils and the 'hidden' and 'informal' curriculum which includes the values demonstrated to pupils through the way the school is run (see Unit 7.2).

Knowledge of educational ends (aims), purposes, values and philosophical and historical influences

This includes the values and priorities which shape the education pupils receive. Teaching is a purposeful activity, both in the sense of short-term goals for a lesson or series of lessons and in the sense of long-term purposes of education. Some would argue that a long-term goal is to produce efficient workers to serve the needs of society well. Others see education as being of intrinsic worth in itself. Aims and purposes tend to be implicit rather than obvious and openly enacted.

Personal subject construct

All of the above aspects of professional knowledge are brought together in your *personal subject construct* (Banks *et al.*, 1999). This takes into account your values and assumptions about your subject which provides the basis of your work as a teacher and for what you understand to be the nature of your subject and how to teach it. For example, how do your political, philosophical, theoretical and religious views shape the version of subject you wish to teach? Within subject areas specific questions may arise. For example, what is the role of sport in physical education? Should creationism be taught alongside evolution and the Big Bang in science lessons? A very different issue is your wider role as a teacher, beyond your subject boundaries. Is it your job to support language development or teach mathematic skills as the need arises in your lessons? As has been said often 'every teacher is a teacher of English'. Such questions have a significant impact on the choices you make as a teacher. You should ensure that personal beliefs and constructs of subject you use in the classroom do not exclude pupils with different views.

Some of the units in this book aim to develop your *general pedagogic knowledge* – your understanding of classroom management and organisation and what makes for effective teaching and deep learning; your *knowledge of learners and their characteristics* and your *knowledge of educational contexts*. Subject-specific pedagogic issues are covered in the subject texts in the *Learning to Teach in the Secondary School* series. Task 1.1.3 asks you to consider pedagogical content knowledge.

Task 1.1.3 **PEDAGOGICAL CONTENT KNOWLEDGE**

Look closely at the forms of pedagogical content knowledge in Table 1.1.2. Consider carefully how you could apply your knowledge in each of the categories identified by Grossman to your work with pupils to make them more reflective learners and to personalise their learning experience.

MANAGING THE LEARNING ENVIRONMENT – A KEY PART OF YOUR GENERAL PEDAGOGIC KNOWLEDGE

An important aspect of your job is managing the learning environment of your classrooms. *Learning to manage the classroom* is similar in many ways to learning to drive. At the outset there seems so much to remember: e.g. using the clutch; brake; changing gear; watching other traffic; looking in the mirror; indicating; obeying the speed limit and so on After a short time, however, such skills become part of subconscious, internalised patterns of behaviour.

Much of what experienced teachers do to manage their classes has become part of their unconscious classroom behaviour. Their organisation of the lesson so that pupils learn is implicit in what they do rather than explicit. So much so, that often teachers find it hard to articulate exactly what it is they are doing or why it is successful. This situation, of course, does not help you as a student teacher. It also gives weight to the spurious notion that teachers are born rather than made and that nobody can tell you how to teach.

Some teachers may well begin teaching with certain advantages such as a 'good' voice or organisational skills. Nevertheless there are common skills and techniques to be learned that, when combined with an awareness of and sensitivity to the teaching and learning contexts, enable you to manage your classes effectively. *Teaching is a continuously creative and problem-solving activity.* Each learner or each group of learners has their own characteristics and group dynamics which the experienced teacher takes into account in planning the relevant learning programme. For example, if there has been recent controversy over environmental issues in the local area effective teachers adapt their approach to the discussion of such matters to make lessons more relevant and to allow pupils to draw on their experience. Although lessons with different groups may have similar content, a lesson is rarely delivered in the same way twice. Variations in interactions between pupils and the teacher affect the teaching strategy chosen.

Discussions about managing the classroom and learning environment are often hijacked by issues of behaviour. Of course, as a teacher you have to develop a set of effective strategies for encouraging behaviour for learning amongst your pupils, i.e. trying to prevent poor behaviour through establishing positive expectations rather than managing it after inappropriate behaviour has begun, *but first and foremost you are there to manage their learning, and that should be your primary emphasis.*

This is not always easy, and a whole range of circumstances come into play. As Rogers (2002:5) identifies:

> Day-to-day school teaching normally takes place in a rather unusual setting: a small room (for what is asked of it), often inadequate furniture and space to move, a 50-minute time slot (or less) to cover set curriculum objectives, and 25 to 30 distinct and unique personalities, some of whom may not even want to be there. Why should there not be some natural stresses and strains associated with a teacher's day-to-day role?

One of your most important roles as a teacher is to bring together the various personalities of your classroom (including your own) and to create from these the best possible context for learning. This means you need to do some careful thinking, planning and preparation. The key to success is to minimise the element of surprise. Of course, issues always arise within the classroom to which you have to react. The majority of events and issues that arise in the classroom are, however, foreseeable and can, therefore, be planned for. It is always better to be proactive than reactive.

When you plan you should think not only of what you are going to teach and how you are going to teach it, but also of the implications of these choices. If, for example, you want your class to watch a DVD, have you checked that the equipment works and that you know the relevant section/chapter? If you want the class to move into groups halfway through the lesson, have you thought about the rationale for your groups, who is going to work with who and how you are going to move the pupils into the groups? How are you going to manage the distribution of books or worksheets in the course of the lesson? Are all pupils working from books with the same page numbering? Such questions may seem small, but failure to think about issues like this can cause interruption and disruption to learning in your lessons, not to mention their potential impact on behaviour. Effective teachers run efficient classrooms, and efficiency maximises the potential for learning and cooperation. Some of the important things for you to consider are:

- timing
- seating plans
- organisation of desks/materials/texts/etc.
- how you plan to use Teaching Assistants (TAs) – meeting with them prior to the lesson is always advisable, where this can be arranged
- pitch/differentiation/extension of work
- range of activity
- likely trouble spots (e.g. using technology, writing on the board, distributing papers, setting homework, moving pupils into groups, etc.).

Skills and strategies can be learned and practised until they become part of your professional repertoire. In Units 4.3, 5.1 and 5.3 we introduce you to theories underpinning educational practice and ideas which can provide a foundation for your development as an effective teacher whatever your subject. But what do we mean by effective teaching?

Effective teaching occurs where the learning experience structured by the teacher matches the needs of each learner, i.e. tasks develop each individual pupil's knowledge, skills, attitudes and/or understanding so that past knowledge lays the foundation for the next stage of learning. A key feature of effective teaching is balancing the pupils' chance of success against the level of difficulty required to challenge them. The units in Chapter 5 provide further information about pupil learning. Understanding about the ways in which learning takes place is essential to your work as a subject teacher and provides the foundations on which to build your professional knowledge and judgement about teaching and learning. The more closely the teaching method matches the preferred learning style of the pupils the more effective the teaching is.

CLASSROOM RIGHTS AND RESPONSIBILITIES

Another important area that you, as a teacher, are responsible for is establishing and managing the rights and responsibilities of the classroom, including your own. It is important that these are clear to everyone and that rights are counterbalanced by responsibility in terms of behaviour and participation. In order for your classroom to run the way you wish it to, in the best interests of everybody, it is important that you establish clearly the framework according to which everyone must operate. Hand in hand with this must be clear and appropriate sanctions for those who do not comply.

The following are useful areas to consider in relation to the rights and responsibilities of your classroom:

■ *Respect*: every pupil has the right to personal respect; everyone should employ respectful language; it is important to respect the views and beliefs of others.
■ *Attention*: every pupil has the right to receive a fair share of the teacher's attention; when addressing the class at the teacher's invitation each pupil has the right to be heard; everyone must pay full attention to the requirements of the lesson; when the teacher speaks all must pay attention.
■ *Learning/teaching*: all pupils have the right to learn; the teacher has the right to teach; everyone has the responsibility of cooperating so that effective teaching and learning can take place.
■ *Safety*: everyone should expect to be safe; everyone must take all reasonable steps to ensure that safety is not compromised.

There may well be other rights and responsibilities that you wish to establish for your classroom. Task 1.1.4 asks you to think now about what these might be and how you are going to establish and maintain them. See Unit 8.3 on your legal responsibilities.

Task 1.1.4 CLASSROOM RIGHTS AND RESPONSIBILITIES

Consider the rights and responsibilities operating in the classroom practice you have observed. Draw up a list for your classes and keep this to refer to as you develop your skills in managing the learning environment.

SUMMARY AND KEY POINTS

So let us return to the question that is the title of this unit. What do teachers do? In the UK, while the curriculum is to a large extent determined centrally, the choice of teaching strategies and materials is largely in the hands of the individual teacher. Your own philosophy of teaching affects the way you approach your work and develops over time as you acquire further professional knowledge and judgement.

As a student teacher you begin to develop a repertoire of teaching styles and strategies and to test these out in the classroom. It may take you considerable time before you can apply the principles of effective teaching to your classroom practice but you can monitor your development through regular evaluation of lessons. In this book we aim to provide a basic introduction to what are complex areas and it is up to you to develop systematically your professional knowledge and judgement through analysing your experience (i.e. through reflection) and wider reading.

As a teacher you have responsibilities to your pupils, their parents and carers, your head of department, your school, your headteacher, the General Teaching Council and others. Being an effective teacher does not mean simply knowing your subject. It also means knowing how to deliver lessons that are intellectually robust, challenging and stimulating; managing the classroom effectively and fairly; assessing and monitoring pupil progress promptly and accurately; modelling in your own behaviour and practice what you expect pupils to do; planning for inclusion and the needs of individual learners; managing the rights and responsibilities of the classroom; upholding school policies and procedures; responding to the pastoral and personal needs of your pupils; completing administrative duties; contributing to the wider life of the school; knowing your legal responsibilities; and so on.

As you progress through your ITE course you develop knowledge, understanding and skills which enable you to fulfil your roles and responsibilities in all of these areas. Through your experiences in school you should move from knowing about skills to a position where you can use them flexibly and appropriately in a range of situations. In other words, you learn to do what teachers do – the school equivalent of plate-spinning – as you balance the many demands of the wonderful job that is teaching.

Check which requirements for your course you have addressed through this unit.

FURTHER READING

Banks, F., Leach, J. and Moon, B. (1999) 'New understandings of teachers' pedagogic knowledge', in J. Leach and B. Moon (eds) *Learners and Pedagogy*, London: Paul Chapman Publishing, pp. 89–110.
This chapter is particularly useful as you begin to think about the subject you are learning to teach and your own relationship with it. It encourages you to think carefully about how knowledge of your subject relates to the practicalities of teaching it in the context of the secondary school.

Grossman, P. L., Wilson, S. M. and Shulman, L. S. (1989) 'Teachers of substance: subject matter knowledge for teaching', in M. C. Reynolds (ed.) *Knowledge Base for the Beginning Teacher*, Oxford: Pergamon Press, pp. 23–36.
This chapter addresses a wide range of issues relevant to teachers at the beginning of their careers. It challenges you to think in detail about what precisely you need knowledge of if you are to be an effective teacher.

Moore, A. (2004) *The Good Teacher: Dominant Discourses in Teaching and Teacher Education*, London: RoutledgeFalmer.

This book offers an insight into the background of a set of key educational issues and provides an overview of key debates.

Shaffer, R. H. (1997) *Making Decisions about Children*, 2nd edn, Oxford: Blackwell Publishers.

Teachers, by the fact they are teachers, have been successful in various learning environments. From a privileged position it is easy to lack understanding of the difficult lives that many pupils lead. This book provides a useful framework to help you understand the different emotional issues which pupils may be facing. It discusses the impact of divorce and marital conflict, the relationship with step parents, the relationship between poverty and psychological development and issues of vulnerability in general.

White, J. (ed.) (2004) *Rethinking the School Curriculum: Values, Aims and Purposes*, London: Routledge Falmer.

This text contains a series of essays discussing the place of each subject in the curriculum in England and giving an overview of curriculum developments within each subject.

Wragg, E. C. (ed.) (2004) *The RoutledgeFalmer Reader in Teaching and Learning*, London: Routledge Falmer.

This book offers an insight into the background of a set of key educational issues and provides an overview of key debates.

RELEVANT WEBSITES

Teacher Support Network: http://www.teachersupport.info

Teacher Support Network is a 24 hour confidential counselling, support and advice service. They also offer support lines in England (tel: 08000 562 561), Wales (tel: 08000 855 088) and Scotland (tel: 0800 564 2270).

Teacher Training Resource Bank: http://www.ttrb.ac.uk

This is a resource designed to give you access to the evidence base underpinning educational practice.

Each of the following units in this book lists the websites specific to the unit. It is recommended that early on in your ITE you become familiar with the central government curriculum websites listed in the units in Chapter 7.

Additional resources for this unit are available on the companion website:
www.routledge.com/textbooks/9780415478724

THE STUDENT TEACHER'S ROLES AND RESPONSIBILITIES

Michael Allen and Rob Toplis

INTRODUCTION

Schools are busy places and teachers are often required to juggle many tasks at once. Unit 1.1 provides some insight into what it is to be a teacher. In this unit we look at what it is to be a student teacher in a secondary school. We consider preparation for school experience, then the school experience itself. We look at your relationships to other people, both staff and pupils, that form part of the busy life of schools, discuss some specific expectations of student teachers on school experience and offer some guidance about your roles and responsibilities. We then link this to an examination of the concept of teacher professionalism. Finally, we discuss how your development as a professional is likely to pass through significant changes over your initial teacher education (ITE) course.

OBJECTIVES

At the end of this unit you should be able to:

- prepare for school experience
- work with other staff and pupils on school experience
- identify expectations, roles and responsibilities of student teachers on school experience
- explain what it means to be a professional
- chart aspects of your development as a teacher over your ITE course and into your future learning and development.

Check the requirements of your course to see which relate to this unit.

PREPARING FOR SCHOOL EXPERIENCE

Before you start any school experience it is important to find out as much as you can about the school and its organisation, as well as the specific department. Ideally, you should visit the school at least once before you start your school experience. On any visit it is helpful to have a list of things you want to find out about the school, the department and the activities in which you are going to be engaged. It is likely that your tutor will have given you a list of information to gather and questions to ask to help you with this. If not, you can find a list on the website accompanying this text (www.routledge.com/textbooks/9780415478724). Task 1.2.1 is intended to help you prepare for school experience.

Task 1.2.1 **PREPARING FOR SCHOOL EXPERIENCE**

As you work through this unit, and as you read other relevant units in the book, make notes about what you need to do to prepare for school experience and what you might do to make the most of your school experience. Compare your notes with those of other student teachers.

During your visit(s) you may be introduced to the headteacher. However, you can expect to talk to the professional tutor and staff with specific areas of responsibility in the school. There are many policy and procedure documents in every school, covering a wide range of subjects, e.g. school uniform; equal opportunities; behaviour management; marking policy; risk management; and health and safety information such as the fire assembly point and how to record accidents. Often these can be found in a staff handbook. You may be issued with a copy of this, or there may be a copy in the staff room or school office. Your tutor may discuss the most relevant sections in the handbook which you can then read in your own time after the visit. This discussion and reading of the handbook provides you with useful practical information about how the school operates and what you need to do to comply with its policies and procedures and routines. The staff handbook should also include a diagram showing the school's management structure and lines of accountability.

You can also expect to talk to the head of department or faculty, your subject tutor and others in your subject department about the curriculum, schemes of work and your teaching timetable. These discussions are likely to include specific aspects of teaching in the department, e.g. safety issues, organisation of equipment and pupils, schemes of work, lesson plans, homework routines, access to texts and resources including information and communications technology (ICT). Some of the information may be in a departmental handbook.

You can gather further information about schools in England from Office for Standards in Education (Ofsted) inspection reports. The school may be able to lend you a copy of the school's last inspection report or you can find it on the Ofsted website. It provides you with a wealth of information about all aspects of the school as it was assessed at the time of the inspection. This also provides you with questions to discuss with staff and areas to follow up as you learn more about the school.

On your visits (and later when you start school experience) be aware of staff-room protocols. Some staff rooms are like lounges where teachers can relax and chat safely away from work and pupils during break and lunch times. Others have an additional function as a workroom (with or without allocated work spaces) where teachers can do marking and lesson preparation during their free periods. There are still some schools where the same staff have sat in the same chairs for ten, twenty or even thirty years! Colleagues may have brought in their own mugs for tea/coffee. There may or may not be a 'tea/coffee club'. Likewise, if you are planning to drive to school, check out the parking facilities and conventions; there may be reserved spots for some staff. If you check these things, you avoid upsetting anyone.

Such visits also enable you to familiarise yourself with the geography of the building. This is particularly important if you are going to teach in a large school, perhaps with several different blocks or operating on more than one site. Secondary schools vary immensely not just in size but also in physical features, ranging from the small rural or special school with under 100 pupils to the very large school with 1,000–2,000 pupils. Some schools are modern, or comparatively modern, while others are old, dating back to the 1880s. Each type of building has advantages and disadvantages. Whichever type of school you are in on school experience, it is important that you locate important facilities such as the office, lavatories and the staff room, before you start. The last thing you need to do on your first day is to get lost!

DURING SCHOOL EXPERIENCE

Work with other staff and pupils on school experience

Despite the fact that teaching involves spending large amounts of time away from colleagues and working autonomously or just with a teaching assistant, you still have to work with other staff and to be a team player. Further, as a student teacher you are likely to be in the classroom with the class teacher, your subject tutor and/or higher education institution (HEI) tutor for much of your time on school experience. Taking on the role of a teacher as a student teacher means forging and managing professional relationships with a range of staff. During the initial days on school experience you introduce yourself to staff you did not meet on visits prior to school experience, including teaching and support staff in your department and key personnel outside the department such as the headteacher, deputies/assistant heads, heads of key stage, heads of year and the special educational needs co-ordinator (SENCO). In addition, you start to build a working relationship with school and HEI tutors who are supporting you and observing your teaching. You try to make a good first impression on all these people. Figure 1.2.1 suggests some perceived attributes that help convey a positive image of a professional and well-prepared student teacher.

■ **Figure 1.2.1** Setting out to create a positive image

In this section advice is presented with regard to developing and managing relationships with specific members of staff who play significant roles in your school experience. Relationships with pupils are also considered.

Relationship with the professional tutor

It is worth remembering that the professional tutor is a key element in your ITE programme, with oversight and management of all student teachers within the school and liaison with your HEI tutor, if appropriate for your course. You may see the professional tutor in a formal context only once or twice a week, but they are senior members of school staff. They may organise sessions on general school issues. Likewise, you should be able to seek their advice on general school issues, if needed. They expect you learn school routines, practices and procedures, including rewards/sanctions and to follow these.

Relationship with your subject tutor (often called subject mentor)

In the early stages of learning to teach your subject tutor is an important person. Your tutor reviews your developing practice and writes reports on observations of your teaching. Your tutor is responsible for giving you a pass/fail on school experience. There are a number of aspects of the relationship which you should consider.

There are likely to be agreed structures for your tutor to give you support, advice and guidance, e.g. written feedback on one or more lessons each week; a weekly tutorial meeting. For other activities, for example jointly preparing lessons or approval of lesson plans by your tutor, seeking advice on planning and preparation for lessons or on aspects of teaching with which you are less familiar, completing the required paperwork for your course, keeping records of pupil attendance, class-work, homework, you should be clear about what your tutor expects of you and then do what is expected.

You should arrange regular meetings and clarify the purpose of those meetings, so that you are prepared for the meeting.

Your tutor is an experienced teacher, from whom you can learn a lot. Do not think you know it all already and either do not seek advice or ignore your tutor's advice. Do not be afraid to ask for advice if you are not sure about anything, but check when is a convenient time – so that you know when to ask and when is inconvenient.

Also check with your tutor your status with support staff, e.g. technical staff, reprographics staff. In some schools you approach them yourself, in others you do so through your tutor. Likewise, discuss with your tutor your attendance at school and departmental staff meetings.

Your attendance and punctuality at school (and at lessons when in school) is important. Your placement school has an agreed procedure if you have an important reason (e.g. an interview) to be off school or you are sick. Let your tutor know of any foreseen absences well in advance. If you find that one morning you are too ill to attend school, try to contact your tutor directly by phone/text message, otherwise speak with the school office staff. (On most courses you are also required to contact the HEI on the day of absence.)

The tutor–student teacher relationship is vital to your success and it is worthwhile taking steps to ensure this remains cordial. However, from time to time problems do occur, and can often be associated with the friction generated when the student teacher fails to seek or to act on the tutor's advice. If your relationship with your tutor breaks down you need to contact your HEI tutor or senior staff member immediately and seek further advice. It is important to be aware that any breakdown in the relationship that ends in the student teacher leaving the placement may subsequently result in course failure.

Relationships with class teachers

You spend the bulk of your time in school in the company of the teachers whose classes you are teaching and so it is important also to establish good working relations with them. Remember that they are going to have to teach the class again after you leave, so discuss with them what they want you to do. Some teachers want you to follow their routines, practices and procedures; others allow you to experiment with what is best for you. Plan your lessons well in advance of when they are going to be taught to allow time for any planning meetings with, or checking by, the class teacher (or your tutor) and any further planning or adjustment to take place. Collate resources well in advance, be flexible and be prepared to change lesson plans at short notice in the light of unexpected events. Avoid the situation where you are chasing the class teacher ten minutes before the start of a lesson for an important resource or piece of information. Arrive early before a lesson. Keep teachers fully informed of any new approaches you are taking in your teaching and events that take place with their classes, particularly behavioural issues that need following up.

Your main task as a student teacher is to ensure the pupils in your classes learn. This is most effective when you are able to treat each pupil as an individual. Learning pupils' names (Buzan (2003) advises how to do this) is a good first step as is getting to know something about their interests. It is important to greet pupils at the beginning of the lesson.

You also need to gain the respect of the pupils you are teaching. This is not usually automatic; it requires a pro-active approach. A general guideline is that if you treat pupils with respect, the feeling is reciprocated (although some pupils may not necessarily respond in this way). For instance, you should be polite when dealing with pupils, and ensure they are polite back to you. At the same time you should clearly define the boundaries of behaviour. Pupils are sensitive to actions they perceive as being unfair, e.g. if one person has been talking, the whole class is kept in for a detention.

Make sure you understand the material you are teaching and have planned and prepared your lesson and your resources. Do not be afraid to admit it if you are asked a question to which you do not know the answer, provided you follow it up in a later lesson.

Planning and preparation are essential for learning and to motivate pupils to learn (see planning in Unit 2.2 and motivation in Unit 3.2). Encouragement is one effective means of keeping pupils on task in your lesson. Motivation and encouragement work best when tailored to the needs of individuals. During the lesson you need to keep pupils on task. To be effective your approach needs to be tailored to each individual. However, this is difficult unless you have some knowledge of individual pupils.

Well-planned lessons support your approach to behaviour for learning (see Unit 3.3). Despite this, you may encounter some behavioural issues in the class, therefore you should also be clear how you are going to deal with any poor behaviour, in line with school behaviour policies. Here are some steps to take to deal with poor behaviour. On occasion you may find yourself involved in a confrontation with a pupil. Sometimes this is unavoidable, as to retreat would be interpreted by onlookers as a loss of your authority. Never be drawn into a public confrontation with a pupil because you may lose your authority, which is difficult to recoup later. In any case, you do need to think of the effect on the rest of the class and also on what the rest of the class are doing when a confrontation is going on. Simply saying 'I will see you later' allows you to choose the time and place to follow up. This enables you to maintain a working relationship with the particular pupil after the event.

Physical contact with pupils should be avoided unless there is an immediate health and safety concern, or is a requirement in a practical lesson such as physical education, e.g. to support a pupil. It is unlikely that you will be called on to make decisions in contexts where physical restraint is necessary, because the supervising class teacher should be available when restraint is the pertinent action. Likewise, any contact with parents/carers in reaction to classroom events, both positive as well as negative, should be undertaken in conjunction with the class teacher. Further, more specific advice on encouraging behaviour to maximise learning is found in Unit 3.3.

A particularly important point to remember is to keep a professional distance in your relationships with pupils. It is easy with some classes to become over-friendly; this is especially the case during the first phase of development (see the section on student teachers' development below). To be the target of an adolescent 'crush' is not unusual for young student teachers, and if this is the case, maintaining an appropriate professional distance is imperative e.g. avoid situations where you are alone or with a small group, stand rather than sit beside a pupil. If you are alone in a room with any pupil (or parent), it is good practice to seek the presence of another member of staff or to leave the door wide open. Similarly, you should avoid texting or e-mailing pupils. False allegations are uncommon but anticipate and avoid situations where they might arise.

Task 1.2.2 presents some scenarios you might have to deal with.

Task 1.2.2 **RELATIONSHIPS WITH PUPILS**

Consider your responses to the following events:

■ there is a struggle between two pupils in the corridor
■ you observe a pupil going through another pupil's desk
■ a pupil hasn't done their homework
■ a pupil says a personal item has gone missing just a few minutes before the end of the lesson.

Discuss your responses with your tutor or another student teacher and record your reflections in your professional development portfolio (PDP). Identify other scenarios to discuss – these may be real events which have taken place in school.

Expectations, roles and responsibilities on school experience

The main expectation of you as a student teacher is that you promote pupils' learning. To achieve this there is a range of *structured teaching* activities which you are likely to engage in. These include:

■ micro-teaching: a short teaching episode where you teach peers or small groups of pupils; it might be video-taped to enable analysis of different aspects of teaching
■ observation of experienced teachers: where you look at specific aspects of teaching in a lesson, e.g. how teachers use questions to promote learning (see Unit 3.1)
■ team teaching: where you share the lesson with others; planning, teaching the lesson and evaluating together
■ whole-class teaching with the class teacher present
■ whole-class teaching on your own. (As a student teacher, you should always have an experienced teacher nearby.)

You should be given feedback on your planning and teaching in each of these situations to enhance your learning. The amount of feedback you get from teachers watching your lessons varies. However, student teachers also have preferences. If you wish to have feedback on every lesson, ask if this can be done. Some student teachers prefer a small amount of very focused feedback, others can cope with more – a page or more of written comments. Written feedback is essential because it provides a record of your progress and ideas for your development. In practice, there are likely to be agreed conventions governing this aspect of your work. These take into account how you are to achieve the requirements to complete your course successfully.

Comments on your teaching divide into those relating to tangible technical issues which can be worked on relatively easily and those relating to less tangible issues relating to pupils' learning. Technical problems such as the quality and clarity of your voice, how you position yourself in the classroom, managing transitions from one activity to another, your use of ICT and/or audio-visual aids, are easy to spot, so you may receive considerable advice on these issues. Problems with these aspects of your work are usually resolved early in your course, whereas less tangible issues which are directly related to the quality of pupil learning require ongoing reflection, attention and discussion, e.g. your approach to the explanation of lesson content, your style of questioning, your evaluation of pupil learning. If you have access to videos of yourself teaching, you are advised to spend some time in the detailed analysis of your performance in these different aspects of teaching. More detailed advice related to the teaching of your specific subject is given in the subject-specific texts in the Routledge *Learning to Teach in the Secondary School* series which accompany this text (see list on p. ii).

Expectations relating to your social skills in developing relationships with staff and pupils and of your teaching are summarised in Table 1.2.1.

Thus, your main roles and responsibilities relate to teaching particular classes. Teachers have other roles and responsibilities such as planning the curriculum and liaising with outside agencies, but these are not usually undertaken by student teachers. You become involved in the wider roles and responsibilities of teachers after completing your ITE course. This is part of your development as a professional. (See also Unit 8.2.)

The roles and responsibilities of teachers, including student teachers, are underpinned by the concept of *professionalism*. It is therefore appropriate to explore some ideas associated with the concept of professionalism and how these have changed over time, often as the result of policies and government initiatives.

TEACHERS AS PROFESSIONALS

Teachers' work changes constantly as a result of government initiatives and agendas about, for example, the curriculum, the care of children, the management of schools, teachers' professional standards and teachers' learning.

Views about other professions, such as doctors and lawyers, can be used as a lens with which to compare teachers' professionalism. The *main attributes of professions* that distinguish them from other groups of workers are: their specialised knowledge – there is a substantial body of knowledge which the professional needs to acquire; that substantial training is required before an individual can be accepted into the profession; a commitment to meeting the needs of clients; a collective identity, and a level of professional autonomy that controls their own practice; the profession is self-governing as well as publicly accountable. At this point it is worth thinking about teacher professionalism in these contexts by undertaking Task 1.2.3.

Task 1.2.3 COMPARE PROFESSIONALS

Using the attributes of profession above, list some examples of the ways in which the practice of teachers is similar and the ways in which it is different. Discuss your findings with other student teachers.

■ **Table 1.2.1** The school's expectations of student teachers

(i) Social skills

You are expected to:

- develop a good relationship with staff and pupils.
- be able to communicate with adults as well as pupils.
- work well in teams.
- learn to defuse difficult situations.
- keep a sense of humour.

(ii) Planning, teaching and evaluating lessons

You are expected to:

- be well organised.
- know your subject.
- plan and prepare thoroughly. Be conscientious in finding out what lesson content is appropriate to the class you are teaching. For some classes you may be teaching material which is new to you or which you last thought about many years ago. You must know the subject matter you are teaching and you are expected to improve your own subject content knowledge. However, you are also expected to ask if you are unsure about the content for a particular lesson.
- share your plans with the class teacher, explaining why you want to do things the way you plan. Discuss any new/different teaching strategies or innovations in your teaching. Evaluate these carefully afterwards.
- check the availability of books and equipment; test out equipment new to you; talk to staff about the work and the pupils' progress; and clarify any safety issues – before the day on which you are teaching the lesson.
- arrive in plenty of time for a lesson in order to arrange the classroom and lay out any equipment or books needed.
- during the lesson learn names of pupils, focus on the learning that is taking place, and ensure that good behaviour is maintained during your teaching and that learning is taking place.
- evaluate the lesson.
- keep good records: have your file of schemes of work and lesson plans, pupil attendance and homework record up to date. Your evaluations of your lessons are best completed on the same day as the lesson; although sometimes you might want to add to this after you have marked pupils' work.

Becoming *a member of the teaching profession* means that you make the following commitments – that you will:

■ *reach an acceptable level of competence and skill* in your teaching by the end of your course. This includes acquiring knowledge and skills which enable you to become an effective teacher and which enable you to understand the body of knowledge about how young people learn and how teachers can teach most effectively.

■ *continuously develop your professional knowledge and professional judgement* through experience, further learning and reflection on your work.

■ *be publicly accountable for your work.* Various members of the community have the right to inspect and/or question your work: the head, governors, parents, inspectors. You have a professional duty to plan and keep records of your work and that of the pupils. This accountability includes implementation of school policies, e.g. on behaviour, on equal opportunities.

■ *set personal standards and conform to external standards* for monitoring and improving your work.

Table 1.2.2 summarises aspects of professionalism.

The General Teaching Councils for England, Scotland, Wales and Northern Ireland were set up as elected regulatory bodies for professional standards. The General Teaching Council for England (GTCE, 2008) states that its overall purpose is to shape the development of professional practice and policy, and to maintain and set professional standards. The statement of professional values and practice of the GTCE for teachers includes the centrality of children and young people to professional practice, recognition of the role of parents and carers to the education of children, the need to work with colleagues, and teachers' own learning and professional, including *reflection* which is one of the important facets of professionalism we discuss next. (See also Unit 5.4.)

Reflection includes reflection both *on* practice and *in* practice (Schön, 1983). Reflection *on* your practice of teaching may involve the familiar lesson evaluations that follow planned, taught and assessed lessons. These reflections may be written and discussed with your tutor (this should be done sooner rather than later if all the important points are to be remembered accurately). An example may be a lesson where some aspect of pupil behaviour affected learning. What was this? Who was involved? What was the real issue? Why did you respond in this way? Did it work?

■ Table 1.2.2 The school's expectations of your professionalism

---

Professionalism

You are expected to:

- dress appropriately (different schools have different dress codes).
- act in a professional manner, e.g. be punctual and reliable; act with courtesy and tact; respect confidentiality of information.
- take active steps to ensure that your pupils learn.
- discuss pupil progress with parents.
- become familiar with and work within school procedures and policies. These include record keeping, rewards and sanctions, uniform, relationships between teachers and pupils.
- be open to new learning: seek and act on advice.
- be flexible, e.g. if there is a change in the timetable on a particular day.
- accept a leadership role. You may find imposing your will on pupils uncomfortable but unless you establish your right to direct the work of the class, you are not able to teach effectively.
- recognise and understand the roles and relationships of staff responsible for your development.
- keep up to date with your subject.
- uphold school policies.

Can/should you respond differently in a similar situation in future? How can you organise your next lesson to reduce or help to eliminate this behaviour? This reflection involves careful analysis, evaluation and subsequent planning; it involves you being self-critical and open to advice from experienced teachers. Reflection *on* practice is an inherent part of learning to teach and should not be ignored or underestimated. Importantly, it is not necessarily a failure on your part. Reflection *in* practice involves your thoughts and the actions you take at the time within classroom contexts. Experienced teachers appear to do this automatically but it is worth remembering that they have built up a stock of intuitive responses to a number of situations over their years of practice. Your reflection *in* action will be limited but you may respond to some everyday examples as you build up your knowledge of pupils and situations. One example may be the timing of pupil activities, where you have allowed ten minutes for the activity, only to find restless pupils after only four minutes; your reflection *in* action may lead you to move the lesson on to the next stage or to use another pupil activity.

It is important to consider the process you go through to become an experienced teacher. We do this next.

A MODEL OF STUDENT TEACHER DEVELOPMENT

The aim of an ITE course is to facilitate your transformation from a student teacher to a competent professional. Plainly, this change is not instantaneous; instead, it proceeds by increments with each little piece of experience contributing to your development. Your perception of yourself as a teacher alters as different aspects become the focus of concern at different points during your ITE course. A major change for you might be assuming the role of the teacher after being a learner, e.g. on a university course. You become one of them (teachers) instead of being one of us (learners). This role reversal requires significant behaviour modifications. Observing other teachers to see how they act in and out of the classroom helps you through these phases of development.

You can expect your expertise and confidence to build over time. Various models of student teacher development have been identified (Fuller and Brown, 1975; Leask and Moorehouse, 2005: 18–31; Furlong and Maynard, 1995). Table 1.2.3 summarises this earlier work and our experience of the phases student teachers go through in the process of becoming competent professionals. Initially your focus will be on how you come across and how you mange the class with a focus on the learning outcomes the class and then individual pupils achieve.

The rest of this section builds on the work of Furlong and Maynard (1995), itself being based on a body of previous work. The model of development (see Table 1.2.3) does *not* assume that everyone passes through a predetermined, invariable linear process during ITE, because individual and contextual aspects differ in many respects, such as the school environment and the course. That said, research (e.g. Calderhead and Shorrock, 1997) has suggested that student teachers have common focuses for their concern at different times during their development.

Remember, your primary role as expected by a prospective employer is to *teach the curriculum*, with the aspiration being every pupil in the class achieves the learning outcomes for each of your lessons. Attainment of the final mature stage in Table 1.2.3 is the aim. With its emphasis on individual pupil learning and the successful achievement of learning outcomes by all pupils, you need to develop aspects of this third phase right at the start of your first school experience. However, in this model your self-percceived role shifts from being the pupil's friend, to a crowd controller, then finally to teacher of subject knowledge.

Other units in this book, for example, classroom management, and planning and differentiation, cover specific issues described in the model in Table 1.2.3. Timing of the stages is difficult to predict because some student teachers progress more quickly than others during their ITE course and because

■ **Table 1.2.3** Stages of development of a student teacher

Stage	Stage 1: Focusing on self development: what am I like as a teacher?	Stage 2: Focusing on whole-class learning: are the pupils learning anything?	Stage 3 Stability and focusing on individual pupils' learning: how can I teach more effectively?
Characteristics	How do I come across? Will they do what I want? Can I plan enough material to last a lesson? Idealism and insecurity. Desire to portray a caring image but also be in control. Pupils as passive learners.	Are the pupils learning? How do I know? Setting boundaries and meeting challenges. Getting through the curriculum. Steady improvements in teaching performance. Beginning to focus on assessment for learning.	Understanding individual behaviour and the different needs of pupils. Opportunities for personal development and experimentation. How effective are my strategies for ensuring all pupils learn? How can I find out? Stability and development. Pupils as active learners

Source: Adapted from Furlong and Maynard (1995); Leask and Moorehouse (2005)

of the individual and contextual differences described above. The three stages may span a single school experience or the whole course; in some cases stage one occurs at the start of the first school experience, with stage two being experienced after a couple of weeks, and with some aspects of stage three appearing right at the end. At the start of the second school experience there may then be a repeat of this process, only the first two phases are shorter. It is important to note that some student teachers who have had difficult and problematic school experiences emerge, after a number of years of qualified experience, as among the best teachers in their schools.

Stage 1: Focusing on self-development

You may begin your first school experience holding certain idealistic views about your role as teacher, partly based on your own memories of school when you were a pupil. Some student teachers adopt an empathetic self-image, seeing themselves identifying with the pupils more than the class teacher, wanting to create a caring persona, being 'there' for the pupils, being popular. You may want to avoid becoming too strict or scary, not wanting pin-drop silence in your classroom, but instead a good-humoured, industrious buzz, so avoiding an atmosphere that negatively affects pupils' emotions. The most important factor determining success is your relationships with pupils, and you feel that if this can be arranged satisfactorily, then accomplishment in other areas naturally follow without a great deal of further effort.

Once you begin your first school experience these idealistic views may begin to evaporate in the face of immediate issues presented to you, and you switch to a more pragmatic stance triggered particularly by the primary need to establish classroom control. At this point, you are developing your understanding of the boundaries you need to establish to create the modern classroom environment. For instance, initially you may be unclear about whether a particular pupil behaviour such as chatting during written work needs challenging. On top of this, you may wonder that if you were to challenge behaviours, whether the pupils would merely ignore you and carry on. Both of these feelings can conspire to make you feel reluctant to assert your authority. Observe other teachers (see Unit 2.1) and discuss this point with them. Student teachers sometimes avoid challenging poor behaviour because they would rather not interrupt the flow of the lesson and because they feel they must keep rigorously to the lesson plan. Pupils actively test your knowledge of these boundaries of behaviour in the classroom, as well as your willingness to act on them. To be seen as a caring friend and equal by the pupils is not appropriate to a working relationship, and is unworkable in practice. Planning issues can also be a cause of anxiety, such as do you have enough work to last the whole lesson, or what if they ask difficult questions? (These points are addressed in Unit 2.3.)

Thus, the first couple of weeks on school experience are likely to be a time of insecurity with respect to self-image and readjustment of some prior idealistic notions, and you may at times feel out of your depth and run off your feet. You may have previously felt comfortable handling small groups or one-to-one situations, but whole-class teaching requires different skills which take time to develop.

Stage 2: Focusing on whole-class learning

With experience, you begin to realise exactly what constitute appropriate boundaries for classroom behaviour, for example, what is a tolerable level of noise, pupil movement around the classroom and what level to pitch your lessons at. Having said that, you are still working to develop strategies that successfully address every one of these issues. You can expect to feel pressure to put on a 'good show' for the significant players in your own assessment as a student teacher, your class teacher, school subject and professional and HEI tutors, and work hard on your creative planning, delivery and especially your discipline, in order to foster these relationships. You develop your skills at assessing pupil work and providing feedback to help them learn.

As a consequence of your hard work in addressing these issues, you begin to experience some successes. The pupils behave better (although perhaps not consistently so), which increases your confidence. Getting to grips with discipline allows you to think more about whether the pupils are achieving learning outcomes, and you begin to adjust your lesson content in the light of this knowledge, although you may not yet be differentiating work for individuals. For most student teachers, these successes are initially inconsistent and largely unpredictable.

This phase is typified by steady improvements in classroom performance. You start to think more about your autonomy as a teacher, about things you would like to do differently, although these desires are tempered by the need to fit in with the clear expectations of your school and HEI tutors.

Stage 3 Stability and focusing on individual pupils' learning:

During the last weeks of your final school experience there is usually a period of stability. Tried and tested methods have brought with them hard-won success, albeit not consistently. You are less anxious about discipline. A common idea held by student teachers is that if pupils have enjoyed a lesson then this shows it was successful. Considerable effort is needed to ensure *all* pupils are achieving the learning outcomes, and this requires differentiation of work for individual pupils. You need to check that the learning outcomes you set don't simply reflect an epistemology of the transmission of concrete knowledge, with an avoidance of the more abstract ideas.

In order to move on from this phase and progress towards becoming a more effective professional, concerted efforts are necessary, often requiring the intervention of others such as class teachers or tutors. Student teachers at this stage of development may not be aware that further improvements are indeed necessary or even possible, so the first step is to become aware of areas where your competence could be further advanced. Critical self-analysis – reflection on your practice (see Unit 5.4) – informs these areas for development, and *you* should recognise the need to make the effort to move on.

The greatest challenge lies in ensuring that each and every member of the class has accessed the learning outcomes; currently, in English state-maintained schools, the view that *Every Child Matters* (Department for Education and Skills (DfES), 2003a) predominates with the emphasis on the inclusion of all pupils in the learning process. The purpose of lessons needs to swing towards the needs of pupils, and away from you as a student teacher, with content focused on learning. The first step is determination of the extent to which pupils have learned during your lessons, which may be indicated by an effective plenary, end of topic test or more formative types of assessment, all of which need to be referenced in your lesson evaluations.

You may in fact already realise that there are certain aspects of your teaching that could obviously be furthered, but the ability to progress is hampered by classroom management issues, for instance you avoid practical work, or you do not feel confident enough to experiment with innovative pedagogies. If this is the case, advice from other members of staff may prove invaluable in moving you on to higher levels of achievement.

BEYOND YOUR ITE COURSE

Teaching is an ever-changing process and learning to teach never stops. The old comment 'I've been teaching this topic like this for the last twenty years' is simply unacceptable: continuing professional development (CPD) is not only a necessity in schools, it is a professional requirement for high quality teaching and learning (Pollard, 2002). CPD is covered in Unit 8.2.

SUMMARY AND KEY POINTS

In this unit we have attempted to provide a background for your multiple, changing roles and responsibilities as a student teacher. In doing so we have considered preparation for school experience, your relationships with other staff and pupils and your roles and responsibilities in school. We have also looked at the concept of teacher professionalism and a model of student teacher development over your ITE course and into your future learning and development.

Check which requirements for your course you have addressed through this unit.

FURTHER READING

The subject-specific texts in this series provided detailed advice about teaching in your subject area (see p. ii).

RELEVANT WEBSITES

Teacher Training Resource Bank: http://www.ttrb.ac.uk
Subject-specific, behaviour management, teaching English as a second language, diversity and special educational needs websites can be found through this website.

TeacherNet: http://www.teachernet.gov.uk
Is a resource for education professionals supported by the Department for Children, Schools and Families.

Qualification and Curriculum Authorities: http://www.qca.org.uk/

Scottish Qualifications Authority: http://www.sqa.org.uk

> For more information about the qualifications and curriculum authorities for England, Northern Ireland, Wales and Scotland see Units 7.4 and 7.5 in this book, and Units 7.6 and 7.7 on the companion website.

Association of Teachers and Lecturers (ATL): http://www.atl.org.uk

National Association of Schoolmasters Union of Women Teachers (NASUWT): http://www.nasuwt. org.uk

National Union of Teachers (NUT): http://www.teachers.org.uk

Voice (previously the Professional Association of Teachers): http://www.voicetheunion.org.uk

> These are the four main teachers' professional associations (unions) in the UK. In addition to offering direct advice and support to members on employment-related matters, the associations produce useful newsletters and publications on a range of topics, offer special concessions, e.g. on car and travel insurance, and training courses.

Ofsted: http://www.ofsted.gov.uk

DCSF: http://www.dcsf.gov.uk

> Provide access to Ofsted reports and other DCSF information.

General Teaching Councils of England: http://www.gtce.org.uk

General Teaching Councils of Northern Ireland: http://www.gtcni.org.uk

General Teaching Councils of Wales: http://www.gtcw.org.uk

General Teaching Councils of Scotland: http://www.gtcs.org.uk

> The General Teaching Councils govern professional standards in teaching. You will be required to register.

Voice Care Network: http://www.voicecare.org.uk

> Is a registered charity with subscribing members. They provide advice and training to help people keep their voices healthy and to communicate effectively. Their booklet More Care for Your Voice is intended to help people whose voice is needed for their work.

Additional resources for this unit are available on the companion website:
www.routledge.com/textbooks/9780415478724

ACKNOWLEDGEMENT

The authors would like to acknowledge the input of Marilyn Leask and Catherine Moorhouse into the first four editions of the unit.

MANAGING YOUR TIME AND STRESS

Susan Capel

INTRODUCTION

Although teaching can be rewarding and exciting, it can also be demanding and stressful. The three main reasons given by teachers as factors which are demotivating and lower morale (e.g. General Teaching Council (GTC), 2003; PriceWaterhouseCoopers, 2001; School Teachers' Review Body, 2002) and by teachers who leave the profession within the first few years (e.g. Spear *et al*., 2000; Wilhelm *et al*., 2000) are related to time and stress. These three factors are: too heavy a workload; work is too pressurised and stressful; and too much administration. These reasons are also causes of concern for student teachers.

You may be surprised by the amount of time and energy you use while on school experience (and later as a teacher), inside and outside the classroom and outside the school day. There is little time within a school day in which you can relax.

Although you may feel as though you have to keep running faster to keep up with all that is required of you as a student teacher (and later as a teacher), this is not going to help you – indeed, it is likely to increase your stress. You need to plan to use your time and energy effectively over the week. You must not spend so much time preparing one lesson that you do not have time to prepare others well (there are, of course, times when you want to take extra time planning one particular lesson, e.g. for a difficult class with whom the last lesson did not go well or if you are less familiar with the material). Likewise, you must use your energy wisely, so that you have enough energy to teach each lesson well.

Undoubtedly you will be tired. Many student teachers have told us that they are so tired when they get home from school that they have to force themselves to stay awake. If your teaching commitment is not to take over your whole life, you need to manage your time and energy and the stress associated with your school experience and teaching.

There has been significant investment nationally to resolve workload issues. This includes transferring some tasks from teachers to teaching assistants. Thus, you should see yourself as part of a team of professionals and paraprofessionals. These changes should help you, as student teachers, although you still need to manage your time and stress effectively, otherwise you are not going to benefit from these changes.

> ## OBJECTIVES
>
> At the end of this unit you should be able to:
>
> ■ identify ways you can use your time effectively in the classroom
> ■ develop ways to manage your time effectively
> ■ identify factors that may cause you stress
> ■ develop methods of coping with stress.
>
> Check the requirements of your course to see which relate to this unit.

MANAGING YOUR TIME

As Amos (1998) emphasised, everyone has the same amount of time. It cannot be lost, increased, saved, delegated, reallocated nor reclaimed by turning the clock back. Time can easily be misused or wasted.

Pupils spend little time in school each year. Assuming six hours contact time per day and 200 days per year, secondary school pupils spend less than 14 per cent of their time in lessons (Arnold (1993) calculated that primary school pupils spend less than 12 per cent of their time in lessons (assuming five hours contact each day)). Over 12 years of compulsory schooling pupils spend about 92 weeks in total in lessons. Calculate how much (or little) time pupils spend in lessons in your subject over a year. It is therefore very important that you use this time effectively. Therefore, it is especially important to consider what you do in your teaching and how you do it.

Managing your time in the classroom

To use classroom time effectively and economically you need to plan to maximise the amount of time available in the lesson, reduce the time it takes for pupils to get to lessons, to settle down and to pack up at the end and to manage pupils' behaviour in the lesson. Ways of doing this include:

■ allocating a high proportion of available time for academic work (sometimes called academic learning time)
■ maintaining a good balance in the use of time on teaching, supervisory and organisational activities
■ spending a high proportion of time in 'substantive interaction' with pupils (i.e. explaining, questioning, describing, illustrating)
■ regularly reviewing the conduct of lessons in terms of effective use of your own and pupils' time
■ devising simple, fast procedures for routine events and dealing with recurring problems
■ eliminating unnecessary routines and activities from your own performance
■ delegating (to teaching assistants or pupils) responsibilities and tasks that are within their capability (adapted from Waterhouse, 1983: 46).

This should enable pupils to:

■ spend a high proportion of their time engaged on learning tasks
■ experience a high degree of success during this engaged time.

These time management principles can be applied in many ways in the classroom, for example:

■ spending time at the start of the first lesson with the pupils (and as a teacher at the start of the academic year) establishing rules and routines. This saves time on organisation and management as you proceed through the year, scheme or unit of work. Pay special attention here to safety issues. See Unit 3.3 for further information about behaviour for learning and Units 1.2 and 2.2 for further information about organisation, rules and routines in the classroom.

■ teaching pupils to seek answers themselves rather than putting their hand up as soon as they get stuck.

■ organising your files and other work so you can easily locate them (throw away paper you do not need again).

■ using teaching assistants or pupils to help give out and collect textbooks, pupils' books or equipment, to mark straightforward homework tests in class, make sure the classroom is left ready for the next class with the chairs tidy, floor clear, board clean and books tidied away.

■ carrying a marking pen with you as you move around the class checking the work that is going on. As you skim pupils' work and comment to them, you can make brief notes on the work. It is easier to pick up mistakes and check work when it is fresh in your mind. This not only provides formative feedback to pupils to promote learning, it saves you having to go back to the work at a later stage which, in itself, wastes time.

■ collecting in books which are open at the page where you should start marking.

■ ensuring that work is dated and that homework is clearly identified so that it is easy for you to check what work has been done and what is missing. Ruling off each lesson's work helps you to check this.

■ keeping one page of your mark book for comments about progress (folding the page over ensures that comments are not seen inadvertently by pupils). As you see pupils' work in class or when you are marking, you can make brief notes which are then immediately at hand for discussions with parents, head of year, report writing, etc.

There are many other ways of managing time effectively in classrooms which you develop as you gain experience.

Task 1.3.1 is designed to help you look at how you spend your time in lessons.

Task 1.3.1 **HOW YOU SPEND YOUR TIME IN LESSONS**

Observe how several experienced teachers use their time effectively in lessons. For example, look at how they divide their time between teaching, supervisory, organisational and management activities; time spent on explaining and questioning, time spent on procedure for routine events such as collecting homework or giving back books, what is delegated to teaching assistants or pupils, time spent managing pupils' behaviour. Ask another student teacher or tutor to observe how you use your time in the classroom in one lesson or over a series of lessons. Discuss with the observer the findings and possible ways of using your's or the pupils' lesson time more effectively and economically. Try these ideas out systematically in your teaching.

Planning outside the classroom

Carefully plan the use of time in each lesson. This planning takes time; indeed, it takes more time when you start out than it does later in your teaching career. Use a time line in your lesson plan, allocating time for each activity, as described in Unit 2.2. Allow time for pupils moving from one part of the school to another for the lesson (and in physical education time for changing), getting the class settled, particularly at the beginning of the day, after a break or lunch. You may find initially that you under- or over-estimate the time needed for each activity, including organisation and management activities. In your reflection and evaluation at the end of each lesson compare the time taken for each activity with that allocated. Although this helps you gradually to become realistic about how long different activities in a lesson take, early in your learning to teach you take longer to organise and manage your classes. It is therefore important that you do not base your planning on the time it takes to organise and manage classes initially; rather, you should work hard to develop routine procedures to reduce this time as much as possible so that you maximise the learning time in the lesson.

Similarly, in planning a series of lessons, allocate time carefully. You have a certain amount of work to cover over a given period of time. If you do not plan carefully, you may find yourself taking too long over some of the content and not leaving yourself with enough time to cover all the content. Pupils' knowledge and understanding develop over a period of time; therefore if they do not complete the content required, their learning may be incomplete. Unit 2.2 provides more information about lesson planning and schemes of work.

In order to use your time outside the classroom effectively you need to plan your use of time and prioritise your work. Keeping records of activities can help with this, for example, keep a file of activities for the week (e.g. lessons to plan, marking to do, assignments for your course, completing specific records of your work, including how you have met certain standards). You may also want to make a list of activities you are going to complete each day. If there are activities left on the list at the end of the week or the day, why is this, e.g. you are spending too much time on each activity, you are unrealistic in how much you can achieve in a day. Also leave time for reflection on your teaching overall and your development as a teacher (what have you learned and how are you going to develop further?).

To help you plan how to use lesson time, complete Task 1.3.2.

Task 1.3.2 PLANNING HOW TO USE LESSON TIME

When planning your lessons, deliberately think about how best you can use the time available. Determine what proportion of time to allocate to each activity and indicate, next to each activity, the amount of time to be allocated to it. When you evaluate the lesson and each activity in it, look specifically at how the time was used. Ask yourself how you can organise pupils and establish routines to make more time available for teaching and learning. Include these in future lesson plans.

Managing your own time effectively

However well you use time in the classroom, you may not be using the time you put into your work and your own time to best advantage. Some people always seem to work long hours but achieve little, whereas others achieve a great deal but still appear to have plenty of time to do things other than work.

One explanation for this could be that the first person wastes time, through, for example, being unsystematic in managing time or handling paperwork, putting off work rather than getting on and doing it, trying to do it all rather than delegating appropriately or not being able to say no to tasks, whereas the second person uses time well by, for example, having clear objectives for work to be done, prioritising work, completing urgent and important tasks first and writing lists of tasks to achieve during the day. Which of these descriptions fits you? To check this, you need to analyse the way you work and, if necessary, try to make changes. Task 1.3.3 is designed to help with this.

Task 1.3.3 PLANNING YOUR USE OF TIME OUTSIDE THE CLASSROOM

Record for one week the amount of time you spend on school work outside the classroom, both at school and at home, e.g. planning, preparation, marking, record keeping, extra-curricular activities, meetings. You might want to use a grid such as the one below.

Day	Work undertaken (along with time for each activity)	Total time
Monday		
Tuesday		
Wednesday		
Thursday		
Friday		
Saturday		
Sunday		
Total time for one week		

Then answer the following questions:

■ Is the time spent outside the classroom and total hours worked during the week reasonable?

■ Are you using this time effectively, i.e. is the balance of time spent on the activities right, e.g. are you spending more time on record keeping than on planning and preparation?

■ Do you need to spend more time on some activities?

■ Could you reduce time on some activities, e.g. can some of the work be delegated to pupils (e.g. mounting and displaying work)?

Compare the time you spend and how you spend it with other student teachers.

If time spent is excessive (48 hours is the maximum working week in the European Union working time directive), plan what action you are going to take to reduce the time spent on school-related work each week. Recheck the use of time outside the classroom by repeating the log for one week to see whether this has worked and in light of the results what further action you need to, and can, take.

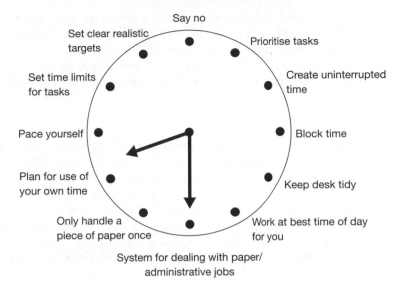

Say no

Set clear realistic
targets

Prioritise tasks

Set time limits
for tasks

Create uninterrupted
time

Pace yourself

Block time

Plan for use of
your own time

Keep desk tidy

Only handle a
piece of paper once

Work at best time of day
for you

System for dealing with paper/
administrative jobs

■ **Figure 1.3.1** Working round the clock

Fontana (1993a) stressed that if we could use our time effectively at work we would be more efficient and more productive, be better able to plan long term, be more satisfied with our work and our job, be less stressed, have more time for ourselves and more opportunity to switch off out of work. There are many different techniques you can use to manage your time effectively. Figure 1.3.1 highlights some of these. Draw your own clock and insert your own techniques to avoid working around the clock and achieve a balance between work and leisure time (a work–life balance).

Now complete Task 1.3.4 on balancing your work and leisure time.

Task 1.3.4 **BALANCING YOUR WORK AND LEISURE TIME**

In Task 1.3.3 you recorded the time spent outside the classroom over the course of a week on school experience. Now do the same for the time spent on, and use of, leisure time. You may want to use a table similar to the one in Task 1.3.3. Looking at both tables, why is the balance between school-related work and leisure time as it is? Is this balance acceptable? If not, is it because of, for example, inefficiency, lack of experience or overload? How can you improve it? Discuss the balance of work and leisure with other student teachers and discuss with teachers how they achieve a work–life balance.

PREVENTING, MANAGING AND COPING WITH STRESS

'Teacher stress may be defined as the experience by a teacher of unpleasant, negative emotions, such as anger, anxiety, tension, frustration or depression, resulting from some aspect of their work as a teacher' (Kyriacou, 2001: 28). Compared to other social welfare professions, teachers experience the highest levels of stress (e.g. Dunham and Varma, 1998; Kyriacou, 2000; Travers and Cooper,

1996). Other studies have suggested that between a quarter and a third of teachers report their job to be (extremely) stressful (e.g. Gold and Roth, 1993; Mills, 1995). It would therefore be surprising if student teachers did not find teaching stressful.

Although it is preferable if you can prevent stress, this is not always possible therefore it is important that you identify causes of stress for you and develop strategies to be able to cope with it.

Causes of stress

Many causes of teacher stress (including time factors and demanding work conditions) have been identified (see for example Benmansour, 1998; Brown and Ralph, 2002). Stokking *et al.* (2003) identified shock and stress when starting to teach as a result of lack of preparation, which they suggested might be due to student teachers having false expectations of the profession (which may be for a range of reasons). Likewise, Terry (1997) found that unrealistic expectations as a result of lack of preparation caused by inadequate training were one source of stress. This can be exacerbated by 'being thrown in at the deep end' or alleviated by gradually growing into more independent roles. On the other hand, research indicates that good preparation for teaching has been found to reduce stress (e.g. Dussault *et al.*, 1997; Terry, 1997). You need to check the causes of stress for you as a student teacher and work to eliminate them.

Student teachers are likely to have different concerns at different stages in learning to teach. Fuller and Brown (1975) classified changes in concerns over time in learning to teach as a three-way process, i.e. concerns about self; concerns about tasks/situations; concerns about impact on pupils (see also Unit 1.2). Thus, being concerned about specific aspects of your teaching or your development as a teacher at specific times is a natural part of learning to teach. As you go through your initial teacher education (ITE) course, reflect on your own development as a teacher, particularly whether your concerns are the same or different at different times of the year.

Studies of stress in student teachers (e.g. Capel, 1996, 1997, 1998; Kyriacou and Stephens, 1999; Morton *et al.*, 1997; Murray-Harvey *et al.*, 2000) have shown that major causes of stress for student teachers include:

■ not being regarded as a real teacher
■ managing the class, control and discipline and dealing with disruptive behaviour
■ motivating pupils and maintaining pupils' interest
■ conflict with pupils
■ coping with the ability range of pupils within a class
■ practical skills of teaching, techniques of lesson preparation and getting the teaching and/or planning right
■ coping with a heavy workload/managing time
■ striking a balance between the practicum and personal commitments
■ having high expectations of own teaching performance
■ disagreement with the tutor
■ observation, evaluation and assessment of teaching by the tutor, particularly receiving the tutor's or class teacher's opinion of classroom competence
■ role ambiguity, role conflict and role overload.

There are, of course, many other aspects of your teaching that may cause you stress or anxiety, e.g. delivering material with which you are not very familiar or reprimanding a pupil. Later units in this book identify practical ways to help you overcome many of these anxieties. The last two in the list above are considered briefly below because they are particularly relevant to student teachers.

When you are being observed, evaluated and assessed, you are 'on show'. You are vulnerable because your developing skills are analysed and criticised constructively. According to Pateman (1994) this may be exacerbated where teachers are involved in assessing the teaching competence of student teachers because student teachers may feel unable to talk freely and openly to teachers about other concerns. Thus, the role of the teacher-tutor in assessment does not take account of student teachers' needs for friendship, counselling and tutoring. This may cause stress for student teachers. This finding is supported in other research on stress in student teachers, e.g. Capel (1994).

Often your role is ambiguous, so you are not quite sure how to perform in the role of a student teacher.

Role conflict can result from doing a number of different activities within your job, each requiring different responsibilities, demands and skills, e.g. teaching, form tutoring, talking to parents, administration (clerical work and committee duties), other tasks within the department, continuing professional development (inter-role conflict) or from trying to meet the different expectations of a number of people with whom you are working, e.g. pupils, your tutor, other teachers, head of department, senior managers, parents (intra-role conflict).

Role overload can occur because there are so many things for you to do as a teacher and too little time in which to do them. Indeed, in a study by Kyriacou *et al.* (2003) less than 10 per cent of student teachers were absolutely certain that they would have enough time to do a good job. Overload can result in not doing a good job, working very long hours to get the task done and not having enough time mentally and physically to relax for work the following day or week. You can help with this by, for example, spending five minutes identifying what you need for the next day (this also helps to save time – see above).

Role ambiguity, conflict and overload may affect student teachers more than qualified teachers for a number of reasons. For example, as student teachers you may, at any one time, be answering to and trying to please a number of people, who expect different things. You may also take longer to prepare each lesson than more experienced teachers. Further, you may be unsure of your role in a lesson, a department or the school as a whole.

It may be that stressors outside work, e.g. tensions of home and family or finances, are brought to and add to stress at work and make a person more vulnerable to stressors at work. Job stress may vary during the year according to the demands of a job, personal circumstances and/or other factors at any one time. A significant stressor at a particular time could account for differences in stress experienced by people at different times of a school year.

In Task 1.3.5 you are asked to look at causes of stress for you.

Task 1.3.5 CAUSES OF STRESS FOR STUDENT TEACHERS

Write a list of factors that cause stress for you – both stressors as a student teacher and stressors outside your ITE course. Compare these with causes of stress identified by another student teacher. Discuss similarities and differences. Use this list for Task 1.3.6.

How can you cope with your stress?

If you cannot prevent stress, you need to be able to cope with it. As there are different causes of stress for different people and for the same person at different times, there is no one way to cope with stress; you have to find out what works for you or for you in particular situations. Different ways of coping with stress, therefore, are appropriate for different people and for the same person at different times. Arikewuyo (2004) classified strategies for coping with stress as:

■ active behavioural strategies (confronting the source of stress and attempting to change these sources by, for example, envisaging that you will get through in any situation whatever the circumstances, becoming more organised and devoting more time and energy to the job)
■ inactive behavioural strategies (behaviours of escape, such as engaging in physical and recreational activities, and avoidance of the source of stress, e.g. those individuals who might create stressful situations)
■ active cognitive strategies (identifying the sources of stress and trying to tackle them by, for example, restructuring priorities, seeking more clarification, identifying strategies to manage and reduce stress)
■ inactive cognitive strategies (conforming to, and trying to meet, expectations of e.g. mentors and tutors by e.g. meeting all duties and deadlines).

Murray-Harvey *et al.* (2000) identified personal and professional coping strategies. There were five types of personal strategies: cognitive; physical; behavioural; emotional; and rational/time organisation strategies. Three specific professional coping strategies were identified: knowledge of the curriculum and what they were expected to teach and knowing the structure, organisation and culture of the school which helped student teachers feel comfortable in that environment; use of self-management skills, such as preparation, planning and organisational skills; and professional qualities.

Specific coping strategies identified below have been drawn from a number of sources (e.g. Cains and Brown, 1998; Cockburn, 1996; Head *et al.*, 1996). These lists are by no means exhaustive and you may find other strategies useful.

■ *Prepare for stressful situations when you are not under pressure*, e.g. prepare lessons before the day on which you are teaching them.
■ *Role-play a situation that is causing you anxiety and/or visualise what you can do to overcome the problem*. This helps you to focus on the problem and can be used to rehearse how you are going to cope.
■ *Actively prepare for a situation*, e.g. if you are anxious about a particular lesson prepare it more thoroughly than normal. Plan thoroughly how you can reduce the likelihood of a problem occurring or deal with a particular problem. This strategy can help you to identify the reasons for a problem and to focus on possible ways of preventing or dealing with it.
■ *Develop effective self-management techniques*, e.g. establish routines so that you can do things automatically, particularly when you are tired.
■ *Recognise and try to develop your strengths as well as your weaknesses* so that you can rely on your strengths as you work on improving any weaknesses.
■ *Identify where you can get help*. You should get regular feedback on your teaching, but identify other people who may be able to help.
■ *Develop support systems which provide a network of people with whom you could talk through problems*, e.g. other student teachers, your tutor, other teachers, a partner or friend. You may want to talk to different people for help with different problems. You may form a group with other student teachers to provide mutual support, talk about your anxieties/concerns, develop a shared understanding of a problem and provide possible alternative solutions and practical help to address a problem, e.g. a lesson being observed then discussed with another student teacher.

■ *Do not worry about incidents that have happened in school and keep problems in proportion.* Try not to take problems home.

■ *Take account of the amount and variety of work you are doing to reduce both role overload and conflict.* This may mean, for example, that you need to try to take work home less often or take on fewer extra-curricular activities. You may need to work on this over a period of time.

See also the strategies for managing your time above.

However, it is important not only to focus on your concerns and fears, but also to pay attention to your aspirations and hopes as a teacher. Conway and Clark (2003: 470) suggested that focusing on resolving immediate concerns can result in 'an unduly pessimistic understanding of teachers and teaching'. You might find it difficult as a student teacher to focus on your development as a teacher, on the positive aspects of learning to teach and on your long-term goals and aspirations as a teacher. However, if you can do this you are likely to have a more balanced view and be able to put things into perspective and therefore reduce your stress (Unit 8.2 is designed to help you think about your continuing professional development). Task 1.3.6 is designed to help you to cope with your stress.

Task 1.3.6 **COPING WITH YOUR STRESS**

In Task 1.3.5 you listed factors that cause stress for you. Now identify ways that you can cope with this stress. Are the same or different methods appropriate for coping with stress, irrespective of the cause? Try out these coping methods as soon as you can and reflect on and evaluate whether these are effective. If they are not totally successful in all or some situations – what other methods are you going to try? Evaluate the effectiveness of these methods and adapt them or try new methods until you find those that work for you to cope with different stressful situations.

SUMMARY AND KEY POINTS

We would be very surprised if, as a student teacher on school experience, you are not tired. Likewise, we would be very surprised if you do not feel as though you do not have enough time to do everything, are not anxious when someone comes in to watch your lessons, particularly if that person has a say in whether you become a qualified teacher, or if you are not worried about other aspects of your teaching and/or school experience. It may help to know that you are not going to be alone in being tired or feeling anxious or worried about your school experience and many of the causes of tiredness, lack of time and stress are the same for other student teachers. Where you are alone is in developing effective techniques for managing your time and for coping with stress. There are no ready answers. Other people can help you with this, but nobody else can do it for you because what works for someone else may not work for you. Finally, you must work at managing your time and stress; there are no short-term, one-off solutions to these problems. However, it helps also to focus on the positive aspects of teaching and why you want to become a teacher.

Check which requirements for your course have been addressed through this unit.

FURTHER READING

Bubb, S. and Earley, P. (2004) *Managing Teacher Workload: Work–Life Balance and Wellbeing*, London: Sage.

This book provides guidance, along with a self-audit tool, on managing your workload including, e.g. how long you are working, what you are spending your time on and whether you are working efficiently.

Capel, S., Heilbronn, R., Leask, M. and Turner, T. (2004) *Starting to Teach in the Secondary School: A Companion for the Newly Qualified Teacher*, 2nd edn, London: RoutledgeFalmer.

Although this book is written for newly qualified teachers, Chapter 2, 'Managing yourself and your workload', provides guidance on managing stress and time which is also appropriate for student teachers.

Child, D. (2007) *Psychology and the Teacher*, 8th edn, London: Continuum.

Chapter 8, 'Human motivation', includes a section on stress in teachers and pupils.

Kyriacou, C. (2000) *Stress-Busting for Teachers*, Cheltenham: Stanley Thornes.

This book aims to help teachers to develop a range of strategies for coping with stress at work. It looks at what stress is; sources of stress; how to pre-empt stress; how to cope with stress; and what schools can do to minimise stress.

Kyriacou, C. (2001) 'Teacher stress: directions for future research', *Educational Review*, 53 (1), 27–35.

This article gives definitions. It looks at research into the prevalence and main sources of teacher stress. It looks at coping with teacher stress – both individually and what schools can do. It concludes by looking at directions for future research.

Nelson, I. (1995) *Time Management for Teachers*, London: Routledge.

This book provides practical ideas for successful time management to help teachers deal with the range of demands made on teachers.

> **Additional resources for this unit are available on the companion website:**
> www.routledge.com/textbooks/9780415478724

USING ICT FOR PROFESSIONAL PURPOSES

An introduction

Richard Bennett and Marilyn Leask

INTRODUCTION

In recent years there has been considerable investment in information and communications technology (ICT) in schools and you will be expected to be able to use various forms of technology in your teaching and in your wider role. In this unit we focus on the use of ICT to support:

- your teaching, through the presentation of information and ideas to pupils (including the use of interactive whiteboards), the preparation of learning materials including those on the school virtual learning environment (VLE)
- pupils' learning of your specialist subject(s), in the development of ICT capability through meaningful contexts
- your wider professional role, including administration including online reporting, record keeping and monitoring of pupil achievement
- your professional development through access to online resources and information sources (see also Unit 8.2).

The British Education and Communication Technology Agency (BECTA, 2004a) defines ICT as:

- broadcast material, DVD or CD-ROM as sources of information e.g. in history
- micro-computers with appropriate keyboards and other devices e.g. to teach literacy and writing
- keyboards, effects and sequencers in music teaching
- devices to facilitate communication for pupils with special needs
- electronic toys to develop spatial awareness and psycho-motor control
- email e.g. to support collaborative writing and sharing of resources
- video-conferencing e.g. to support the teaching of modern foreign languages
- Internet-based research e.g. to support geographical enquiry
- integrated learning systems (ILS) e.g. to teach basic numeracy
- communications technology e.g. to exchange administrative and assessment data.

The ideas in the unit are based on practice in innovative schools. You may wish to take your *European Computer Driving Licence (ECDL)*. This provides a structured way of developing your skills supported by material freely available from the British Computing Society. In England student teachers are required to pass an online test in ICT.

Inspection reports (e.g. Office for Standards in Education (Ofsted), 2004a, 2004b) and research projects (e.g. Harrison *et al.*, 2003; BECTA, 2007c) have shown that effective use of ICT resources in schools to support subject learning and teaching is patchy. At best, teachers are making highly successful use of ICT-based resources and activities to stimulate and extend the quality of their pupils' learning experiences. At worst, teachers avoid the use of ICT-based resources which could contribute positively to the development of their pupils' learning. In the world outside school, pupils are surrounded by ICT-based information sources and most are well versed in the use of ICT-based technologies in their daily lives. It has been estimated by the Sector Skills Development Agency (2004) that more than 90 per cent of new jobs require ICT skills and it is now expected that schools will ensure pupils' ICT skills are developed. There is considerable convincing research evidence (e.g. BECTA 2002, 2003a; Organisation for Economic Cooperation and Development (OECD), 2007) to indicate that, when ICT is effectively deployed, pupil motivation and achievement are raised.

OBJECTIVES

At the end of this unit you should be able to:

■ understand a range of ways in which ICT can be deployed for educational purposes

■ have identified specific applications relevant to your subject area and used those available to you

■ have audited your ICT skills and knowledge and developed an action plan for improving these.

Check the requirements of your course to see which relate to this unit.

THE SKILLS AND KNOWLEDGE REQUIRED OF STUDENT TEACHERS

As part of your course you can expect to have to demonstrate that you can appropriately integrate forms of ICT into your teaching both to support learning as well as to develop pupils' basic ICT skills. In Task 1.4.1 you are asked to audit your strengths and weaknesses in basic ICT skills.

Task 1.4.1 AUDITING YOUR BASIC ICT SKILLS AND KNOWLEDGE WRITING YOUR OWN ACTION PLAN

Check the requirements of your course for ICT skills and use and that you can carry out the basic functions identified in Table 1.4.1 such as word processing, use of spreadsheets and databases, email, the production of PowerPoint slides.

Write an action plan setting out what you need to learn by when and how you are going to do that.

■ **Table 1.4.1** Basic ICT skills. Put a tick beside each skill indicating your level of competence/ confidence (0 = no confidence, 3 = very confident).

General skills	0	1	2	3
Choosing appropriate software to help solve a problem				
Dragging and dropping				
Having more than one application open at a time				
Highlighting				
Making selections by clicking				
Moving information between software (e.g. using the clipboard)				
Navigating around the desktop environment				
Opening items by double clicking with the mouse				
Printing				
Using menus				
How to change the name of files				

Word processing skills	0	1	2	3
Altering fonts: font, size, style (bold, italic, underline)				
Text justification: left, right and centre				
Using a spellchecker				
Moving text within a document with 'cut', 'copy' and paste				
Adding or inserting pictures to a document				
Counting the number of words in a document				
Adding a page break to a document				
Altering page orientation – (landscape, portrait)				
Using characters/symbols				
Using find and replace to edit a document				
Using styles to organise a document				
Using styles to alter the presentation of a document efficiently				
Adding page numbers to the footer of a document				
Adding the date to the header of a document				
Changing the margins of a document				

E-mail skills	0	1	2	3
Recognising an e-mail address				
Sending an e-mail to an individual				
Sending an e-mail to more than one person				
Replying to an e-mail				
Copying an e-mail to another person				
Forwarding an incoming e-mail to another person				
Adding an address to an electronic address book				
Filing incoming and outgoing e-mails				
Adding an attachment to an e-mail				
Receiving and saving an attachment in an e-mail				

Database skills (not now required but useful)	0	1	2	3
Searching a database for specific information				
Using Boolean operators (and/or/not) to narrow down searches				
Sorting database records in ascending or descending order				
Adding a record to a database				
Adding fields to a database				
Querying information in a database (e.g. locating all values greater than 10)				
Filtering information in a database (e.g. sorting on all values greater than 10)				
Categorising data into different types (numbers, text, and yes/no (Boolean) types)				

Web browser skills	0	1	2	3
Recognising a web address (e.g. www or co.uk, etc.)				
Using hyperlinks on websites to connect to other websites				
Using the back button				
Using the forward button				
Using the history				
Understanding how to search websites				
Using Boolean operators (and/or/not) to narrow down searches				
Creating bookmarks				
Organising bookmarks into folders				
Downloading files from a website				

Spreadsheet skills	0	1	2	3
Identifying grid squares in a spreadsheet (e.g. B5)				
Inserting columns into a spreadsheet				
Inserting rows into a spreadsheet				
Sorting spreadsheet or database columns in ascending or descending order				
Converting a spreadsheet into a chart				
Labelling a chart				
Adding simple formulae/functions to cells				
Applying formatting to different types of data including numbers and dates				

Presentation skills	0	1	2	3
Inserting text and images on a slide				
Inserting a slide in a presentation				
Adding a transition between slides				
Adding buttons to a presentation				
Using timers in a presentation				

Table 1.4.1 continued

An editable version of Table 1.4.1 is available on the companion website:
www.routledge.com/textbooks/9780415478724

■ **Table 1.4.2** Elements of ICT in various subject areas

Art and design		Maths	
Finding things out	Surveys (e.g. consumer preferences), web galleries, online artist/ movement profiles	Finding things out	Databases, surveys, statistics, graphing, calculators, graphical calculators, dynamic geometry, data logging/ measurement (e.g. timing), web-based information (e.g. statistics/ history of maths)
Developing ideas	Spreadsheets to model design specs	Developing ideas	Number patterns, modelling algebraic problems/probability
Making things happen	Embroidery CAD/CAM	Making things happen	Programming – e.g. LOGO turtle graphics
Exchanging and sharing information	Digital imagery/CAD/ multimedia for students' design portfolios	Exchanging and sharing information	Formulae/symbols, presenting investigation findings, multimedia
Reviewing, modifying and evaluating	Real world applications – e.g. commercial art	Reviewing, modifying and evaluating	Comparing solutions to those online, online modelling and information sources

Business and commercial studies		Technology	
Finding things out	Pay packages, databases, online profiling,	Finding things out	Product surveys, consumer preferences, environmental data
Developing ideas	Business/financial modelling	Developing ideas	CAD, spreadsheet modelling
Making things happen	Business simulation	Making things happen	CAM, simulations (e.g. environmental modelling), textiles, embroidery, control
Exchanging and sharing information	Business letters, web authoring, multimedia CVs, e-mail	Exchanging and sharing information	Advertising, product design and realisation, multimedia/web presentation
Reviewing, modifying and evaluating	Commercial packages, dot.com, admin. systems	Reviewing, modifying and evaluating	Industrial production, engineering/electronics

Performing arts

Finding things out	Online information sources, surveys
Developing ideas	Planning performance/choreographing sequences
Making things happen	Lighting sequences, computer animation, MIDI, multimedia presentations
Exchanging and sharing information	Video, audio, digital video, web authoring, multimedia, animation, DTP posters/flyers/programmes e-mail
Reviewing, modifying and evaluating	Ticket booking, lighting control, recording/TV studios, theatre/film industry

Physical Education

Finding things out	Recording/analysing performance, internet sources (e.g. records)
Developing ideas	Planning sequences/tactics
Making things happen	Modelling sequences/tactics, sporting simulations
Exchanging and sharing information	Reporting events, posters, flyers, web/multimedia authoring, video, digital video
Reviewing, modifying and evaluating	Website evaluation, presentation of performance statistics, event diaries, performance portfolios

English

Finding things out	Surveys, efficient searching/keywords, information texts, online author profiles, readability analysis
Developing ideas	Authorship, desktop publishing (balancing text and images)
Making things happen	Interactive texts/multimedia/web authoring
Exchanging and sharing information	Exploring genres (e.g. writing frames), authoring tools, text/images, scripting, presenting, interviewing (audio/video)
Reviewing, modifying and evaluating	Website evaluation, online publishing, e-mail projects

Modern foreign languages

Finding things out	Class surveys, topic databases, web searching/browsing
Developing ideas	Concordancing software, interactive video packages, DTP and word processing
Making things happen	Online translation tools, interactive multimedia
Exchanging and sharing information	Word processing, DTP, web/multimedia authoring, e-mail projects, video/audio recording, digital video editing
Reviewing, modifying and evaluating	Internet communication, website/CD ROM language teaching evaluation, translation software

Humanities		Science	
Finding things out	Surveys, databases, Internet searching, monitoring environment (e.g. weather), census data etc.	Finding things out	Data recording and analysis, spreadsheets and graphing packages, Internet searching (e.g. genetics info.)
Developing ideas	Multimedia, DTP, modelling (spreadsheets/ simulations)	Developing ideas	Modelling experiments/ simulations
Making things happen	Simulations, interactive multimedia/web authoring	Making things happen	Datalogging, modelling experiments, simulations (what if...?)
Exchanging and sharing information	Web authoring, e-mail projects	Exchanging and sharing information	Communicating investigation findings (DTP, web/multimedia authoring, DV
Reviewing, modifying and evaluating	Weather stations, satellite information, website/CD ROM evaluation, archive information	Reviewing, modifying and evaluating	Accessing information (evaluating for bias on issues, e.g. nuclear power)

There will be various opportunities for you to use ICT to support specific aspects of teaching and learning in your subject area. Task 1.4.2 asks you to identify these and make plans to develop your skills in the use of ICT in your subject area.

Task 1.4.2 **IDENTIFYING ICT RESOURCES FOR YOUR SUBJECT AREA**

Find out what ICT resources are available to you to support your subject area. Table 1.4.2 and the material at the end of the unit provide guidance for different subjects. This is likely to include areas new to you and you will need to take an active role in your own professional learning and development. Add your subject specific learning goals to the action plan from Task 1.4.1.

Discuss the application of ICT subject resources to the lessons you are taking with your tutor and fellow student teachers. Use these resources in lessons and evaluate your success in achieving the learning objectives you set.

ICT AND PUPIL LEARNING

ICT should only be used where its use is justified as a method of achieving the stated learning outcomes for any lesson. From a learning perspective, BECTA (2001) state that the effective use of ICT can lead to benefits in terms of:

- greater motivation
- increased self-esteem and confidence
- enhanced questioning skills
- promoting initiative and independent learning
- improving presentation
- developing problem-solving capabilities
- promoting better information handling skills
- increasing time 'on task'
- improving social and communication skills.

More specifically, ICT can enable pupils to:

- combine words and images to produce a 'professional' looking piece of work
- draft and redraft their work with less effort
- test out ideas and present them in different ways for different audiences
- explore musical sequences and compose their own music
- investigate and make changes in computer models
- store and handle large amounts of information in different ways
- do things quickly and easily which might otherwise be tedious or time-consuming
- use simulations to experience things that might be too difficult or dangerous for them to attempt in real life
- control devices by turning motors, buzzers and lights on or off or by programming them to react to changes in things like light or temperature sensors
- communicate with others over a distance.

They go on to define pupil ICT capability as characterised by an ability to use effectively ICT tools and information sources to analyse, process and present information, and to model, measure and control external events. More specifically, a pupil who has developed ICT capability should:

- use ICT confidently
- select and use ICT appropriate to the task in hand
- use information sources and ICT tools to solve problems
- identify situations where the ICT use would be relevant
- use ICT to support learning in a number of contexts
- be able to reflect and comment on the use of ICT they have undertaken
- understand the implications of ICT for working life and society.

Pupils are expected to be given opportunities to develop and apply their ICT capability in the context of all curriculum subjects.

In addition to the features outlined above, ICT has been shown to provide specific support for pupils with special educational needs (SEN). Peacey (2005) provides detailed advice about particular forms of ICT which support pupils with SEN. Teachers interested in SEN and inclusion issues will find further information and support in the inclusion section of the BECTA website.

The most obvious and hence the most common use of ICT to support teaching is in the production of paper-based resources such as worksheets, template documents, handouts, information leaflets and pupil booklets. Increasingly, teaching resources are being provided online through websites such as the Teacher Resource Exchange (TRE), which is government sponsored and is aimed at enabling teachers and other education professionals to share their handiwork and ideas with others. Several other organisations, such as TeacherNet and Schoolzone provide similar resources and some commercial companies market their services or provide adaptable paper-based resources on CD-ROM. Also, the websites for subject organisations (e.g. the Maths Association, Geography Association) provide educational materials and links to other resources. However, as a teacher you must balance the time taken to search for and modify others' resources to meet your pupils' learning needs against the time it might take to produce your own from scratch.

Interactive whiteboards

In many schools you will find data projectors and interactive whiteboards which you can use to support interactive whole-class teaching. Interactive whole-class teaching was developed particularly by educators in the Pacific Rim countries whose pupils demonstrated the impressive learning gains in basic subjects identified in international comparisons of pupils' performance (see OECD, 2001). The basic principles of interactive whole-class teaching require teachers to build shared understanding in pupils through careful questioning and the presentation of information and ideas which expose and challenge learners' misconceptions. The teacher channels and develops pupils' thinking interactively, rather than presenting information, solutions and ideas didactically (see Dickenson, 1999; Department for Education and Skills (DfES), 2004Z).

The use of data projectors and interactive whiteboard technology further enhances this approach by giving teachers and pupils flexible access to resources, including the Internet, and provides teachers with opportunities to present challenging information and ideas through the use of text, images, animations, sound and video – i.e. through multi-sensory (multimedia) approaches, thereby enabling pupils to access information in accordance with their preferred learning styles.

To facilitate the use of such approaches, as a teacher you need to develop your skills in finding, accessing, cataloguing and presenting information in a range of formats. Furthermore, interactive whiteboards offer a series of tools, such as the ability readily to highlight and manipulate text and images, to translate handwriting into text and to save and replay screens and sequences of on-screen actions. A BECTA information booklet outlines some of the strategies for effective use of interactive whiteboard teaching technology (BECTA, 2004b). Further subject specific information, guidance and support can be found and downloaded by visiting the Teachernet publications website and accessing the *Embedding ICT @ Secondary* documents on interactive whiteboards.

Learning platforms

The government's e-learning strategy (DfES, 2004Z; BECTA, 2007b) requires every maintained school in England to have its own learning platform (a Virtual Learning Environment (VLE) or Managed Learning Environment (MLE)). You may be familiar with VLEs or MLEs from your university courses. These provide the means by which teaching staff can provide online information, resources and activities for learners. The complexity of a learning platform can vary from the basic presentation of handouts, presentations and information covered in taught sessions, to resources which exploit Web 2.0 technologies such as online assessment tasks, blogs, wikis and user-controlled network spaces. The most sophisticated learning platforms integrate data about individual pupils' attendance, performance and progress, with learning resources produced by teachers. Staff, pupils and parents can log into the system to check on a child's progress and monitor their access to online tools and resources.

Your use of a school's learning platform as a teacher is likely to be influenced heavily by your own experience of online learning as a student. It is important to find out about the school's learning platform and how your department makes use of it when joining the school as levels of sophistication and adoption vary not only from school to school but also from department to department within schools. More information on developments in this *Harnessing Technology* agenda can be found by accessing information on the BECTA website.

Developing pupils' ICT skills

In addition to using ICT resources to support or enhance your role as a teacher, there is an expectation you will contribute to the development of pupils' ICT capabilities by giving pupils hands-on experience of ICT. In most cases, your prime motivation in making use of ICT is to address learning objectives related to your subject. However, with a slight shift of emphasis you could modify an activity to develop concomitantly aspects of pupils' ICT knowledge and skills. For example, you might produce a word-processed paper-based writing frame for your pupils to structure the presentation of some information associated with your subject. In this case, you have used your ICT skills to research and produce the worksheet, but the children are making no use of ICT to complete it. The activity could be modified as follows:

■ The writing frame is presented on-screen for completion by a pair of pupils working collaboratively, using the features of a word processor to enhance the communication of information (e.g. by reworking sections of text and by selecting and incorporating appropriate images).
■ Relevant information is located and copied from two or three websites you have identified and pupils edit the pasted text, shifting the focus of the information or targeting a specific audience. (Pupils need to be aware that they should acknowledge the sources they use.)
■ Groups of pupils search websites (or a CD-ROM encyclopaedia) to create information leaflets for each other on different aspects of a topic – or to present controversial information from different viewpoints.

Consider the different levels of cognitive demand (i.e. the decision-making) in each of the above examples. Whilst the subject learning in each case might be broadly similar, shifting the ICT focus for the task not only supports the application and development of the pupils' ICT capabilities, it also deepens their learning by helping them engage more fully with the subject matter.

Educational resources provided through the Internet support:

■ access to information
■ the use of interactive tools and resources
■ participation in and/or creation of online projects
■ communication with 'experts' and other learners
■ publishing and sharing information and ideas with a potential world audience.

Information sources dominate the Internet and most pupils already know how to access them before commencing their secondary education. However, they may not have learned how to search for information efficiently or to discriminate between sources in terms of accuracy, reliability, plausibility and the currency of the information. When using Internet-based information sources it is important to help pupils learn how to search for and locate the most appropriate information for the task in hand. Similarly, it is essential that they adopt safe surfing habits to avoid unsuitable materials and protect them from unwanted attention. The most useful information on 'Internet safety' is provided online by BECTA and also by the Child Exploitation and Online Protection Centre (CEOP).

A fun teaching approach which uses the Internet more systematically for educational purposes is the setting up, a 'webquest' (see Figure 1.4.1).

Alternatively, you might decide to participate in an online project with pupils from other schools, maybe in other parts of the world. The European Schoolnet e-Twinning site provides a partner finding service.

If you are interested in undertaking such projects then starting with something small and achievable enables you to develop strategies which work for you in your particular subject. For example, a survey on a specific topic carried out by pupils in two countries can be done over a very short time span, perhaps a couple of weeks. This could enable you to avoid problems of timetabling or clashes of holidays. Funding for collaborative work may be available from UK or European Union sources. Table 1.4.3 provides some guidelines for running e-mail projects. For example, Lord Grey School in Bletchley has undertaken sustained ICT curriculum projects across subjects and involving many countries. Holy Cross Convent School in Surrey has undertaken innovative cross-curricular

---

A 'webquest' is a framework made by teachers for pupils for stimulating educational adventures on the web and to help pupils in the acquisition of problem-solving and searching skills. Taking account of curricular goals, the teacher sets up a few guidelines, with a simple structure:

■ introduction – context information related with the task/problem/adventure/questions to be completed by the pupils;
■ the task/problem/adventure/questions – what has to be done by the pupils;
■ internet resources – location of internet resources like websites, databases, live video cameras for educational purposes (e.g. vulcanology . . .);
■ reporting results and final discussions.

For further information and examples visit the Webquest website www.webquest.org

---

■ **Figure 1.4.1** An example of a 'webquest'

■ **Table 1.4.3** Checklist for planning ICT projects with other schools

1 What learning outcomes do you want the pupils to achieve in terms of knowledge/concepts, skills, attitudes?
2 What is the time scale of the project and how does that fit with school holidays and other events in the partner school?
3 What languages can you work in? (Don't forget that parents, other schools and the local community may be able to help here.)
4 What resources – staff, equipment, time – are involved?
5 Does anyone need to give their permission?
6 How are you going to record and report the outcomes?
7 Do staff need training?
8 Can you sustain the project within the staff, time and material resources available to you?
9 What sorts of partners are you looking for?
10 How are you going to find the partners?
11 How are you going to evaluate the outcomes?

An editable version of Table 1.4.3 is available on the companion website:
www.routledge.com/textbooks/9780415478724

videoconferencing projects with a school in Japan. This work is described further by Lawrence Williams, the director of studies (Leask and Williams, 2005). Further examples of projects with other schools, e.g. virtual field trips, virtual art galleries, are given in Leask and Pachler (2005).

FINDING PARTNERS FOR E-MAIL/VIDEO-CONFERENCING/ INTERNET-BASED PROJECTS

There are a number of ways of finding partners for inter-country projects. These include:

■ using existing contacts, through, for example, exchanges or through the local community and teachers in the school
■ e-mailing schools direct; various sites provide lists of schools' e-mails, e.g. European Schoolnet and ePals
■ advertising your project, e.g. by registration on a site such as those mentioned below
■ searching sites listing school projects and finding projects which seem to fit with your curriculum goals.

Sites such as the Global School Net in the USA and Internet Scuola in Italy provide all three of the last options.

Assessment and reporting

It is too early to predict the extent to which teaching processes and assessment methods are likely to change in response to the opportunities discussed above. In the UK at secondary level, change would accelerate if the examination boards incorporated ICT-based work into assessment requirements. Clearly, pupils have to be taught skills needed for the critical appraisal of material but good teachers will be doing this already. Issues related to plagiarism as pupils download sections of text and incorporate these into assignments are likely to be more problematic for teachers. Whilst teacherless classrooms are unlikely to occur, certainly the positive motivation which some learners feel when using technology is not to be underestimated, but this does depend on the context for learning which the teacher establishes. You may come across schools using ICT for pupils to take assessments when they are ready. This kind of practice is expected to expand as ways of personalising the curriculum develop further.

Schools in England are expected to move to online reporting to parents and you can expect to be engaged in working in this way.

Motivation and classroom management

The integration of technology into lessons requires confidence and competence in the management of the resources and pupils on your part. Whilst it has been shown that ICT-based tasks are often motivating for pupils, the use of a computer does not automatically make a poorly conceived lesson interesting. As with all teaching, if a computer-based lesson is not carefully planned, it could result in poor motivation, lack of involvement (or off-task activity) and very little learning benefit: Cox (1999) gives the following advice on the teaching of word processing, regardless of the subject context:

■ tasks must be relevant
■ pupils should be prepared for their tasks before being assigned computers
■ don't let pupils sit at computers while you are talking to them at the introduction of the lesson

- don't leave pupils for the whole lesson just working on their task with no intervention to remind them of the educational purpose
- don't expect pupils to print out their work at the end of every lesson
- end each lesson by drawing pupils together to discuss what they have achieved
- don't rely on the technology to run the lesson.

The availability of wireless technology gives teachers flexibility in their use of ICT.

USING ICT TO SUPPORT YOUR WIDER PROFESSIONAL ROLE

Using ICT for administration and monitoring

You can expect the school to use ICT for recording and analysing pupils' results, and to create reports for parents on pupils' progress. In Task 1.4.3 you are asked to become familiar with the systems in your school.

Task 1.4.3 **USING ICT FOR ADMINISTRATION AND MONITORING**

Schools and teachers use a variety of systems for recording, monitoring and analysing pupil progress and teacher effectiveness against targets, and predictions. Find out what systems are in use in the school in which you are placed and compare these if possible, with those used in other schools by talking with other student teachers.

Using ICT for professional development

The Internet is the most prominent source of information for professional development and lends itself particularly to communication at a formal and informal level. However, it should be remembered that broadcast television and radio, and video, CD-ROM or DVD-based materials provide information and offer opportunities for professional development. The Teachers' TV website provides guidance for teachers on the effective use of a range of technologies and a range of clips which you can incorporate into your teaching.

The Internet provides educators with the following:

1   Access to a huge range of *free and high quality information sources* including the rapid and inexpensive publication of the latest research findings from researchers around the world in all disciplines, as well as access to museums, galleries, newspapers, radio stations and libraries. These resources are often available in a variety of languages. In the UK, *Teachernet* and the *Teacher Training Resource Bank* (TTRB) are intended to provide information and resources for teachers and those intending to become teachers. In addition, various government-supported sites have information of potential use to teachers.

2   *Teaching and learning resources*, in the form of lesson plans, worksheets and computer-based learning materials. By browsing through the work of others who are tackling the same sorts of problems and issues (e.g. differentiating tasks for pupils), you not only gain material support, you have to opportunity to appraise others' teaching approaches. For examples, refer to the Teacher Resource Exchange and the Schoolzone websites.

3   For teachers and schools *the opportunity to publish and share information about their work.* School websites and virtual learning environments (VLEs) provide opportunities for publishing material for a range of purposes. For example, pupils are sometimes set projects to publish material which they have researched themselves. In doing this, both pupils and teachers are developing their knowledge about the use of this technology. Parents can be kept informed more fully about the work their children are doing through website publications and, of course, parents who are seeking schools for their children may find such sites of value in guiding their choice. In addition, the school VLE can provide a useful resource as colleagues pool ideas and use the VLE as a form of departmental filing cabinet for resources.

4   *Synchronous (e.g. video conferencing and online chat/discussion groups) and asynchronous communication (e.g. email)* with single or multiple audiences, e.g. with other teachers, pupils, parents and experts in particular fields regardless of their location. Some schools tap the expertise of parents and local companies to provide experts on line for short periods. These are specialists who are able to answer pupils' questions in areas relevant to their expertise. Teachers are using these facilities for a range of purposes, both curriculum based and for professional development. For example, joint curriculum projects with classes in other countries can be easily maintained through the use of email. Results of such collaboration can also be posted on the school website for participants in both countries to see.

Tasks 1.4.4–1.4.6 provide examples of tasks which could lead to accreditation at M level. Tasks at this level normally require you to undertake considerable critical analysis and to have a wider understanding of the role ICT might play in your subject and in the learning of your pupils.

Task 1.4.4 **INCREASING THE COMPLEXITY OF PUPILS' WORK**

Reflect on an ICT-based activity which you have observed or taught and, using the definition of ICT capability shown in the unit, discuss how the activity could have been amended to increase the level of ICT demand for more experienced pupils. Justify your amendments by reference to relevant curricular documentation, inspection evidence and background reading.

Task 1.4.5 **ICT AND PUPIL LEARNING**

Identify aspects of work within your subject for which ICT is not appropriate and some areas of study for which ICT will make a significant difference or which will transform pupils' learning experiences beyond that which would be possible without ICT. Use reading and research to critically analyse the learning gains associated with ICT within the context of your subject.

Task 1.4.6 **SOCIAL CONSTRUCTIVIST APPROACHES AND ICT**

Using the information provided in Unit 5.1 and further background reading, evaluate the extent to which ICT-based activities within your subject could be structured to support social constructivist approaches to learning and teaching

SUMMARY AND KEY POINTS

This unit has touched on only some of the classroom practice and professional development opportunities available through ICT. We recommend that you extend your understanding beyond the guidance here by reading more widely in this area, by experimenting with different types of software of particular use in your subject, by spending some time searching the Internet to identify high-quality resources and educational websites which are specifically relevant to your interests and by talking to teachers and student teachers who are themselves exploring the possibilities offered by new technologies. Make sure you know what your subject association website offers.

However, it is important to remember that ICT use in the classroom should be directly related to the achievement of specified learning outcomes. Using ICT in your classroom provides no guarantee that learning takes place.

Check which requirements for your course have been addressed through this unit.

FURTHER READING

The subject-specific texts in this Routledge *Learning to Teach* series all contain units about the use of ICT in the specific subject area. You may find further ideas for the application of ICT in your subject areas in these texts. The Routledge text *Learning to Teach Using ICT in the Secondary School* (Leask and Pachler, 2005) provides more detailed guidance on the topics raised in this unit.

RELEVANT WEBSITES

British Education and Communication Technology Agency (BECTA): http://www.becta.org.uk
Provides the most recent information about ICT in education. Information relating to special educational needs is scattered across the Internet so a starting point that gathers many of them together is useful. Such a site can be accessed through the teachers' area of the BECTA website: http://schools.becta.org.uk.

British Educational Research Association (BERA): http://www.bera.ac.uk
Access through websites to research papers from conferences around the world, including information about research into the use of ICT for education purposes.

Child Exploitation and Online Protection Centre (CEOP) website Think U Know: http://www.thinkuknow.co.uk
Reinforcing the message about safe internet use is the responsibility of all teachers. This website contains useful information for teachers, children and parents.

National Association for Gifted Children: http://www.nagcbritain.org.uk
Further help and advice for teachers working with gifted pupils.

British Computing Society (BCS): http://www.bcs.org
The BCS supports the European Computer Driving Licence. Details can be found on the website.

ICT Skills Test for QTS on the TDA website: http://www.tda.gov.uk/skillstests/ict.aspx

European Schoolnet: http://www.eun.org/portal/index.htm
Allows access to European government supported educational websites.

Global SchoolNet: http://www.globalschoolnet.org/

Webquest.Org: http://www.webquest.org

24 Hour Museum: http://www.24hourmuseum.org.uk
Provides links to over 3000 museum and gallery websites.

Tramline: http://www.field-guides.com
  Virtual field trips.

Teachers' TV: http://www.teachers.tv
  Download clips to incorporate into your own presentations.

Mathematical Association: http://www.m-a.org.uk and the Association for Teachers of Mathematics (ATM): http://www.atm.org.uk
  Both provide information, ideas and resources for ICT-based activities.

Geography Association: http://www.geography/org.uk
  Another subject association providing resources for teachers.

Additional resources for this unit are available on the companion website:
www.routledge.com/textbooks/9780415478724

2 BEGINNING TO TEACH

The previous chapter was concerned with the role and responsibilities of the teacher and how you might manage those. In this chapter, we look first at how you might learn from observing experienced teachers and then move on to consider aspects of planning and preparing lessons.

For most pupils there is a period during which you observe other teachers working, take part in team teaching and take part of a lesson before taking on a whole lesson. During this period, you use observation and critical reflection to build up your professional knowledge about teaching and learning and your professional judgement about managing learning. Unit 2.1 is therefore designed to focus your attention on how to observe the detail of what is happening in classrooms.

It is difficult for a student teacher to become fully aware of the planning that underpins each lesson as planning schemes of work (long-term programmes of work) is usually done by a team of staff over a period of time. The scheme of work then usually stays in place for some time. The extent of the actual planning for each lesson may also be hidden – experienced teachers often internalise their planning so their notes for a lesson are brief in comparison with those that a student teacher needs. Unit 2.2 explains planning processes. Unit 2.3 combines much of the advice of the first two units in an analysis of the issues you probably need to be aware of before taking responsibility for whole lessons.

The quality of lesson planning is crucial to the success of a student teacher in enabling the pupils to learn. Defining clear and specific learning objectives and learning outcomes for pupils' learning in a particular lesson is one aspect of planning that many student teachers initially find difficult. The following story (from Mager, 1990: v) reinforces this need to have clear objectives and outcomes for lessons:

> Once upon a time a Sea Horse gathered up his seven pieces of eight and cantered out to find his fortune. Before he had travelled very far he met an Eel, who said, 'Psst. Hey, bud. Where 'ya goin'?'
>
> 'I'm going out to find my fortune,' replied the Sea Horse, proudly.
>
> 'You're in luck,' said the Eel. 'For four pieces of eight you can have this speedy flipper and then you'll be able to get there a lot faster.'
>
> 'Gee, that's swell,' said the Sea Horse and paid the money and put on the flipper and slithered off at twice the speed. Soon he came upon a Sponge, who said, 'Psst. Hey, bud. Where 'ya goin'?'
>
> 'I'm going out to find my fortune,' replied the Sea Horse.
>
> 'You're in luck,' said the Sponge. 'For a small fee, I will let you have this jet-propelled scooter so that you will be able to travel a lot faster.'

So the Sea Horse bought the scooter with his remaining money and went zooming thru the sea five times as fast. Soon he came upon a Shark, who said, 'Psst. Hey, bud. Where 'ya goin'?'

'I'm going to find my fortune,' replied the Sea Horse.

'You're in luck. If you take this short cut,' said the Shark, pointing to his open mouth, 'you'll save yourself a lot of time.'

'Gee, thanks,' said the Sea Horse and zoomed off into the interior of the Shark and was never heard from again.

The moral of this fable is that if you're not sure where you're going, you're liable to end up someplace else.

We hope that by the end of this chapter, you will be able to plan lessons in which both you and the pupils know exactly what they are meant to be learning.

Explicitly sharing your learning objectives with pupils provides them with clear goals and potentially a sense of satisfaction from your lesson as they achieve the goals set.

Readings for Learning to Teach in the Secondary School: A Companion to M Level Study brings together essential readings to support you in your critical engagement with key issues raised in this textbook.

READING CLASSROOMS

How to maximise learning from classroom observation

Susan Heightman

INTRODUCTION

How do you actively read a classroom rather than simply watch a teacher at work? You need to answer this question if you are going to make the most of opportunities to observe. In your school placements you can expect to observe experienced teachers or fellow student teachers. Whatever the case, there is so much to learn from every observation, therefore you should not prejudge what observations are likely to be of greatest value to you professionally. Equally you should not consider observation to be an activity reserved for student teachers. Lesson observation, and the improved practice that can be gained from it, is an important continual professional development (CPD) activity for all teachers. It is also a key tool for reflective practice and practitioner research (see Unit 5.4).

Qualified teachers expect to be observed in the context of appraisal, inspections and preparations for inspections. They may be less used to being observed by people who are concerned to analyse what is happening in the lesson rather than measure against preconceived ideas about what should be happening. It is important that the teacher experiences your presence and any follow-up discussion positively. Thus, before you undertake any observation in a classroom you should ask the teacher first; agreeing areas for observation and what you plan to do with the data. To achieve this you need to express appreciation for being able to observe and ask the teacher's guidance on the nature of the role they wish you to adopt. They may suggest you participate or alternatively just observe.

Remember that observation is a fundamental research activity; seeking to know what is happening, why it is happening and what the impact is likely to be. It is not passive. It is perceptive rather than judgemental; it does not seek to confirm judgements. Take an open, positive approach to observation. After the lesson you should thank the teacher. It is always a good idea to discuss your data with the teacher if possible since a deeper understanding emerges from the discussion. In the discussion you should avoid negative comments. Teachers themselves are most likely to begin any discussion of the lesson with reference to the things that went wrong or were not in the plan. Always focus the discussion on the agreed observations and aspects of the lesson that they were pleased with because it is from this positive discussion that you learn most about their professional craft.

The form in which you record your observations varies according to the selected focus. Many schools possess their own observation proformas or you may be given a general proforma by your tutors. Alternatively you can draw up a range of schedules for yourself. There are advantages to having blank proformas to work with because they facilitate the quick recording of information and prompt you to consider specific aspects, which in the flow of the lesson, you might otherwise forget. There are also great advantages to sharing an observation activity with another student teacher to share perceptions and engage in a discussion about the significance of what you have observed.

OBJECTIVES

At the end of this unit you should be able to:

■ define the focus of your observations to achieve specific learning purposes
■ use different strategies for recording your observations in forms that lend themselves to subsequent analysis
■ use observation to analyse teaching strategies and pupil learning behaviours to enhance your ability to plan and teach your own lessons
■ understand the teaching and learning process and have an insight into how you wish to teach.

Check the requirements of your course to see which relate to this unit.

WHO SHOULD YOU BE OBSERVING AND WHY?

The majority of your observations are of classes being taught by their established teacher prior to you working more directly with the class yourself.

Here you are naturally concerned to study how the pupils interact with each other and with the teacher and to see these observations as of particular value in preparing you for teaching this specific class. However, you should not be observing the teacher with a view to mimicking their teaching style. Each teacher is unique and has a unique relationship with the class that cannot be replicated. If you attempt to mimic, the pupils are likely to reject this because you have no shared background or relationship with the class. The teacher has established rules and routines in the first weeks of the term and these are often reinforced by just a single gesture or signal from the teacher, e.g. a raised eyebrow if a pupil speaks out of turn. Instead, as a student teacher, you should be seeking to understand and record for analysis what and how pupils are learning and the routines and teaching strategies of the teacher. This analysis should enable you to select those aspects that you wish to adapt in developing your own teaching style. You are also familiarising yourself with the subject content in action and the resources the pupils are using in order to learn.

Alternative forms of observation may be of small groups of pupils working with a special needs teacher or teaching assistant. Here you are learning a particular range of teaching skills and have the opportunity to understand the learning needs of individual pupils in greater depth. In addition, it is valuable to follow a class or pupil across the curriculum and observe teachers working in subject disciplines other than your own. All opportunities to observe teachers working with pupils are learning opportunities. Teachers in assembly, on duty at break or working with pupils in an extra curricular activity are all professionally engaged in their work and how they work in these contexts directly relates to their work in the classroom. It is therefore a good idea to carry a small notepad with you at all times to record significant observations and any questions that occur to you at the time. These can be followed up later with the teacher or your tutor. Effective observation is a matter of intellectual curiosity and being alert.

Table 2.1.1 gives examples of research questions you might address through observation, for example you may be observing a class prior to teaching them on your school experience. Add questions of your own. The tasks are designed to help you achieve effective professional learning from a range of observation contexts.

WHAT DO CLASSROOMS LOOK LIKE?

The classroom or teaching space is more than a room with chairs, desks and a teaching board of some kind. It is a learning environment. It should both promote and support pupil learning.

It should also express the values and ethos of the school. As part of lesson observation, it is useful to 'read' the impact of the appearance and layout of the room on pupil and teacher learning and performance. You may find the planned approach in Task 2.1.1 useful.

Task 2.1.1 **THE LEARNING ENVIRONMENT**

Consider a classroom or teaching space you are going to be teaching in on school experience and how you can use this effectively. Record your observations in your professional development portfolio (PDP).

- Is this space a specialist room or a general classroom? Is the space used mainly by one teacher?
- Sketch the layout of the space and the seating arrangements. Identify the light source and other features of note such as the board and the teacher's desk/display boards.
- Note your perceptions of the advantages and limitations of the room layout to pupil and teacher learning and teaching.
- Describe any displays. Note the different proportions of pupil work and teacher/published material displayed. Are the displays colourful and well cared for? Do the displays prompt pupils to value their own work and the work of other pupils more highly?
- Comment on whether you would like to be taught in this room and whether the environment promotes the subject and pupil learning. Give reasons for your response.

HOW LESSONS BEGIN

As with all relationships and activities, how a lesson begins is significant to its success or otherwise. Experienced teachers are very aware of this and it is important that you spend some time focusing upon the different ways in which teachers manage lesson beginnings to enable you to begin to establish your own set of routines. Invariably schools have their own school-wide policies on this, for example lining up outside the classroom in single file (so as not to block the corridor) until the teacher arrives. As a student teacher you should always know the school and department policies in these respects and follow them. Task 2.1.2 is designed to help you to analyse the beginning of a lesson.

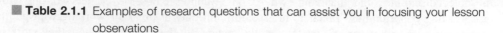

Table 2.1.1 Examples of research questions that can assist you in focusing your lesson observations

Briefing and preparing for observation

Note the date, time and place of the lesson.

Are you briefed on the topic being taught in the lesson and the composition of the class?

Have you agreed the role you are to take during observation, a participant or non-participant?

Have you agreed the form in which you are going to note down your observations and your major areas of focus?

Have you agreed how you are going to feed back to the teacher and any future use you may use of the notes you have made?

Teaching and learning questions

What was the plan/structure/shape of the lesson?

How was the lesson introduced? How did the pupils know the learning objectives and outcomes planned for the lesson?

What were the different learning activities that the pupils undertook?

Was there group or pair work during the course of the lesson?

Did any pupils receive different work, degrees of help or resources during the lesson?

What were the different ways that pupils recorded or presented their learning?

How did the teacher direct the pace of the lesson?

What form of question and answer sessions did the teacher initiate?

How were pupils encouraged to ask and answer questions?

What resources were used to assist in learning?

How did the teacher provide visual, auditory and kinaesthetic learning resources and opportunities in the lesson?

Was ICT a dimension in the lesson?

Pupil and class management dimensions

What were the teacher's expectations about pupil behaviours?

Were there established routines or codes of conduct? What were they?

Were issues of health and safety referred to during the lesson?

Did the teacher use any assertive behaviour management techniques? Were any sanctions used during the lesson?

How did the teacher use seating plans in the lesson and for what purpose?

How did the teacher use voice and gesture in the lesson to manage pupil response?

How did the teacher assess pupils work during the lesson and how did they feedback to the pupils?

Other professional issues

Was a teaching assistant or special educational needs teacher also in the room? What was their role and how did they work with the pupils and the teacher?

Have you identified any gaps in your subject knowledge through watching this lesson? How will you fill them?

How has this observation made you reconsider your future professional practice as a teacher?

Task 2.1.2 **ANALYSING THE BEGINNING OF A LESSON**

There are three stages to the beginning of a lesson:

1   outside the classroom
2   entrance of pupils and settling
3   introduction to the lesson and possibly a starter activity.

Useful prompt questions are:

Outside the classroom

■   What procedures were used for pupils gathering outside the classroom?
■   Were pupils free to enter as they arrived or did they have to line up?
■   Did the teacher wait for the class at the classroom door – were they welcomed on arrival outside the classroom or did the teacher stay inside the classroom until the class were directed to enter?

Settling into place

■   Did the pupils sit where they pleased or did they have their own places? Did they wait to greet the teacher standing before they were told to sit down?
■   Was a register taken and in what manner?
■   What signals did the teacher use to indicate that the lesson had begun?

The beginning of the lesson

■   How did the teacher explain the learning objectives and expected learning outcomes (see Unit 7.4) of the lesson?
■   How long was it before the lesson proper began?
■   What problems or issues did the teacher have to deal with before the lesson began? How did they do this?
■   In settling the class, what praise or reprimands did the teacher use and how did pupils respond?

If possible discuss this list of questions with another student teacher and add to them.

Then undertake the observation. In carrying out this task you should arrive at least five minutes before the beginning of the lesson.

Using a similar checklist to that drafted below, record your observations in your PDP.

Observer name . . . .

Teacher name . . . .

Class name and subject . . . . Date . . . .

Real Time	Place	Pupil Actions	Teacher Actions	Pupil Talk	Teacher Talk	Other Notes

After the lesson discuss with the teacher what you have noted to check for any misunderstandings and to discuss further the strategies you have observed.

THE STRUCTURE OF A LESSON AND TRANSITIONS

The structure of a lesson is very important to its effectiveness (see Unit 2.3). When an experienced teacher is delivering a lesson they work to a plan but equally they deviate from the plan when a new learning need becomes apparent. Good teaching is flexible and responsive. Experienced teachers use *transitions* in a lesson to summarise the learning at key points before moving on to the next learning activity. Figure 2.1.1 is a flow diagram of a lesson about communications where the transitions are highlighted in bold.

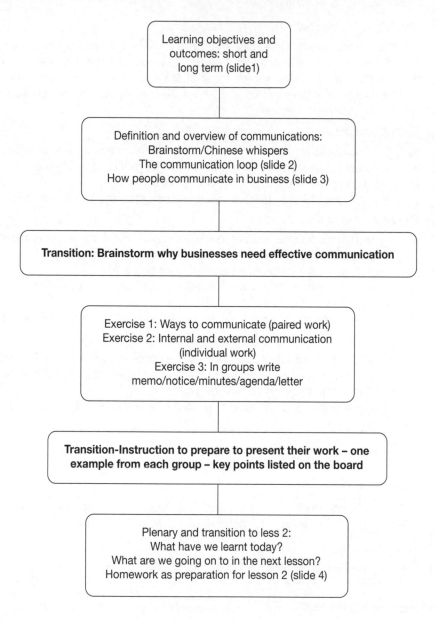

■ **Figure 2.1.1** Flow diagram of a Year 10 Business Studies lesson which is the first lesson of a double lesson on communication for business

TEACHER TALK AND ORAL FEEDBACK

It is useful to focus on teacher talk and feedback to pupils during a lesson to enable you to estimate its impact on learning and behaviour. There are various ways of doing this, including tape recordings and video recordings of a lesson. The use of a digital camera for detailed observation work is covered later in this unit. However for most occasions you will use simple non-technical procedures to capture this kind of data. Task 2.1.3 is designed to help you analyse the way teachers talk to pupils.

Task 2.1.3 **ANALYSIS OF TEACHER TALK**

Complete the following checklist for a section of the lesson or the entire lesson.

Oral Feedback	Examples observed in the lesson	Learning impact on pupils
Giving information		
Correcting errors or misapprehensions		
Praising		
Questioning to check understanding		
Questioning to deepen understanding		
Asking pupils to focus on specific aspects		
Summarising learning		
Encouraging pupil reflection		
Coaching in skills		
Answering pupil questions		
Correcting poor behaviour		
Guiding pupils back on task		
Outlining next learning tasks		

Discuss your observations with the teacher after the lesson and record both the checklist and discussion in your PDP.

PUPIL TALK AND INTERACTION

It is equally interesting and highly relevant to analyse pupil talk and interaction which, if you follow a class or group to different lessons, often changes in significant ways.

It is enlightening simply to use a class list and place a tick in red next to each pupil's name as they ask for information and a tick in black against each pupil's name as they answer a question or offer information. This gives you a clear indication of the degree to which pupils are engaging in pupil talk during a lesson.

A more detailed observation with a small group of pupils, one pupil or the entire class can be undertaken using the checklist in Task 2.1.4 over a twenty-minute period, recording minute by minute. A three-minute exemplar record is shown in the example.

Task 2.1.4 **ANALYSIS OF PUPIL TALK**

Complete the following checklist to analyse individual, group or whole-class pupil talk over a twenty-minute period.

Real time – minute intervals	Pupil(s) initiated questions of teacher	Pupil(s) answering teacher questions	Off task(s) discussion with peers	On task discussion with peers	Other notes about the class activity or other events
9.10	John B Mary Y				JB not paying attention Mary Y confused
9.11					
9.12		Mark S	John B		Good recall by Mark
9.13				Mark S Mary Y	

Discuss your observations with the teacher after the lesson and record both the checklist and discussion in your PDP.

FOCUS UPON PUPIL LEARNING

Often the instinct of student teachers in the classroom is to focus upon the teacher teaching rather than upon how the pupils are learning. This is not surprising. When we first 'go back to school' our memories are those of a pupil who watched and listened to the teacher. However, you need to shift your focus and observe what and how the pupils are learning. There are many ways this can be done; some of which are given in Task 2.1.5.

OBSERVING PUPIL MANAGEMENT AND ENCOURAGING PUPIL BEHAVIOURS WHICH MAXIMISE LEARNING

Before beginning lesson observations you should familiarise yourself with the school and department's behaviour policies. These are important documents that staff rely upon to achieve a common way of developing appropriate pupil behaviour. It is essential that you work within

Task 2.1.5 **ANALYSIS OF PUPIL LEARNING**

Before beginning your analysis write down the learning objectives and expected learning outcomes of the lesson (see Units 2.3 and 7.4) and identify the task/activities on which you are going to focus.

If you are a participant observer, you are able to make notes of key information as the lesson unfolds, look over the work of pupils and be actively engaged with coaching and guiding their learning activities. This enables you to begin to appreciate what strategies and resources are working effectively for their learning and why.

As a non participant observer, your task becomes more subtle. If you are free to move around once pupils are involved in an activity, you can oversee their task completion. If it is appropriate, you can ask them brief questions but you must not disengage them from their task. Equally, do not be tempted to do the work for them. Always lead them to think through the task with your help.

In highly active lessons there may be considerable chat and activity but there may equally be considerable learning taking place. Quiet lessons where the pupils seem attentive to the teacher are not necessarily lessons where learning is occurring.

The most flexible way to record pupil learning is to make bullet points as you perceive the evidence; a proforma is not appropriate because pupil learning is so complex. Once the lesson is over, ask the teacher if you may review the work the pupils have completed.

Discuss your observations with the teacher after the lesson and record both your analysis and discussion in your PDP.

these policies even if you sometimes notice that some teachers do not follow them. (See also Unit 3.3. Task 3.3.4 asks you to record observations of unacceptable behaviour by one pupil you are teaching.)

OBSERVING ASSESSMENT FOR LEARNING

A common weaknesses identified in school inspections is a lack of focus upon assessment for learning (see also Unit 6.1). Ensure that you know your school's assessment policies in relation to both teaching and learning as well as national assessments and those of the subject department. Task 2.1.6 (p. 74) asks you to focus on assessment for learning.

HOW DOES THE TEACHER USE LEARNING RESOURCES AND AIDS DURING THE LESSON?

It is interesting to map the resources and aids teachers use during lessons to help pupils with different learning styles to learn and to differentiate by matching the lesson activities to the differing abilities of pupils. Some pupils learn most easily through activities such as experiments or making things, others through visual media or through listening.

Task 2.1.7 (p. 74) provides an example of a proforma you might use to observe learning resources to help pupils learning.

Task 2.1.6 ASSESSMENT FOR LEARNING

Observe a class and during the lesson write detailed notes about any activities the teacher initiates to assess pupils' understanding of the work. The teacher may use many strategies but the most common include:

■ direct questions
■ discussion
■ asking pupils to present their work to their partner, the whole class or a group
■ reviewing work on a computer, whiteboard or in exercise books
■ through role play or display activities
■ setting another task to test understanding
■ posing a problem to solve to evaluate and deepen understanding.

Also note down what the teacher does to correct misconceptions and misunderstandings or to deal with total lack of understanding and to advance the learning. How does the teacher reassure and motivate? How does the teacher consolidate the learning? Discuss your observations with the teacher.

Task 2.1.7 LEARNING RESOURCES AND PUPIL LEARNING

Identify the teaching/learning resources and aids used in activities in a lesson you observe and analyse the pupil learning benefits.

The resource	Learning activity	Learning benefit
Text book		
Prepared study guide or worksheet		
Pictures, mind maps, graphics		
Video, CDs, DVDs		
Computer programs including Internet		
Tape recording		
Television programme or film		
Experiment		
Games, puzzles, models and activity cards		
Whiteboard		
Electronic whiteboard and digital camera		

Discuss the lesson with the teacher and selected pupils. File the data in your PDP.

SUBJECT CONTENT FOCUSED OBSERVATION

All teachers during the early stages of their career have to work hard to fill gaps in subject content knowledge and as schemes of work and the curriculum change, experienced teachers also need to learn new content. You may want to focus an observation on the content of a lesson on a topic in which there is a gap in your own subject content knowledge (see also Unit 1.1).

USING DIGITAL CAMERAS OR VIDEO CAMERAS IN LESSON OBSERVATION

Lesson observation and the subsequent analysis is a very challenging activity. In five minutes of a lesson so much can happen that is of significance and worthy of discussion. Recording part of a lesson for detailed analysis with the teacher is an ideal way of learning.

Digital cameras have made a significant difference to the ease with which lessons can be recorded. They are small, easy to use, with good sound quality and work well in normal classroom lighting. Observation, if it is not to disturb the natural flow of a lesson, needs to be subtle so that the pupils forget that observation is taking place. Many schools use regular video recording of lessons as part of their programme of CPD for all teachers and in these circumstances pupils learn to ignore the cameras. Software is available to annotate videos in real time so that key aspects of the lesson e.g. questions asked, explanations given, can be easily grouped for playback and analysis.

COLLABORATIVE TEACHING AS A FORM OF OBSERVATION

Collaborative teaching opportunities, where the planning and presentation of a lesson are shared with another student teacher or teacher, also provide opportunities to observe. You should seek every opportunity during your school experience to be involved in collaborative teaching because it is an active and powerful training experience where two-way observation and feedback can be very constructive.

FURTHER INQUIRIES AND OBSERVATIONS: SOME SUGGESTIONS

Many other aspects of teaching can be observed; some examples are given in Tasks 2.1.8 and 2.1.9. You should be able to adapt these for your own use.

Task 2.1.8 **TEACHERS' QUESTIONS**

Questions are often classified into 'closed' and 'open'; focusing on recall of fact or using prior knowledge to speculate about events or anticipate new ideas (see Unit 3.1 for further information about questioning). What types of question do teachers ask? Are they simple questions with one-word answers or are they more complex involving explanation? Investigate the frequency of different types of questions. The following questions may help to focus your observation. Does the teacher:

- ask mainly closed questions?
- ask both open and closed questions according to purpose and circumstance?
- accept only right answers?
- dismiss wrong answers?
- give enough time for pupils to give an answer?
- encourage pupils to frame a reply?

How does the teacher respond to right and wrong answers given by pupils?

Discuss your responses and your interpretation of them with the class teacher or your tutor.

Task 2.1.9 **WHERE IS THE TEACHER DURING A LESSON?**

The movement of teachers in the classroom may say a lot about their relationship with pupils, about how they keep an eye on activity and behaviour and about their interest in the pupils.

Draw an A4 map of the classroom in which you are observing. Mark on key points: teacher's desk, pupil desks, whiteboard, projector, etc. Have several copies of the map available. At regular intervals throughout the lesson, e.g. every minute or so, mark on your map where the teacher stands and where they have moved from, to build up a picture of position and movement. At the same time record the time and what is going on in the lesson. This enables you to relate teacher movement to lesson activity. Analyse your map and discuss the following:

■ Where is the teacher most often positioned during the lesson? What possible reasons are there for this: writing on the board; explaining with a projector; helping pupils with written work?

■ Does the teacher keep an eye on all events in the room and, if so, how?

■ Is it done by eye contact from the front or does the teacher move around the room?

■ How did the teacher know that pupils were on task for most of the lesson?

■ Were some pupils given more attention than others? What evidence do you have for this? What explanations are there for this?

■ Was teacher movement related to pupil behaviour in any way? Examine this idea and look for the evidence.

■ Did the nature of the subject matter dictate teacher movement? How might movements change in different subject lessons? Give an example.

■ Some teachers use their desk and board and equipment as a barrier between them and pupils; others move in among pupils and desks. Are there 'no-go areas' which the teacher does not visit? Are there similar spaces for teachers which the pupils do not visit?

Summarise your findings in your PDP and reflect on what information your 'map' gives you about 'teacher territory' and 'pupil territory'. Share your information with other student teachers.

SUMMARY AND KEY POINTS

Teachers and pupils set up a working relationship in which both parties know the rules, the codes of behaviour and their boundaries. In most classes boundaries are kept and teachers work smoothly with the class, apparently without great effort. Beneath that order there is a history of carefully nurtured practice by the teacher in establishing an appropriate atmosphere.

Sometimes these boundaries break and you may have seen ways in which the teacher restores a working atmosphere. Each teacher has their own way of dealing with this problem. Watching the ways other teachers deal with such problems helps you widen your own repertoire of skills. You should focus on how teachers handle behavioural problems for at least one of your classroom observation tasks.

The class you work with during school experience is someone else's; you are unlikely to break the relationships set up between them. In this respect your job of learning to teach is made more difficult; do you break the established pattern of behaviour or not? Remember that school experience is a learning exercise and not one in which you are expected to take on the whole class as if it is yours for the year.

The tasks in this unit were designed to help you organise your enquiries and get the most out of observation; more importantly, they should enable you to organise other inquiries of interest as you seek solutions to problems arising in your own teaching.

Observing can include looking, listening, recording, analysing and selecting; after evaluation some of what you have seen and heard may be incorporated into your own teaching. Unless observing moves on from you merely being in the classroom, letting the events wash over you, observation becomes boring and more importantly, of little value. So you need to focus your observations on events, strategies, circumstances; then do something with the results. Observing other people's lessons is also about feelings: your own because of the task ahead of you; the teachers' because they are under scrutiny; and the pupils' because they are wondering what you are there for and what you might be like next week. So too is the observer observed!

What you see and hear you have to interpret. You are likely to be familiar with classrooms. You have spent many years of your life, literally hundreds of hours, in classrooms as a pupil and student. You have a good idea of what you think makes a good classroom. However, you are biased because of that experience. You are not a neutral observer or one that brings a fresh eye to teaching. You need to unlearn most of what you know about the classrooms of your adolescence and of your undergraduate days before you can start to understand today's classrooms. Your experience then was not only that of a learner but also that of a successful learner; otherwise you would not be in the position you are in now. In addition you were probably a keen rather than reluctant learner. All these features of your background may have to be carefully examined and their usefulness evaluated for the task in hand, that of teaching today's pupils in today's classrooms.

Observation is in one sense a research exercise. It enables you to gather data on teacher performances and pupil learning. By analysing that data, you can begin to identify factors that contribute to effective teaching and learning and so place them in the framework of your own emerging skills.

Observation is affected by past experience and interpretation and is open to bias. Similarly, each pupil responds differently to you as a result of their experiences. Teaching is a personal activity (as is learning), leading to a personal style of teaching. Observation of events in classrooms helps you evaluate your teaching contributing to your emerging style. Your skills as a teacher develop through both observation and experience throughout your career. During your course your tutor helps you evaluate your development in the context of your course requirements.

You must prepare for observation by identifying both focus and purpose. You should share these with the class teacher. Evaluation is enhanced by discussion afterwards, bearing in mind that the teacher has been under scrutiny and your appreciation made clear. In the same spirit, you should welcome observation by experienced teachers. In these ways you experience education as a process where pupils and teachers transform knowledge through inquiry into improved understanding.

Check which requirements for your course have been addressed through this unit.

FURTHER READING

Altrichter, H., Feldman, A., Posch, P. and Somekh, B. (2008) *Teachers Investigate Their Work: An Introduction to Action Research across the Professions*, 2nd edn, London: Routledge.
This comprehensive text provides detailed advice about observations.

Education Broadcasting Services (EBS) (2004) *Looking at Learning: Tactics of Questioning,* (CD-ROM), London: Teacher Training Agency.
This is available in the libraries of all teacher education institutions in England or from the Training and Development Agency for Schools (TDA) and it demonstrates questioning approaches which support learning.

Marsden, E. (2009) 'Observing in the classroom', in S. Younie, S. Capel and M. Leask (eds) *Supporting Teaching and Learning in Schools: A Handbook for Higher Level Teaching Assistants,* London: Routledge; Wragg, E. C. (1999) *An Introduction to Classroom Observation,* London: Routledge.
Both these give more detail and more examples of how to observe in classrooms.

Additional resources for this unit are available on the companion website:
www.routledge.com/textbooks/9780415478724

UNIT 2.2

SCHEMES OF WORK AND LESSON PLANNING

Jon Davison and Marilyn Leask

INTRODUCTION

> Our lesson observations revealed that in classes run by effective teachers, pupils are clear about what they are doing and why they are doing it. They can see links with their earlier learning and have some ideas about how it could be developed further. The pupils want to know more.
>
> (Hay McBer, 2000: para.1.2.4)

If your time with the pupils is to be used effectively, you need to plan carefully for each lesson – taking account of how pupils learn, the requirements of the curriculum, the most appropriate methods of teaching the topic and the resources available as well as the evaluations of previous lessons.

Two levels of planning are particularly relevant to your work in the classroom – the *scheme of work*, which outlines a group of lessons around a particular topic, and the *lesson plan*, for each individual lesson. You will quickly gain experience of planning as you plan lessons and schemes of work on your school experience. Because planning is integrally linked to evaluation and development, evaluation of plans for a specific situation may point to the need to change or develop your plans so planned activities do not have to be followed through rigidly and at all costs.

OBJECTIVES

At the end of this unit you should be able to:

- explain what is meant by the terms: 'aims', 'learning objectives', 'learning outcomes', 'progression', 'differentiation'
- construct schemes of work (also known as programmes or units of work)
- construct effective lesson plans.

Check the requirements of your course to see which relate to this unit.

PLANNING WHAT TO TEACH AND HOW TO TEACH IT

The factors influencing *what* should be taught (content of lessons) are discussed in Unit 1.1, but how much you teach in each lesson and *how* you teach it (teaching strategies) are the teacher's own decisions.

Content of lessons

The knowledge, skills, understanding and attitudes appropriate for a young person entering the world of work in the twenty-first century are vastly different to those which were considered appropriate even fifteen years ago. Ideas about what teachers should teach change regularly and the curriculum is under constant scrutiny by those responsible for education.

As a student teacher, you are usually given clear guidelines about what to teach and the goals for pupils' learning within your subject. These goals are in part usually set out in government produced documents, e.g. the National Curriculum documents, school documents and syllabuses prepared by examination boards. You need to become familiar with the curriculum requirements and the terminology relevant to your subject. However, before you plan individual lessons you need an overall picture of what learning is planned for the pupils over a period of time. This overall plan is called a scheme of work and most departmental schemes of work cover between half a term's work and a couple of years' work.

Teaching strategies or methods

However constraining the guidelines on content are, the decision about which teaching strategies to use is usually yours (see Unit 5.3). As you become more experienced as a teacher, you acquire your own personal teaching style. But as people learn in different ways and different teaching strategies are suitable for different types of material, you should become familiar with a range of ways of structuring learning experiences in the classroom. For example, you might choose to use discussion, rote learning, discovery learning, role play and so on to achieve particular learning objectives. Unit 5.3 gives you detailed advice on teaching styles and strategies which help you achieve different learning objectives. Task 2.2.1 asks you to reflect on your preferred approaches to learning.

Task 2.2.1 HOW DO YOU LEARN?

Spend a few minutes making notes of the methods which you use to help you learn and the teaching strategies used by teachers from whom you felt you had learned a lot (see Unit 5.3). Then make notes about those situations from which you did not learn. Compare these notes with those of other student teachers. People learn in different ways and different areas of learning require different strategies. You need to take account of such differences in planning your lessons and to demonstrate that you can use a range of teaching methods in order to take account of such differences (see Unit 4.1 on differentiation).

SCHEMES OF WORK AND LESSON PLANS

There are two main stages to planning for pupil learning:

1   Preparing an outline of the work to be covered over a period – *the scheme of work*.
2   Planning each individual lesson – *the lesson plan*.

A number of formats for both schemes of work and lesson plans are in use. We suggest you read the advice given for the teaching of your subject in the subject specific texts in this *Learning to Teach* series. However, whilst the level of detail may vary between different approaches, the purpose is the same – to provide an outline of the work to be completed either over an extended period (scheme of work) or in the lesson (lesson plan) so that the planned learning objectives and learning outcomes can be achieved. Try different approaches to planning in order to find those most appropriate to your situation. The best plans are ones which support you in your teaching so that your pupils learn what you intend them to learn. The illustrations in this unit are intended to provide examples with which you can work and later modify.

The scheme of work

This might also be called the 'programme of work' or the 'unit of work'. Different terms may be used in your school or in your subject but the purpose is the same – to devise a long-term plan for the pupils' learning. So a scheme of work sets out the long-term plans for learning and thus covers an extended period of time – this could be a period of years, a term or half a term or weeks, e.g. for a module of work. A scheme of work should be designed to build on the learning which has gone before, i.e. it should ensure continuity of pupil learning.

Schemes of work should be designed to ensure that the knowledge, skills, capabilities, understanding and attitudes of the pupils are developed over a particular period in order to ensure progression in learning. The term 'progression' means the planned development of knowledge, skills, understanding or attitudes over time. In some departments, schemes of work are very detailed and include teaching materials and methods as well as safety issues.

Using a scheme of work

Usually on school experience you are given a scheme of work. In putting this together, the following questions have been considered:

1   What are you trying to achieve? (Aims for the scheme of work and learning objectives for particular lessons – see the definitions below.)
2   What has been taught before?
3   How much time is available to do this work?
4   What resources are available?
5   How is the work to be assessed?
6   How does this work fit in with work pupils are doing in other subjects?
7   What is to be taught later?

The scheme itself may be quite brief (Figure 2.2.1 shows a proforma used by student teachers on one course) but it will be based on the above information.

Each of these areas is now discussed in turn.

1 *What are you trying to achieve?* The aims of a scheme of work are general statements about the learning that should take place over a period.

Learning objectives are specific statements which set out what pupils are expected to learn from a particular lesson in a way that allows you to identify if learning has occurred. Learning objectives are prepared for each lesson and further detail is included under lesson planning later in this unit.

In devising each scheme of work a small aspect of the whole curriculum will have been taken and a route planned through this which provides the best opportunities for pupils to learn. Progression in pupil learning should be considered and built into schemes of work.

2 *What has been taught before?* This information should be available from school documentation and from staff. In the case of pupils in their first year of secondary education, there is usually a member of staff responsible for liaising with primary schools who may have this information.

Scheme of work for x topic

Area of work		Ref.

Class	No in class	Age	Key stage

No of lessons	Duration	Dates	

Aims (from the National Curriculum programmes of study)

(Objectives are listed in each lesson plan)

Framework of lessons	NC reference

Assessment strategies

Other notes (safety points)

■ **Figure 2.2.1** Scheme of work proforma

An editable version of Figure 2.2.1 is available on the companion website:
www.routledge.com/textbooks/9780415478724

3 *How much time is available to do this work?* The number and length of lessons devoted to a topic are decided by the department or school in which you are working. Don't forget that homework has a valuable role to play in enhancing learning and that not all the lessons you expect to have are available for teaching. Some time is taken up by such things as tests, revision, fire drill, special events, lateness.

4 *What resources are available?* Resources include material resources as well as human resources and what is available depends on the school where you are working. You need to find out the procedures for using resources in the school and what is available. You may find there are resources outside the school to draw upon – parents, governors and charities. Many firms provide schools with speakers on current topics. There may be field studies centres or sports facilities nearby. You need to check if there are any safety issues to consider when choosing appropriate resources.

5 *How is the work to be assessed?* Teaching, learning and assessment are interlinked. Most of the work you are doing with pupils is teacher assessed although some is externally assessed. A key purpose of teacher assessment is *formative*, assessment *for* learning – to check and guide pupils' progress, e.g. in relation to learning objectives. Teacher assessment may also be *summative* – undertaken at the end of an extended period to assess the level achieved. In any case, you should keep good records of the pupils' progress (homework, classwork, test results) in your own record book as well as providing these in the form required by the school or department. Unit 6.1 focuses on assessment issues.

Task 2.2.2. asks you to consider how you might record the outcomes of your assessment of pupils.

Task 2.2.2 **RECORD KEEPING AND ASSESSMENT**

Ask staff in your department how they expect pupil assessment records to be kept and what forms of assessment you should use for the work you are doing. Make notes and compare practice in your school with that in the schools your fellow student teachers are working in.

---

6 *How does this work fit in with work the pupils are doing in other subjects?* There are many areas of overlap where it is useful to discuss pupils' work with other departments. For instance, if pupils are having difficulty with measurement in technology, it is worth checking if and when the mathematics department teaches these skills and how they teach them. Cross-curricular dimensions to the curriculum (see Unit 7.2) will have been considered by the school and responsibilities for different aspects shared out among departments. Ask staff in your department what responsibilities the department has in this area.

7 *What is to be taught later?* Progression in pupil learning has to be planned and a scheme of work is drawn up for this purpose. From the scheme of work you know what work is to come and the contribution to pupil learning that each lesson is to make.

Task 2.2.3 asks you to draw up a scheme of work.

Task 2.2.3 **DRAWING UP A SCHEME OF WORK**

In consultation with your tutor, draw up a scheme of work to last about six to eight lessons. Focus on one particular class which you are teaching. Use the format provided for your course (or the one we provide in Figure 2.2.1) or one which fits in with the planning methods used in the department.

The lesson plan

The lesson plan provides an outline of one lesson within a scheme of work. In planning a lesson, you are working out the detail required to teach one aspect of the scheme of work. To plan the lesson you use a framework and an example of a lesson planning framework is given in Figure 2.2.2.

The following information is required for you to plan effectively.

1 *Overall aim(s) of the scheme of work and the specific learning objectives for this lesson* Defining learning objectives and associated learning outcomes which clarify exactly what learning you hope will take place is a crucial skill for the effective teacher (see Unit 7.4). These help you to be clear about exactly what pupils should be achieving and help the pupils understand what they should be doing. However, drawing up effective objectives and specifying and planning for outcomes require considerable thought.

At this stage in your career, if you ensure that your lesson objectives focus on what should be achieved from the lesson in terms of pupils' learning, then you have made a good start.

Listing learning objectives after the following phrase 'By the end of this lesson, pupils will be able to . . .' may help you to devise clear goals and to understand the difference between aims (general statements), learning objectives (statements about specific goals e.g. demonstrate an understanding of a mathematical formula) and learning outcomes (specific lesson outputs, e.g. the accurate completion of six exercises applying the formula).

Date .... Class ....
Area of work ....
Aim...
Learning objectives..
Learning outcomes ....

TIme	Teacher activity	Pupil activity	Notes / equipment needed
0–5 min	Class enter and settle	Coats and bags put away	
5–10 min	Homework discussed / recap of work so far / task set / new work explained		
10–25 min	Teacher supports groups / individuals	Pupils work in individual groups to carry out the task	
… and so on			
Ending	Teacher summarises points / sets homework		

Evaluation: Were objectives achieved? What went well? What needs to be addressed next time? How are individuals responding?

■ **Figure 2.2.2** Planning a lesson: one possible approach

An editable version of Figure 2.2.2 is available on the companion website:
www.routledge.com/textbooks/9780415478724

Words that help you be precise are those such as state, describe, list, identify, prioritise, solve, demonstrate an understanding of. These words force you to write statements which can be tested. If you think your learning objectives are vague, ask yourself whether the objectives can be measured and if the learning outcomes make it clear what the pupils must do to achieve the objectives. When you tell the pupils what learning outcomes are expected from the lesson do they understand what is expected of them? Objectives may be related to knowledge, concepts, skills, behaviours and attitudes (see Unit 7.4 for more advice about setting learning objectives and outcomes setting).

Task 2.2.4 challenges you to set learning objectives, specify learning outcomes and then analyse the learning which may result.

Task 2.2.4 **WRITING LEARNING OBJECTIVES AND LEARNING OUTCOMES**

Learning objectives refer to the observable outcomes of the lesson, i.e. to what pupils are expected to be able to do. Specifying the expected learning outcomes for the lesson will help you clarify your learning objectives. Discuss the writing of learning objectives with other student teachers and your tutor. Choose a particular lesson and, as a group, devise appropriate learning objectives which relate to changes in pupils' learning or behaviour. Pay particular attention to the quality and type of objectives you are setting – are they focused on the pupils' learning? Then identify the learning outcomes related to the learning objectives. How might learning be demonstrated?

2 *Range of abilities of the pupils* As you develop as a teacher, you are expected to incorporate *differentiation* into your planning. This refers to the need to consider pupils' individual abilities when work is planned so that both the brightest pupils and those with lesser ability are challenged and extended by the work. Differentiation can be achieved in different ways depending on the material to be taught. Differentiation may, for example, be achieved by **outcome**, i.e. different types or qualities of work may be produced, or by *task*, i.e. different tasks may be set for pupils of differing abilities, or by *teacher input* (Unit 4.1 provides further information). You provide continuity of learning for the pupils by taking account of and building on their existing knowledge, skills, capabilities and attitudes.

3 *Time available* On the examples of a lesson plan provided, a time line is drawn down the left-hand side. If you refer to this in the lesson, you are able to see easily if it is necessary to adapt the original plan to fit the time available.

4 *Resources available* Staff usually go out of their way to help pupils have the appropriate resources. But don't forget that others may be needing them so ask in good time for the resources you require. Check how resources are reserved in your department.

5 *Approaches to classroom management* These should be suitable to the topic and subject (see Units 4.1 and 5.2).

6 *Teaching strategies and the learning situations* These should be set up as appropriate to the work being covered (see Unit 5.3). Modeling, explaining and questioning are three key skills which you should work to improve. It is a good idea to write out questions in advance which you may want to use to test the pupil's grasp of the topic and which develop thinking. Phrasing appropriate questions is a key skill for a teacher (Unit 3.1 has further details).

7 *Assessment methods* Decide which ones to use in order to know whether your learning objectives have been achieved (see Unit 6.1).

8 *Any risks associated with the work* Safety is an important issue in schools. In some subjects, the assessment of risk to the pupils and incorporation of strategies to minimise this risk are a necessary part of the teacher's planning. Departmental and national guidelines are provided to ensure the safety of pupils and should be followed. As a student teacher you should consult your head of department, class teacher or tutor for guidance on safety issues. If you are in doubt about an activity and you cannot discuss your worries with the class teacher or your tutor, do not carry out the activity.

9 *What do the pupils know now?* As your experience of the curriculum and of pupils' learning develops, you will find it easier to answer this question. You need to consider what has been taught before as well as the experience outside school which pupils might have had. It may be appropriate to do some form of testing or analysis of knowledge, skills, attitudes and understanding or to have a discussion with pupils to discover their prior experience and attitudes to the work in question. As a student teacher you should seek advice from the staff who normally teach your classes.

Lessons have a structure and a rhythm to them. As you read this next section, think about the overall pattern to a lesson and the skills you use at each stage.

CONSTRUCTING A LESSON

Initially, you might find it difficult to see exactly how teachers manage their classes. In order to help you see the underlying structure of a lesson, we have divided the lesson and its planning into five key stages: preparation; beginning; moving on; ending, evaluation. Figure 2.2.3 (The structure of a lesson) illustrates this rhythm. Each stage is discussed below.

Task 2.2.5 asks you to summarise key points about lesson planning.

Task 2.2.5 **PLANNING AND GIVING LESSONS**

As you read about the five stages of a lesson, make notes in your professional development portfolio (PDP) to remind you of key points to pay attention to when you are planning and giving lessons. Unit 2.3 provides more details.

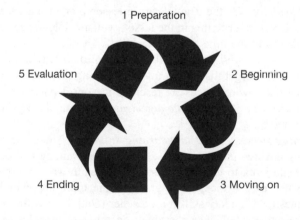

1 Preparation

5 Evaluation

2 Beginning

4 Ending

3 Moving on

■ **Figure 2.2.3** The structure of a lesson

Preparation

The most successful lessons are thoroughly planned and structured beforehand and you manage a class more effectively if you carefully consider in advance how to organise yourself and the pupils.

Make sure you have enough of the necessary materials, equipment and resources. Know the exact number of the items you are using so that you know if something has been lost and can take steps to find it immediately. Most departments have developed their own systems of stock control, e.g. a useful technique for textbooks is to number them and when you give them to pupils, record the textbook number in your mark book.

Ensure that you know how to operate any equipment you plan to use (e.g. computers or subject-specific equipment) and that it is in working order. If you are carrying out a science experiment, you should do it yourself before the lesson. This enables you to anticipate problems pupils might encounter.

Plan a variety of appropriate teaching and learning activities (see Unit 5.2). Remember, the concentration span of adults is about twenty minutes and that of most pupils is shorter. Plan extra, related activities in case your chosen activity does not work or pupils complete activities more quickly than you anticipate.

Give advance warning to pupils of any books, materials, etc., that they need for the lesson. If you have asked them to collect particular items or materials, do not rely on them remembering – bring enough yourself in order for the lesson to proceed just in case or have alternative plans.

Beginning

A good beginning is a crucial part of a successful lesson as it sets the tone, motivates pupils and establishes your authority. There are a number of key points to be kept in mind when you think about beginning your lessons.

Be in the classroom before the pupils arrive and ensure equipment is ready. Undoubtedly, the school you are in has established rules about pupil movement around the school and entry to classrooms. However, in the lower years in particular, it is common to line up pupils outside your teaching room and to usher them inside in an orderly manner.

Settle the class as quickly as possible and ensure that all pupils are facing you – even when they are seated in groups around tables – and are listening in silence before you begin the lesson. Do not begin the lesson when any pupil is talking, but wait calmly, confidently and expectantly for quiet. You will get it! Do not press on until you have established quiet. It is worth taking the time to do so. When observing other lessons, note the techniques teachers use for managing transitions from one activity to another in the lesson (see Unit 3.1).

Class management is much easier when you know the pupils by name. So make a determined effort to learn pupils' names as quickly as possible. It does not happen by osmosis, so you have to work at it. Seating plans are useful, as is the practice in the early stages of asking pupils to raise their hands when you register them. Although it might appear time consuming, giving out exercise books to pupils individually quickly allows you to put a face to a name (see Buzan (2003) for techniques for remembering names).

If you are unable to address pupils by name address them by their class/form designation. For example, 'Right 7G, I want everyone looking this way'. This is far better than 'Right girls/boys/ladies/lads etc.' Never resort to 'Oi you, blondie!' or some equally unprofessional outburst. Similarly, impersonations of deflating a balloon through continued 'Sshh-sshh-ing' do nothing to enhance your authority.

Pupils like to know what is expected of them. They relax and have a far more positive approach if you explain the learning objectives and what you plan to do in the lesson, with a brief rationale of how it fits in with previous and future work and if you let them know what you want them to achieve in the lesson. Some teachers display their lesson plan on the wall to help the pupils understand what is to be covered, why and how.

Establish a crisp, but not rushed, pace from the beginning. Never stand in one place in the room for more than a matter of a few minutes; some teachers suggest that as a student teacher, you do not sit at the teacher's desk during the lesson except in extremis. Use eye contact, vary the pace and tone of your voice (see Unit 3.1) and monitor pupil reaction continually.

Moving On

Smooth, seamless transitions between one part of the lesson and the next are vital if there is to be overall continuity and coherence. Having introduced the lesson, you need to explain the purpose of the first (and thereafter any subsequent) pupil task. Be very clear about what you want the pupils to do and tell them exactly how long they are to spend on the activity. They then have an idea of the pace at which they need to work, how much you expect them to 'produce', and what quality of work you require.

Before they begin the activity, check that all pupils understand exactly what they are expected to do. Deal with any queries before the class begins work. This saves endless repetition of the task to individuals.

Have a definite routine for distributing books and materials. Will you give out equipment? Will pupils come out to collect it row by row, table by table? Will one pupil per table/row collect it? In any event, it is essential that this activity is carried out in a controlled and orderly manner in any classroom. Moreover, if you are teaching a physical education or science subject, the safety aspect of this area of class management is of unequalled importance. (the subject specific books in this series provide specific advice).

When the pupils are engaged in the activity, move around the room monitoring pupil progress and dealing with questions; but do not interfere unnecessarily. Let them get on with the activity. Effective class management depends upon your active involvement. Key skills are: circulation; monitoring progress; the use of proximity to pupils; sensitivity to and awareness of pupil needs. Even when the whole class is engaged in a task, it is rarely appropriate to sit at the teacher's desk and 'switch off'.

Give one or two minutes' warning of the end of the activity. Be vigilant about keeping to the time limit you imposed at the beginning of the activity. Do not let things 'slide'. Be aware that not every pupil will finish the task set. Use your judgement in assessing that, while a few may not have finished within your deadline, most are ready for the next stage. If, however, it becomes clear that the whole class needs longer than you anticipated for an activity, be flexible enough to adjust your planning.

At the end of the activity, settle the class and expect all pupils to be sitting quietly, facing you before you proceed to the next stage of the lesson. Be sure to maintain your business-like manner and the crisp pace you established earlier.

Ending (sometimes called the 'plenary')

It is important that any learning experience is rounded off, that pupils experience a sense of completion. Similarly, pupils need some mental space between lessons. They need to 'come down' from one lesson in order to prepare themselves for the next. Remember, depending upon the timetable, pupils may need to negotiate the conceptual intricacies of between four and eight subjects in a day. Your lesson, therefore, needs to be completed in an organised manner.

Plan enough time at the end of the lesson to: sum up what has been achieved with reference to the planned learning objectives and outcomes; set homework where appropriate; give a brief idea of what the next lesson will comprise and (if necessary) explain what pupils need to bring to it.

As with the distribution of materials (see the section on beginning), have a definite, orderly routine for collection.

Before pupils leave, make sure the classroom is neat and tidy and remember that the pips or bell are signals for you, not the pupils. Dismiss the pupils by table or row and ensure that they leave the room in a quiet, controlled fashion. Enforcing a quiet orderly departure also adds to the pupils' experience of the standards you expect, i.e. that your classroom provides an orderly and calm learning environment. Take a well-earned ten-second breather before beginning the whole process again with the next class!

Evaluation and planning future lessons

As soon as you can after the lesson, evaluate its success. Were the learning outcomes achieved? What went well? What didn't go well? What evidence do you have which allows you to answer with some degree of certainty? (See Unit 5.4.) What should you change next time on the basis of this evaluation and how does this fit in with the scheme of work? If you develop the practice of reflecting on your work as a matter of course, then modifying future practice on the basis of this reflection becomes second nature. In this way, you use your experience systematically to build up your professional knowledge and to develop your professional judgement.

SUMMARY AND KEY POINTS

You should now be able to explain the following terms – aims, learning objectives, learning outcomes, progression, differentiation – and have considered how to construct schemes of work and lesson plans which are comprehensive and useful.

Check which requirements for your course have been addressed through this unit.

FURTHER READING

See the subject texts in this series as listed on p. ii.

Hay McBer (2000) *Research into Teacher Effectiveness,* **DfEE: London.**
This comprehensive report into effective teaching proposes a model of teacher effectiveness comprising teaching skills and professional characteristics. The early sections are particularly useful in relation to the preparation and planning of lessons.

Kyriacou, C. (2007) *Essential Teaching Skills,* **3rd edn, Cheltenham: Nelson Thornes.**
An excellent and readable overview of the key skills which underpin effective teaching.

Mager, R. (2005) *Preparing Instructional Objectives,* **3rd edn, Atlanta, GA: Center for Effective Performance.**
Making the desired learning objectives and learning outcomes clear to students helps ensure effective learning. This text provides useful information about objective setting.

Learning and Teaching Scotland (2007) *Teaching for Effective Learning,* **Dundee (http://www. ltscotland.org.uk/Images/TEFL%20complete_tcm4-435808.pdf).**
This is a very readable booklet on the principles of effective teaching and learning.

Warburton, N. (2003) *Thinking from A-Z*, 2nd edn, London, RoutledgeFalmer.
This slim text will help you refine your critical thinking abilities. It provides an easily accessible dictionary of terms describing different types of thinking. The ideas may help you in your teaching of pupils to reason.

RELEVANT WEBSITES

Learning and Teaching Scotland: http://www.ltscotland.org.uk/
Qualifications and Curriculum Authority: http://www.qca.org.uk
TeacherNet: http://www.teachernet.gov.uk
Teacher Training Resource Bank: http://www.ttrb.ac.uk
These sites provide a range of useful teaching materials and advice.

Additional resources for this unit are available on the companion website:
www.routledge.com/textbooks/9780415478724

UNIT
2.3

TAKING RESPONSIBILITY FOR WHOLE LESSONS

Marilyn Leask and Mike Watts

INTRODUCTION

This unit draws attention to issues which have particular relevance to you when you are just starting to take responsibility for whole lessons. It focuses on particular aspects of planning and teaching which initially cause many student teachers problems.

Recall the iceberg image of a teacher's work from Unit 1.1. The delivery of the lesson in the classroom represents the tip of the iceberg, while the bulk of the teacher's work for a lesson – routines, preparation, subject knowledge, professional knowledge and judgement, previous lesson evaluations – is hidden.

OBJECTIVES

At the end of this unit you should be able to:

- plan routines for good class management
- evaluate how your personal attributes contribute to your effectiveness
- undertake lesson preparation
- plan to avoid common problems
- work effectively with other adults in the room
- review what you are doing.

Check the requirements for your course to see which relate to this unit.

ROUTINES FOR CLASS MANAGEMENT

Routines for class and lesson management provide a structure so that learning can take place within a classroom where the rules are understood by all. In time, these routines become instinctive for you. Establishing rules decreases the likelihood of having to use lesson time to discipline pupils at a later stage.

But your routines are not established in a vacuum. The pupils you teach have been in schools for at least seven years – they expect the teacher to establish 'norms' for classroom work, talk and movement and most pupils are conditioned to accept such classroom routines. This doesn't mean

91

that they won't resist you when you insist on certain types of behaviour but it does mean that they have certain expectations that you will set the rules. Three types of *routines* in operation are:

■ for managing work and movement
■ for managing relationships and reinforcing expectations of attitudes and behaviour
■ for gaining attention – for both the pupils and the teacher.

Routines for managing work and movement

For your early lessons, one of your main goals is to get the pupils down to work fairly promptly by providing them with clear learning objectives, tasks and clear instructions about the tasks to be done and the learning outcomes you expect. Your concern is to establish yourself as an organised teacher who sets – and shares – clear outcomes for a lesson and provides work that allows pupils to achieve those outcomes. At this stage in your development as a teacher it makes sense to fit in with established routines. Task 2.3.1 complements Task 1.2.2 in asking you to draw up rules and routines for your classroom.

Task 2.3.1 **CLASSROOM NORMS**

Make a list of your expectations for the presentation of work, talk and movement for pupils in your classroom. Find out through observation and discussion what the expectations of the experienced teachers are, especially the teachers of the classes you are taking. Update and amend your list as you gain experience with what works for you. When you are teaching ensure pupils understand your expectations and that you apply these consistently.

Routines for managing relationships and reinforcing expectations

Adopting a *firm, fair, friendly* approach helps you develop good relationships with pupils. Pupils have certain expectations of you. They expect the teacher to be consistent and fair in applying rules. They expect those who do well to be rewarded and/or acknowledged, e.g. through praise, even just a quiet word at the end of the lesson. There are many occasions when you are not the only adult in the room and, while they are there to help and assist, they too are looking to you for clear expectations. Teaching assistants (TAs), Learning Support assistants (LSAs) and technicians, tutors, even the classroom teacher, may all be in the room while you are working.

Pupils who do not abide by your rules expect to be reprimanded. A quiet individual reprimand may be sufficient to establish your authority with many pupils. Confrontation in front of the whole class is generally to be avoided. Remember, the role of routines is to make your lessons run smoothly – everyone should know what your expectations are for classroom behaviour.

It takes time for the student teacher and for any teacher new to a school to find out about influences on classroom relationships which come from the community. Information about the range of group 'norms' of behaviour for teenagers in the local area and background information about other social relationships (e.g. which pupils are cousins, siblings or stepsiblings) may help you understand more easily your pupils and their expectations. For example, the school has policies about bullying, systems for reprimand, and ways of excluding and dealing with troublesome pupils. It is important that you read these – the pupils most likely know most of these even if you do not.

Experienced teachers can often sense that trouble is brewing between pupils and defuse the situation. They use their voice sparingly – drawing on a range of other controls, e.g. placing themselves near pupils who need more encouragement to stay on task and using non-verbal gestures to remind pupils to keep working. If needed, there are a number of sanctions which all teachers can use. Establishing your authority can be achieved by using one of a range of means:

■ the pupil is required to apologise (or face sanctions)
■ a verbal warning is given – a brief reprimand or keeping the pupil for a moment at the end of the lesson to indicate your displeasure
■ a couple of pupils tidy up after the others have left (teachers are wise to protect their professional reputations by not remaining alone in closed classrooms with individual pupils)
■ additional work is given
■ another adult in the room is asked for assistance, perhaps to work closely with a particular pupil.

In Task 2.3.2 you are asked to establish what you are going to do to reward behaviour.

Task 2.3.2 **REWARDS AND SANCTIONS**

Find out about the policies on rewards and sanctions at the school where you are teaching. Make notes of the key issues which affect your work. Check your understanding of the application of these policies with experienced staff. Write down the approach you are going to take with respect to rewards and sanctions and discuss this with fellow student teachers and/or your tutor.

Routines for gaining attention

Getting the attention of the whole class at points during the lesson is a skill which experienced teachers practise effortlessly. First, act as though you believe the pupils will obey you. One pattern teachers often practise is to call for attention ('Stop what you're doing and just look here for a minute'). They then follow this with a focus on an individual ('Paul, that means you too'), which acts as a reminder to all pupils that if they don't want to be the focus of the teacher's attention they need to stop what they are doing. The time to call the class to attention is not when they're all working well but when the work is flagging and they need to be spurred on or they've come to a difficult point – unless, for example, you wish to draw their attention to a point on safety.

One of the fundamental rules of the classroom is that pupils should not speak when the teacher is speaking. Spending a few minutes in a lesson waiting for silence until you speak saves a lot of time later as pupils know what you expect. Pupils may need reminding of your expectations and you probably need to reinforce the idea that this is one of your basic rules. You must be able to get the class's attention when you require it. When observing classes, the following questions may help you see some of the strategies used by teachers to establish this aspect of their authority:

■ What verbal cues does the teacher use to establish quiet? Key phrases such as 'Right then', 'Put your pens down now' establish that the teacher requires the class to listen. Some student teachers make the mistake of thinking the words 'quiet' or 'shush' repeated over and over will gain the required effect. Experienced teachers tend to use more subtle or strident methods, e.g. QUIET! said once with great emphasis. Units 3.1, 3.2 and 3.3 provide further advice.

■ What non-verbal cues does the teacher use to gain attention? Look at the way teachers use gestures using eyes, face, arms and hands to establish that they require the class to listen. They may stand still and just wait. Their pupils know that if they keep their teacher waiting they will be penalised.

Unit 3.1 contains more ideas on establishing authority.

There are also routines related to the way pupils gain the teacher's attention. The usual routine is that pupils put up their hands and don't call out. Again, we suggest you find out what the current practice is for the classes you are teaching. If you decide to change established practices then you have to put in considerable effort to establish the new rules.

YOUR PERSONAL ATTRIBUTES

Body language plays an important role in your communication with others and is an aspect of the way you present yourself which you should consider. Some personal attributes which may interfere with your teaching may reveal themselves only once you are teaching. For this reason, it is worth keeping this aspect of your interaction with pupils and staff under review.

Try to establish early in your teaching experience:

■ whether your voice can be heard at the back of a classroom
■ whether you are comfortable talking directly to individuals and groups
■ whether you have any particular habits which may interfere with the developing of a relationship with a class, e.g. do you rattle coins in your pocket as you speak; do you play with your hair; are you able to use facial expressions effectively to indicate enthusiasm; do you speak in a monotone; do you look at people when you speak to them; what you communicate through your smiling (some people inadvertently smile when they are angry)
■ what messages your positioning, posture and your movement in classrooms and corridors convey
■ what gestures you normally use when speaking.

For example, part of your role is to comment on individual pupils' work as the class is in progress. It is not always easy to be critical while at the same time supporting and motivating pupils. They will respect fair comment, eye contact, genuine advice on their work. They will be suspicious if you show discomfort or evasiveness. We suggest you ask for feedback occasionally on these aspects in order to check whether you are inadvertently presenting yourself in an unfavourable manner. If you are able to ask other adults in the room, they may be willing to help you in this. Observing other teachers can be very revealing: make a note of what seems to make them personally effective. Having some lessons recorded on video provides you with evidence of your performance on which you can build. Unit 3.1 on communicating with pupils provides more detail.

An indispensable quality to cultivate is patience, for example, when having to answer the same question several times. The expression 'to try your patience' is apt: pupils may wittingly or unwittingly test your patience. You may want to show your impatience, but always keep reserves of patience to call on.

LESSON PREPARATION

I spent days preparing my first lesson on my first teaching experience – geography with a group of 15-year-olds who weren't exactly enamoured with the subject. Educationally it was a disaster! I was so nervous that I rushed through my carefully prepared 40-minute lesson and at the end of 10 minutes I had nothing more to say. I panicked and told them to draw a map – any map – for the rest of the lesson.

This true story, from a (now) very successful teacher, highlights the nervousness which many student teachers experience when faced with their first lessons. Such nervousness is natural. You are assuming an unfamiliar role as a teacher but there are no set lines which you can learn to carry you through the scene. Over time, you reflect on and learn from such events and thus build your professional knowledge about teaching and learning (pedagogy) and your professional judgement about how to manage the work in the classroom so that the situation described above does not arise. In the meantime, you are learning from each situation you face.

When you are spending hours planning for your first lessons, you may wonder whether you've made a sound choice of career but in many ways, learning to teach is similar to learning to drive. In time, many aspects of driving become automatic; so it is with some features of teaching. In your first lessons, it is a good idea not to try to do anything too ambitious. Limited success is better than unlimited disaster! We suggest you work through Units 2.2, 5.2 and 5.3 as they provide ideas about the basic teaching skills and planning approaches that you need to employ in your first lessons. You should, of course, be building on your experience of group work, micro-teaching and on the observations you have made.

Following the steps outlined in Table 2.3.1 should ensure that you start your first lessons from a position of being well prepared.

Task 2.3.3 directs you to reflect on and develop your approach to lesson planning.

■ **Table 2.3.1** Planning lessons

---

1   Plan the lesson and ask for advice about your plan. Be clear about what you want to achieve. Have your plan to hand at all times.

2   Think about your presentation needs – are you using worksheets, pictures, cue-cards, video, text-books, PowerPoint, the interactive whiteboard, OHP transparencies? Some are easier and safer to use than others, and you may need back-up if something fails.

3   Check that you have adequate extension and alternative work. Anticipate that additional work may be needed. You may find that equipment you had planned to use stops working or the specialist in your subject is not available to supervise you.

4   Link tasks to earlier work and set *authentic* tasks: ones that are relevant to the learners lives and locality.

5   Know the class if possible through your observations and have a strategy for using and learning names. Try to learn names quickly: making notes beside the names in the register may help you remember. Drawing up a seating plan can help; pupils may always, or at least usually, sit in the same places. In any case, you can ask them to sit in the same seats until you know their names.

6   Think about how you are going to assess pupils' work during the lesson. You should give them comment and feedback on what they are doing.

7   Check which other adults will be in the room, and the need for any materials to help them with particular pupils. It might help, for example, to give them a copy of your lesson plan so that they are aware of what you are trying to do.

8   Keep track (in your head, at least) of what activities 'worked' with the class and why.

Task 2.3.3 **CHECKING YOUR LESSON PLAN**

Consider the plan for one of your forthcoming lessons. Are the pupils clear about the learning objectives for the lesson and the types of outcomes expected? Are the tasks 'authentic' tasks? Are you making your expectations of the pupils clear at each stage? Are the pupils actively engaged at each point or are they wasting time waiting for you to organise books or equipment? Are you expecting them to concentrate on your talking for too long; to take in too much new information without the chance to discuss it and assimilate it? Is there scope for pupils to feed back to you what they've learned this lesson, e.g. through question and answer? As a trial run, try to explain the main points of the lesson to another student teacher. The quality of explanation you are able to give affects the learning which takes place as does the nature of questions you ask. Ask a colleague observing your lesson to give you feedback afterwards on these points.

Learning names is an important skill. For some people it comes easily, for others – unfortunately – it takes time. Unit 2.2 provides some advice on learning names, for example, try to observe the class beforehand and note how the teacher manages potentially noisy or very quiet pupils. Quickly knowing a pupil's name can have a very positive effect: it is personal, affirming and cautionary, all at the same time.

AVOIDING COMMON PROBLEMS

By this point of your course, you know the routines you will use, your lesson is planned. You have also given some thought to where you will stand, when and how you will move around the room. You know to keep scanning the class and, how and when you talk to children, not to have your back to most of the class.

Judging the timing during a lesson is one of the most difficult problems initially and following a time line on your lesson plan can help you see at a glance how the lesson is progressing in relation to the time allowed. Always keep an eye on the clock and keep your lesson plan to hand. Judging the timing and pace of a lesson takes considerable time to develop.

Unavoidable incidents will occur to interrupt the flow of your carefully prepared lesson but other incidents can be anticipated or at least dealt with effectively if you are prepared. It is as well to anticipate problems so that you are not too distracted from the lesson you planned to deliver. We discuss below some of the more common incidents and possible solutions so that you are not taken by surprise.

One or more pupils won't settle to the work

When some pupils are being disruptive, it is essential to get the bulk of the class working, preferably on work which requires less input from you than normal. This allows you time to deal quietly and firmly with those resisting your authority and thus to establish your authority over them. Ignoring deliberately provocative remarks such as 'This is boring' can help you avoid confrontation. Try to motivate uninterested pupils by linking the work with their interests if possible. Letting them feel you are interested in them as people can promote positive relationships but you still should expect them to work. Ask your experienced colleagues for advice if particular pupils constantly cause you trouble. It is likely that they are also causing some other staff difficulties.

You are asked a question and you don't know the answer

This is bound to happen. You can admit you don't know – 'What an interesting point, I've not thought of it that way'; 'I just can't remember at the moment'. It is possible to celebrate the moment: 'Jane, that is a really good question. Where might we go for an answer to that?' Make arrangements for the answer to be found. The pupil can follow it up for homework, use the library to look for the answer or write to those who might know. You may also be able to find out from other teachers, student teachers or your subject association.

You are asked personal questions

At some point you will be asked 'Have you got a boyfriend/girlfriend?', 'Are you a student?' (ask the school if you are to be introduced as a member of staff or a student), 'Have you ever done this before?', 'How old are you?' or comments may be made about your car (or lack of one) or what you are wearing.

Don't allow yourself to be distracted from the work in the lesson. You can choose whether or not to answer personal questions but do set boundaries beyond which you won't go. Often a joke deflects the questioner – 'Mine is the Rolls parked around the corner'. Some questions will probe your tastes in, say, clothes, music, food, football teams. Don't be personal, and don't take things personally. If you decide to share personal information with the pupils then choose the time and place. This is rarely at the start of a lesson or in the middle of an activity, say, on flowering plants, but may be in the last few moments as they are leaving the classroom.

A pupil swears

As a student teacher, you cannot solve all the problems of the pupils and the school. Usually if a pupil is asked to repeat what they said, they omit the offensive word and feel sufficiently rebuked. You have indicated that swearing is unacceptable.

What you do need to do is to establish a line about what is acceptable and stick to it. Make it clear to your classes what your rule is and link it to school policy which should be 'no swearing'.

Swearing at teachers or abusing other pupils are serious offences and you must take action. There are different ways in which you might react – depending on the pupils, the context, the school. You may require an apology or you may wish to take the matter further. Take advice from experienced teachers. Act in haste and repent at leisure is good advice for a student teacher. Take a little time to decide on the response. Letting a pupil know an act was unacceptable and that you are thinking about how to respond can be more effective than an ill-considered response from you at the time. Consistency in your approach to discipline is an important facet of establishing your reputation. You want the pupils to know that if they do X, which is unacceptable, some form of action, Y, always follows. (This approach is an application of behaviourist learning theories (see Unit 3.3) – you are teaching the pupils to understand that a certain negative action on their part always gets a certain negative response from you, regardless of what they think about it. Thus it is perfectly clear to them how to avoid a negative response.) Work through Task 2.3.4 to develop your approach to dealing with incidents.

Task 2.3.4 **SWEARING**

Discuss the following two scenarios with the teachers with whom you're working. What is an appropriate response for you in each case?

1   You overhear a pupil use swear words in conversation with another pupil. The word is not used in an abusive way.
2   A child swears at another child or at you! What are the routine responses for dealing with these incidents in your school?

Pupils are not properly equipped to do the work – they lack PE kit, pens, books, maths equipment

You should aim to get most of the class working so that you can then direct your attention to those who require individual attention. Many departments have systems in place for dealing with pupils' lack of kit and equipment. In the early days of your teaching, it can be less disruptive to your lesson for you simply to supply the missing item (pencil/paper) so you can keep the flow of the lesson going. But make sure you retrieve what you have loaned and indicate firmly that you expect pupils to provide their own. Task 2.3.5 develops your approach to dealing with poorly equipped pupils.

Task 2.3.5 **PROCEDURES FOR DEALING WITH POORLY EQUIPPED PUPILS**

Find out whether there is a system in the department in which you are working for dealing with pupils who are not properly equipped. Plan how you can avoid this problem interfering with the smooth running of your lesson and review and revise your plans as necessary if incidents occur.

Equipment doesn't work

You must check equipment beforehand and, in any case, have a back-up planned if your lesson is dependent on equipment, like information and communications technology (ICT) software, working. It can be a disaster if the video, DVD or computer program you want to use won't load, it can take the heart out of the lesson. One back-up might be that you have revision quizzes to hand, another back-up may be to have a paper copy of the PowerPoint slides you were using, and you can work from those. See also the advice below on what to do where pupils finish earlier than you had planned.

You have too much material

Pupils have to get to their next lesson on time and to have their break on time. So you must let them go on time! Five minutes or so before the end of a lesson (more if they have to change or put equipment away), draw the lesson together, reminding them of what's been achieved against the learning objectives and outcomes, what's expected in the way of homework and perhaps what's coming next. They then pack away and are ready to go at the correct time.

The pupils finish the work earlier than you had anticipated

Inevitably there are occasions when you have time with a class which you didn't expect. Table 2.3.2 provides ideas of what to do if you have unplanned time with a class.

MORE GENERAL CONCERNS

There are a number of more general concerns which most student teachers feel at some point or another. We discuss these here so that by anticipating problems and posing solutions, you may be better prepared for dealing with them.

■ **Table 2.3.2** Back-up plans in case pupils finish work quickly

---

It is essential that you have additional work available in case pupils finish the work early or some expected equipment or resources are not available. Examples of what you might do are as follows:

* Have questions prepared relating to recent work in the area under study.
* Recall what assessment plans you have. Go round and look at pupils' work, give constructive comment and share this with the class if appropriate. Do a quick test of the issues covered in the lesson or a spelling test of new words.
* Use your lesson objectives to devise questions about the work.

Ask pupils to work in pairs or teams to devise questions to be put to the rest of the class or to other teams. Answers can be written in rough books with the pupils swapping books to mark them.

* Work for subsequent lessons can be introduced so that pupils can see the purpose in what they are doing now – remember repetition can be an aid to learning. Introducing concepts briefly in one lesson means they will be more familiar when you go over them in depth later. (This is an example of constructivist learning theory in action – new knowledge can be fitted into a broader understanding of the issues.)
* Pupils' existing knowledge on the next topics could be discussed through question and answer. (Learning is more certain where you, as the teacher, build on pupils' existing knowledge and experience.)
* You may take the opportunity to check the pupils' ability to apply a range of study skills. There are excellent books on this topic. Plan together with the pupils ways of learning the work you've been covering, e.g. developing a spider diagram for summarising the key points in a topic in history or geography, producing a mnemonic to aid the recall of key issues.
* Homework (either past or just set) can be discussed in more detail. You may allow the pupils to discuss this together.
* In practical subjects you may ask pupils to repeat a sequence they've been working on in PE, perhaps extending it to incorporate another skill; to observe each other performing the sequence and to comment on the performance; or to demonstrate what is coming up in the next lesson.
* In history you might ask pupils to consider the consequences of different outcomes for major historical events.

With experience, you acquire the skill of fitting work to the time available so the problem ceases to cause you anxiety.

Maintaining behaviour for learning

This is cited as an area of concern by student teachers more often than any other area and Units 3.2 and 3.3 provide more detailed guidance. See also the Behaviour 4 Learning website (http://www.behaviour4learning.ac.uk). Pupils are influenced by your confidence, the material, the demands of the work and your ability to enforce rules. The first lesson may be a honeymoon period, where the pupils are sizing you up or, on the other hand, they may test you out. If you insist (in a quiet firm way) that you are in charge of what happens in the classroom, then the vast majority of pupils give way – as long as you are seen as fair and reasonable. Whilst it is important not to see the class as 'them' against 'you', adopting a 'divide and rule' strategy can pay dividends: praise those who work well and reward them, e.g. with privileges or house points or merit slips, using the systems established within the school. Do not expect to win over all pupils immediately; some may take months; a few may never be won over. Discuss any difficulties you have with other teachers – they may have effective strategies for dealing with the pupils who are giving you concern.

Defusing situations

Inexperienced teachers tend to reprimand pupils much more frequently than experienced teachers, probably because they have less well developed subtle control mechanisms, e.g. body language. Table 2.3.3 provides further advice.

Task 2.3.6 directs you to collecting ideas from observing the practice of experienced teachers.

Task 2.3.6 **DEFUSING DIFFICULT SITUATIONS**

We suggest that when you are observing experienced teachers you look specifically at how they defuse situations so that reprimands are not required. Note these for future reference.

■ **Table 2.3.3** Techniques used by teachers to defuse situations

- Anticipating changes of mood and concentration and moving the lesson on and perhaps increasing the pace of the lesson, e.g. 'Right, let's see what you've understood already . . .'
- Scanning the class regularly, even when helping individuals or groups so that potential problems are prevented: 'That's enough, Julia' or 'Have you got a problem over there?' is usually sufficient to remind pupils to keep on task. Pupils are impressed if you can see what they are doing without them realising you are looking at them. Standing at a pupil's desk perhaps towards the back of the room allows you to monitor the work of pupils close to you as well as to scan the rest of the class without them seeing you.
- Using humour to keep pupils on task – a knowledge of adolescent culture and local activities is useful: 'You're too busying thinking about what you'll be doing on Friday night to concentrate.'
- Using a whole range of non-verbal cues: posture, facial expressions, gestures, positioning in the classroom to reinforce your authority. The children recognise these and you need to recognise them too (see Unit 3.1).

Notice that none of these techniques require the teacher to shout or to be angry in order to keep the pupils on task.

Retrieving situations

You may have a poor lesson with a class. This doesn't mean that all lessons with that class will be like that. What it does mean, however, is that you must analyse the situation and put into place strategies for ensuring that the next lesson is better. Experienced colleagues should be able to give you advice. Observing an experienced teacher, even if in another discipline, teaching a group that you have difficulty with can be eye opening and can provide you with ideas for the way forward. Discuss what you've seen with the teacher. Ask another adult in the room, or a colleague to watch you teach the class with whom you are having difficulty and ask for suggestions about how you can improve.

Personal vulnerability, lack of self-belief and confidence

In becoming a teacher, you are more vulnerable than when being educated for many other professions as you are exposed to a discerning audience (the class) early on. So much of your performance in the classroom depends on your own personal qualities and your ability to form good relationships with pupils from a wide range of backgrounds (see Unit 1.2). Your performance is analysed and commented on by those who observe your teaching. You are forced to face your own strengths and weaknesses as a result of this scrutiny. This can be stressful particularly when you may be given apparently conflicting advice from different observers. As you become more experienced and you develop more analytical skills for use in appraising your performance, you should build your self-belief and confidence.

Dealing with your feelings

Incidents occur which leave you feeling deflated, unsure or angry. Try to adopt a problem-solving reflective approach to your work so that you maintain some objectivity and can learn from any difficult experiences you have. One group of student teachers was asked, at the end of their year of initial teacher education, what advice they would give new student teachers. Above all else, they said, keep in touch with other student teachers so that you can discuss your concerns with others in the same situation. It is likely that your concerns are also the concerns of other student teachers.

The challenge to your own values

Most people mix with people who hold similar values and attitudes. As a teacher, you are dealing with children from different backgrounds and with different expectations about education and different values to your own. You need to consider how you can best provide equal opportunities in your classroom and what strategies you might use to motivate disaffected pupils. Chapter 4 provides further advice in these areas. See also the Multiverse, the initial teacher education website on diversity (http://www.multiverse.ac.uk).

Loss of books or equipment and breakages

Schools have different approaches to dealing with loss and breakage of equipment by pupils. Seek advice from those with whom you are working. Anticipating and thus avoiding problems makes your life easier. The simple strategy of managing your lesson so that there is sufficient time at the end to check that equipment and books are returned saves you time in the long run. See also Unit 8.3 on health and safety.

Having a ready answer

There are a number of routine situations which can throw you off balance in the lesson, e.g. 'Someone's taken my pen/book', 'Sir, she did it too', 'Miss, he started it', 'But Miss, you let her go to the toilet', 'Do you like . . .' (and here they name a pop group about which they all know but of which you've never heard). Discuss these situations with other student teachers and make notes for yourself about how you might deal with them. See how other teachers deal with these.

Time and stress management

These are important enough issues that Unit 1.3 is devoted to them. Here we want to raise three points:

■ giving the lesson is only one part of a teacher's job
■ preparing your first lessons takes you a long time
■ if you skimp on lesson preparation, then the stress level you experience in the lesson will be high as you will not feel in control.

■ **Table 2.3.4** Lesson preparations checklist

- Set clear, simple learning objectives for the lesson that are likely to be achieved.
- Plan the lesson carefully and have extension work ready.
- Be clear about the learning outcomes you want.
- Know pupils' names and obtain pupil lists.
- Check the room layout: are things where you want them? What about safety issues?
- Know the school, class and lesson routines.
- Be on time.
- Prepare board work beforehand if possible (check that it won't be rubbed off if there is a lesson before yours) or use a pre-prepared overhead projector transparency, or computer presentation.
- Act as though you are in charge although you probably won't feel that you are.
- Know the subject and/or make crib notes and put key points on the board, transparency or computer presentation if you're unsure.
- Plan the rhythm of the lesson to give a balance between teacher talk and pupil activity.
- Include a timeline in your lesson plan so that you can check during the lesson how the plan is working. Try not to talk too quickly.
- Be prepared to clamp down on misbehaviour. It is easier to reprimand one pupil who is misbehaving than to wait until they have goaded other pupils into following suit or retaliating.
- Visualise yourself being successful.
- Have a fallback plan for the lesson (see Table 2.3.2).

SUMMARY AND KEY POINTS

Your first encounters with the pupils are important in setting the tone for your relationships with them. It is worth carefully considering the image you wish to project in these early lessons and planning your work to help reinforce this image. If you want to create the image that you are a disorganised teacher who doesn't know what the lesson is about any more than the pupils do, then this is relatively easy to achieve. Your image is something you should create deliberately and not just allow to happen.

Most student teachers have to work on controlling their nerves and developing their self-confidence. Covering the points in Table 2.3.4 in your preparation should prevent some of the difficulties you would otherwise encounter.

From the observations you've done, you should have established how other teachers deal with minor infringements of school rules – remember there are a number of types of reprimand you can use before you give out detentions (see Table 2.3.3).

One of your major problems may be believing that you are indeed a teacher. This is a mental and emotional transition which you need to make. The pupils, parents and staff usually see you as a teacher, albeit a new one and expect you to behave as such.

Check which requirements for your course have been addressed through this unit.

FURTHER READING

The subject-specific texts in the *Learning to Teach* series provide you with further advice. Some teachers join chat rooms or discussion groups on the Internet to discuss issues related to their teaching. You should be aware that these discussions may be public and may be archived.

Buzan, T. (2003) *Use Your Memory*, **London: BBC Books.**
This is just one of Tony Buzan's books which are packed with ideas for improving memory. Why not draw them to the attention of your pupils?

Millar, R., Leach, J., Osborne, J. and Ratcliffe, M. (2006) *Improving Subject Teaching,* **London: RoutledgeFalmer.**
This text is part of an Improving Learning Series and whilst this one is specifically around science teaching, more texts in the series will be published. They report findings from a major national research initiative into teaching and learning – the ESRC Teaching and Learning Research (TLRP) programme (http://www.tlrp.org/).

Westwood, P (2007) *Commonsense Methods for Children with Special Educational Needs,* **5th edn, London: RoutledgeFalmer.**
Many of the approaches used with pupils with SEN work well with all pupils. This text outlines approaches to effective teaching and provides research to back up different strategies.

RELEVANT WEBSITE

Teacher Training Resource Bank: http://www.ttrb.ac.uk
This resource bank includes sections from all the major subject associations where they publish what they consider is best practice in teaching in the subject. These resources provide useful foundations for your work as a teacher.

Additional resources for this unit are available on the companion website:
www.routledge.com/textbooks/9780415478724

CLASSROOM INTERACTIONS AND MANAGING PUPILS

Effective classroom management is essential to effective learning. Classroom management refers to arrangements made by the teacher to establish and maintain an environment in which learning can occur, e.g. effective organisation and presentation of lessons so that pupils are actively engaged in learning. Classroom management skills and techniques are addressed throughout this book in a number of different chapters and units. This chapter includes three units about different aspects of classroom management related to interacting with pupils. Together they give an insight into the complex relationships which are developed between teachers and pupils, and emphasise the need for well-developed skills and techniques that you can adapt appropriately to the demands of the situation. They reinforce the fact that, although you must plan your lessons thoroughly, not everything you do in the classroom can be planned in advance, as you cannot predict how pupils will react in any situation on any given day.

One commonality of teachers from whom we have learned a lot is their ability to communicate effectively with pupils to enhance their knowledge, skills and understanding. Most of us tend to think we communicate well. However, communication is a complex process. Unit 3.1 is designed to help you communicate effectively in the classroom. The unit looks first at verbal communication including using your voice, the language you use and the importance of active listening. It then considers aspects of non-verbal communication, e.g. appearance, gesture, posture, facial expression and mannerisms, particularly in relation to how you present yourself as a teacher.

Some pupils are motivated to learn and maintain that motivation, others are inherently motivated to learn but various factors result in them losing motivation, others may not be inherently motivated to learn, but their motivation can be increased. A study of motivation therefore is crucial to give you some knowledge and insight into how you can create a motivational climate that helps to stimulate pupils to learn. Unit 3.2 looks at what motivation is, presents a number of theories of motivation and considers how these can inform your teaching and pupils' learning, looks at the motivational learning environment in your classes and how this influences pupil motivation and identifies some specific methods to motivate pupils extrinsically (e.g. the use of praise and punishment, feedback) in order to encourage the development of intrinsic motivation.

We recognise student teachers' concerns about managing behaviour and misbehaviour. Unit 3.3 looks at 'behaviour for learning', a positive approach to behaviour which focuses on positive relationships with pupils and a positive classroom climate in which all pupils can learn effectively. This approach is more consistent with an inclusive schooling approach. Thus, the focus of the unit is on preventing misbehaviour as far as possible rather than on a reactive approach which focuses on 'discipline' for misbehaviour.

Readings for Learning to Teach in the Secondary School: A Companion to M Level Study
brings together essential readings to support you in your critical engagement with key issues raised
in this textbook.

COMMUNICATING WITH PUPILS

Paula Zwozdiak-Myers and Susan Capel

INTRODUCTION

We can all think of teachers who really understand their subject but cannot communicate it with others, as well as teachers from whom we have learned a lot. These teachers may have very different personalities and teaching styles, but they all have in common the ability to communicate effectively with pupils.

Communication is a complex two-way process involving the mutual exchange of information and ideas that can be written, verbal and non-verbal. Clear and effective communication includes not only delivering but also receiving information, which involves listening, observation and sensitivity. The quality of communication between pupil and pupil in lessons is very important, as it can enhance or hinder learning. Pupils can learn from communicating with each other, e.g. through discussion or by talking about a task. Equally, such communication can be irrelevant to, and interfere with the progress of, the lesson, therefore detracting from pupils' learning. However, the focus of this unit is the quality of communication between a teacher and pupils, which is critical to effective learning.

Most of us tend to think we communicate well. However, when we study our communication skills systematically, most of us can find room for improvement. You cannot predict how pupils will react to an activity, a conversation or a question asked. Your response, both verbal and non-verbal, in any classroom situation influences the immediate and, possibly, long-term relationship with the class. In order to respond appropriately you need well-developed communication skills, combined with sensitivity to pupils and 'where they are at' in relation to their understanding (Dillon and Maguire, 2001) and judgement.

We first consider aspects of verbal communication, including using your voice (volume, projection, pitch, speed, tone, clarity and expressiveness), the language you use and the importance of active listening. We then consider aspects of non-verbal communication, e.g. appearance, gesture, posture, facial expression and mannerisms, particularly in relation to how you present yourself as a teacher. Further aspects of communication are addressed in Unit 5.2.

OBJECTIVES

At the end of this unit you should be able to:

■ appreciate the importance of effective verbal and non-verbal communication skills
■ vary your voice consciously to enhance your teaching and pupils' learning
■ appraise your use of language and use questioning more effectively as a teaching and learning tool
■ understand the relationship between verbal and non-verbal communication
■ be aware of and have control over your own self-presentation in order to present yourself effectively.

Check the requirements for your course to see which relate to this unit.

VERBAL COMMUNICATION

Gaining attention

You need to establish procedures for gaining pupils' attention at the beginning of a lesson and also when you want the class to listen again after they have started an activity. This latter skill is especially important if there is a safety risk in the activity. Before you start talking to a class, make sure that all pupils can see and hear you, that you have silence and that they are paying attention. Establish a means of getting silence, e.g. say 'quiet please', clap your hands, blow a whistle in physical education or bang on a drum in music, and use this technique with the class each time to ensure consistency of approach. Wait for quiet and do not speak until there is silence. Once you are talking, do not move around. This distracts pupils, who may pay more attention to the movement than to what you are saying.

Using your voice

A teacher's voice is a crucial element in classroom communication. It is like a musical instrument and if you play it well, then your pupils will be an appreciative and responsive audience. Some people have voices that naturally are easier to listen to than others. Certain qualities are fixed and give your voice its unique character. However, you can alter the volume, projection, pitch, speed, tone, clarity and expressiveness of your voice to use it more effectively and to lend impact to what you say. These verbal dynamics are important elements of paralanguage and your voice, if sensitively tuned, can become a powerful agent of expression and communication.

The most obvious way you can vary your voice is by altering the *volume*. It is useful to have the whole volume range available, from quiet to very loud, but it is rarely a good thing to be loud when it is not needed. Loud teachers have loud classes. If you shout too much, you may get into the habit of shouting all the time – sometimes people know somebody is a teacher because of their loud voice. Also, if you shout too much, you may lose your voice every September! Of course, you have to be heard, but this is done by projection more than by volume.

You *project* your voice by making sure it leaves your mouth confidently and precisely. This needs careful enunciation and breath control. If your voice is projected well, you are able to make a whisper audible at some distance. Equally, good projection brings considerable volume to your ordinary voice without resort to shouting or roaring.

Each group of words spoken has its own 'tune' that contributes to the meaning. A person may have a naturally high or low voice but everybody can vary the 'natural' *pitch* with no pain. Generally speaking, deep voices sound more serious and significant; high voices sound more exciting and lively. To add weight to what is being said the pitch should be dropped; to lighten the tone the pitch should be raised. A voice with a lower pitch can create a sense of importance as it comes across as more authoritative and confident than a high-pitched voice. It can also be raised more easily to command attention, whereas raising a naturally high-pitched voice may result in something similar to a squeak, which does not carry the same weight.

Speed variations give contrast to delivery. You can use pause to good effect. It shows confidence if you can hold a silence before making a point or answering a question. Having achieved silence, do not shout into it. Equally, have the patience to wait for a pupil to respond. Research (e.g. Muijs and Reynolds, 2005) suggests that a reasonable time for any such pause is three seconds or slightly longer although up to fifteen seconds might be required for open-ended, higher level questions. Speaking quickly can be a valuable skill on occasion, however, this needs concentration and careful enunciation.

To use your voice effectively these factors need to work together. For example, you do not communicate effectively if the pitch of your voice is right, but you are not enunciating clearly or the volume is wrong, e.g. you are shouting at a group or pupils at the back cannot hear what you are saying. It is also important to put feeling into what you say to engage pupils so that your voice does not sound dull and monotonous. Often, pupils respond to *how* you say something rather than *what* you say. If you are praising a pupil sound pleased and if you are disciplining a pupil sound firm. If you deliver all talk in the same way without varying the verbal dynamics do not be surprised if pupil response is undifferentiated. Now complete Task 3.1.1.

Task 3.1.1 **THE QUALITY OF YOUR VOICE**

Record your voice either reading from a book or a newspaper or in a natural monologue or conversation. Listen to the recording with a friend or another student teacher. If you have not heard yourself before, the experience may be a little shocking! Your voice may sound different from the way you hear it and a common response is to blame the recording equipment. This is probably not at fault. Remember that normally you hear your voice coming back from your mouth. Most of your audience hear it coming forward. As you become used to listening to yourself, try to pick out the good points of your voice. Is it clear? Is it expressive? Is the basic pitch pleasant? When you have built up your confidence, consider areas for improvement. Do you normally speak too fast? Is the tone monotonous?

Repeat the task, but this time trying to vary your voice. For example, try reading at your normal speed, then faster, then as quickly as you can. Remember to start each word precisely and to concentrate on what you are saying. Then try varying the pitch of your voice. You will be surprised at how easy it is. Ask another student teacher to listen to the tape with you, comment on any differences and provide helpful advice for improving. Try these out in your teaching.

Language of the teacher

Teaching involves communicating with pupils from a variety of backgrounds and with different needs. Pupils who are learning English as an additional language (EAL) for example, may have very limited experience in talking, listening, reading and writing in English and need particular support in developing their language and communication skills (Department for Education and Skills (DfES), 2002a). In England the teaching of English is considered to be 'the responsibility of all teachers' (School Curriculum and Assessment Authority (SCAA), 1996: 2), which implies that all teachers are teachers of language. The National Literacy Strategy (NLS) in England explicitly states that 'Good oral work enhances pupils' understanding of language and of the way in which language can be used to communicate' (Department for Education and Employment (DfEE), 1998: 3).

In order to develop pupils' language skills, a teacher's language must be accessible. There is no point in talking to pupils in language they do not understand. That does not mean subject-specific vocabulary cannot be introduced, rather that you gradually introduce your class to the language of the subject. To do this you must not assume that everybody knows the words or constructions that you do, including simple connecting phrases, e.g. 'in order to', 'so that', 'tends to', 'keep in proportion', etc. Start with a simple direct language that makes no assumptions.

It is easier for pupils to understand a new concept if you make comparisons or use examples, metaphors or references to which they can relate. Where appropriate, use a variety of words or explanations that ensures the meaning of what you intend to convey is understood by all pupils. As a teacher your language must be concise. When you are speaking, you stress or repeat important words or phrases. Placing an accent on certain syllables of the words you use gives rise to rhythmic patterns that affect the meaning of your message. These are important techniques in teaching. If they help learning, repetition, accentuation and elaboration are valuable, but filling silence with teacher talk is generally unproductive. You take longer to deliver the same information and pupils' time may not be used most effectively. However, it is generally accepted that pupils understand something and learn it better if they hear it a number of times and if it is explained in different ways. Therefore, as the Chinese proverb says, you should:

▪ tell them what you are going to tell them
▪ tell them
▪ then tell them again what you have told them.

Task 3.1.2 focuses on the language of your subject.

Task 3.1.2 THE LANGUAGE OF YOUR SUBJECT

Compile a list of specialist words and phrases used in your subject or in a particular topic that you may be teaching. How many of these might be in the normal vocabulary of an average pupil at your school? In your lesson planning how might you introduce and explain these words and phrases? How might you allow pupils opportunities to practise their use of the words in the lesson? Tape a lesson that you are teaching then replay the tape and consider your use of language, including words and phrases identified above. It can be particularly helpful to listen to this with a student teacher learning to teach another subject who does not have the same subject knowledge and language and who therefore may be nearer to pupils' experience of the subject. How might you improve your use of language in future lessons?

As well as conveying content, a teacher's language is also used to create individual relationships with pupils that make them more interested in learning. Using pupils' names, 'saying something positive to every pupil individually over a period of time and thanking pupils at the end of a good lesson' (DfES, 2004a: 18), showing interest in their lives outside the classroom, valuing their experience, are all important in building mutual respect and creating a positive atmosphere for classroom learning (see also Unit 3.2).

Teachers also use language to impose discipline. Often, negative terms are used for this. This is not inevitable and a positive approach may have more success. For example, can you suggest a constructive activity rather than condemning a destructive one? Could earlier praise or suggestion have made later criticism unnecessary? Hughes and Vass (2001) provide guidance on types of language that teachers can use to positively enhance pupils' motivation and learning (see also Units 3.2 and 3.3).

TYPES OF COMMUNICATION

There are many different ways in which verbal communication is used in teaching. Explaining, questioning and discussion, are considered briefly below.

Explaining

Teachers spend a lot of time explaining to pupils. In some teaching situations it can be the main form of activity in the lesson, thus being able to explain something effectively is an important skill to acquire. Pupils learn better if they are actively engaged in the learning process and a good explanation actively engages pupils and therefore is able to gain and maintain their attention. You must plan to involve pupils, e.g. mix an explanation with tasks, activities or questions, rather than relying on long lectures, dictating notes or working out something on the board. (See also Unit 5.2.)

Explaining provides information about what, why and how. It describes new terms or concepts or clarifies their meaning. Pupils expect teachers to explain things clearly and become frustrated when they cannot understand an explanation. A good explanation is clear and well structured. It takes account of pupils' previous knowledge and understanding, uses language that pupils can understand, relates new work to concepts, interests or work already familiar to the pupils. Use of analogy or metaphor can also help an explanation.

Table 3.1.1 identifies a range of features that characterise effective explanations. You might find this checklist and the sample questions useful when analysing and reviewing both your own and another student teacher's explanations.

Teachers often reinforce verbal explanations by providing pupils with a visual demonstration, or model. Modelling is an effective learning strategy that allows pupils to ask questions about and hear explanations related to each stage of the process as it happens as the teacher can, for example:

■ 'think aloud', making apparent and explicit those skills, decisions, processes and procedures that would otherwise be hidden or unclear
■ expose pupils to the possible pitfalls of the task in hand, showing how to avoid them
■ demonstrate to pupils that they can make alterations and corrections as part of the process
■ warn pupils about possible hazards involved in practical activities, how to avoid them or minimise the effects if they occur.

(DfES, 2004c: 3)

■ **Table 3.1.1** Characteristic features of explanations

Characteristic feature	Sample questions to ask yourself
Clear structure	Is the explanation structured in a logical way showing how each part links together?
Key features identified	What are the key points or essential elements that pupils should understand?
Dynamic opening	What is the 'tease' or 'hook' that is used at the start?
Clarity – using voice and body	Can the voice or body be used in any way to emphasise or embellish certain points?
Signposts	Are there clear linguistic signposts to help pupils follow the sequence and understand which are the key points?
Examples and non-examples	Are there sufficient examples and non-examples to aid pupils' understanding of a concept?
Models and analogies	What models might help pupils understand an abstract idea? Are there any analogies you could use? Will pupils understand the analogy? How might you help pupils identify the strengths and weaknesses of the analogy?
Props	What concrete and visual aids can be used to help pupils understand more?
Questions	Are there opportunities to check for pupils' understanding at various points, and to note and act on any misconceptions or misunderstandings? Are there opportunities for pupils to rehearse their understanding?
Connections to pupils' experience	Are there opportunities, particularly at the start, to check pupils' prior knowledge of the subject and to link to their everyday experiences?
Repetition	Are there a number of distinct moments in the explanation when the key points that should be learned are repeated and emphasised?
Humour	When and how might it be appropriate to use humour?

Source: DfES (2004b: 11)

Showing learners what to do while talking them through the activity and linking new learning to old through questions, resources, activities and language is sometimes referred to as scaffolding (see Unit 5.1). The idea is that 'learners are supported in carrying out a task by the use of language to guide their action. The next stage in scaffolding is for the learner to talk themselves through the task. Then that talk can, in turn, become an internalised guide to the action and thought of the learner' (Dillon and Maguire, 2001: 145–146). Combining verbal and visual explanations can be more effective than using verbal explanations exclusively, particularly with pupils who prefer a visual learning style (see Units 5.1, 5.3), are learning EAL or who have special educational needs.

Questioning

One technique in the scaffolding process for actively involving pupils in their learning is questioning. Teachers use a lot of questions; indeed 'every day teachers ask dozens, even hundreds of questions, thousands in a single year, over a million during a professional lifetime' (Wragg and Brown 2001: 1).

Asking questions effectively

Effective use of questioning is a valuable part of interactive teaching. However, if not handled effectively, pupils misunderstand and/or become confused. To be able to use questioning effectively in your lessons requires planning (see Unit 2.2 on lesson planning). To use questioning effectively you need to consider:

■ why you are asking the question(s)
■ what type of question(s) you are going to ask
■ when you are going to ask question(s)
■ how you are going to ask question(s)
■ of whom you are going to ask a question, how you expect the question answered, how you are going to respond if the pupil does not understand the question or gives an inappropriate answer, and how long you are going to wait for an answer.

However, you cannot plan your questioning rigidly; you must be flexible, adapting your plan during the lesson to take account of the development of the lesson.

Asking questions is not a simple process. Questions are asked for many reasons, e.g. to gain pupils' attention or check that they are paying attention, to check understanding of an instruction or explanation, to reinforce or revise a topic, to deepen understanding, to encourage thinking and problem solving, or to develop a discussion. Wragg and Brown (2001: 16–17) classified the content of questions related to learning a particular subject, rather than procedural issues, as one of three types: *empirical questions* requiring answers based on facts or on experimental findings; *conceptual questions* concerned with eliciting ideas, definitions and reasoning in the subject being studied; and *value questions* investigating relative worth and merit, moral and environmental issues. These broad categories often overlap and some questions may involve elements of all three types of questions.

Another classification that can be used to help you plan questions with specific purposes in mind is Bloom's (1956) 'taxonomy of educational objectives' through which questions can be arranged into six levels of complexity and abstraction. Lower-level questions usually demand factual, descriptive answers whereas higher-level questions are more complex and require more sophisticated thinking from pupils. Research indicates that pupils' cognitive abilities and levels of achievement can be increased when they are challenged and have regular access to higher-order thinking (Black and Harrison, 2001; Muijs and Reynolds, 2005; Wragg and Brown, 2001). Table 3.1.2 links the hierarchical levels in Bloom's taxonomy with what pupils might be expected to do and the types of question that would help them to realise those tasks. Exemplars of possible question stems are provided for each cognitive objective that you could draw upon when planning questions to ask pupils in your lessons.

Black and Wiliam (2002) and others have studied the use of Bloom's taxonomy in questioning and you might like to refer to this literature.

There are a number of other ways in which questions can be categorised.

Closed and open questions

The most common reason for asking questions is to check that pupils have learned what they are supposed to have learned or that they have memorised certain facts or pieces of information. These are questions like: what is the capital of Brazil? What is Archimedes' Principle? How many kilometres are there in a mile? What does the Latin expression 'Veni, vidi, vici' mean, who first used it and when? How do you spell 'emphatic'? These are called *closed* questions. There is only one correct answer; pupils recall information. The pupil either knows the answer or not, no real thought is required. Closed questions might be given to the whole class, with answers coming instantaneously. A short closed question–answer session might reinforce learning, refresh pupils' memories or provide a link to new work.

■ **Table 3.1.2** Linking Bloom's (1956) taxonomy to what pupils need to do, thinking processes and possible question stems

Cognitive objective	What pupils need to do	Use of questioning to develop higher order thinking skills	Links to thinking	Possible question stems
Knowledge	Define Recall Describe Label Identify Match	To help pupils link aspects of existing knowledge or relevant information to the task ahead.	Pupils are more likely to retain information if it is needed for a specific task and linked to other relevant information. Do your questions in this area allow pupils to link aspects of knowledge necessary for the task?	Describe what you see . . . What is the name for . . .? What is the best one . . .? Where in the book would you find . . .? What are the types of graph . . .? What are we looking for? Where is this set?
Comprehension	Explain Translate Illustrate Summarise Extend	To help pupils to process their existing knowledge.	Comprehension questions require pupils to process the knowledge they already have in order to answer the question. They demand a higher level of thinking and information processing than do knowledge questions.	How do you think . . .? Why do you think . . .? What might this mean . . .? Explain what a spreadsheet does . . . What are the key features . . .? Explain your model . . . What is shown about . . .? What happens when . . .? What word represents . . .?
Application	Apply to a new context Demonstrate Predict Employ Solve Use	To help pupils use their knowledge to solve a new problem or apply it to a new situation.	Questions in this area require pupils to use their existing knowledge and understanding to solve a new problem or to make sense of a new context. They demand more complex thinking. Pupils are more likely to be able to apply knowledge to a new context if it is not too far removed from the context with which they are familiar.	What shape of graph are you expecting? What do you think will happen? Why? Where else might this be useful? How can you use a spreadsheet to . . .? Can you apply what you now know to solve . . .? What does this suggest to you? How does the writer do this? What would the next line of my modelled answer be?

Analysis	Analyse Infer Relate Support Break down Differentiate Explore	To help pupils use the process of inquiry to break down what they know and reassemble it.	Analysis questions require pupils to break down what they know and reassemble it to help them solve a problem. These questions are linked to more abstract, conceptual thought which is central to the process of enquiry.	Separate . . .(e.g. fact from opinion) What is the function of . . .? What assumptions are being made . . .? What is the evidence . . .? State the point of view . . . Make a distinction . . . What is this really saying? What does this symbolise? So, what is the poet saying to us?
Synthesis	Design Create Compose Reorganise Combine	To help pupils combine and select from available knowledge in order to respond to unfamiliar situations.	Synthesis questions demand that pupils combine and select from available knowledge to respond to unfamiliar situations or solve new problems. There is likely to be a great diversity of responses.	Propose an alternative . . . What conclusion can you draw . . .? How else would you . . .? State a rule . . . How do the writers differ in their response to . . .? What happens at the beginning of the poem and how does it change?
Evaluation	Assess Evaluate Appraise Defend Justify	To help pupils compare and contrast knowledge gained from different perspectives as they construct and reflect upon their own viewpoints.	Evaluation questions expect pupils to use their knowledge to form judgements and defend the positions they take up. They demand complex thinking and reasoning.	Which is more important/moral/logical . . .? What inconsistencies are there in . . .? What errors are there . . .? Why is . . . valid . . .? How can you defend . . .? Why is the order important? Why does it change?

Source: DfES (2004d: 13–14)

By way of contrast, *open* questions have several possible answers and it may be impossible to know if an answer is 'correct'. These questions are often used to encourage divergent thinking and to develop understanding. Examples of open questions are: how could we reduce our carbon footprint? Should Western governments intervene in Middle Eastern politics? What words could you use to describe a family reunion?

These questions are much more complex than closed questions. They are designed to extend pupils' understanding of a topic. To answer them the respondent has to think and manipulate information by reasoning or applying information and using knowledge, logic and imagination. Open questions cannot usually be answered quickly. Pupils probably need time to gather information, sift evidence, advance hypotheses, discuss ideas and plan answers.

An example from a science lesson shows the difference in purpose between closed and open questions. 'What is the chemical formula of carbon dioxide?' is a closed question requiring factual knowledge but 'How does carbon combine with oxygen during the respiration process?' requires understanding. Further, a question such as 'Do you think that reducing our carbon footprint can slow down global warming' requires a deeper level of reflection and research by pupils.

You can ask closed or open questions or a combination of the two as *a series of questions*. The questions in the series can start with a few relatively easy closed questions and then move on to more complex open questions. A series of questions takes time to build up if they are to be an integral part of the learning process. They must therefore be planned as an integral part of the lesson not as a time filler at the end of a lesson where their effect is lost. Questions at the end of the lesson are much more likely to be closed-recall questions to help pupils remember what they have been taught in the lesson. There are implications for assessment of closed and open questions (see also Unit 6.1).

There are other aspects of questioning that are important to consider. Questions can be asked to the whole class; to groups; or to specific named individuals. Questions can be spoken, written on a whiteboard, or given out on printed sheets. Answers can be given at once or produced after deliberation, either spoken or written. For example, you may set a series of questions for homework and either collect the answers in to mark or go through them verbally with the class at the start of the next lesson.

Effective questioning is a skill you must develop as a teacher. It requires you to be able to ask clear, appropriate questions, use pauses to allow pupils to think about an answer before responding and use prompting to help pupils who are having problems in answering a question. Some key tactics identified by Wragg and Brown (2001: 28) for asking questions include: structuring; pitching and putting clearly; directing and distributing; pausing and pacing; prompting and pacing; listening and responding to replies; sequencing. Muijs and Reynolds (2005) identify three types of prompts to help pupils answer questions: *verbal prompts* (cues, reminders, tips, references to previous lessons or giving part of a sentence for pupils to complete); *gestural prompts* (pointing to an object or modelling a behaviour); and *physical prompts* (guiding pupils through motor skills).

Follow-up questions can be used to probe further, encourage pupils to develop their answers, extend their thinking, change the direction of the questioning and distributing questions to involve the whole class. Non-verbal aspects of communication such as eye contact, gesture, body language, tone of voice, humour, smiles and frowns are important in effective questioning because they go with the words that are used.

Wragg and Brown (2001: 28) identified common pitfalls or 'errors' in questioning by student teachers as:

- asking too many questions at once
- asking a question and answering it yourself
- asking questions only of the brightest or most likeable pupils
- asking a difficult question too early in the sequence of events
- asking irrelevant questions
- always asking the same types of questions (e.g. closed ones)
- asking questions in a threatening way
- not indicating a change in the type of question
- not using probing questions
- not giving pupils the time to think
- not correcting wrong answers
- ignoring pupils' answers
- failing to see the implications of pupils' answers
- failing to build on answers.

Errors of presentation, e.g. not looking at pupils when asking a question, talking too fast, at the wrong volume or not being clear, were identified as the most common errors. One reason for this may be the ease of detection of these errors (Wragg, 1984). The second most common type of error was the way student teachers handled responses to questions, e.g. they only accepted answer(s) to open-ended questions that they wanted or expected. Open questions are likely to prompt a range of responses, which may be valid but not correspond to the answer expected. You need to respond appropriately to these. You must avoid the guessing game type of question-and-answer session where the teacher has a fixed answer in mind and is not open to possible alternative answers. Pupils then spend their time guessing what the teacher wants.

Other errors identified in this study were pupils not knowing why particular questions were being asked, pupils not being given enough background information to enable them to answer questions, teachers asking questions in a disjointed fashion rather than a logical sequence, jumping from one question to another without linking them together and focusing on a small group of pupils and ignoring the rest of the class. Student teachers tended to focus on those pupils sitting in a V-shaped wedge in the middle of the room.

Some aspects of questioning were not identified as common errors, e.g. whether the vocabulary is appropriate for the pupils' level of understanding or whether the questions are too long, complex or ambiguous. Wragg (1984) suggested that one reason for this might be that they are difficult to detect and correct. It is as important to think about and develop these aspects of questioning as it is those that are most obvious.

The use of questioning in a lesson should be considered in relation to the use of other teaching techniques rather than in isolation. For example, you can encourage pupils to participate actively in questioning by listening and responding appropriately to answers, praising good answers, being supportive and respecting answers and not making pupils feel they will be ridiculed if they answer a question incorrectly (see also Units 3.2, 5.2, 5.5).

Discussion

Questioning may lead naturally into discussion in order to explore a topic further. Although pupils generally have more control over the material included in, and direction of, a discussion than in many teaching situations the teacher is still in charge. As with all other aspects of your teaching, discussion should be planned. Seating arrangements are important to develop a less structured atmosphere for a discussion, which can encourage as many pupils as possible to contribute to the discussion. You also need to plan how you are going to stimulate the discussion and how you are going to respond if a discussion drifts off its main theme. By interjecting suggestions or key questions you can keep a discussion on the topic.

> For a fruitful discussion which allows pupils some significant say over what is discussed, whilst at the same time covering ground that teacher's judge to be important, it is best to think of questions that may be perplexing, intriguing or even puzzling to pupils. Skilfully chosen encouraging, broad questions are often effective in sparking off animated conversations. The process may begin with recall questions to extend and activate knowledge and then thought questions to lift the discussion.
>
> (Wragg and Brown, 2001: 44)

To maximise pupils' learning through discussion you need to be able to chair a discussion effectively. Before you use discussion in your classes, it is wise to observe another teacher use this technique in their teaching. See also Unit 4.5 and the Appendix to Unit 4.5, 'Handling Discussion with Classes'.

Listening

For effective communication, *being able to listen effectively and take account of the response* is as important as being able to send the message effectively. Learn to recognise and be sensitive to whether or not a message has been received properly by a pupil, e.g. you get a bewildered look or an inappropriate answer to a question. Be able to react appropriately, e.g. repeat the same question or rephrase it. However, also reflect on why the communication was not effective, e.g. was the pupil not listening to you? If so, why? For example, had the pupil 'switched off' in a boring lesson or was the question worded inappropriately? Do not assume that pupils have your grasp of meaning and vocabulary (see 'Language of the teacher' above). It is all too easy to blame a pupil for not listening properly, but it may be that you had a large part to play in the breakdown of the communication. It is also important that you listen effectively to what pupils are communicating to you.

Wragg and Brown (2001: 34) identify four types of listening:

Skim listening – little more than awareness that a pupil is talking (often when the answer seems irrelevant); *Survey listening* – trying to build a wider mental map of what the pupil is talking about; *Search listening* – actively searching for specific information in an answer; *Study listening* – a blend of survey and search listening to identify the underlying meaning and uncertainties of the words the pupil is using.

It is too easy to ask a question and then 'switch off' while an answer is being given, to think about the next question or next part of the lesson. This lack of interest conveys itself to the pupil. It is distracting to the pupil to know that the teacher is not listening and not responding to what is being said. Also, you may convey boredom or indifference, which has a negative impact on the tone of the lesson. Effective listening is an active process, with a range of non-verbal and verbal responses that convey the message to the pupil speaking that you are listening to what is being said. Effective listening is associated with conveying enthusiasm and generating interest, by providing reinforcement and constructive feedback to pupils. These include looking alert, looking at the pupil who is talking to you, smiling, nodding and making verbal signals to show you have received and understood the message or to encourage the pupil to continue, e.g. 'yes', 'I see what you mean', 'go on', 'Oh dear', 'mmmm', 'uh-huh'.

NON-VERBAL COMMUNICATION

Much teacher–pupil communication is non-verbal (e.g. your appearance, gestures, posture, facial expression and mannerisms). Non-verbal communication supports or detracts from verbal communication, depending on whether or not verbal and non-verbal signals match each other; for example, if you are praising someone and smiling and looking pleased or if you are telling them off and looking stern and sounding firm, you are sending a consistent message and are perceived as sincere. On the other hand, if you are smiling when telling someone off or are looking bored when praising someone, you are sending conflicting messages that cause confusion and misunderstanding. Robertson (1996: 94) expresses this well: 'When non-verbal behaviour is not reinforcing meaning . . . it communicates instead the speaker's lack of involvement. Rather than being the message about the message, it becomes the message about the messenger.'

However, non-verbal communication can also have a considerable impact without any verbal communication, e.g. looking at a pupil slightly longer than you would normally communicates your awareness that they are talking or misbehaving. This may be enough to make the pupil stop. You can indicate your enthusiasm for a topic by the way you use gestures. You can probably think of a teacher who stands at the front of the class leaning against the board with arms crossed waiting for silence, the teacher marching down between the desks to tell someone off or the teacher who sits and listens attentively to the problems of a particular pupil. The meaning of the communication is clear and there is no need to say anything. Thus, non-verbal communication is important for good communication, classroom management and control.

Effective communication therefore relies not only on appropriate content, but also on the way it is presented. Studies by Mehrabin (1972) indicate that 93 per cent of the meaning behind verbal messages is received through non-verbal channels: notably, 55 per cent through gesture; 38 per cent through tone of voice; and 7 per cent from the words actually used.

PRESENTING YOURSELF EFFECTIVELY

There might seem to be some contradiction in discussing ways of presenting yourself as it could indicate that there is a correct way to present yourself as a teacher. However, the heading clearly refers to you as an individual, with your own unique set of characteristics. Herein lies one of the keys to effective teacher self-presentation: while there are some common constituents and expectations, it is also the case that every teacher is an individual and brings something of their own unique personality to the job.

Initial impressions are important and the way you present yourself to a class on first meeting can influence their learning over a period of time. Having prepared the lesson properly, the pupils' impressions of the lesson, and also of you, are important. An important part of the impression created is your appearance. Research by Sage (2000) indicates that pupils 'value adults' in school who take pride in their appearance and are well dressed. Pupils expect all teachers to wear clothes that are clean, neat and tidy and certain teachers to wear certain types of clothes, e.g. it is acceptable for a physical education teacher to wear a tracksuit but not a history teacher. Thus, first impressions have as much to do with non-verbal as with verbal communication, although both are important considerations.

How teachers follow up the first impression is equally important, e.g. whether you treat pupils as individuals, how you communicate with pupils, whether you have any mannerisms such as constantly flicking a piece of hair out of your eyes or saying 'er' or 'OK' – which reduce or prevent effective communication (pupils tend to focus on any mannerism rather than on what is being said and they may even count the number of times you do it!). It is generally agreed that effective teaching depends on and is enhanced by self-presentation that is *enthusiastic*, *confident* and *caring*. Why are these attributes important? How can you work towards making these attributes part of your self-presentation as a teacher?

Enthusiasm

One of the tasks of a teacher is to enable pupils to learn to do or to understand something. Before many pupils will make an effort to get to grips with something new, the teacher needs to 'sell' it to them as something interesting and worthwhile. However, your enthusiasm should be sustained throughout a lesson, and in relation to each activity – not only when you are presenting material but also when you are commenting on a pupil's work, particularly perhaps when a pupil has persevered or achieved a goal.

Your enthusiasm for your subject is infectious. However, there could be a danger of 'going over the top' when showing enthusiasm. If you are over-excited it can give a sense of triviality, so the enthusiasm has to be measured.

There are perhaps three principal ways in which you can communicate enthusiasm both verbally and non-verbally. The first is via *facial expression*:

An enthusiastic speaker will be producing a stream of facial expressions which convey his excitement, disbelief, surprise or amusement about his message. Some expressions are extremely brief, lasting about one fifth of a second and may highlight a particular word, whereas others last much longer, perhaps accompanying the verbal expression of an idea. The overall effect is to provide a running commentary for the listener on how the speaker feels about the ideas expressed. In contrast, a speaker who is not involved in his subject shows little variation in facial expression. The impression conveyed is that the ideas are brought out automatically and are failing even to capture the attention of the speaker.

(Robertson, 1996: 86)

The second way is via the *use of your voice*. The manner in which you speak as a teacher gives a clear indication of how you feel about the topic under debate and is readily picked up by pupils. Your voice needs to be varied and to indicate your feelings about what you are teaching. As you are engaged in something akin to a 'selling job' your voice has to show this in its production and delivery – it has to be persuasive and occasionally show a measure of excitement. A monotone voice is hardly likely to convey enthusiasm. 'Enthusiastic teachers are alive in the room; they show surprise, suspense, joy, and other feelings in their voices and they make material interesting by relating it to their experiences and showing that they themselves are interested in it' (Good and Brophy, 2000: 385).

A third way to convey enthusiasm is via your *poise and movement*. An enthusiastic speaker has an alert posture and accompanies speech with appropriately expressive hand and arm gestures – sometimes to emphasise a point, at other times to reinforce something that is being described through indicating relevant shape or direction, for example an arrangement of apparatus or a tactical move in hockey. If you are enthusiastic you are committed and involved, and all aspects of your posture and movement should display this.

Think back to teachers you have worked with and identify some whose enthusiasm for their subject really influenced your learning. How did these teachers convey their enthusiasm? How do you convey your enthusiasm?

Confidence

It is very important that as a teacher you present yourself with confidence. This is easier said than done because confidence relates both to a sense of knowledgeable mastery of the subject matter and to a sense of assurance of being in control.

There is an irony in pupils' response to teacher confidence. Expression of authority is part of the role pupils expect of a teacher, and where exercised with confidence, pupils feel at ease and reassured. In fact, pupils prefer the security of a confident teacher. However, if they sense at any time that a teacher is unsure or apprehensive, it is in young people's nature to attempt to undermine authority (for further information see Robertson, 1996).

Of course, in many cases it is experience that brings confidence but sadly pupils seldom allow that to influence their behaviour. Although the key to confident self-presentation is to be well planned, both in respect of material and in organisation, without the benefit of experience, all your excellent plans may not work and you may have no alternative 'up your sleeve'. Whatever happens you need to cultivate a confident exterior, even if it is something of an act and you are feeling far from assured inside.

Confidence can be conveyed verbally in clear, purposeful instructions and explanations that are not disrupted by hesitation. Instructions given in a direct and business-like manner, such as 'John, please collect the scissors and put them in the red box', convey a sense of confidence. On the other hand, the same instruction put in the form of a question, such as 'John will you collect the scissors and put them in the red box?' can convey a sense of your being less assured, not being confident that, in fact, John *will* co-operate. There is also the possibility of the pupil saying 'No'! Your voice needs to be used in a firm, measured manner. A slower, lower, well-articulated delivery is more authoritative and displays more confidence than a fast, high-pitched method of speaking. Use of voice is particularly important in giving key instructions, especially where safety factors are involved and in taking action to curtail inappropriate pupil behaviour. This is perhaps the time to be less enthusiastic and animated and more serious and resolute in your manner.

Non-verbally, confidence is expressed via, for example, posture, movement and eye contact, both in their own right and as an appropriate accompaniment to verbal language. There is nothing agitated about the movement of confident people. They tend to stand still and to use their arm gestures to a limited extent to reinforce the message being conveyed.

Eye contact is a crucial aspect of conveying confidence to pupils. A nervous person avoids eye contact, somehow being afraid to know what others are thinking, not wanting to develop a relationship that might ultimately reveal their inability or weakness. Clearly it is your role as a teacher to be alert at all times to pupil reaction and be striving to develop a relationship with all pupils that encourage them to seek your help and advice. Steady, committed eye contact is usually helpful for both of these objectives. You must also recognise that the use of eye contact is regarded differently by people of different cultures, e.g. some members of some cultures avoid use of eye contact. You should therefore take into account cultural sensitivities. This also applies to other aspects of non-verbal communication, such as gesture, touch and spatial proximity to another person. For further information about cultural differences take advice from your tutor, a staff member of that culture, staff at the local multicultural centre or the Commission for Racial Equality (additional guidance on cultural diversity can be accessed from http://www.multiverse.ac.uk).

Caring

It is not surprising perhaps that pupils feel that a caring approach is important in developing an effective relationship with teachers. Wentzel (1997) described caring teachers as those who demonstrate a commitment to their teaching, recognise each pupil's academic strengths and needs and have a democratic style of interaction. Wragg (1984: 82) reported that many more pupils preferred teachers who were 'understanding, friendly and firm' than teachers who were 'efficient, orderly and firm' or 'friendly, sympathetic and understanding'. It is interesting to note that firmness is also a preferred characteristic.

Notwithstanding pupils' preferences, interest in pupils as individuals and in their progress is surely the reason most teachers are in teaching. Your commitment to pupils' well-being and learning should be evident in all aspects of your manner and self-presentation. While this attitude goes without saying, it is not as straightforward as it sounds as it demands sensitivity and flexibility. In a sense it is you as the teacher who has to modify your behaviour in response to the pupils, rather than it always being the pupil who has to fall into line with everything asked for by the teacher. There is a potential conflict, and balance to be struck, between firm confidence and flexible empathy. It is one of the challenges of teaching to find this balance and to be able to respond suitably at the appropriate time.

A caring approach is evident in a range of features of teaching, from efficient preparation through to sensitive interpersonal skills such as listening. Those teachers who put pupils' interests above everything have taken the time and trouble to prepare work thoroughly in a form appropriate to the class. Similarly, the classroom environment shows thoughtful design and organisation. In the teaching situation, caring teachers are fully engaged in the task at hand, observing, supporting, praising, alert to the class climate and able to respond with an appropriate modification in the programme if necessary. Above all, however, caring teachers know pupils by name, remember their work, problems and progress from previous lessons and are prepared to take time to listen to them and talk about personal things as well as work. In other words, caring teachers show a real sensitivity to pupils' individual needs. They communicate clearly that each pupil's learning and success are valued.

Now complete Tasks 3.1.3 and 3.1.4.

Task 3.1.3 **COMMUNICATING EFFECTIVELY**

Select in turn each aspect of verbal and non-verbal communication identified above; your use of voice, language, explaining/questioning/discussion, listening, presenting yourself effectively, enthusiasm, confidence, caring. Prepare an observation sheet for your tutors or another student teacher to use when observing a class. Use this as a basis for evaluation and discussion about how you can further develop this aspect of your teaching.

Task 3.1.4 **STUDYING ONE ASPECT OF COMMUNICATION IN DEPTH**

Select one specific aspect of communication to study in greater depth. Review the literature on that aspect of communication. Design and conduct a piece of action research on that specific aspect of communication with one of your classes. Report critically on your research, identifying issues related to the assumptions underlying the methods of data collection and analysis. Critically analyse the outcomes of the study and reflect on your learning about improving your communication.

SUMMARY AND KEY POINTS

Good communication is essential for developing good relationships with pupils, a positive classroom climate and effective teaching and learning. This unit has aimed to help you identify both the strengths and weaknesses in your verbal and non-verbal communication and in your self-presentation, to provide the basis for improving your ability to communicate. Your developing professional knowledge and judgement should enable you to communicate sensitively and to best advantage.

Check which requirements for your course have been addressed through this unit.

FURTHER READING

Good, T. and Brophy, J. (2000) *Looking in Classrooms*, 8th edn, New York: Addison-Wesley.
Chapter 9 discusses research relating teacher behaviour to pupil achievement and considers its implications for the role of the teacher in actively presenting information to pupils, such as the effectiveness of demonstrations and questioning techniques.

Muijs, D. and Reynolds, D. (2005) *Effective Teaching: Evidence and Practice*, 2nd edn, London: Paul Chapman (Sage).
Chapter 2 considers the important relationship between interactive teaching and pupils' learning. Elements of effective questioning techniques are identified and then reviewed in relation to class discussion.

Wragg, E. C. and Brown, G. (2001) *Questioning in the Secondary School*, London: RoutledgeFalmer.
This book combines relevant research with practical resources that enable you to reflect upon your use of questions; develop approaches to preparing, using and evaluating your own questions; and explore ways in which pupils may be encouraged to question and to provide answers.

Additional resources for this unit are available on the companion website:
www.routledge.com/textbooks/9780415478724

ACKNOWLEDGEMENT

The authors would like to acknowledge the significant input of Roger Strangwick and Margaret Whitehead to the first three editions of this unit.

UNIT 3.2

MOTIVATING PUPILS

Susan Capel and Misia Gervis

INTRODUCTION

Pupils' attitudes to school and motivation to learn are a result of a number of factors, including school ethos, class climate, past experiences, future expectations, peer group, teachers, gender, family background, culture, economic status and class. However, the link between motivation and educational performance and achievement is complex.

Some pupils have a more positive attitude to school and to learning, e.g. it is valued at home or they see a link between education and a job. For example, if pupils see a relationship between success at school and economic success, they are more likely to work hard, behave in the classroom and be more successful. Many pupils want to learn but depend on teachers to get them interested in a subject. Even though some pupils may not be inherently motivated to learn, the school ethos, teachers' attitudes, behaviour, personal enthusiasm, teaching style and strategies in the classroom can increase their motivation to learn (see also, for example, Unit 5.3). On the other hand, pupils who do not feel valued at school are, in turn, unlikely to value school. Therefore, although some pupils may be inherently motivated to learn, they may become demotivated or have low motivation because of a learning environment that does not meet the needs of their learning style or does not stimulate them, or a task being too difficult or a negative impact of factors such as those identified above. Pupils for whom the motivational climate is not right are more likely to become disinterested and misbehave. If the teacher does not manage the class and their behaviour effectively, the learning of all pupils in the class can be negatively affected.

Thus, a central aim for you as a teacher is to create a motivational climate that helps to stimulate pupils to learn. There are a range of techniques you can use to increase pupils' motivation to learn, e.g.:

- showing your enthusiasm for a topic, subject or teaching
- treating each pupil as an individual
- providing quick feedback by marking work promptly
- rewarding appropriate behaviour.

In order to use such techniques effectively you need to understand why each technique is used. A study of motivation therefore is crucial to give you some knowledge and insight into ways of motivating pupils to learn. There is a wealth of material available on motivation. This unit tries to draw out some of the material we feel is of most benefit to you as a student teacher. However, the further reading list at the end of this unit, plus other reading in your library, will help you to develop your ability to motivate pupils further.

> ## OBJECTIVES
>
> At the end of this unit you should be able to:
>
> ■ understand the role and importance of motivation for effective teaching and learning and classroom management;
> ■ appreciate some of the key elements of motivation for effective teaching
> ■ understand how to motivate pupils effectively.
>
> Check the requirements of your course to see which relate to this unit.

WHAT IS MOTIVATION?

Motivation 'consists of internal processes and external incentives which spur us on to satisfy some need' (Child, 2007: 226). There are three key elements to motivational behaviour, which it is helpful for you to understand as they can help you to interpret the behaviour of your pupils. These three elements are:

■ direction (what activities people start)
■ persistence (what activities people continue)
■ intensity (what effort people put in).

These three key elements determine the activities that people start (direction) and continue (persistence) and the amount of effort they put into those activities at any particular time (intensity).

Motivation can be intrinsic (motivation from within the person, i.e. engaging in an activity for its own sake for pleasure and/or satisfaction inherent in the activity, e.g. a sense of achievement at having completed a difficult piece of work) or extrinsic (motivation from outside, i.e. engaging in an activity for external reasons, e.g. to receive a reward, such as praise from a teacher for good work, or to avoid punishment). Research has found that a person intrinsically motivated in an activity or task is more likely to persist and continue with that activity than a person extrinsically motivated. This can be illustrated by some (intrinsically motivated) pupils succeeding at school/in a subject despite the quality of the teaching, whereas other (extrinsically motivated) pupils succeed because of good teaching. Therefore, intrinsic motivation is to be encouraged in learning. A teacher's job would certainly be easier if all pupils were motivated intrinsically. However, pupils are asked to do many activities at school which are new to them, which are difficult, at which they may not be immediately successful or which they may perceive to be of little or no relevance to them. In order to become intrinsically motivated, pupils need encouragement along the way, e.g. written or verbal praise for effort, making progress or success, feedback on how they are doing or an explanation of the relevance of the work. For example, the quality of the feedback that pupils receive can directly impact on their self-confidence. Teachers can deliberately plan extrinsic motivators (see below) into their lessons with a view to enhancing both self-confidence and intrinsic motivation (there is a cyclical relationship between self-confidence and intrinsic motivation, such that high self-confidence increases intrinsic motivation whereas low self-confidence decreases intrinsic motivation). Task 3.2.1 asks you to reflect on what motivates you as a learner.

> ### Task 3.2.1 **WHAT MOTIVATES YOU AS A LEARNER?**
>
> Reflect on your own experiences at school. What was it that motivated you in the subject that you are now learning to teach? Identify another subject in which you were less motivated and reflect on why this was the case. Discuss your reflections with another student teacher. Also identify anything you can learn from these experiences to build into your own teaching.

THEORIES OF MOTIVATION

There are a number of theories of motivation. In addition, we adopt our own, often unconscious, theories. Examples of theories of motivation, along with some of their implications for you as a teacher in determining learning activities, are given in Table 3.2.1 and in the text below.

What motivates people?

It is often difficult for a teacher to identify the exact reason for a particular pupil's behaviour at a particular time, and therefore what is motivating them. Likewise, it is often difficult for a pupil to identify exactly what is motivating them. As a teacher you can often only infer whether or not pupils are motivated by observing their behaviour. Although there may be other reasons for a pupil not listening to what you are saying, talking, looking bored or staring out of the window, one reason may be that the motivational climate is not right and therefore the pupil is not motivated to learn.

Some of the factors which have been found to be motivating include: positive teacher–pupil relationships; supportive peer relationships; a sense of belonging; pupils' beliefs about their abilities; pupils' beliefs about the control they have over their own learning; pupils' interest in the subject; and the degree to which the subject or specific tasks are valued. Such factors have been categorised as:

- achievement (e.g. completing a piece of work which has taken a lot of effort)
- pleasure (e.g. getting a good mark or praise from a teacher for a piece of work)
- preventing or stopping less pleasant activities/punishment (e.g. avoiding getting a detention)
- satisfaction (e.g. feeling that you are making progress)
- success (e.g. doing well in a test).

The need for individuals to achieve can be encouraged by creating a learning environment in which 'the need for achievement in academic studies is raised' (Child, 2007: 254). Each individual sets themselves a standard of achievement, according to their level of aspiration. It is therefore important to raise pupils' levels of aspiration. Pupils who are challenged are more likely to improve their performance than those who are not challenged. Thus, setting tasks which are challenging but achievable for each individual pupil, i.e. individualised tasks, can be used to raise aspirations. Tasks on which pupils expect to achieve approximately 50 per cent of the time are the most motivating. However, this means that pupils are likely to fail on the task approximately 50 per cent of the time, so it is important to plan to reduce loss of motivation when pupils fail (e.g. by praising effort, giving feedback on performance, etc.).

Intrinsic motivation in pupils is related to interest in the activity and to effort (hard work), which leads to deep learning (Entwistle, 1990; see also deep and surface learning in Units 5.2 and 5.3). Deep learning means that learners try to understand what they are doing, resulting in greater understanding of the subject matter. This is a prerequisite for high-quality learning outcomes, i.e. achievement. According to Achievement Goal Theory, Covington (2000) reported deep learning as being associated with task (or mastery) goals, whereas surface (superficial or rote) learning is associated with outcome (or performance) goals. In line with other studies, Lam *et al.* (2004) found that pupils with an outcome orientation were more likely to focus on better performance than on mastering a task.

Pupils' perceptions of the motivational context, i.e. the goal orientation of the classroom and school, has been found to be important for their motivation and adjustment to school. Controlling environments (teachers attempt to guide pupils' thinking by providing specific guidelines for their academic and personal behaviours in class) have been found to have a negative effect on perceived competence and participation, which results in decreased intrinsic and self-motivation. On the other hand, pupils are motivated by teachers who know, support, challenge and encourage them to act independently from each other and from the teacher. An autonomy–support environment is one in which the teacher gives increasing responsibility to pupils, e.g. for choices/options about what they want to do; encourages pupils' decision-making by spending less time talking, more time listening, making less directive comments, asking more questions, and not giving pupils solutions; allows pupils to work in their own way; and offers more praise and verbal approval in class. Such an environment supports pupils' academic and social growth by increasing intrinsic and self-motivation to succeed at school, self-confidence, perceived competence and self-esteem. Research by Manouchehri (2004) has found a relationship between the motivational style adopted by teachers of mathematics and their commitment to implementing new teaching methods. Those adopting an autonomy–support style of motivation increase pupils' participation and engagement by, for example, creating more opportunities for pupils to examine and develop their understanding of mathematical ideas, to listen to the arguments, and to ask questions of other pupils.

Figure 3.2.1 illustrates the link between teachers' actions in creating the motivational climate and pupils' responses which influences their intrinsic motivation.

Further, pupils' goal orientations may be influenced by the motivational climate created by what teachers do and say. Research in physical education has found a relationship between pupils' perceptions of a lesson as being task orientated and adaptive motivational responses, including increased intrinsic motivation. In contrast, a relationship has been found between pupils' perceptions of a lesson as being outcome orientated and maladaptive motivational responses. Such a lesson focuses on individual achievement and competition between individuals. This may foster extrinsic motivation, discourage hard work and effort to achieve success by pupils who fail to achieve the outcome.

Although it changes with age, it is generally accepted that pupils are more likely to try harder if they can see a link between the amount of effort they make and success in the activity. Indeed, there is a link between achievement goal theory and attribution theory in that pupils with a high need to achieve attribute their success to internal causes (e.g. aptitude and effort), whilst they attribute failure to lack of effort. On the other hand, pupils with a low need to achieve attribute their failure to external factors, e.g. bad luck, or to lack of ability. Therefore, as a teacher, you should design activities which encourage pupils to attribute success or failure to effort. However, this is not always easy. Postlethwaite (1993) identified the difficulty of determining how much effort a pupil has made on a piece of work (especially that done at home) and hence the problems of marking the work. You can no doubt think of occasions where one person has made a lot of effort on a piece of homework, but missed the point and received a low mark, whereas another person has rushed through the homework and managed to achieve a good mark. In 'norm-referenced' marking a certain percentage of the class are given a designated category of mark, no matter how good each individual piece of work. Thus, each pupil's mark for a piece of work is given solely for their performance

■ **Table 3.2.1** Important theoretical perspectives and their implications for you as a teacher

Theory	Source	Main points	Implications for teachers
Theory *x* and theory *y*	McGregor, 1960	*Theory x* managers assume that the average worker is lazy, lacks ambition, is resistant to change, self-centred and not very bright. *Theory y* managers assume that the average worker is motivated, wants to take responsibility, has potential for development and works for the organisation. Any lack of ambition or resistance to change comes from experience.	Your treatment of pupils may be related to whether inherently you believe in theory *x* or theory *y*. A *theory x* teacher motivates pupils externally through a controlling environment, e.g. by directing and controlling pupils actions, persuading, rewarding and punishing them to modify their behaviour. A *theory y* teacher encourages intrinsic motivation by allowing pupils to develop for themselves. This may be through an autonomy-support environment (see below).
Achievement motivation	Atkinson, 1964; McClelland, 1961	Motivation to perform an achievement-orientated task is related to the: (i) need to achieve on a particular task; (ii) expectation of success on the task; and (iii) strength of the incentive after the task has been completed successfully. This results in individuals setting themselves standards of achievement.	Create a learning environment which raises the need for achievement in academic studies by raising levels of aspiration. Plan tasks that are challenging but attainable with effort. Work should be differentiated according to individual needs.
Achievement goal theory	Ames, 1992a, 1992b; Dweck, 1986; Dweck and Leggett, 1988; Nicholls, 1984, 1989	A social-cognitive perspective which identifies determinants of achievement behaviour; variations of which result from different achievement goals pursued by individuals in achievement situations. In achievement settings, an individual's orientation towards one of two incompatible goals by which to judge success underpins how they strive to maximise their demonstration of ability. In a task (or mastery) orientated setting the focus is on skill learning and exerting effort to succeed and success is judged by self-improvement, mastery of a task. In an outcome (ego or performance) orientated setting individuals compare their performance and ability with others and judge success by beating others with little effort to enhance social status.	Pupils' goal orientation may be influenced by the motivational climate created by what teachers do and say. Therefore, plan a task orientated learning environment that encourages pupils to improve their performance by trying hard, selecting demanding tasks and persisting when faced with difficulty; rather than an outcome orientated learning environment that encourages pupils to select easy tasks which they can achieve with minimum effort and on which they are likely to give up when facing difficulties.

Attribution theory	Weiner, 1972	Success or failure is attributed to ability, effort, difficulty of task or luck, depending on: (i) previous experience of success or failure on the task; (ii) the amount of work put in; or (iii) a perceived relationship between what is done and success or failure on the task.	Reward effort as well as success, as pupils are more likely to try if they perceive success is due to effort, e.g. can give two marks for work, one for the standard of the work, the other for effort. Use teaching and assessment which is individualised rather than competitive.
Expectancy theory	Rosenthal and Jacobson, 1968	A range of cues is used by one person to form expectations (high or low) of another. That person then behaves in a way that is consistent with their expectations. This influences motivation, performance and how the other person attributes success or failure. The other person performs according to the expectations, thus creating a self-fulfilling prophecy.	In order to avoid the self-fulfilling prophesy of pupils performing according to the way teachers expect them to perform (by forming expectations (high or low) based on a range of cues and conveying these expectations), do not prejudge pupils on their past performance. Rather, encourage pupils to work to the best of their ability all the time.
Hierarchy of needs theory	Maslow, 1970	Hierarchy (highest to lowest): 1. Self-actualisation (need to fulfil own potential); 2. Self-esteem (need to feel competent and gain recognition from others); 3. Affiliation and affection (need for love and belonging); 4. Need for physical and psychological safety; 5. Physiological needs (e.g. food, warmth). Energy is spent meeting the lowest level of unmet need.	If basic needs, e.g. sleep, food, warmth, are not met, a pupil concentrates on meeting that need first and is unlikely to benefit from attempts by teachers to meet higher level needs. Try to create a classroom environment to fulfil basic needs first, e.g. rules for using dangerous equipment provide a sense of physical safety, routines give a sense of psychological security, group work can give a sense of belonging (affiliation) (Postlethwaite, 1993).
Behavioural learning theories	Skinner, 1953	Activity or behaviour is learned and maintained because of interaction with the environment. An activity or behaviour reinforced by a pleasurable outcome is more likely to be repeated.	Positive reinforcement (reward), e.g. praise, generally increases motivation to learn and behave. This has a greater impact if the reward is relevant to the pupils, they know how to get the reward and it is given fairly and consistently (there are, however, exceptions: see 'Praise', below).

compared to that of the rest of the group. This encourages success or failure to be attributed to ability or luck. In 'criterion-referenced' marking, all pupils who meet stated criteria for a particular category of mark are marked in that category. Thus, pupils are given a mark which reflects how closely the criteria for the assessment have been met, irrespective of the performance of other pupils. Although this overcomes some of the disadvantages of norm-referenced marking, it does not reflect how much effort the pupil has put into the work. Postlethwaite went on to say that effort can best be judged by comparing different pieces of the same pupil's work, as the standard of work is likely to reflect the amount of effort put in (i.e. ipsative assessment). Giving two marks for the work, one

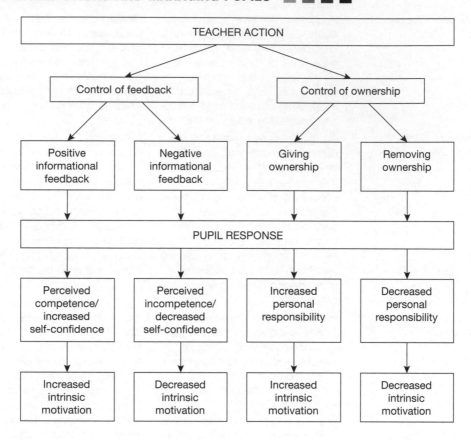

■ **Figure 3.2.1** Creating a motivational climate

for content and standard of the work and one for effort and presentation, can encourage effort. Thus, even if the content and standard are poor, it may be possible to praise the effort. This praise can motivate the pupil to try harder, especially if the pupil values the mark for effort. He suggested that another way of encouraging pupils to attribute success to effort is to ask them to write about the way they tackled the task (see also Units 6.1 and 6.2 on assessment).

According to expectancy theory, a teacher forms an impression of a pupil on which they base their expectations of that pupil; the teacher's verbal and non-verbal behaviour is based, consciously or unconsciously, on those expectations; the pupil recognises, consciously or unconsciously, the teacher's expectations of them from their behaviour and responds in a way that matches these expectations. Thus, there is a self-fulfilling prophecy. It is generally accepted that if a teacher expects high achievement and good behaviour, pupils perform to the best of their ability and behave well. Murdock (1999) found that where teachers held high expectations of pupils, they were engaged more academically. If, on the other hand, teachers have low expectations of pupils' achievement and behaviour, pupils achieve little and behave badly. In the same way, teachers can develop stereotypes of how different groups of pupils perform or behave; stereotypes can direct expectations. (See also Unit 4.4 for further information.)

One aspect of the organisation of a school that may particularly influence teachers' expectations of pupils is the way pupils are grouped. Pupils streamed by ability remain in the same group throughout the year, whatever their ability in different subjects. Whatever the labels attached to each stream, pupils are perceptive and judge their abilities by the stream they are in. This may be partly because teachers' verbal and non-verbal behaviour communicates clearly their expectations. Teachers expect pupils in the 'top' stream to do well; therefore they behave accordingly, e.g. actively encouraging pupils, setting challenging work. Teachers do not expect pupils in the 'bottom' stream to do as well; therefore they behave accordingly, e.g. constantly nagging pupils, setting easy work (or none at all). Both groups of pupils tend to fulfil the expectations of teachers. No doubt many of you have heard of the notorious 'bottom' stream in a school. Setting (or banding) pupils for different subjects can overcome problems of streaming, i.e. recognising pupils' ability in different subjects and changing the grouping of pupils according to their ability in a specific subject. The problem can also be overcome by grouping pupils in mixed ability classes and providing differentiated work to enable pupils of different abilities to work alongside each other on tasks that are challenging but achievable for each pupil (see achievement motivation above). For further information about differentiation see Unit 4.1.

Task 3.2.2 asks you to think about the hierarchy of needs theory and Task 3.2.3 asks you to think about the application of these theories to your teaching.

Task 3.2.2 **HIERARCHY OF NEEDS**

Consider some of the home conditions likely to leave pupils with unmet needs which prevent effective learning at school. Discuss with your tutor or another student teacher what can be done in the school and what you can do in your lessons that may help pupils to meet these basic needs to provide a foundation for effective learning. Discuss when and to whom you should report if you suspect pupils' most basic needs are not being met, as this may require the skills of other professionals.

Task 3.2.3 **USING THEORIES OF MOTIVATION IN YOUR TEACHING**

Review a range of literature on theories of motivation. Write a reflective commentary on these.

Work in a group with other student teachers. Each select one theory of motivation. Use the literature and the implications for teachers in Table 3.2.1 and in the text above as a basis for identifying practical implications of the theory for your lessons. Use this as the basis for planning for motivating pupils in one of your lessons based on this theory of motivation. Ask your tutor to observe the lesson and give you feedback on the effectiveness of your motivation. Record your own reflections. Meet with your group and discuss the theory, your lesson and reflection, then identify ways you can improve. Repeat the cycle of planning, teaching the lesson, reflecting and evaluating your effectiveness. Record the outcomes of your own and other student teachers work in your professional development portfolio (PDP).

You might then like to try this using other theories of motivation.

Some specific factors which influence pupils' motivation to learn

Extrinsic motivation is sometimes used to encourage intrinsic motivation. However, it is important to recognise that motivating pupils extrinsically can have a detrimental effect when a pupil is already motivated intrinsically.

PERSONAL ACHIEVEMENT (SUCCESS)

Personal achievement (generally called success in an outcome orientated learning environment) is generally motivating in itself. Some pupils struggle to succeed, whereas others succeed much more quickly. There are many ways to help pupils succeed, e.g. using a technique often called whole–part– whole teaching. In this, pupils are shown the whole activity first so that they know what they are trying to achieve. The activity is then broken down into small, self-contained, achievable parts, which allow pupils to receive reinforcement for each small, successful step. The separate parts of the activity are gradually put together until the whole activity has been built up. Pupils are given appropriate feedback at each stage (see below for more information about giving feedback), therefore when they attempt the whole, they are most likely to succeed. You may relate to this by thinking about when you learned (or tried to learn) front crawl in swimming. You probably practised your arms, legs and breathing separately before you tried to put it all together (see Task 3.2.4). What other techniques can you use to help pupils succeed?

Task 3.2.4 WHOLE–PART–WHOLE TEACHING

As part of your normal lesson planning with a class, select one activity which you can break down into small, self-contained, achievable parts, which can be put together to build up gradually to the whole. Ask your tutor or another student teacher to observe you teaching this activity using whole–part–whole teaching. At the end of the lesson discuss with some of the pupils how this went. Discuss with the observer how the pupils responded and how well they learned the task. Evaluate the lesson yourself and build on this experience in future lessons.

REWARDS

Although personal achievement or success is motivating in itself, pupils may not be immediately successful on activities they undertake at school, therefore they may need external rewards (positive reinforcements) to motivate them. Bull and Solity (1987) identified four types of rewards, listed below in the order in which they are used most often:

- social rewards (social contact and pleasant interactions with other people, including praise, a smile to recognise an action or achievement or to say thank you, encouraging remarks or a gesture of approval)
- token rewards (house points, grades, certificates)
- activity rewards (opportunities for enjoyable activities)
- material rewards (tangible, usable or edible items).

Task 3.2.5 focuses on these four types of reward.

PRAISE

Generally, pupils respond more positively to praise and positive comments about their work or behaviour than to criticism and negative comments. This, in turn, may produce a motivational learning environment in which pupils work harder and behave better. If pupils misbehave in a classroom in which there is a positive motivational learning environment, Olweus (1993: 85) suggested that the

Task 3.2.5 **USING REWARDS**

Develop an observation schedule with sections for the four types of reward listed above. Observe a class and mark in the appropriate category any reward used by the teacher in the class. Discuss with the teacher the variety and frequency of use of the different possible methods of reward as well as why a particular type of reward was used to achieve a particular purpose. Ask your tutor or another student teacher to undertake the same observation on one of your lessons. Discuss the differences in variety and frequency of reward used. As you plan your lessons consider how you might use reward. Ask the same person to observe a lesson a couple of months after the first one and see if you have changed your use of reward in your lessons. Relate this to what you know about behavioural learning theories.

use of praise makes pupils feel appreciated, which may make it easier for them to accept criticism of inappropriate behaviour and to attempt to change.

However, the Office for Standards in Education (Ofsted, 1993) reported that teachers give relatively little praise and that their vocabulary is generally more negative than positive. Praise is given more often for academic than social behaviour and social behaviour is more likely to be criticised than praised. One reason for this may be that teachers expect pupils to behave appropriately in the classroom.

Some teachers use very few different words to praise pupils, e.g. 'good', 'well done', 'OK'. What other words can you use to praise someone or give feedback? Try to develop a list of such words because if you use the same word to praise pupils all the time, the word loses its effect. The range of words must be accompanied by appropriate non-verbal communication signals (see non-verbal communication in Unit 3.1), see Task 3.2.6, p. 134.

Although it is generally accepted that praise aids learning, there are dangers in using praise. There are times when it may not be appropriate. For example, pupils who become lazy about their work as a result of complacency may respond by working harder if their work is gently criticised on occasion. If praise is given automatically, regardless of the work, effort or behaviour, pupils quickly see through it and it loses its effect. Praise should only be used to reward appropriate achievement, effort or behaviour.

Some pupils do not respond positively to praise, e.g. they are embarrassed, especially if they are praised in front of their peers. Others perceive praise to be a form of punishment, e.g. if they are teased or rejected by their peers for being 'teacher's pet' or for behaving themselves in class. Thus, conforming to the behaviours and values promoted in school results in negative social consequences. Therefore, although pupils know that they will be rewarded for achievement, effort or behaviour, they may also be aware of the norms of the peer-group which discourage them from achieving academically, making an effort or behaving well.

Other pupils do not know how to respond to praise because they have not received much praise in the past; e.g. because they have continually received low marks for their work or because they have been in the bottom stream. They have therefore learned to fail. Some of these pupils may want to attribute failure to not caring or not trying to succeed. One way they may do this is by not making an effort with work, another is misbehaving in the classroom.

Thus, pupils respond differently to praise. In the same class you may have some pupils working hard to receive praise from the teacher or a good mark on their homework, while others do not

Task 3.2.6 **THE LANGUAGE OF PRAISE**

Use the observation schedule below (or develop a similar one of your own with categories for praise and negative comments given to an individual, a group or the whole class, for both academic work and behaviour).

Tick each time praise or negative comment is given in each category					
Praise to individual for academic work					
Praise to group for academic work					
Praise to whole class for academic work					
Praise to individual for behaviour					
Praise to group for behaviour					
Praise to whole class for behaviour					
Negative comment to individual for academic work					
Negative comment to group					
Negative comment to whole class for academic work					
Negative comment to individual for behaviour					
Negative comment to group for behaviour					
Negative comment to whole class for behaviour					

Observe a class taught by an experienced teacher. Sit in a place where you can hear everything that is said. Record the number of times the teacher gives praise and makes negative comments to individuals, groups and the whole class in relation to academic work and behaviour. Observe the same experienced teacher in another lesson. This time write down the different words, phrases and actions the teacher uses to give praise and negative comments in each of these categories and the number of times each is used.

Ask someone to conduct the same observations on your lessons. You might be surprised to find that you use a phrase such as 'good' or 'OK' very frequently in your teaching. Discuss the differences with your tutor and, if appropriate, develop strategies to help you improve the amount of praise you give and the range of words, phrases and actions you use to give praise. Record these strategies in your PDP and gradually try to incorporate them into your teaching.

respond well to praise or are working hard at avoiding praise. You have to use your judgement when giving praise; e.g. if you praise a pupil who is misbehaving to try to encourage better behaviour, you may be seen to be rewarding bad behaviour, thereby motivating the pupil to continue to misbehave in order to get attention. If you are not immediately successful in your use of praise, do not give up using it, but consider whether you are giving it in the right way, e.g. would it be better to have a quiet word, rather than praise pupils out loud in front of their peers? As your professional knowledge and judgement develop you become able to determine how best to use praise appropriately to motivate pupils in your classes.

PUNISHMENT

As well as using praise, teachers also use punishment to try to change behaviour. However, reward, most frequently in the form of praise, is generally considered to be more effective because it increases appropriate behaviour, whereas punishment decreases inappropriate behaviour. If pupils are punished, they know what behaviour results in punishment and therefore what not to do, but may not know what behaviour avoids punishment.

However, there are times when punishment is needed. At such times, make sure that you use punishment to best effect; e.g. avoid punishing a whole class for the behaviour of one or a few pupil(s), always make it clear which pupil(s) are being punished for what behaviour, always give punishment fairly and consistently and in proportion to the offence. Also, make sure that the punishment does not include the behaviour that you want exhibited (e.g. do not punish a pupil by requiring them to run round the football pitch if that is what you had wanted them to do). This sends mixed messages and is likely to put the pupil off that activity. Do not make idle threats to pupils, by threatening them with punishment that you cannot carry out. In order to increase appropriate behaviour, identify to the offender any positive aspects of the behaviour being punished and explain the appropriate behaviour (see also Units 3.3 and 4.5).

FEEDBACK

It may be that pupils who do not respond positively to praise are underperforming and have been doing so for a long time. You may be able to check whether they are underperforming by comparing assessment data over a period of time to measure current achievement with past achievement. The achievement of all pupils, including underperforming pupils, can be enhanced by receiving feedback on their work. Feedback is a formative process which gives pupils information about how they are doing and whether they are on the right track when learning something. This motivates them to make an effort and to continue.

A pupil is more likely to learn effectively or behave appropriately if feedback is used in conjunction with praise. A sequence in which feedback is sandwiched between praise, i.e. praise–constructive feedback–praise, is designed to provide encouragement and motivation, along with information to help the pupil improve the activity or behaviour. Giving praise first is designed to make pupils more receptive to the information and, afterwards, to have a positive approach to try again. Try combining feedback with praise in your teaching.

By observing pupils very carefully you are able to spot small changes or improvements, which allows you to provide appropriate feedback (Unit 2.1 looks at observation techniques and Unit 6.1 focuses on assessment for learning). Feedback can be used effectively with the whole–part–whole teaching method (see above). If you give feedback about how a pupil has done on each part, this part can be improved before going on to the next part. If you give feedback immediately (i.e. as an attempt is being finished or immediately after it has finished, but before another attempt is started), pupils can relate the feedback directly to the outcome of the activity. Thus, pupils are more likely to succeed if they take small steps and receive immediate feedback on each step. This success can, in turn, lead to increased motivation to continue the activity.

One problem with giving immediate feedback is how you can provide feedback to individual pupils in a class who are all doing the same activity at the same time. There are several methods which you can use to provide feedback to many pupils at the same time, e.g. getting pupils to work through examples in a book which has the answers in the back, setting criteria and letting pupils evaluate themselves against the criteria or having pupils assess one another against set criteria (Unit 5.3 covers teaching strategies and styles, including the reciprocal teaching style of Mosston and Ashworth, 2002). If they have been properly prepared for it, pupils are generally sensible and constructive when given responsibility for giving feedback. In Task 3.2.7 the focus is on pupils giving feedback to each other.

Task 3.2.7 **PUPILS GIVING FEEDBACK**

As an integral part of your lesson planning, select one activity in which pupils can observe each other and provide feedback. Devise a handout with the main points/criteria to be observed. Plan how you are going to introduce this activity into the lesson. Discuss the lesson plan with your tutor. Ask your tutor to observe the lesson. Discuss the effectiveness of the strategy afterwards, determining how you can improve its use. Also try to observe teachers who plan for pupils to observe and give feedback to each other. Try the strategy at a later date in your school experience. Think of other ways in which you can get more feedback to more pupils when they are doing an activity. Include these in your lesson plans, as appropriate (see also teaching styles in Unit 5.3).

However, it is not always appropriate to give immediate feedback.

Not all feedback comes from another person, e.g. the teacher or another pupil; feedback also comes from the activity itself. The feedback from an activity may be easier to identify for some activities than others; e.g. a pupil gets feedback about their success if an answer to a mathematics problem matches that given in the book or the wicket is knocked down when bowling in cricket. In other activities, right or wrong, success or failure, is not as clear-cut, e.g. there is often no right or wrong answer to an English essay. In the early stages of learning an activity pupils find it hard to use the feedback from the activity, e.g. they may notice that they were successful at the activity, but not be able to identify why. Normally, therefore, they need feedback from another person. This immediate, external feedback can be used to help pupils become more aware of what they are doing, how they are improving, why they were successful or not at the activity and therefore to make use of feedback from the activity. Later in the learning, e.g. when refining an activity, pupils should be able to benefit from feedback from the activity itself and therefore it is better to encourage this internal feedback by, for example, asking appropriate questions, e.g., how did that feel? In this situation the teacher should not give immediate feedback.

Finally, to be effective, feedback should be given about pupils' work or behaviour, not about the pupils themselves. It must convey to pupils that their work or behaviour is satisfactory or not, not that they are good (or bad) per se.

Motivating individuals

As the discussion above has highlighted, there is no one correct way to motivate pupils to learn. Different motivation techniques are appropriate and effective in different situations, e.g. pupils of different ages respond differently to different types of motivation, reward, punishment or feedback. Likewise, individual pupils respond differently. Further, any one pupil may respond to the same motivator differently at different times and in different situations.

Pupils need to feel that they are individuals, with their needs and interests taken into account, rather than just being a member of a group. Pupils need to be given opportunities to take ownership of the tasks in which they are engaged. If pupils are not motivated or bored, do not let them avoid doing an activity, but try to find ways of motivating them; e.g. by relating it to something in which they are interested. You can motivate pupils most effectively by using motivation techniques appropriate for a particular pupil in a particular situation. Therefore you need to get to know pupils as individuals.

Thus, you need to try to find out what motivates each pupil in your class. Learning pupils' names quickly gives you a start in being able to motivate pupils effectively (Unit 2.2 provides advice on learning pupils' names). You also need to get to know pupils as individuals. Observation of pupils, talking to them, discussing a pupil with the form tutor or other teachers all help. As you get to know pupils you can identify what motivates them by finding out what activities they enjoy, what they choose to do and what they try to avoid, what types of reward they work for and to what they do not respond.

The sooner you can relate to pupils individually, the sooner you can manage a class of individuals effectively. However, this does not occur at an early phase of your development as a teacher (see phases of development in Unit 1.2). As a student teacher you are at a disadvantage here because you can only know what motivates each pupil and what rewards they are likely to respond to if you know your pupils well and know something about their needs and interests. As a student teacher, you do not usually spend enough time in one school to get to know the pupils well and therefore you can only try to motivate individual pupils by using your knowledge and understanding of pupils of that age.

SUMMARY AND KEY POINTS

This unit has identified some theoretical underpinnings, general principles and techniques for achieving an appropriate motivational climate in your lessons and therefore to increase pupils' motivation to learn. However, you need to be able to use these appropriately. For example, if you praise a group for working quietly while they are working you may negatively affect their work. It is better in this situation to let the group finish their work and then praise them. In addition, pupils are individuals and therefore respond differently to different forms of motivation, reward, punishment and feedback. Further, the same pupil responds differently at different times and in different situations. To motivate each pupil effectively therefore requires that you know your pupils so you can anticipate how they will respond. Motivation is supported by good formative assessment techniques (see Unit 6.1). Your developing professional knowledge and judgement enables you to combine theory with practice to motivate pupils effectively in your classes, which raises the standard of their work.

Check which requirements for your course have been addressed through this unit.

FURTHER READING

Chalmers, G. (ed.) (2001) *Reflections on Motivation*, London: Centre for Information on Language Teaching and Research (CILT).
> By blending theoretical and practical classroom applications, this book provides activities, and a rationale for their use, for motivating learners in modern foreign languages – although these could be used in other subjects.

Child, D. (2007) *Psychology and the Teacher*, 8th edn, London: Continuum.
> Chapter 8 provides in-depth consideration of motivation in education. It starts by considering three broad types of theories of motivation, then looks specifically at how some of the theories of motivation impact on you as a teacher and on your pupils.

Gilbert, I. (2002) *Essential Motivation in the Classroom*, London: RoutledgeFalmer.
> This book covers strategies, ideas and advice to help teachers understand how to motivate pupils and how pupils can motivate themselves.

Kyriacou, C. (2007) *Essential Teaching Skills*, 3rd edn, Cheltenham: Stanley Thornes.
> This book contains chapters on lesson management and classroom climate, both of which consider aspects of motivation, e.g. whether lesson management helps to maintain pupils' motivation and whether the opportunities for learning are challenging and offer realistic opportunities for success.

Additional resources for this unit are available on the companion website:
www.routledge.com/textbooks/9780415478724

BEHAVIOUR FOR LEARNING

A positive approach to classroom management

Philip Garner

INTRODUCTION

This unit is designed to enable you to enhance your knowledge and skills in classroom management and especially to support the development of positive approaches to behaviour. The unit takes account of the most recent shifts in thinking and policy regarding the management of pupil behaviour in classrooms. There is a move away from reactive approaches, characterised by a preoccupation with 'discipline' being something which the teacher imposes on pupils to the concept of *behaviour for learning*, which is consistent with a quest to develop inclusive schooling for all learners. It emphasises that teachers are integral to creating an appropriate climate in which all pupils can learn effectively. The approach promoted by *behaviour for learning* encourages you to link pupil behaviour with their learning, via three interlinked relationships: how pupils think about themselves (their relationship with themself); how they view their relationship with others (both teachers and fellow pupils); and how they perceive themself as a learner, relative to the curriculum (their relationship with the learning they are undertaking). A recognition of these three relationships is seen as the basis of a preventative approach; one in which pupils themselves have a key role to play in learning to manage their own behaviour.

This unit does not provide detailed accounts of what to do, nor commentaries on individual behavioural needs. There is now an extensive literature relating to the practical aspects of behaviour management (see, for example, the Behaviour 4 Learning website (http://www.behaviour4learning. ac.uk); Dix, 2007; Haydn, 2006; Roffey, 2004). This unit examines the behaviour for learning approach, and the way that it should be woven into your teaching, including some key, underpinning principles of behaviour for learning. It:

- is a positive description, emphasising teacher expectations; it does not focus on behaviours that the teacher does not want
- emphasises the centrality of effective relationships
- puts a value on behaving in ways which enable and maximise pupil learning
- places an emphasis on setting targets that are reachable
- is relevant to all pupils, irrespective of their stage of learning.

The crucial factor in all these principles is the manner in which a positive climate for learning is established. This is directly under your control in the classroom setting by selecting approaches which are more likely to increase the learning behaviour of pupils. Research strongly suggests that these approaches are characterised by the promotion of positive relationships and the development of an appropriate emotional climate in the classroom (Evans *et al.*, 2003).

In England national policies regarding effective classroom management are fully embedded in and sympathetic to a behaviour for learning approach. The role of schools in supporting a movement towards greater social and educational inclusion implies that all teachers, not simply those charged with responsibility for special educational needs (SEN), should have a stake-holding in this aspect of pupil development. And so it is vital that, as a student teacher, you both understand and practise the core principles of a behaviour for learning approach as a routine aspect of your teaching. It is important that the three relationships enshrined within it are applicable to all pupils, thus making behaviour for learning an ideal way of developing inclusive practices in your classroom.

OBJECTIVES

At the end of this unit you should be able to:

■ recognise the policy context for promoting pupil learning in classrooms
■ interrogate a definition of the term 'unacceptable behaviour' and understand the significance of its underlying causes
■ recognise the importance of a behaviour for learning approach and its core principles
■ develop positive approaches to unacceptable behaviour which are based on relationships with pupils.

Check the requirements for your course to see which relate to this unit.

THE POLICY CONTEXT

There is increasing emphasis upon the inclusion of a greater diversity of learners in schools (Department for Education and Employment (DfEE), 1999a; Department for Education and Skills (DfES), 2001b, 2003a). The underpinning ideology of educational inclusion is that the needs of all pupils in schools should be met, irrespective of their level of achievement or the nature of their social behaviour. Thus, you are likely to encounter a range of learner needs. This is certainly the case with pupils who exhibit social, emotional and behavioural difficulties (SEBD), who represent over 20 per cent of all pupils who have SEN (Department for Children, Schools and Families (DCSF), 2008a). Thus, the progress of some pupils in your classes is being supported by the special educational needs co-ordinator (SENCO), and by other key workers in the school. Provision for this group of pupils is formally set out in the SEN Code of Practice (DfES, 2001c).

SEBD is a term which refers to a continuum of behaviours, from relatively minor behaviour problems to serious mental illness (Department for Education (DfE), 1994a). Here we discuss pupil behaviours which are viewed as low level unacceptable behaviours. However, you may encounter pupils who present more challenging behaviours in your classes, including some who may abuse drugs and other substances; pupils with mental health needs and pupils who experience behaviour-related syndromes, such as Attention Deficit/Hyperactivity Disorder (ADHD) or Autistic Spectrum

Disorders (ASD). All of these behaviours, including those which are sometimes intense and very challenging to teachers, can be more effectively managed if you build proactive, positive strategies into your teaching. You can also expect to have school strategies and support to address these more extreme behaviours.

Behaviour and attendance initiatives now emphasise a no-blame approach, built around the development of a positive classroom ethos. Emphasis is placed on teachers developing a repertoire of knowledge, skills and understanding about classroom management. You may find lead behaviour teachers and other professionals in your school who provide practical support in positively managing behaviour and attendance (DCSF, 2008b). Further support is provided by teachers in other educational settings (pupil referral units (PRUs)), special schools for pupils experiencing SEBD, teaching assistants and local authority (LA) personnel who have a specific brief for work in SEBD (Walker, 2004).

The Key Stage 3 (KS3) Behaviour and Attendance Strategy (DCSF, 2008b) is the key guidance document in England in this area. The strategy aims to help schools and individual teachers to promote positive behaviour and to support them in tackling issues of low-level unacceptable behaviour. It also emphasises that attendance is a matter of concern to teachers, in that pupils who choose not to attend a lesson, or school as a whole, are demonstrating important signals of disengagement from formal education.

The core objective of the Behaviour and Attendance Strategy is to ensure all schools have the skills and support they need to maintain creative and positive learning environments for all pupils. The strategy promotes the development of positive behaviour throughout the school and assists teachers to develop or refine proactive policies on behaviour and attendance. The emphasis, then, is upon establishing an appropriate climate in classrooms, based on the development of positive teacher–pupil relationships, which help to insulate pupils from those factors (discussed later in this unit), which might cause them to behave inappropriately and thereby promote active engagement in learning.

Teachers also have responsibility for tackling bullying in schools. Like other unacceptable behaviours, bullying varies in its type and intensity. From September 1999, headteachers of maintained schools in England have had a duty to draw up measures to prevent all forms of bullying among pupils. A pack entitled *Bullying: Don't Suffer in Silence* (including a video aimed at pupils) (DCSF, 2008c) considers many aspects of bullying, including the importance of tackling homophobic abuse and bullying and bullying by mobile phone text messages.

Each of these initiatives recognises that social and emotional aspects of learning (SEAL) need to be developed by all pupils, and especially by those who, from time to time, present behaviour which is unacceptable or challenging. Goals for teachers which support their efforts to establish an appropriate classroom climate in which learning can take place, and motivate learners, are that pupils are able to:

- be effective and successful learners
- make and sustain friendships
- deal with and resolve conflict effectively and fairly
- solve problems with others or by themselves
- manage strong feelings such as frustration, anger and anxiety
- recover from setbacks and persist in the face of difficulties
- work and play co-operatively
- compete fairly and win and lose with dignity and respect for competitors.

Establishing these as guiding principles will pay rich dividends as you progress in your teaching career. Discuss with your tutor how you can do this.

Moreover, the concept of SEAL includes an awareness of emotional intelligence (Goleman, 1996) and enables pupils to improve their behaviour. Weare (2004: 63) suggests that 'generally a punitive approach tends to worsen or sometimes even create the very problems it is intended to eradicate . . . punishment alienates children from their teachers and does nothing to build up trust that is the bedrock of relationships'. Weare examined several systematic reviews of research which have looked at programmes designed to promote mental health in schools in the USA. For example, Wells *et al.* (2003) concluded that many programmes had clear and positive effects on behaviour. Successful programmes taught emotional and social competences and focused on the whole-school environment, not just on an individual behaviour.

The Behaviour 4 Learning website (http://www.behaviour4learning.ac.uk) provides a valuable set of resources to assist you, as student teachers and in your first years of teaching, in developing a positive approach in managing behaviour.

Now complete Task 3.3.1.

Task 3.3.1 **WHAT IS BEHAVIOUR FOR LEARNING?**

You are gathering new information almost daily which assists you in developing your professional skills. Respond to each of the questions listed under each of the three relationships central to behaviour for learning.

Relationship with self

■ Why do I want to teach?
■ How do I feel about my general progress so far?
■ How confident 'in myself' am I about the career I am embarking upon?
■ What factors motivate me to succeed?
■ What issues do I view as impacting negatively on my current situation?
■ Where do I want to be in 1/5/10 years' time?

Relationship with others

■ How well do I interact with other course members?
■ Do I empathise with the feelings/views of other student teachers?
■ How well do I communicate my own views and opinions?
■ What personal attributes enable me to work as part of a team or group?
■ Are there any social/interactional skills which I feel I need to enhance?
■ Do I feel comfortable in both 1:1 and group settings?

Relationship with the curriculum

■ Am I learning an appropriate range of skills to equip me for my job?
■ Am I sufficiently aware of the relevant theory underpinning my practice?
■ Am I coping with the range and amount of work?
■ Are there any aspects of my ITE course about which I feel less certain?

Reflect on how your responses might compare with those of the pupils you come in contact with. Record your responses in your professional development portfolio (PDP).

WHAT IS UNACCEPTABLE BEHAVIOUR?

Current policy and guidance on behaviour management in England emphasises developing appropriate, positive behaviour. Such an approach has significant benefits for all pupils (Harker and Redpath, 1999) and invites teachers to be clear about what behaviour they want pupils to engage in, and modelling this as part of their teaching.

However, the term behaviour has traditionally been taken to mean unacceptable behaviour. The Elton Report (Department of Education and Science (DES), 1989) refers to misbehaviour as behaviour which causes concern to teachers. The term is one which can variously be replaced by a range of other expressions that teachers use to describe unwanted, unacceptable behaviour by pupils. Disruptive, challenging, anti-social, emotional and behavioural difficulties (EBD) and SEBD are terms which are widely used, according to the personal orientation of the teacher concerned and to the type of problem behaviour being described.

The term unacceptable behaviour, and its companion descriptors (see above), is often used as a catch-all expression for pupil behaviours that span a continuum (DfE, 1994a). The so called EBD continuum ranges from low level unacceptable behaviour at one end (such as talking out of turn, distracting others, occasionally arriving late in class) to more serious, sometimes acting out behaviour at the other (such as non-attendance, verbal or physical aggression, wilful disobedience, bullying). This confusion was recognised by DfE (1994a) who describe EBD as all those behaviours which comprise a continuum from normal though unacceptable to mental illness; confusion which has been increased as the term EBD has recently incorporated the social difficulties into the spectrum to become SEBD (DCSF, 2008a).

The SEN Code of Practice (DfES, 2001a: 93) defines 'children and young people who demonstrate features of emotional and behavioural difficulties' as those who are 'withdrawn and isolated, disruptive and disturbing, hyperactive and lacking concentration'. The definition also includes those who display 'immature social skills and those who present challenging behaviours arising from other complex special needs'.

A major difficulty in defining what inappropriate behaviour constitutes is that it varies according to the perception, tolerance threshold, experience and management approach of individual teachers. Thus, what might be an unacceptable behaviour in your own classroom may be viewed in another context, or by another (student) teacher, as quite normal. Alternatively, what you accept as normal may be seen as unacceptable in another context or by another (student) teacher. This leads to confusion in the mind of pupils, and to potential tension between individual teachers in a school or between a student teacher and tutor or other experienced teacher. So it is important to recognise that: (a) pupil behaviour needs to be described explicitly in terms of observable actions; and (b) responses to it have to take full regard of a school's policy concerning behaviour. When you describe a behaviour, you should always ensure that your definition is of the behaviour itself and not of the pupil as a whole. This avoids any likelihood of labelling the pupil as a disruptive pupil or a problem pupil.

The Qualifications and Curriculum Authority (QCA, 2001a) usefully identified 15 behaviours by which a pupil's emotional and behavioural development might be defined and assessed. These were divided into learning behaviours, conduct behaviours and emotional behaviours. Each of these groupings is subdivided into sets of criteria, depicting desirable and undesirable behaviours (see Figure 3.3.1). Now complete Task 3.3.2.

Identifying or defining unacceptable behaviour is important if you are going to develop strategies to deal with it in ways which promote learning. You need to describe exactly what any unwanted behaviour actually comprises in order to give a precise and objective description of what has occurred; importantly, you need to describe the behaviour itself, not the pupil, otherwise there may be unwarranted negative labelling of the pupil. Task 3.3.3 helps you to develop a definition of unacceptable behaviour.

Task 3.3.2 **SCHOOL POLICY ON PUPIL BEHAVIOUR**

Familiarise yourself with your placement school's whole-school policy on behaviour and attendance. Discuss it with another student teacher who is placed in a different school. Consider both the similarities and differences in the two policies. What are the implications of the document for you as a student teacher?

Record your reflections in your PDP.

Task 3.3.3 **WHAT IS UNACCEPTABLE BEHAVIOUR?**

It is important that you arrive at a personal definition of what comprises unacceptable behaviour. Divide a blank sheet of paper into three. Head the left-hand section 'Totally unacceptable' and the right-hand section 'Acceptable'. The middle section is reserved for 'Acceptable in certain circumstances'. Now examine your own classroom teaching, and complete each section. Remember, behaviour is as much about positive, learning behaviour as it is those pupil actions which you regard as unacceptable or challenging. Reflect on your responses and discuss with your tutor and record in your PDP. Should the opportunity arise, you might wish to undertake this exercise with your pupils, in order to gather their thoughts. Comparing your list to theirs is likely to prove very revealing!

SCOPING THE CAUSAL FACTORS

As Ayers and Prytys (2002: 38) noted, 'The way in which behaviour is conceptualised will determine the treatment of emotional and behavioural problems'. There are a number of causal factors that assist in explaining unwanted behaviour, disaffection and disengagement amongst some pupils; these are often multivariate and overlapping. The attribution of a cause can frequently result in the acquisition of a negative label by the pupil. Understanding and recognising these gives you clues as to what might be successful strategies. A brief outline of causal factors is given below. There is more exhaustive coverage in a variety of other sources (e.g. Clough *et al.*, 2004).

Factors which may cause unacceptable behaviours

You should recognise that, for some pupils, their unacceptable behaviour is caused by several of the factors identified below:

INDIVIDUAL FACTORS

■ A pupil believes that the work is not within their grasp and as a result feels embarrassed and alienated and lacks self-esteem as a learner.
■ A pupil may well experience learning difficulties.
■ A pupil may have mental health, stress and possible drug misuse issues, all of which are important factors explaining under-achievement and inappropriate behaviour in adolescence.

Desirable behaviour	Undesirable behaviour
L1. Attentive/interested in schoolwork	
• attentive to teacher, not easily distracted • interest in most schoolwork· starts promptly on set tasks/motivated • seems to enjoy school	• verbal off-task behaviours • does not finish work/gives up easily • constantly needs reminders • short attention span • negative approach to school
L2.Good learning organisation	
• competent in individual learning • tidy work at reasonable pace • can organise learning tasks	• forgetful, copies or rushes work • inaccurate, messy and slow work • fails to meet deadlines, not prepared
L3. Effective communicator	
• good communication skills (peers/adults) • knows when it's appropriate to speak • uses non-verbal signals and voice range • communicates in 1:1 or group settings	• poor communication skills • inappropriate timing of communication • constantly talks • lack of use of non-verbal skills
L4. Works efficiently in a group	
• works collaboratively • turn-takes in communication/listens • takes responsibility within a group	• refuses to share • does not take turns
L5. Seeks help where necessary	
• seeks attention from teacher when required • works independently or in groups when not requiring help	• constantly seeking assistance • makes excessive and inappropriate demands • does not ask 'finding out' questions
C6. Behaves respectfully towards staff	
• co-operative and compliant • responds positively to instruction • does not aim verbal aggression at teacher • interacts politely with teacher • does not deliberately try to annoy or answer the teacher rudely	• responds negatively to instruction • talks back impertinently to teacher • aims verbal aggression, swears at teacher • deliberately interrupts to annoy
C7. Shows respect to other pupils	
• uses appropriate language; does not swear • treats others as equals • does not dominate, bully or intimidate	• verbal violence at other pupils • scornful, use of social aggression (e.g. 'pushing in') • teases and bullies • inappropriate sexual behaviour
C8. Seeks attention appropriately	
• does not attract inappropriate attention • does not play the fool or show off • no attention-seeking behaviour • does not verbally disrupt • does not physically disrupt	• hums, fidgets, disturbs others • throws things, climbs on things • calls out. eats, runs around the class • shouts and otherwise attention seeks • does dangerous things without thought

■ **Figure 3.3.1** Desirable and undesirable behaviour

Source: Adapted from QCA (2001a)

C9. Physically peaceable	
• does not show physical aggression • does not pick on others • is not cruel or spiteful • avoids getting into fights with others • does not have temper tantrums	• fights, aims physical violence at others • loses temper, throws things • bullies and intimidates physically • cruel/spiteful
C10. Respects property	
• takes care of own and others' property • does not engage in vandalism • does not steal	• poor respect for property • destroys own or other's things • steals things
E11. Has empathy	
• is tolerant and considerate • tries to identify with feelings of others· • tries to offer comfort • is not emotionally detached • does not laugh when others are upset	• intolerant • emotionally detached • selfish • no awareness of feelings of others
E12. Is socially aware	
• understands social interactions of self and peers • appropriate verbal/non-verbal contacts· • not socially isolated • has peer-group friends; not a loner • doesn't frequently daydream • actively involved in classroom activity • not aloof, passive or withdrawn	• inactive, daydreams, stares into space • withdrawn or unresponsive • does not participate in class activity • few friends· not accepted or well-liked • shows bizarre behaviour • stares blankly, listless
E13. Is happy	
• smiles and laughs appropriately • should be able to have fun • generally cheerful; seldom upset. • not discontented, sulky, morose	• depressed, unhappy or discontented • prone to emotional upset, tearful • infers suicide • serious, sad, self-harming
E14. Is confident	
• not anxious • unafraid to try new things • not self-conscious, doesn't feel inferior • willing to read aloud, answer questions in class • participates in group discussion	• anxious, tense, tearful • reticent, fears failure, feels inferior • lacks self-esteem, cautious, shy • does not take initiative
E15. Emotionally stable/self-controlled	
• no mood swings • good emotional resilience, recovers quickly from upset • manages own feelings • not easily flustered or frustrated • delays gratification	• inappropriate emotional reactions • does not recover quickly from upsets • does not express feelings • frequent mood changes; irritable • over-reacts; does not accept punishment or praise • does not delay gratification
Key: L = learning behaviour; C = conduct/behaviour; E = emotional behaviour	

■ **Figure 3.3.1** *continued*

CULTURAL FACTORS

■ Adolescence can be a period of rebellion or resistance for many young people.
■ Possible tension between societal expectation and the beliefs and opinions of the pupil.
■ Group/peer pressure can result in various forms of alienation to school.
■ Negative experience of schooling by parents, siblings or other family members.

CURRICULUM RELEVANCE FACTORS (LINKED TO BOTH INDIVIDUAL AND CULTURAL)

■ The curriculum may be seen by a pupil to be inaccessible and irrelevant.
■ The school may give academic excellence more value than vocational qualifications or curriculum options.

SCHOOL ETHOS AND RELATIONSHIPS FACTORS

■ Some schools can be 'deviance provocative' – their organisational structures and procedures are viewed by pupils as oppressive and negative.
■ Some schools are less inclusive, both academically and socially to pupils who behave 'differently'.

EXTERNAL BARRIERS TO PARTICIPATION AND LEARNING FACTORS

■ Family breakdown or illness usually impacts negatively on a pupil's mental health, and often on their sense of priority.
■ Poverty and hardship can mean that a pupil's physiological needs are not met – such pupils may be tired, hungry and consequently easily distracted (see also Maslow (1970) in Unit 3.2).
■ Sibling and caring responsibilities may mean that some pupils arrive late in your lesson – or not at all.

One aspect of causality which is directly related to behaviour for learning is the three interlinked relationships first identified above; what Bronfenbrenner (1979) called the ecosystemic theory of relationships. In the case of a pupil who is consistently behaving inappropriately, it is suggested that there has been a breakdown in one (or more) of these three relationships:

■ pupils' relationship with themselves (how pupils feel about themselves, their self-confidence as learners and their self-esteem)
■ pupils' relationship with others (how they interact socially and academically with all others in their class and school)
■ pupils' relationship with the learning they are undertaking (the curriculum) (how accessible they feel a lesson is, how best they think they learn).

The interrelationship between these are shown in Figure 3.3.2.

All three relationships need to be taken into account when planning your strategy to tackle unacceptable behaviour. The emphasis upon positive relationships is an integral component of a behaviour for learning approach. Establishing good relationships with individual pupils as well as whole classes, from your first encounter with a group of pupils, enables you to establish a climate in which learning can flourish.

Task 3.3.4 links causes to possible teaching strategies.

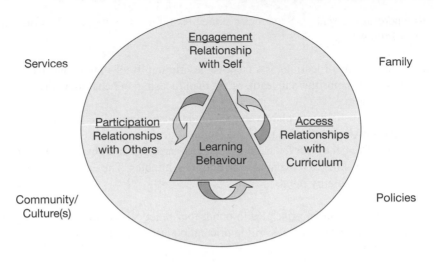

■ **Figure 3.3.2** The behaviour for learning model

Source: After Tod and Powell (2004)

Task 3.3.4 **LINKING CAUSES TO POSSIBLE TEACHING STRATEGIES**

Consider one pupil you are teaching who sometimes exhibits behaviour(s) unacceptable to you. Write a brief description of each of the behaviours, making sure that your language is clear and describes clearly observable pupil actions. Taking each behaviour in turn and, referring to the causal factors identified above, assess which factors you feel might underlie that particular behaviour. Consider how amenable to change each of the causal factors you have identified is. Also identify any other teachers (for example, the SENCO) who might be able to provide you with advice and support. Finally, reflect on how your interpretation of cause might inform the way in which you choose to address the behaviour(s) shown.

Make notes in your PDP and discuss your responses with your tutor.

KEY PRINCIPLES OF BEHAVIOUR FOR LEARNING

As has already been implied in this unit, a behaviour for learning approach accepts that most social and emotional aspects of learning (SEAL) are learned. Evidence strongly suggests that the most successful strategies for developing a positive learning environment are those which incorporate the promotion of positive relationships (Burnett, 2002).

Each of the three (interlinked) relationships is important in developing a positive learning environment in the classroom and as a teacher you are at the very heart of orchestrating them. Although some pupils have relatively advanced skills when they arrive at school, others might need support and the direct teaching of the specific skills they have not yet learned. So your task is to focus on helping to develop appropriate skills which enable each pupil to learn within a variety of learning contexts. This can be in whole-class or small group situations in the classroom and elsewhere in the school. Some basic principles inform the way in which this can be done. These are as follows:

- Behaviour for learning is a positive description. It tells pupils what you want them to do and why this helps them to learn, rather than focusing on behaviours that you do not want in your classroom.
- It requires that you place value on (and praise appropriately) pupil behaviour which enables and maximises learning.
- Effective behaviour for learning strategies can range from high-level listening and collaborative learning skills to remaining seated for two minutes. The emphasis is upon setting targets that are reachable by pupils.

With these broad principles in mind, four practical features of this approach, each of which is central to a behaviour for learning orientation are explored below.

Leadership in the classroom

Most pupils come to your classroom wanting to learn, although there are times when some are either unable or unwilling to learn on account of some of the factors described above. So, you need to develop certain classroom leadership skills which contribute to your being able to establish an appropriate climate for learning and forge positive relationships with pupils, and thus establish an ethos which allows pupils to demonstrate positive behaviour.

First, and in more general terms, you have to take the lead in promoting three broad elements which help to define the ethos of your classroom. These are:

- motivation: you need to provide time at the start of each lesson to tell pupils what they are learning and why. Pupils need to be involved at every stage in assessing whether these learning intentions have been met (Unit 3.2 looks at motivation in more depth).
- emotional well-being: to help reduce pupil anxiety you should share the lesson structure with pupils at the start, so they know what is going to happen during the lesson.
- expectations: you need to give time at the start of the lesson and before each new activity to make clear what behaviours are needed for this piece of learning to be successful.

These three underpinning elements are embedded in more specific teacher actions, which allow you to demonstrate your role as the classroom leader to your pupils. These include:

- good communication between yourself and your pupils (see Unit 3.1)
- secure subject knowledge
- providing lively, well-paced lessons
- understanding and meeting the learning needs of all pupils in your class
- acting on your reflections and evaluations of previous lessons (feedback loop) (see also Unit 2.1)
- demonstrating confidence and direction in managing pupils
- modelling desired behaviours yourself.

It is unlikely that, as a student teacher, everything clicks into place straight away; some of these develop with experience. Nevertheless it is worth noting that research has shown that student teachers who display confidence in managing their classes are less likely to encounter problem behaviour by pupils (Giallo and Little, 2003).

Building positive relationships in classrooms

A behaviour for learning approach places emphasis on the relationships you form with your pupils. Moreover, a positive teacher–pupil relationship is a significant factor in encouraging attendance. Ineffective interventions are usually the product of unsatisfactory relationships with individual pupils. These interventions, even though they are ultimately unsuccessful, take up valuable teaching time and impact negatively on the learning of an individual pupil, the rest of the class, and also on your own confidence. Most interventions should take the form of positive actions that fit somewhere on a continuum from positive reinforcement through to positive correction. The actions you select should be those that enable learning to continue. They usually include eye contact, use of pupil name, description of the appropriate behaviour you would like to see, praise and affirmation (see also Units 3.1 and 3.2). For example:

- modelling appropriate behaviour
- positive reinforcement and the use of praise; positive correction
- consistent and firm application of rules
- use of verbal and non-verbal communication
- listening to pupils and respecting their opinions
- remaining vigilant (pre-empting unacceptable behaviour)
- dealing positively with lateness and non-attendance.

By assimilating these into your professional repertoire you are more likely to forge meaningful and positive relationships with your pupils. In sum, effective relationships mean that there is common ground between pupil and teacher. This is as vital in securing appropriate conditions for learning as it is for managing those behavioural issues which may be potentially problematic.

Structuring the lesson, for positive behaviour and attendance

The design of effective lessons is fundamental to high-quality teaching and learning. This in turn promotes and supports behaviour for learning in the classroom. At its heart is effective lesson design irrespective of the level of achievement of the learner, or the subject or skill being learned. Your teaching should be characterised by:

- focus and structure so that pupils are clear about what is to be learned and how it fits with what they know already
- actively engaging pupils in their learning so that they make their own meaning from it; developing pupils' learning skills systematically so that their learning becomes increasingly independent (see also Unit 5.2)
- using assessment for learning to help pupils reflect on what they already know, reinforce the learning being developed and set targets for the future (Units 6.1 and 6.2 discuss assessment); having high expectations of the effort that pupils should make and what they can achieve (see also Unit 3.2)

- motivating pupils by well-paced lessons, using stimulating activities matched to a range of learning styles which encourage attendance
- creating an environment that promotes learning in a settled and purposeful atmosphere.

You can further reinforce a behaviour for learning approach by building individual teaching sequences within an overall lesson. The lesson (or a sequence of lessons) needs first of all to be located, in the mind of the pupil, in the context of (a) a scheme of work, (b) pupils' prior knowledge, and (c) their preferred learning styles. It also stresses the importance of identifying clear learning outcome(s) for, and making them explicit to, pupils. It is helpful, then, to structure lessons as a series of 'episodes' by separating pupil learning into distinct stages or steps and then planning how each step should be taught. You can secure overall coherence by providing (a) a stimulating start to the lesson, (b) transition 'signposts' between each lesson episode which reviews pupil learning so far and launches the next episode, and (c) a final plenary session that reviews learning. (Lesson planning is covered in Unit 2.2.)

Rights, responsibilities, routines and rules

A framework for promoting positive classroom behaviours is commonly constructed around rights, responsibilities, rules and routines; the 4Rs (Hook and Vass, 2000). You should recognise that such a focus operates best within an overall context of a fifth, over-arching 'R' of relationships, which are crucial to the successful implementation of any activity in the classroom. Of course, as a student teacher, you need to be very sensitive to the existing protocols of any class you take – these have been established over a longer period of time by the permanent class teacher. But you can begin by being conscious of how each of these 'Rs' can have a positive impact on your teaching.

RIGHTS AND RESPONSIBILITIES

Rights and responsibilities are inextricably linked. They refer to both teacher and pupils and are the basis on which classroom relationships, teaching and learning are built.

- Teacher's responsibilities – you must seek to enable all pupils to learn, to seek out and celebrate improvements in learning, to treat pupils with respect and to create a positive classroom environment in which pupils feel safe and able to learn.
- Teacher's rights – you must be allowed to teach with a minimum of hindrance, to feel safe, to be supported by colleagues and to be listened to.
- Pupils' responsibilities – pupils must be willing to learn, to allow others to learn, to co-operate with teaching and other staff and peers and to do their best at all times.
- Pupils' rights – pupils should be treated with respect, be safe, be able to learn and be listened to.

RULES

These are the mechanisms by which rights and responsibilities are translated into adult and pupil behaviours. They are best constructed collaboratively, so that the views of all pupils are taken into account.

ROUTINES

These are the structures which underpin the rules and reinforce the smooth running of the classroom. The more habitual the routines become the more likely they are to be used. Pupils who behave inappropriately often do so because they are unsure of what is happening in the classroom at a given time.

CHOICES

Pupils should be encouraged to make choices about their behaviour and thus take responsibility for their own actions. Choice is guided by their responsibilities and leads to positive or negative consequences according to the choice made by the pupil.

CONSEQUENCES

Pupils know the consequences of sensible or inadvisable choices. Responsible choices lead to positive consequences; conversely, a choice to behave inappropriately leads to a known negative consequence.

Now complete Tasks 3.3.5 and 3.3.6 on praise.

Task 3.3.5 **MONITORING YOUR USE OF PRAISE AND ENCOURAGEMENT IN THE CLASSROOM**

A useful starting point to promote the notion of positive approaches to behaviour is to examine the ways in which you provide encouragement, positive feedback and praise to your pupils. You can assess this by developing a log of praise and encouragement to use as a tool for measuring these positive interactions.

Add to the list of positive pupil behaviours identified below which you can use to give praise. Underneath each one, note the words or actions you might use to convey to the pupil that your recognition carries value and meaning.

1   Queuing sensibly and quietly to enter the classroom
2   Allowing another pupil to go first
3   Lending an item of equipment to another pupil
4   Putting waste paper in the bin
5   Supporting another pupil's learning
6
7
8
9
10

During your observation of a lesson taught by a more experienced teacher in your placement school, record other ways in which that teacher acknowledges positive behaviour by pupils. Compare your notes on this topic with another student teacher working in a different setting.

Task 3.3.6 **THE IMPACT OF PRAISE**

Undertake a small-scale project, designed to establish the impact of 'praise' on pupils in your class. In doing this you should: (a) develop one or more research questions, so that your data-collection has a focus; (b) identify a small, but recent and relevant, set of literature which contributes to a theoretical understanding of the issue; (c) define and provide a rationale for your methodology (including coverage of any ethical issues which might emerge in such a study); (d) gather and analyse data; (e) consider the relevance of your findings to your practice.

Amongst the possible research questions you might wish to consider are:

■ Do boys prefer different kinds of praise and encouragement than girls?
■ Does the nature and type of praise change according to age?
■ What types of praise do pupils prefer?
■ Is praise evenly distributed amongst your teaching group?
■ Does personal praise link closely with a whole-school approach?

In Task 3.3.7 you are asked to explore the links between behaviour and learning.

Task 3.3.7 **EXPLORING FURTHER THE LINKS BETWEEN BEHAVIOUR AND LEARNING**

To explore further the links between behaviour and learning, select one or more pupils who you currently teach in your placement school and who present you with a particular challenge on account of their unacceptable behaviour. You should explore the learning-and-behaviour interface by responding to the following key questions.

■ Does the educational achievement of this pupil vary from one curriculum subject to another?
■ What are the characteristics of those curriculum subjects in which the pupil appears to perform more effectively?
■ Has the pattern of educational achievement been inconsistent over time? Are there any logical explanations for this?
■ Do you know anything about the pupil's preferred learning style?
■ What are your own views about the capabilities of this pupil?
■ What do other subject teachers say about the educational achievements of this pupil?

Each of the above questions can form the basis of a small-scale classroom enquiry. For each you could: (a) gather evidence from the school's pupil data; (b) obtain information from key personnel (for instance, the pupil's form tutor, or the SENCO); and (c) secure inputs from the pupil directly (subject to the appropriate permissions).

On the basis of what you discover, try to formulate a theoretical model for both the unacceptable behaviours displayed and their relationship with more positive aspects of this pupil's school performance.

Task 3.3.8 asks you to consider different definitions of unacceptable behaviour.

Task 3.3.8 INTERPRETATIONS OF UNACCEPTABLE BEHAVIOUR

Interpretations of 'unacceptable behaviour', and the ways in which it has been defined, have changed over time. In spite of this, the educational literature is replete with material (research papers, books, official reports and guidance documents) looking at ways in which schools and teachers can manage behaviour more effectively. Two examples of this, separated by nearly 20 years, are the Elton Report (1989) and the Steer Report (2005). Using the links provided below, consider the similarities and differences in the recommendations of each report. What does the content of these documents tell you about official policy on pupil behaviour? Are there many commonalities regarding the practical advice that these reports offer to classroom teachers? Are you able to draw any inferences from the generic commentaries given concerning the nature and extent of pupil behaviour in schools?

Write up your analysis and discuss with your tutor or other student teachers.

Elton Report: http://www.dg.dial.pipex.com/documents/docs1/elton.shtml
Steer Report: http://www.behaviour4learning.ac.uk/ViewArticle2.aspx?
    ContentId=12597

You should now look at a selection of the 'Behaviour Scenarios' accessible via http://www.behaviour4learning.ac.uk. Using these examples you should try to (a) identify any apparent theories which inform the preferred response to the example of inappropriate behaviour illustrated and (b) considering your own work situation, reflect on how the behaviour illustrated would be managed within existing school policy and your own thoughts on this.

SUMMARY AND KEY POINTS

This unit has focused on the current policy emphasis on positive approaches to managing behaviour. More specifically, it has afforded you the opportunity to examine some individual elements which should give you a more complete understanding of the changed national policy context for promoting pupil learning in classrooms, the difficulties in arriving at a definition of the term 'unacceptable behaviour', some of the key underlying causes of a pupil's unacceptable behaviour, the characteristics of a behaviour for learning approach and an overview of its core principles, a consideration of a number of practical skills which you can develop or refine, in order to promote pupil learning.

At the heart of this approach is a recognition that old-fashioned notions of authoritarian discipline are no longer viable or efficacious (if ever they were). This way of working tends to be unsatisfactory for both pupil and teacher. The former is imbued with a greater sense of resistance and alienation, while the latter spends an increasing amount of time controlling particular individuals in the class rather than providing learning opportunities.

In choosing to place emphasis on positive approaches to managing behaviour you need to place a premium on your relationships with pupils. It is axiomatic, such is the centrality of relationship building, that you need to recognise that positive relationships do not happen by chance. Indeed, the opposite is the case: you need to plan for positivity in your interactions with pupils, while at the same time invoking the '4Rs rule' to orchestrate your strategy. Above all, you need to bear two things in mind. First, pupils expect you to assume a leadership role in making 'good behaviour' happen in your classroom. And, second, unacceptable behaviour, by its very nature, is complex and ever changing and establishing a broad set of personal principles to inform the way you view it is vital groundwork for your continuing professional development (CPD).

Check which requirements for your course have been addressed through this unit.

FURTHER READING

Berryman, M., Glynn, T. and Wearmouth, J. (2006) *Perspectives on Student Behaviour in Schools: Exploring Theory and Developing Practice*, London: Routledge.

This comprehensive text considers the causes of disruptive behaviour, assessment issues and the development of effective intervention strategies that are of practical use to teachers. It does this by considering a range of perspectives: psychodynamic, behavioural and socio-cultural theory, whilst retaining a highly practical orientation.

Clough, P., Garner, P., Pardeck, T. and Yuen, F. (eds) (2004) *The Handbook of Emotional and Behavioural Difficulties*, London: Sage.

This book provides a systematic, comprehensive and up-to-date overview of a series of themes which underpin a study of emotional and behavioural difficulties. It is divided into four sections, dealing with (i) contexts of problem behaviour, (ii) roots and causes, (iii) strategies and interventions, and (iv) points of tension and development. A range of well-known authors contribute to this resource, providing informative and challenging perspectives and commentaries. Moreover, the international authorship reveals that 'unacceptable behaviour' is not simply an issue that is encountered in a single school or individual country. It has a global dimension, from which all teachers can learn.

Weare, K. (2004) *Developing the Emotionally Literate School*, London: Paul Chapman Publishing.

This is a practical and up-to-date account of how schools can use emotional literacy to increase learning and improve relationships between teachers and pupils. The term 'emotional literacy' is discussed in great detail, and its significance to teachers is made clear. It outlines the research base on which the efficacy of 'emotional literacy' is premised, and emphasises that the concept is as applicable to teachers as it is to pupils.

RELEVANT WEBSITE

Behaviour for Learning online: http://www.behaviour4learning.ac.uk

The Behaviour for Learning website is supported by the Training and Development Agency for Schools (TDA). It functions as a 'collecting point' for information on aspects of promoting a behaviour for learning approach. It contains information to enable you to enhance your understanding of pupil behaviour and to refine approaches which enable pupils to learn. The site relates much of its resources (which include case studies, research papers, video extracts and subject-specific materials) to both the standards for qualifying to teach and for induction. From this website you can link to many additional behaviour-related websites, including the DCSF's Behaviour and Attendance site. Registration is free, though not compulsory.

Additional resources for this unit are available on the companion website:
www.routledge.com/textbooks/9780415478724

MEETING INDIVIDUAL DIFFERENCES

It is a truism to say that each pupil in your class is different but from time to time it is important to remind ourselves of this fact. A class of same-age pupils is likely to contain individuals at different stages of development arising from differences in physical and mental development or cultural experiences, or some combination of all three. Significant differences arise in the achievements of members of a class of pupils, especially in a mixed ability class. Other differences arise from the cultural, religious and economic backgrounds of your pupils which may strongly affect their response to schooling. Some pupils respond to academic challenge while others see no point in such demands. Some pupils are gifted and need special attention, as do many pupils with learning or behavioural difficulties. Some pupils are at ease with adults whilst others find the experience less comfortable

This chapter, comprising six units, invites you to consider several aspects of the background and development of your pupils. In practice the features discussed interact, giving rise to the complex and varied behaviours which characterise human beings. For ease of discussion, some factors are discussed separately; we hope this approach helps you subsequently better to integrate your understandings of pupils and their learning and develop your relationships with your pupils.

One response of schools in recent years has been to acknowledge the differences between pupils in their response to school subjects and their associated achievements. Unit 4.1 addresses ways in which pupils are grouped by schools for teaching and learning in the context of differentiation and progression. Central to successful differentiation is the identification of pupil needs; thus case studies invite you to inquire more deeply into the background and response of individual pupils and planning differentiated work. You may want to return to this unit after dipping into other units.

Unit 4.2 focuses on the physical characteristics of pupils as they develop and mature in adolescence and young adulthood and draws attention to the range of 'what is normal'. We address issues of diet and health of young people, the ways in which schools contribute to healthy eating both academically and socially. Attention is drawn to the increase in obesity in pupils and more widely in the population.

Unit 4.3 addresses the issue of cognition and cognitive development. Logical reasoning is one important aspect of cognitive development, along with others such as problem solving, developing expertise and creative thinking. The notion of intelligence is introduced including the current theory of 'multiple intelligences'. Some examples of teaching material from secondary school curricula are discussed in terms of their cognitive demand on pupils. You are invited to look specifically at the differences in performance between pupils of different ages on similar tasks.

Through set tasks there are opportunities for you to work with pupils and to see for yourself how pupils respond to different demands; we address too the importance of teaching pupils how to learn and to think about their own learning.

In Unit 4.4 the cultural background of pupils is considered, including class, gender and ethnicity. These factors are discussed separately while recognising that for every pupil these factors combine in different ways. We highlight some differences in performance of groups of pupils from various backgrounds, using research evidence and to speculate on the causes of those differences. We discuss equal opportunities policies in schools and issues of access to the curriculum and career opportunities.

Unit 4.5 links the development of values in young people in the context of the current curriculum structure for schools in England, together with more recent curriculum changes including the subject of Citizenship and the growing importance of personal, social and health education (PSHE) programmes. The emphasis lies in the way schools contribute to values education through both the overt and hidden curricula. We identify some ways in which a school contributes to the development of values through its ethos and practices. While not stressing differences between pupils, the focus of the chapter does acknowledge the range of values and beliefs in our society and how schools not only have to respond to such differences but also to contribute to the spiritual, moral and cultural development of pupils as well as their mental and physical development.

Inclusion and special needs education is the focus of Unit 4.6 and addresses the ways in which those pupils with special physical, behavioural and learning needs may be supported by the *Special Educational Needs Code of Practice* (DfES, 2001a) for schools in England. This later document replaces 1994 Code of Practice (DfE, 1994). The unit draws attention to significant changes in recent years in the way in which pupils with SEN are supported and emphasises the importance of the classroom teacher in the identification of need and response to need. There is a brief survey of some physical disabilities, mental impairment and behavioural difficulties pointing to sources of support and guidance, including the support of other professionals.

Readings for Learning to Teach in the Secondary School: A Companion to M Level Study brings together essential readings to support you in your critical engagement with key issues raised in this textbook.

PUPIL GROUPING, PROGRESSION AND DIFFERENTIATION

Hilary Lowe and Tony Turner

INTRODUCTION

In your development as student teachers you are required to focus on a number of principles for planning and teaching. These include:

- setting suitable learning challenges
- responding to pupils' diverse learning needs
- overcoming potential barriers to learning.

You should also understand the principles of inclusion and work towards ensuring effective teaching of all pupils through matching the approaches used to the pupils being taught. As student teachers you also need:

- knowledge, understanding and be able to use/adapt a range of teaching, learning and behaviour management strategies including how to *personalise learning* and provide opportunities for all learners to achieve their potential
- to understand how children and young people *develop*
- to know how to make effective *personalised provision* for those you teach
- to be able to plan for *progression* across the age and ability range, designing effective learning sequences within lessons and across series of lessons
- to be able to manage the *learning of individuals, groups and whole classes*, modifying your teaching to suit the stage of the lesson
- to assess the learning needs of those you teach in order to set challenging learning objectives.

If you are learning to teach in England you can see the specific standards in the relevant Training and Development Agency for Schools (TDA) document (2007b: 7–11).

There is an increasing acknowledgement of the need for a comprehensive and multi-faceted approach to how we meet pupils' diverse learning needs. This is based on a more complex and inclusive view of ability, of how children learn, of the factors that affect learning and the effectiveness of particular teaching approaches. In the last decade, too, there has been a major shift in expectations about what all pupils can and should achieve.

In this unit we consider how different grouping arrangements have been used and are currently used as a means of coping with differences in pupils' performance. We also consider how progression for all pupils can be achieved through teaching and learning approaches that ensure that account is taken of a range of learning needs. Strategies for developing differentiated units of work are provided, building on the subject specialist focus of the reader. See also Units 2.2 and 3.2.

OBJECTIVES

At the end of this unit you should be able to:

■ understand the links between progression, differentiation and pupil grouping
■ evaluate the implications of learning in a range of pupil grouping arrangements
■ discuss definitions of differentiation and their relationship to effective teaching
■ discuss teaching methods which allow for differentiation
■ begin to apply principles of differentiated approaches to learning in lesson planning.

Check the requirements for your course to see which relate to this unit.

GROUPING PUPILS ACROSS THE SCHOOL

Schools have traditionally sought to cope with differences in pupil performance either through streaming, banding and setting or by setting work at appropriate levels for pupils in wide, or mixed, ability classes. *Streaming* places the best performers in one class for all subjects, the least able performers in another class, with graded classes in between. *Banding* places pupils in broad performance groups for all subjects and tries to avoid producing classes comprising only pupils showing low attainment or unwillingness to learn. *Setting* describes the allocation of pupils to classes by attainment in each subject, i.e. streaming or banding for each subject. Broad streaming and banding support a notion of a general intelligence whereas setting acknowledges that pupil aptitude and attainment may be different across subjects and contexts. The recent focus on the needs of more able or gifted and talented learners has led to renewed interest at a policy and school level in *acceleration* or *fast-tracking*, that is, moving a pupil or groups of pupils into a class with an older age group for some or all subjects. The effects of acceleration is one of the most researched aspects of educational intervention with 'gifted' pupils; for further reading see Brody (2004).

Prior to the Education Reform Act (ERA) of 1988, many state schools grouped their pupils in wide ability classes for teaching purposes. The backgrounds, aptitudes and abilities of pupils, coupled with differences in interest and motivation, leads to large differences in achievement between pupils which, by age 11, are substantial and widen even further as pupils grow older. Recognising these differences without prematurely labelling pupils as successes or failures was regarded as an essential prerequisite for organising the teaching of secondary pupils.

Following the move in many state comprehensive schools to mixed ability teaching in the 1960s, strong arguments in favour of, and opposing, the practice arose. The effects of different pupil groupings became the subject of research; this work has recently been reviewed, e.g. Hallam (2002); Kutnick *et al.* (2005). One finding of these reviews is that all pupils gained socially from working in wide ability groups. Such groupings allowed pupils from a variety of backgrounds, as well as abilities, to work together, strengthening social cohesion. Another finding was support for some form of setting in mathematics and other subjects where learning is dependent on a more linear acquisition of skills and knowledge, e.g. in modern languages and science (Harlen, 1997; Ireson *et al.*, 2002). A more recent comparative study of pupils in two comprehensive schools has rigorously documented their differences in knowledge and understanding of mathematics and their motivation and attitude towards the subject (Boaler, 1997). This study identified advantages and disadvantages to both setting and mixed ability teaching and differences in approach to the teaching of mathematics by the subject staff. One conclusion from the study was that pupils in mixed ability classes did as well as pupils taught in ability sets. However, the latter were taught in a more traditional way, through rule-learning then application whereas the former linked mathematics to the everyday life of pupils and used more open-ended project work as part of their teaching strategies. From this study the evidence in favour of grouping pupils one way rather than another is not clear. More recent research paints a similar conflicting picture of the advantage of setting over mixed ability. For example at Key Stage 3 the advantage of a particular type of setting depended on the subject while at General Certificate of Secondary Education (GCSE) level setting appeared to offer little advantage over other forms of grouping. For more detailed information see Ireson *et al.* (2005).

Since the introduction of the ERA (1988) wide ability grouping has been in retreat and successive governments have advocated a return to grouping by ability, together with increased whole-class teaching. A recent review of attainment and pupil grouping across schools does not support the contention that setting alone contributes to success or that setting improves the standards of those not achieving adequately (Kutnick *et al.*, 2005: 6). For example, at Key Stage (KS) 3 the review identifies:

■ no significant difference between setting and mixed ability teaching in overall attainment outcomes
■ ability grouping does not contribute much to the raising of standards of all pupils
■ lower achieving pupils show more progress in a mixed ability classes
■ higher achieving pupils show more progress in set classes (Kutnick *et al.*, 2005: 5).

The advantages of grouping pupils one way rather than another is not clear cut; the evidence suggests that in some circumstances setting may promote the learning for some pupils in some subjects. The downside of setting may be some reduction in positive attitudes by pupils towards their peers, of heightened sense of anxiety of some pupils in higher sets; of reduced motivation of those labelled 'not bright' and diminished cross-cultural mixing.

It is said to be less demanding on teachers to prepare lessons for setted groups. However, 'top sets' may contain not only very able pupils, who may be unchallenged by the work expected of the majority, but may also contain a wide spread of ability and motivation, given that top sets are often large groups. Organising learning and teaching to maximise the potential of all pupils requires teachers to acknowledge that their classes, however grouped, are 'mixed ability'.

The best ways to group pupils remains a vexed question. The Secondary National Strategy in England (Department for Education and Skills (DfES), 2001c) has led to less emphasis on how pupils are grouped and more on strategies which target individual achievement (e.g. the use of data, Assessment for Learning, booster classes, more challenge for some) and at the same time on approaches to learning which can maximise achievement for all pupils (emphasis on thinking processes, learning to learn, intervention via questioning, collaborative learning and literacy across the curriculum). The government's *Five Year Strategy* for England (DfES, 2004l) placed great emphasis on what it terms 'personalised learning' which is intended to allow for greater tailoring of the curriculum to individual needs. Downloadable support material for the strategy is available for KS 3 and KS 4 (Department for Children, Schools and Families (DCSF), 2008g).

The reform of the 14–19 curriculum proposed an approach to pupil progression through a flexible curriculum and qualifications framework and greater emphasis on differentiated approaches to learning and independent enquiry (DfES, 2004d). At the whole school and class level flexible groupings have a very direct link to how we differentiate to meet individual differences in learning and performance. A common theme in the conclusions to the best evidence studies of research into pupil grouping is that what goes on in the classroom, the pedagogic models and the teaching strategies used, is likely to have more impact on achievement than how pupils are grouped.

Finally, all the research reviews point to the long-lasting effects of particular grouping arrangements not only on pupil self-esteem but also on their future engagement in learning. Task 4.1.1 invites you to investigate pupil grouping in your school.

Task 4.1.1 HOW ARE PUPILS GROUPED IN YOUR SCHOOL?

Find out the ways in which pupils are grouped in your school experience school, the reasoning behind the grouping and how it works in practice. Grouping arrangements often change after pupils have been in the school for a term, or a year. Different policies may apply often at the transition of KS3 to KS4 as well as between subjects.

Are primary school records, e.g. National Curriculum (NC) achievement levels used to group pupils? Or are other tests used, such as Yellis or the Cognitive Abilities Test (National Foundation for Educational Research (NFER)-Nelson) used to assess pupils and assign them to groups? See Chapter 6 for more about these tests.

Write a summary of your findings and discuss it with your tutor. Revise and file in your professional development portfolio (PDP).

Grouping within class

Current thinking about effective teaching and learning sees the use of flexible groupings in class as an aid to learning and as a form of differentiation. The reviews on the effects of particular types of pupil grouping on pupils' achievement point to the positive effects of within-class groupings which may include grouping by:

- ability
- mix of ability
- gender
- expertise
- friendship
- age.

The National Strategy for England guidance gives examples of learning activities which make use of flexible and different forms of grouping, such as paired tutoring, jigsawing and rainbowing examples of which may be found under subject headings (DfES, 2004e).

In jigsawing, the class is split into groups to study a topic. Jigsaw is a co-operative learning strategy that enables each pupil of a 'home' group to specialise in one aspect of a learning unit. Pupils meet with members from other groups who are assigned the same aspect, and after mastering the material, return to the 'home' group and teach the material to their group members.

Rainbowing is similar to jigsawing. For example, pupils work in groups of four and discuss a problem or task. Each group member has a colour. Only four colours are used. After discussion new groups are formed, getting together students with the same colour. This means that in a class of 28, new groups of 7 members are formed and the findings or ideas from each group of 4 can be shared. Pupils can then return to their original groups of 4 armed with new ideas. You could of course begin with 4 groups of 7, each with a colour of the rainbow and then form smaller groups of 4 in the same way.

PROGRESSION AND DIFFERENTIATION

By far the greatest challenge to teachers is to ensure progression in the learning of all pupils in their class. Each pupil is different, whether in streamed, banded or wide ability classes. Each pupil brings to school unique knowledge, skills and attitudes formed by interaction with parents, peers, the media and their everyday experience of their world. Pupils are not blank sheets on which new knowledge is to be written. Many pupils may have skills of which the school is not aware: some pupils care for animals successfully; others play and adapt computer games; yet others may work with parents in the family business. Some pupils may know more arithmetic than we dream of as the following parody of stock market practice suggests:

> *Teacher:* 'What is two plus two, Jane?'
> *Jane:* 'Am I buying or selling, Sir?'

Your classroom is a reflection of your pupils' diversity of background and culture that interacts with their potential for learning. Each pupil responds to the curriculum in a different way. Some parents and their children may value a vocational, relevant education more highly because it is immediately applicable to earning a living and may not subscribe to the values placed by the school on a broad, largely academic education.

The teacher must take account of personal interest, ability and motivation to design learning which challenges and interests pupils but, at the same time, ensures for each a large measure of success. Planning and teaching for progression in learning is the core business of teachers. For example, the attainment levels in each subject for the NC in England shows progression. The Qualifications and Curriculum Authority (QCA) suggest that progression in geography involves:

- an increase in the breadth of studies
- an increasing depth of study associates with pupils' growing capacity to deal with complexities and abstractions
- an increase in the spatial scale of what is studied
- a continuing development of skills to include specific techniques and more general strategies of enquiry, matched to pupils' developing cognitive abilities
- increasing opportunity for pupils to examine social, economic, political and environmental issues.

(QCA, 2004a)

How might these characteristics relate to teaching in your subject?

Facilitating progression: approaches to differentiation and personalised learning

Planning learning for pupils, choosing learning objectives and learning outcomes based on knowledge of the pupils and of what constitutes progression in particular curriculum areas, are critical in ensuring pupils' acquisition of the knowledge and skills underpinning progress.

Progression and *differentiation* are therefore two sides of the same coin. The use and success of differentiated approaches depend on teachers knowing their pupils, being secure in their own subject knowledge and having access to a range of teaching strategies. A straightforward way of thinking about planning for progression is the following:

1   What is it you want your pupils to know, understand and do? This might be at the lesson level, the module level or by the end of year.
2   What is it that pupils know, understand and can do at the start of the topic?
3   What sequence of learning activities may help pupils progress from their present state to your objective(s)?
4   How do you know when pupils have reached where you want them to go? (Levinson, 2005a: 100)

In planning for progression you need to start from where the pupils are, not where you would like them to be. Not all pupils are at the same level so some degree of differentiation must be built into your programme for most pupils if you expect to achieve points 1–4.

There is no one right way to differentiate for pupils. Effective differentiation is a demanding task and is about raising the standards of all pupils in a school, not just for those underachieving, with learning difficulties or the gifted. The purpose of a differentiated approach is to maximise the potential of the pupil and to improve learning by addressing the pupil's particular needs. But what exactly is meant by differentiation and how is it achieved?

Consider the following definitions. Differentiation is:

■   a planned process of intervention by the teacher in the pupil's learning
■   the matching of work to the differing capabilities of individuals or groups in order to extend their learning
■   about entitlement of access to a full curriculum
■   'shaking up' what goes on in the classroom so that pupils have multiple options for taking in information, making sense of ideas, and expressing what they learn (Tomlinson, 1999).

How would you describe differentiation? In Task 4.1.2 an aspect of the NC for England is interrogated for its implications for differentiation.

However differentiation is defined, the challenge begins with its implementation and practice, which in turn is affected by teachers' beliefs about the ability of their pupils, by expectations of particular groups of pupils, and by understanding of how we learn and of optimal learning environments. The nature of the subject itself and the kind of learning it involves may also affect differentiation.

Differentiation has sometimes had a bad press because at its worst it has implied an unrealistic and daunting demand on teachers to provide consistently different work and different approaches at the level of the individual pupil or has, unwittingly perhaps on the part of teachers, placed a ceiling on achievement for some pupils.

At its best, given what we know about effective approaches to learning, the influence of high expectations and the potential of all pupils, differentiation may be said to combine a variety of learning options which tap into different levels of readiness, interests, ability and learning profiles with more individualised support and challenge at appropriate times and in appropriate contexts. As we know relatively little about the potential of each pupil, differentiation should be used sensitively and judiciously.

Task 4.1.2 DIFFERENTIATION IN THE NATIONAL CURRICULUM

In the guidance to teachers for the NC for England (General Teaching Requirements: Inclusion) there are statements which identify actions teachers might take to promote a 'secure learning environment for all pupils', just one aspect of 'Inclusion'. These statements include:

- using teaching approaches appropriate to different learning styles
- using, where appropriate, a range of organisational approaches, such as setting, grouping or individual work, to ensure that learning needs are properly addressed
- varying subject content and presentation so that this matches their learning needs
- planning work which builds on their interests and cultural experiences
- planning appropriately challenging work for those whose ability and understanding are in advance of their language skills
- using materials which reflect social and cultural diversity and provide positive images of race, gender and disability
- planning and monitoring the pace of work so that they all have a chance to learn effectively and achieve success (QCA, 2008a).

In a group with other subject student teachers discuss which of these statements are statements of differentiation. To help you, use the descriptions of differentiation mentioned above this task.

Later, write a working definition of differentiation and discuss it with your tutor.

This perspective on differentiation, as liberating rather than constraining, relies on a number of broader principles informing classroom learning and teaching:

- a focus on key concepts and skills
- opportunities for problem solving, critical and creative thinking
- ongoing assessment for learning
- a balance between flexible groupings and whole-class teaching
- identifying pupils as active learners with whom learning goals and expectations are shared
- collaborative and co-operative learning
- achievable but challenging targets
- motivating and interesting learning activities
- supportive and stimulating learning environments.

Managing differentiation

Differentiation starts with a clear view about what you want your pupils to achieve and what individual pupils may need as a particular learning goal, using and acting upon what you know about pupils' previous learning and achievement using assessment data from a range of sources. Then begins the consideration of appropriate differentiation strategies and how the process can be managed; a useful guide is Kerry (2002). It is unrealistic to expect one teacher to plan differentiated work separately for each pupil; it is perhaps better to identify groups of pupils who can work to a given set of objectives using methods suitable to those pupils and the topic in question.

Differentiation always needs to be included as part of your day-to-day lesson planning. Your lesson planning proforma should include a reminder about differentiation. See also lesson planning in Unit 2.2

It may be helpful to have a framework or steps in which to plan work:

Step 1   Your aims and short-term objectives must be broad enough to apply to most pupils in your class. There are often a number of ways of achieving the same goal.

Step 2   Consider which activities to give pupils, linking them to what the pupil already knows and then identifying outcomes. Achievable outcomes are one way of ensuring motivation but must set pupils a challenge, i.e. not be too easy. By identifying achievable outcomes for different groups of children, the process of differentiation is begun.

Step 3   The selection of an activity or more than one. As well as factors in step 2, check the availability of resources, the back-up needed, e.g. instructions are pupil friendly, the language is suitable for your pupils.

Step 4   Planning must include assessment. This can be achieved in a number of ways, for example by question-and-answer sessions, taking part in small group discussions, responding to queries in class, asking questions of pupils working on an activity, listening to pupils discussing their work as well as marking books or short tests. The information gained helps you identify the next steps for the pupil. Assessment must reflect your objectives and be aligned with your learning outcomes.

Some lesson plans or schemes of work plan for differentiation by identifying different priorities for activities, such as:

■   must/should/could
■   core/support/extension.

For example, an activity is selected which all pupils must attempt; it may contain the core idea of the lesson.

However, differentiation models should also recognise that pupils' learning needs may not be fixed or permanent and may relate to the learning context or topic at hand. Differentiation may therefore involve support or challenge being given to different pupils at different times, for example sometimes to:

■   a whole group
■   a targeted group
■   those who work at speed.

Differentiation strategies: stimulus–task–outcome

The outcome of any particular task depends on the way it is presented to the pupil and how they respond. Teaching methods can be restricted by our own imagination; we are inclined to present a task in just one way with one particular learning outcome in mind, rather than to look for different ways to achieve our goals or to accept a range of sensible responses. One important, but limited, teaching goal is to ensure pupils remember things which often involves rote learning, for example, learning Mark Antony's speech on the death of Caesar. Very simply this exercise is:

STIMULUS Play the role of Mark Antony in a class presentation of excerpts from *Julius Caesar*.
TASK Learn by heart the relevant text.
OUTCOME Complete oral recall.

Much learning depends on recall methods: learning the names of element symbols in science; preparing vocabulary in a language lesson; recalling formulae or tables from mathematics; learning to spell. Recall is necessary, if unexciting.

If we wish to help pupils recall and *use* knowledge then we move up a level, to consolidate and widen understanding. This situation opens opportunities to use a variety of contexts including ones directly appropriate to pupils' needs, i.e. differentiation by choice of stimulus. For example, to consolidate pupils' understanding of punctuation you could:

1   ask pupils to punctuate a piece of text from which the punctuation has been removed
2   as 1 above using a written report of an interview
3   as 1 or 2 above but read it through first with the pupils
4   engage in a taped discussion with pupils and ask pupils to write a short report of what was said, with verbatim examples
5   ask pupils to interview other pupils, or staff, about a topic and write a report which includes a record of some interviews
6   ask pupils to write a scene for a play.

Thus for different stimuli all pupils consolidate their understanding of punctuation but the level of outcome is different according to the difficulty of the task and ability of the pupils, i.e. differentiation and progression.

Now complete Task 4.1.3.

Task 4.1.3 **LESSON PLANNING FOR DIFFERENTIATION: DIFFERENT TASKS FOR A SIMILAR LEARNING OUTCOME**

Choose a specific learning outcome for a topic you have taught or are about to teach. Identify two or three different tasks that allow pupils to achieve the identified learning outcome. In which ways are the tasks the same and different? Use the example above, of developing understanding of punctuation, to help you.

Discuss the tasks with your tutor. If possible try out the tasks with your class and review the outcomes of the lesson

---

By contrast with the task above, Task 4.1.4 (p. 166) invites you to discuss and identify ways in which one stimulus might be used for different outcomes.

Differentiation through teacher input and support

Differentiation also takes place at the *point of contact with the group or individual*. The level and nature of your response to pupils is itself an act of differentiation and includes:

■   checking that they understand what they are supposed to do
■   listening to a discussion and prompting or questioning when needed
■   helping pupils to brainstorm an idea or problem
■   asking questions about procedure or techniques; suggesting further action when difficulties arise or motivation flags

Task 4.1.4 **LESSON PLANNING FOR DIFFERENTIATION: ONE STIMULUS WITH DIFFERENT TASKS AND OUTCOMES**

You have a set of photographs showing the interiors of domestic kitchens covering the period 1850 to the present. Describe two or more ways in which you could use these photographs to teach your subject. Confine your discussion to a class you teach covering one to two lessons. For each example, identify

■ how you use the photographs
■ the activities you set your pupils
■ the objectives and learning outcomes
■ how you assess outcomes
■ the ways in which the activities are differentiated.

Analyse your plan in terms of task and outcome for the differentiated approaches you develop. If you do not like the choice of photographs, choose your own stimulus, e.g. an astronaut working in space lab; a Salvador Dali painting or an environmental activist at work.

Discuss your plan with your tutor. Identify how differentiation can be achieved for your pupils and why your choices of task, learning outcomes and assessment are appropriate.

■ giving pupils supporting worksheets or other written guidance appropriate to the problem in hand; the guidance might explain the topic in simpler terms or simpler language
■ checking pupils' notebooks and noting progress
■ marking pupils' work
■ encouraging pupils by identifying success
■ setting targets for improvement
■ increasing the demand of an existing task
■ noting unexpected events or achievements for a plenary session.

Now try Task 4.1.5.

Task 4.1.5 **DIFFERENTIATION: CLASS–TEACHER INTERACTION**

Discuss the above list of teacher support strategies with other student teachers and identify those strategies appropriate to the teaching of your subject. Add to the list of responses for your teaching.

Knowing how to set differentiated tasks depends on how well you know your pupils. The activity needs to be challenging yet achievable. Other ways in which *activities can be differentiated* include:

■ their degree of open-endedness
■ the pupils' degree of familiarity with the resources

■ whether the activity is a complete piece of work or a contributory part of a larger exercise
■ the amount of information you give pupils
■ the language level at which it is presented
■ whether the activity is set orally or by means of written guidance
■ degree of familiarity with the concepts needed to tackle the activity
■ the amount of guidance given to pupils; for example, in science lessons, the guidance given on making measurements, recording data or drawing a graph.

The activity suggested in Task 4.1.4 could be discussed in terms of these criteria. Task 4.1.6 invites you to appraise this list for your own subject teaching.

Task 4.1.6 **DIFFERENTIATION: HOW THE TASK IS PRESENTED AND SUPPORTED**

Discuss the list above with other student teachers in your subject. In groups, say of two people, rewrite the list using strategies appropriate to your subject and the context of your teaching.

Share your list with the group and go on to revise your own list.

Differentiation by outcome and how the activity is assessed

Differences in outcome may be recognised by the amount of help given to pupils and how the activity was set and supported. This aspect of differentiation was referred to in Task 4.1.6. In addition, your expectations of what counts as a satisfactory response to the activity lies in your assessment criteria. Your assessment strategy reflects your criteria. These criteria might include:

■ the extent to which all aspects of the problem have been considered
■ the adoption of a suitable method of approaching the activity
■ the use of more difficult concepts or procedures in planning
■ the recognition of all the factors involved in successful completion of the activity and limiting the choice appropriately
■ thoroughness and accuracy of recording data in a quantitative exercise
■ appropriateness and selection of ways to present information and the thoroughness and depth of analysis
■ use of appropriate ideas (or theory) to discuss the work
■ accuracy and understanding of conclusions drawn from an activity, e.g. are statements made appropriate to the content and purpose of the activity
■ distinction between statements supported by evidence from speculation or opinion
■ the way the activity is written about, such as the selection of appropriate style for the target audience
■ the ability of pupils to express themselves in an increasingly sophisticated language
■ the use of imagination or insight
■ the selection of appropriate diagrams, sketches or pictures
■ sensible use of information and communications technology (ICT) to support a task
■ recognition of the limitations of the approach to a problem and awareness of ways to improve it.

By choice of assessment criteria you differentiate the work set. Statements such as those listed help you construct your assessment strategy.

Differentiation through curriculum design

Moving from differentiating your teaching in a lesson to one embracing the curriculum, one model of a differentiated curriculum suggests that the curriculum needs to be organised around the following core elements:

- *learning environment or context*, e.g. changes in where learning takes place; open and accepting classroom climate
- *content*, e.g. greater levels of complexity, abstraction
- *process*, e.g. promotion of higher-level skills, greater autonomy, creative thinking
- *product*, e.g. encouraging the solving of real problems, the use of real audiences (Maker and Nielson, 1995; DCFS, 2008f).

You might refer back to the ideas in the list above, 'differentiation by outcome'.

Differentiation can therefore include different or enriched learning experiences that take place outside the classroom or even the school, the environmental factor. The NC exemplar schemes of work give examples of additional learning opportunities for each subject area (DfES, 2005d). The government guidance on teaching gifted and talented pupils also gives guidance on enrichment and extension beyond the classroom, many examples of which are relevant for most pupils (DCFS, 2008f; QCA, 2008b).

Differentiation is good teaching and requires that you know your pupils. This knowledge enables you to judge the extent to which pupils have given an activity their best shot or settled for the easy option. Your role is to motivate your pupils and give support. Some pupils may present a greater challenge than others and examples are given in case studies below. It is important to remember that you are unlikely to be successful with all pupils all the time. Read these case studies and then address Task 4.1.7. See Rose (2004) for further discussion of differentiation of pupils with special educational needs (SEN).

CASE STUDIES OF PUPILS

Peter

Peter is a popular member of his group and has an appealing sense of humour. He can use this in a disruptive way to disquiet teachers while amusing his peers.

He appears very bright orally but when the work is of a traditional nature, i.e. teacher led, he often avoids the task in hand; it is at such times that he can become disruptive. His disruption is not always overt; he employs a range of elaborate avoidance tactics when asked to settle to work and often produces very little. His written language and numeracy attainments are significantly lower than those he demonstrates orally.

When given responsibility in groups, Peter can sometimes rise to the challenge. He can display sound leadership ability and, when he is motivated and interested in a group project, can encourage his peers to produce a good team effort. His verbal presentations of such work can be lively, creative, humorous and full of lateral thinking. At such times Peter displays an extensive general knowledge.

Peter's tutor is concerned about Peter's progress. He fears that Peter will soon begin to truant from those subjects in which teaching is traditional in style. He is encouraging Peter's subject teachers to provide him with as much problem-solving work as possible.

Filimon

Filimon arrived a year ago from Ethiopia via the Sudan. He had not been at school for at least a year due to his country's war. He speaks Sunharic at home, as well as some Arabic, but knew no English on arrival. Eight months of the year he has spent at school here have been a 'silent period' during which time he was internalising what he was hearing. Now he is starting to speak with his peers and his teacher. He has a reading partner who reads to him every day and now Filimon is reading these same stories himself.

Joyce

Joyce is a very high achiever. She always seems to respond to as much extension activity as she can get. She puts in a lot of effort and produces very well presented work (e.g. capably using ICT), and amply demonstrating her ability to understand, evaluate and synthesise. Joyce's achievements are maximised where she is able to work on her own or in a pair with one of a couple of other girls in the class. In other groups she tends to keep herself to herself. Some teachers are concerned that she is not developing her social and leadership potential.

Joyce's parents put a lot of pressure on her and are keen for Joyce to follow an accelerated programme wherever this is possible. Should she achieve her ambitions for higher education, Joyce will not be the first in her family to make it to Oxbridge.

These case studies were provided by Paul Greenhalgh, adapted by him from Greenhalgh (1994). You may find it instructive to select one of the pupils described above and consider how their presence in your class would modify your lesson planning.

Task 4.1.7 **WRITING YOUR OWN CASE STUDY**

Prepare a short case study of two pupils in one of your classes. Identify two pupils for whom further information would be helpful to you in lesson planning and use the examples of case studies above to help you identify the information you need to collect. Do not use the pupil's real name in any report you make or discussion outside the school.

Collect information from the class subject teacher and the form teacher. The form teacher can give you background information about the pupils, as much as is relevant to your study.

After collecting the information and writing your report ask the class teacher to read it and comment on it. Finally, use the information to amend Task 4.1.6 or plan a new lesson.

If there are other student teachers in your school share your case studies with them. Use the case studies to identify some learning needs of these pupils and plan teaching strategies to take account of these needs. The study can contribute to your professional portfolio.

Finally we return to the topic of pupil grouping and the use of the teaching strategies of jigsawing and rainbowing. Task 4.1.8 invites you to explore one of these strategies over a small number of lessons and evaluate the experience.

Task 4.1.8 **IN-CLASS GROUPING AND TEACHING STRATEGIES**

Develop a plan to try out either jigsawing or rainbowing with a class you teach. Read the following two papers before embarking on the task.

1   *Grouping Pupils for Success* (DfES, 2006a); note particularly pp. 6–8
2   Group work *Pedagogy and Practice: Teaching and Learning in Secondary Schools. Unit 10 Group Work* (DfES, 2004e). Read the whole file before moving on.

Further support

3   Advice and strategies on group working including rainbowing and jigsawing. Download the file 'Using interactive whiteboards in secondary mathematics' from http://www. eriding.net/maths/tl_resources_sec.shtml. Then scroll down to 'mini-whiteboards' for the download document (accessed 19 December 2008).
4   Jigsawing in a design and technology (D&T)classroom to develop design skills. The file shows how jigsawing was used in a D&T classroom. Access the file 'Design and technology; framework and training materials' at http://www.standards.dfes.gov.uk/ secondary/keystage3/respub/design/foreword/.
    Select 'Teaching the sub-skills of designing (TRM 4)', in *Appendices* and download this file (accessed 25 September 2008).

Suggested procedure

Try out *one* of strategies identified in 2 and 3 above with a class you teach, e.g. one which would respond to new grouping arrangement; or with a class with whom you are not making expected progress.

Select a topic suitable for sub-group working, draft a rough plan for a lesson, including a set of aims and discuss the plan with your tutor or class teacher. Redraft and try out the plan on a small scale over one lesson, to test out the groupings you can make, the instructions you give and how to move pupils around groups, and how to round of the lesson. Evaluate the lesson and decide whether to continue with this class or choose another.

Develop a set of lesson plans covering three lessons, including aims, learning objectives and learning outcomes for each lesson.

Lesson 1 Identify the task, provide background for the pupils, including resources and perhaps homework; brief your class for the second lesson.
Lesson 2 Introduce the activity, indicate how the groups will form and re-form and what pupils are expected to do in each group. Allow time for pupils to consolidate what they gained from the activity, e.g. by a plenary session and introducing the final lesson.
Lesson 3 Consolidates the work done, leading to the final product.
    Check your plans with your tutor/class teacher and teach the lessons.

You may find it helpful for your evaluation to keep a pocket notebook handy to jot down events that occur, the good things and bad things that happen, including pupil responses and comments. Flesh out these notes after each lesson.

Evaluate your teaching against your aims, including implicit aims such as enjoyment, enthusiasm, greater participation, improved behaviour pattern.

Identify the pupil gains gain from the strategy, including knowledge and skills both cognitively and affectively, the advantages and drawbacks of the strategies.

Discuss your evaluation report with your tutor and file in your PDP. The report may contribute to your coursework.

SUMMARY AND KEY POINTS

This unit has discussed the ways in which pupils can be grouped for teaching, from mixed-ability classes to streaming, setting and banding. The relative merits of each strategy are mentioned in the light of research. The advantages of any one way of grouping pupils are not clear-cut and other factors strongly influence the achievement of pupils, such as the expectations of the teacher. The developments of In-class differentiation of work for pupils and personalised learning have moved the discussion away from grouping.

Differentiation is addressed and several examples are provided. The ways in which tasks are selected, set, supported and assessed are each susceptible to modification to meet the needs of different pupils. Differentiation is addressed from a number of aspects starting from a simple model of planning using stimulus, task and outcome. The importance of the role of the teacher in supporting and guiding their pupils is emphasised. Strategies for the management of differentiation are addressed, emphasising that in all classes there are pupils with different needs irrespective of the way the pupils are grouped for teaching. Differentiating your teaching is a responsibility of all teachers not just for those teachers addressing pupils with special needs.

We have suggested that tasks should be related to the experience of the learner whenever possible and further, that the outcomes should be achievable. You may find the discussion about Vygotsky and Piaget in Unit 5.1 helpful in developing those ideas. For those wishing to go further and explore lesson planning in a different way, using a constructivist approach (Unit 5.1), see Biggs and Moore (1993) who developed a model described as 'constructive alignment'.

The skills of teaching this way are acquired with experience and, importantly, better understanding of your pupils. While acknowledging that many student teachers move schools at least once in their course it is important you begin to understand differentiated approaches to teaching and learning.

Check which requirements for your course have been addressed through this unit.

FURTHER READING

DfES (Department for Education and Skills) (2006a) *National Strategies: Grouping Pupils for Success*: http://www.standards.dfes.gov.uk/secondary/keystage3/all/respub/ns_grp_pup_succ.

This document discusses the arguments for and against different ways of grouping pupils but moves on to explore more flexible ways of grouping, beyond rigid streaming and setting models and drawing upon evidence from research. It does not give subject specific advice about grouping but does identify contexts in which different ways of grouping pupils may be beneficial.

DCSF (Department for Children, Schools and Families) (2008l) *Key Stage 3 National Strategy Materials*: http://www.standards.dfes.gov.uk/secondary/keystage3/.

The Standards site offers a wide range of supporting material for teachers and can be explored for materials on ability grouping and differentiation as well as other teaching resources.

Ireson, J. M. and Hallam, S. (2001) *Ability Grouping in Education,* London: Paul Chapman.

This book provides an overview of ability grouping in education and considers selective schooling and ability grouping within schools, such as streaming, banding, setting and within-class grouping. It addresses the implications of ability grouping for teachers, managers in education and the wider community. It emphasises the complexity and far-reaching implications of ability grouping. Later research by these authors addresses the effects of ability grouping on GCSE attainment (Ireson, J., Hallam, S. and Hurley, C., 2005).

Kerry, T. (2002) *Learning Objectives, Task-setting and Differentiation*, London: Nelson-Thornes.
This book, part of a series addressing professional skills for teachers, clarifies each of these skills, explains their purpose and explores issues around, and the consequences of, the implementation of these skills. Practical application is discussed supported by examples and activities. It also encourages readers to assess their own implementation and progress by analysing the tasks against standards.

Rose, R. (2004) 'Towards a better understanding of the needs of pupils who have difficulties accessing learning', in S. Capel, R. Heilbronn, M. Leask and T. Turner (eds) *Starting to Teach in the Secondary School: A Companion for the Newly Qualified Teacher*, Abingdon: RoutledgeFalmer. This chapter addresses lesson planning for pupils who have difficulty learning. The author suggests an inquiry based approach to teaching alongside co-operative teaching and learning.

Wiliam, D and Bartholomew, H., (2004) 'It's not which school but which set you're in that matters: the influence on ability-grouping practices on student progress in mathematics', *British Educational Research Journal,* 30 (2), 279–294.

Additional resources for this unit are available on the companion website:
www.routledge.com/textbooks/9780415478724

ADOLESCENCE, HEALTH AND WELL-BEING

Margaret Jepson, Barbara Walsh and Tony Turner

INTRODUCTION

Adolescence is a transitional stage of growing up that changes a child into an emerging adult (teenager or young person) and involves biological, social and psychological changes including dramatic changes to the body. The expectation of different societies and/or families influences young people, especially in the school setting. Important factors that may affect adolescents' self-image and their schooling relates to the family, including the socio-economic status, employment history and family harmony. Negative features in any one of these may portend poor career prospects (Child, 2004).

Adolescents begin to develop an independence from their parents which, on its own, can cause many problems and often these can be reflected on their behaviours in school, especially because their parents' opinions become less important and this is sometimes seen as them being disrespectful and ignorant. It is important for you as a student teacher to understand these changes; how a pupil is feeling on the inside often affects how they cope and feel in the external environment.

Most pupils want to be normal, to conform to what they see in others of their peer group. The idea of 'fitting in' is extremely important to them at this stage. You can begin to see the friction that can occur when pressure from their peers contradicts expectations of parents and teachers. Conforming, in part, concerns appearance; personal appearance becomes a highly sensitive issue during adolescence, one because of the notion of normality, shape, size, etc. but also sexuality and emerging relationships. In these respects the place of a balanced diet is important; school has an important part to play in countering obesity and anorexia, both aspects of self-image as well as areas of concern nationally.

Some pupils carry the weight of expectation of their parents, such as choosing subject options and career pathways. Under pressure some pupils begin to think they can't do well in school anymore and the more they think this, the worse they do. Teachers and other adults begin to demand more of secondary pupils, to stretch their minds. Many adults expect pupils to think more like adults than children. Scientists have found that the human brain also goes through a growth spurt during adolescence. This process and the pressure of expectation are difficult for some pupils to deal with as they start to think and act differently than hitherto.

Adolescence is full of changes and full of challenges. There are many changes that happen that can affect how pupils can react in different situations. It is an area that you may find frustrating to deal with, especially the mood swings that are synonymous with this stage in the growth cycle;

adolescents can be excited one minute and depressed the next. It is a time when pupils are searching for a personal identity that often gives them a feeling of insecurity.

Schools have a vital part to play in this developmental period. School may be, for some, the only place where they find consistent messages and a place where their personal autonomy can grow. Schools are crowded places and require a strong discipline for all; these conditions are not necessarily compatible with those for the emergence of the autonomous individual.

OBJECTIVES

At the end of this unit you should be able to:

■ describe and understand some of the physical differences between pupils during adolescence

■ appreciate the effect of external pressures and influences on pupils' behaviour and identify some implications of these differences for teaching and learning

■ discuss healthy eating and the role of the school in promoting this ideal.

Check the requirements for your course to see which relate to this unit.

ABOUT DEVELOPMENT AND GROWTH

Young people tend to have growth spurts, particularly after puberty, the point at which the sex glands become functional. Most girls mature physically earlier than most boys. There are differences in growth rates between boys and girls at the onset of puberty; some girls showing a growth spurt at an earlier age than most boys. However, there is little difference, for example, in mean height of boys and girls up to the age of 13 but after the age of 16 boys on average are over 13cm taller than girls. Height increases appear earlier than weight increases and this has implications for physical activity. The differential rate of height and weight development is the origin of clumsiness and awkwardness of some adolescent pupils. As well as obvious gender differences between pupils in a coeducational context, the differences between individuals within a group of boys, or a group of girls, can be quite large and obvious. These differences in development can be worrying for the individual and may affect pupils' attitudes and performance to academic work. For example, it can happen that some pupils who have developed physically earlier than their peers may dominate activity in a class, causing a number of pupils to reduce their involvement for fear of being ridiculed by more 'grown-up' members of the class.

Another feature of physical development is the onset of puberty. The age at which this occurs varies quite widely between both individuals and cultures, as does the period of puberty. Adolescence can begin at age 10 for some, for others much later and may finish around the age of 19, not like the defined teenage years of 13–19. This means that a 12-year-old girl may be in a pre-pubertal, mid-pubertal or post pubertal state. A 14-year-old boy is similarly placed. Thus it is not sensible to talk to a 14-year-old cohort of pupils as though they are a homogenous group. The onset of puberty is affected too by environmental factors, including diet.

There is evidence that environmental factors affect growth (Tanner, 1990). These factors include:

■ the size of the family; many larger families have children of below average height

■ where the pupil is raised; urban-reared children are often taller and heavier than those raised in a rural society

■ the socio-economic status of the family (parents); lower social class, defined by the employment status of the parents correlates with having shorter children

■ prolonged unemployment has a similar effect to socio-economic factors.

There may be evidence in your school of a link between 'growth and development' and socio-economic factors such as free school meal provision. The up-take of free school meals is used, for example, by the Office for Standards in Education (Ofsted) in school inspections, as a proxy measure of deprivation. Childhood poverty continues to have a significant impact on young people's well-being and education.

Managing your classes

The variation in physical development of pupils shown, for example, in any year cohort has implications for your management of classes. These differences are particularly apparent in Years 7–9 and may stand out in activities which prosper on physical maturity or physical control. Boys in early adolescence who develop late often cannot compete with their peers in games; and girls who mature earlier than their friends can be also be advantaged in physical education and games but, at the same time feel embarrassed. Thus competitive activities such as running or throwing or physical confrontation games such as football, hockey and rugby favour faster-developing pupils. Equally important is physical control, the ability to co-ordinate hand and eye, and control tools and equipment properly and safely. In the past, some adolescents have been regarded as clumsy which may be related to growth spurts, described above. Activity in subjects such as physical education, art and design, technology, science and computing depend, in part, on good co-ordination and psychomotor skills. If growth in weight and height are behind the norm this may have a different impact at that time.

Now try Task 4.2.1, see p. 176.

Some research suggests that pupils physically maturing faster score better on mental tests than pupils developing more slowly. Girls develop physically and mentally faster than boys on average and the results of standard assessment tasks (SATs) and General Certificate of Secondary Education (GCSE) results are a reflection of this.

Large differences in performance in school subjects which, taken together with differences in physical development, has raised the question of whether pupils should be grouped in classes by age, as they are now, or whether some other method should be used to group pupils for teaching purposes, for example, by achievement. Some other educational systems require pupils to reach a certain academic standard before proceeding to the next year, leading to mixed-age classes. Thus under-performing pupils are kept back a year to provide them with an opportunity to improve their performance. This practice has a big impact on friendship, self-confidence and self-esteem.

When pupils feel good about themselves they feel confident and ready to experience new things, but the opposite feeling often means that they take every small setback as destroying their confidence. It is important for you to differentiate your tasks so every pupil achieves some success; personalised learning helps give them belief in their own abilities and the confidence to take on more challenging tasks. It is sometimes difficult with every group you teach to establish which pupils to extend further so they do not plateau and become bored, and which to gently cajole so you do not destroy that finely tuned confidence. Giving your pupils responsibility in your lessons, making sure they achieve some success and giving them positive constructive feedback helps them feel more confident about themselves and improve their self-belief. As a student teacher in your

Task 4.2.1 **A PROFILE OF A CLASS**

Select a class you teach and find out as much as you can about the background of your pupils. Then shadow the class for a day and try to relate your findings to the ways pupils respond to teachers and different subjects.

Discuss your plan with your tutor who can direct you to appropriate sources of information such as the form tutor. The school physical education staff may well be able to provide information on physical development. There may be special provisions for some pupils in your school that provide additional information, e.g. homework club or other provision for pupils unable to work at home. See also the notes below this task.

When you visit classrooms get permission from the teachers, tell them what you are doing, what is to happen to the information and what is expected to emerge. Be prepared to share your findings with them.

Write a short report for your tutor. Respect the confidentiality of information you acquire in any written or oral report. Reports should not quote names. The report may contribute to your Professional Development Portfolio (PDP). Record in your diary your personal response to this work and any implications it has for you.

Notes to help with Task 4.2.1
Some of the information you might gather from the form tutor you are working with includes:

- the names and the numbers of boys and girls
- the ethnicity of pupils; check the way the school reports ethnicity
- the religious or cultural background of pupils
- recent immigrants or children of families seeking asylum
- patterns of absences and whether absences are supported by notes from parents or guardians
- the regularity of completing homework and its quality (the class teacher should have such a record).
  Gather data about:
- the height and weight of pupils; are any pupils deemed overweight or obese (see later in this unit)
- pupils who have statements of special need and the reason for this
- do they have a support teacher and why
- pupils who do not have a support teacher but need one
- pupils who have been identified as 'gifted and talented' in the school
- pupils with specific learning difficulties, e.g. dyslexia.

first school experience it is very difficult to achieve balance between extending pupils further *and* maintaining their personal self-belief; it is an area where your tutor can help to guide you.

Everybody has feelings, it is impossible to be human and not have them and during adolescence emotions are particularly strong. It is often easy for adolescents to feel helpless and overwhelmed by emotions. During this time they may say things in anger or hit out only to regret it later. It is important for you to gauge the situation, if they do shout out, act out of character or tell you some personal details. Some pupils are happy to share their feelings whilst others hide them, if they are suppressing anger, sadness or bitterness it might manifest itself in them blowing small situations totally out of proportion. The side effects of this could be headaches, lethargy and disaffection towards their work and as a teacher it is important to be aware of these changes. (See Unit 3.3).

The school at which you were a pupil may be quite different from your school experience school. Task 4.2.2 is intended to reveal some of those differences and provide you with an opportunity to discuss their implications for your teaching.

Task 4.2.2 PUPIL BACKGROUNDS AND SOME IMPLICATIONS FOR YOUR TEACHING

In what ways do the experiences of school of pupils in your school resemble or differ from pupils from your own school days; or perhaps your expectations prior to your initial teacher education course? In addition to the information gained in Task 4.2.1 you might consider collecting information about:

■ family sizes, and extended families in the school
■ socio-economic classes into which most of your pupils' families may fit
■ employment rates of parents
■ achievement, as measured by standardised national tests; e.g. SATs levels or tests of cognitive ability.

Other factors which may help in describing the pupils in your school include:

■ how pupils are dressed and adherence to school uniform
■ self-confidence, willingness to talk to teachers and to each other
■ attitude to authority, including respect for other pupils and for teachers.

Collect your own impressions of the pupils in your school, for discussion with other student teachers and your tutor. You could limit the study to one class you teach, pooling information with other student teachers.

Repeat this exercise after a period in the school and see how familiarity with the pupils and school has altered first impressions. Keep a record of your inquiry in your PDP. Record in your diary your personal response to this work

---

We have discussed the physical development of pupils and drawn attention to the differences in development both within a gender group and between boys and girls. A large influence on physical development is diet, lifestyle and attitude to exercise and games. There is concern about the dietary habits of some young people, in part about risk of disease, in part about the level of fitness of many young people and issues of overweight and obesity. Yet others draw attention to the increased use of computers in entertainment and the accompanying sedentary habits this entails. Thus we turn to consider diet, development and the curriculum. The advent of the concept of 'healthy schools' in England has required teachers to address 'health' issues including health and well-being (including emotional health), healthy eating, physical activity as well as Personal, Social and Health Education (PSHE). This development has raised the profile of adolescent eating habits and recommendations for physical activity. This is the subject of the next section.

DIET, HEALTH AND WELL-BEING

Background

In the past hundred years the average height and weight of children and adults have increased and the age at which puberty arrives has decreased. Such average changes are due in part to increased nutritional standards, better conditions of health and sanitation, as well as better economic circumstances for the majority. However the increasingly sedentary lifestyle of young people and the rise in obesity has given rise to increased concern about their diet.

Many adolescents in the UK are not eating healthy diets and or meet the recommendations for exercise. The National Diet and Nutrition Survey carried out by the Food Standards Agency (FSA) examined the diets of British schoolchildren aged 4–18 years (FSA, 2000). It found that adolescents ate more than the recommended level of sugar, salt and saturated fats. The most frequently consumed foods were white bread, savoury snacks, biscuits, potatoes and chocolate confectionery. The results of this type of diet and insufficient exercise are an increase in weight, leading to overweight and obesity. The Royal College of Physicians Report (RCP) described the rapid increase in the prevalence of overweight and obesity in all age groups as 'the obesity time bomb' (see Figure 4.2.1). The estimated cost to the National Health Service of an increasing number of overweight and obese adults is about £0.5 billion in terms of treatment alone and possibly in excess of £2 billion in the wider economy (RCP, 2004).

If current trends continue, at least one-third of adults, one-fifth of boys and one-third of girls will be obese by 2020. If the proportion of obese children continues to rise, a whole generation may have a shorter average life expectancy than their parents (National Heart Forum, 2007: 12).

Why is there this increase in overweight and obesity over the last 20 years? A simple answer is to do with energy balance. People are eating too much for the amount of physical activity that they do (see Table 4.2.1).

Energy from food can be measured in Calories (where 1C = 1000c) or joules (4200 J = 1 Calorie). It is estimated that the average adult whose daily energy intake is just 60 Calories more than their energy output will be come obese within 10 years.

The immediate consequences of overweight and obesity in adolescence are social and psychological. Evidence from a longitudinal study in America on self-esteem and obesity shows that, at ages 9–10, there were no significantly different scores between obese and non-obese children but by 13–14 years of age significantly lower levels of self-esteem were observed in obese boys, obese Hispanic girls and white girls compared with their non-obese counterparts (Strauss, 2000). This decrease in self-esteem was associated with significantly increased feelings of sadness, loneliness and nervousness and with an increased likelihood to smoke and drink alcohol. There is

---

Obesity in 2–4-year-old children almost doubled (5%–9%) in 10 years (1989–1999)
Obesity in 6–15-year-olds trebled (5%–16%) in 11 years (1990–2001)
Obesity in adult women nearly trebled (8%–23%) in 22 years (1980–2002)
Obesity in adult men nearly quadrupled (6%– 22%) in 22 years (1980–2002)

■ **Figure 4.2.1** Obesity in England 1980–2002: some summary facts
Source: RCP (2004: 4)

■ **Table 4.2.1** Energy balance equations

Energy from food = energy expended in exercise = maintenance of body weight
Energy from food > energy expended in exercise = addition in body weight
Energy from food < energy expended in exercise = loss in body weight.

also an element of social marginalisation. When asking adolescents to name their friends, it was found that whilst obese adolescents nominated similar numbers of friends as normal weight adolescents they received significantly fewer friendship nominations from others than were received by normal weight adolescents (Strauss and Pollack, 2003). Obese adolescents were also more likely to receive no friendship nominations than were normal weight adolescents. Obese children are believed by their peers to be lazy, dirty, stupid, ugly, cheats and liars. Those who are overweight are often seen as an easy target for bullying, with little peer pressure occurring to prevent it. Such marginalisation may contribute to the social and psychological effects of obesity.

The later health consequences are also serious. Increasing fatness is closely correlated with the development of type 2 diabetes that used to be diagnosed in middle to later life but is now increasingly seen in young adults and children. Childhood obesity that continues into adult life increases the risk of early death from all disease related causes, including cancers and cardiovascular disease (RCP, 2004).

The government has issued information enabling pupils and parent to calculate if they are obese and guidelines for healthy eating. The current measure of body fatness is the body mass index (BMI). This is defined as a person's weight in kilogrammes divided by the square of their height in metres. Obesity in adults is defined as a BMI of 30 or more and overweight as between 25–29.9. BMI varies with age and there are age and gender specific standards used by the medical profession.

The guidelines for healthy eating are:

■ Base your meals on starchy foods.
■ Eat lots more fruit and vegetables.
■ Eat more fish.
■ Cut down on saturated fat and sugar.
■ Try to eat less salt – no more than 6g per day.
■ Get active and try to be a healthy weight.
■ Drink plenty of water.
■ Don't skip breakfast.

Developing strategies to address obesity

Schools should provide two hours of physical education (PE) per week and in 2007 some 85 per cent of schools were meeting this target. Government advice is that children and young people should achieve 60 minutes of at least moderate intensity physical activity every day but by the age of 16 only 50 per cent of girls achieved this. About 70 per cent of boys achieved this target but the activity levels dropped between the ages of 13–15 years (National Heart Forum (NHF), 2007). A survey of Liverpool schoolchildren researching health and fitness found that more than 25 per cent of Year 5 pupils did not participate in any, or low levels, of physical activity (equivalent to a slow walk) during the morning and lunchtime break. For Year 7 boys the figure was 40 per cent and for girls 60 per cent (Liverpool SportsLinx Project, 2003). Break times and lunchtimes for about half of all Year 7 pupils are therefore spent either sitting or doing no more than slow walking.

A simple answer to the problem of obesity is therefore that adolescents should eat less and exercise more. But this is far too simplistic. Adolescents today are not greedier or have less will power than previous generations. There many factors that contribute to what has become known recently as an 'obesogenic environment', a term now used to describe environments which encourage and promote high energy intake and inactivity. 'Obesity is linked to broad social developments, and shifts in values, such as changes in food production, motorised transport and work/home lifestyle patterns' (Foresight, 2007).

Eating habits

Ofsted have found that most pupils have a good understanding of what constitutes healthy eating through food technology lessons and other subjects particularly personal, social and health education (PSHE), PE and science but 'pupils' knowledge had too little bearing on the food they chose (in school)' (Ofsted, 2007a: 6). Clearly other factors played a greater part in food choice than simply a knowledge of healthy eating. New nutritional standards for school meals in England came into force in September 2006 and in 2007 new standards for all other school food and drink, e.g. breakfast clubs, vending machines and water, were introduced. The Ofsted report above found that whilst the meals provided met the new standards the number of pupils eating school meals had fallen. In the worst cases less than 50 per cent of pupils ate a school lunch (Ofsted, 2007: 6). The reasons were varied but included cost, particularly for low income families who were not eligible for free school meals. Those who were on free school meals were often deterred from taking them where there was a visible means of payment. Schools where all pupils used a credit-type card, with credit placed either by parents or the school, encouraged a higher take up of free school meals.

A 'Food in Schools' pilot project found that pupils' perceptions of the dining room were influential in the take-up of school meals. For example, 'students care as much about sitting with their friends as they do about what they actually eat. So the dining room is seen as a social environment' (Department of Health (DoH), 2007). When queues are long and the dining room is crowded, the take up of school meals is lower; this condition limits participation in extracurricular activities or the opportunity to sit with friends who bring packed lunches.

Schools that had most impact on encouraging healthy food choices had a close partnership with staff, pupils and their families; another favourable factor is where a senior member of staff gives a high profile to this work. In secondary schools where the pupils were involved in making decisions about the types of food offered the take up was higher (Ofsted, 2007a). If the fall in take up of healthier schools meals is not reversed then government's food policies will have limited effect.

The school lunch may be the first meal of the day for many pupils since 18 per cent of 16-year-old boys and 29 per cent of 16-year-old girls do not have breakfast before coming to school. School breakfast clubs are said to improve attendance, punctuality, concentration levels, problem solving abilities and creativity (British Nutrition Foundation (BNF), 2003). Breakfast clubs provide a range of healthy foods and it is reported that this had been a successful first step in engaging pupils on healthy choices. Schools that provide tuck shops are offering healthy snacks and many pupils bought food for their lunch to avoid queuing later in the day (Ofsted, 2007a).

Nearly half of all 14–18 year olds bring in a packed lunch from home. A recent lunch box survey found an over-representation of fatty and sugary foods and an under-representation of fruit, vegetables and starchy vegetables (FSA, 2004). The most common items were a cheese or ham sandwich made with white bread, a packet of crisps and a chocolate bar or cake with a high frequency of sweetened fizzy drinks. Schools that have installed vending machines as a means of raising funds must now have healthy drinks such as water, milk and milk-based drinks rather than sugar-based fizzy drinks. Those vending machines that sold sugary or high salt snacks have been removed (FSA, 2007).

The environment outside school, in terms of influence on food habits, has changed in recent years. There has been, for example:

■ increased consumption of pre-prepared foods and carbonated drinks
■ more 'eating out' in restaurants or 'eating in' through takeaways
■ an increase in snacking, often high in saturated fats, sugar or salt
■ increased pocket money for children; crisps and savoury snacks are the most popular after-school snack.

The advertising and promotional campaigns by the food industry to encourage the sale of their particular brands of food are mainly targeted at children. The top ten advertised food brands in the UK are dominated by the market leaders in the food industry and represent relatively unhealthy food options that are aimed at children. The food industry gave confidential commercial evidence to the House of Commons Committee on Obesity whose Report (HCR) concluded 'It is clear advertisers use their increasingly sophisticated knowledge of children's cognitive and social development, and careful consumer research into their motivations, values, preferences and interests, to ensure that their messages have maximum appeal' (HCR, 2004, para. 105). For example, McDonald's fast-food chain spent over £41 million pounds on advertising in 2002 and the other top ten food brands spend between £6 million and £16 million each in the same period (A. C. Nielson cited in HCR, 2004). By contrast the annual budget for the government's Five-a-Day campaign to encourage eating of fruit and vegetables is £5 million. Some fast food outlets are now offering more healthy fruit and salad options, a welcome move.

There is a link between eating habits, social class and income. Whilst pupils from higher socio-economic groups tend to have a more varied diet, those from lower socio-economic groups have more restricted choice of food because parents purchased only food that is sure to be eaten. Many adolescents from low-income families eat less fruit and vegetables and more foods high in fat, sugar and salt.

The British Medical Association (BMA) point out that almost a quarter of 15 and 16 year olds in the UK smoke at least once a week and over a fifth of this age group report having used drugs in the last month (BMA, 2003). The amount of alcohol consumed by young people in this country is one of the highest in Europe, these problems in their own right cause physical, mental, emotional and social problems which only exacerbates how they feel about themselves.

Physical activity

Inactivity starts with getting to school. In 2005, 44 per cent of secondary pupils walked to school and only 2 per cent cycled, with few schools offering safe storage for cycles (Department for Transport (DfT), 2006). Children who walk or cycle to school are likely to be fitter than those who journey by car and are therefore more likely to enjoy and benefit from sport.

Physical education and school sport form less than 2 per cent of school life but are useful in fostering habits of activity that can last into adult life. Active children tend to be less overweight and tend to achieve more academically. The House of Commons Committee on Obesity received evidence that organised school sport seems to alienate many children and there is ample evidence to suggest that much bullying begins in the changing room (HCR, 2004, para. 256). Other activities such as dance or aerobics could be used to broaden the opportunities for physical activities in school. Again there are fewer opportunities available in terms of access to swimming pools and leisure centres for those in lower socio-economic groups than those in higher socio-economic groups.

An international study showed that young people watch more than four hours of TV a day with even higher levels at weekend, a figure higher than many other countries in the survey (BHF, 2007: 5).

By contrast there is a parallel culture that values thinness. Images of men and women in the media, advertising and popular culture emphasise beauty, youth and thinness. Some adolescent boys but in particular many girls may compare themselves to extremely thin models working in the fashion industry and perceive themselves to be 'fat' in comparison, rather than healthy and attractive. This promotion of the 'ideal' thin body undermines self-confidence. Body image and self-esteem are closely connected. Adolescents with high self-esteem see themselves not necessarily better than others but not worse, recognise limitations and expect to grow and improve. Adolescents with low self-esteem are dissatisfied with themselves and wish that they were different.

Moving forward

The National Curriculum in England addresses the issues of healthy living in a variety of ways. Nutrition is quite firmly based in the science programme of study and all pupils in Key Stage (KS) 3 have an entitlement to 24 hours of practical cookery over the Key Stage. This can be delivered through the teaching of food technology, where pupils develop skills in preparing and cooking meals and other food products. In PSHE classes developing a healthy safer lifestyle is a part of the KS3 curriculum. In PE pupils are to be taught knowledge and understanding of fitness and health.

Now complete Task 4.2.3.

Task 4.2.3 **FOOD AND HEALTH**

Access the National Curriculum for England documents (Qualifications and Curriculum Authority (QCA), 2007a) (or related documents in the country where you are learning to teach) and identify those areas of your subject area, and other curriculum areas, that address issues of healthy eating and healthy lifestyles. In England address the Programmes of Study at KS3 first.

Discuss with your tutor whether there is a co-ordinated approach in your school experience school between the subjects and PSHE.

The *National Curriculum in Action* (QCA, 2004b) and the Department of Health/ Department for Education and Science (DoH/DES,) 'Food in Schools' website (http://www. foodinschools.org/) are useful sources for whole school approaches. Resources for the teaching of nutrition are available on the British Nutrition Foundation website (http://www. nutrition.org.uk/).

Given all these teaching and learning opportunities why is there still a problem with overweight and obese adolescents? Perhaps a major reason is that schools do not exist in a vacuum.

There are those who would argue that advertising foods high in fat, sugar and salt should be banned during children's television shows. The Office of Communications (Ofcom) undertook research in 2004 and found that TV advertising had a moderate, direct effect on pupils' food choices and that indirect effects were likely to be larger. Teenagers were influenced by advertisements that were witty or had subtle messages or featured celebrities as role models (Ofcom, 2006). The sponsorship of food products by companies for schools may raise money for books or sports equipment but can give conflicting messages to pupils if the targeted food products are the 'less healthy' variety, e.g. crisps, chocolates.

There is a range of initiatives in schools in England to try to find a means of improving the health of pupils. The National Healthy Schools Strategy is a project where schools can work with pupils, teachers, families and the local community to actively promote the physical, social and mental well-being of all, and 75 per cent of schools are expected to have healthy schools status by 2009. It is an attempt by the government in England to bring together a multi-faceted approach to the broad problem of improving the obesogenic environment of modern society. The five objectives of the campaign are:

1   Being healthy.
2   Staying safe.
3   Enjoying and achieving.
4   Making a positive contribution.
5   Economic well-being.

National Healthy School Programme
(http://www.clusterweb.org.uk/Children/hs_national.cfm)

Now complete Task 4.2.4.

Task 4.2.4 **THE SCHOOL ENVIRONMENT**

The definition of an obesogenic environment is one that encourages and promotes high-energy intake and inactivity. Use the answers to the following questions and your observations to evaluate the environment of your school experience school.

1   How many of your pupils walk or cycle to school?
2   Is there safe storage for cycles?
3   Is there a school tuck shop or vending machine and what is sold?
4   Is there a breakfast club?
5   How would you describe the environment of the dining room at lunchtime?
6   What proportion of the school curriculum is given to PE and sports activities?

In addition observe the:

■   activities of pupils during break and lunchtime on two different weekdays
■   the food availability and food choices made at lunchtimes on one or two days.

Write a short account of this aspect of the school environment and discuss it with your school tutor. File the account in your PDP.

Schools need to empower adolescents by the provision of knowledge and understanding of diet and a healthy lifestyle and confidence in themselves to make the choices that control their lifestyle. You may wish to design a small-scale research project to identify the factors that contribute to a whole-school approach to adolescence, health and well-being (see Task 4.2.5).

Task 4.2.5 **ADOLESCENCE, HEALTH AND WELL-BEING: A WHOLE-SCHOOL APPROACH**

The task is in two parts.

Part 1

Write a literature review which focuses on a whole-school approach to adolescence, health and well-being. You may need to focus on recent research, say in the last ten years in order to manage the task, on what has been published in the three areas. Your purpose is to convey the knowledge and ideas that have been established on these three areas and their strengths and weaknesses (not just a descriptive list of the material available, or a set of summaries).

Part 2

This can be followed by writing a proposal for a small-scale research study in your school that can be evaluated against current thinking. The proposed research should seek to explore the whole school policy for 'health and well-being' for your pupils in your placement school and to be evaluated in the light of the evidence from your literature review.

We suggest that you discuss your review and the proposed research plan with your tutor as you proceed. The final document may be used as a basis for coursework and filed in your PDP.

SUMMARY AND KEY POINTS

Adolescence involves physical, mental and emotional changes leading towards maturity and presents dramatic physical changes in young people. These changes cause nervous introspection: 'am I growing normally, am I too tall, too short, too fat? Will I be physically attractive to others?' Comparison with others becomes the main yardstick of development. Personal appearance assumes a growing importance and causes sensitivity. Girls mature physically earlier than boys, but the range of development of both sexes is wide. The range of physical differences between pupils means that, at the same age, pupils react quite differently to tasks and situations in school.

Young people are taller and heavier than previous generations, in part due to improved diets. But the obesogenic environment of modern society, the more sedentary lifestyle and increased consumption of unhealthy foods can lead to overweight and obesity. There is growing concern about the rising numbers of overweight and obese pupils and adults and the physical, social and psychological effects this may have on the individual.

The social and psychological effects on young people can be as damaging as the health risks. A number of issues affecting the health of adolescents have not been raised including smoking, drinking and drug use, mental health and sexual health. These are discussed further in the report on *Adolescent Health* (BMA, 2003).

Schools have a big role to play in helping pupils through adolescence with the minimum of disruption, to understand the changes in their bodies, to be comfortable with themselves as they are and how they look. Schools play an important part in ensuring that pupils have access to a healthy diet and that they understand its importance to them now and in the future. This knowledge and understanding is achieved through a whole-school approach, including teaching and learning in PSHE, citizenship and other areas of the National Curriculum. Involving pupils in understanding and learning about themselves through active participation encourages confidence and supports a positive self-image (see also Unit 5.2). The pupil who feels valued for their contribution is a pupil who is likely to have good self-esteem. All teachers have the opportunity to contribute to the healthy development of their pupils and engender the self-confidence in young people to take control of this aspect of their lives and to resist fashion and peer pressure.

Check which requirements for your course have been addressed through this unit.

FURTHER READING

British Nutrition Foundation: http://www.nutrition.org.uk.
This site provides reliable information on diet and health. Particularly useful are the 'Teachers Centre' and 'Pupils Centre' which provide a wide range of resources for teachers, including activities for pupils that are downloadable free of charge. It also has research papers relating to diet and health and an excellent resource for those teaching about diet and health.

Food Standards Agency: http://www.eatwell.gov.uk.
Government advice on food is given in this section of the Food Standards Agency site. This site has up-to-date, easy-to-read references with some resources for teachers. The Food Standards Agency main site is: http://www.food.gov.uk.

Tanner, J. M. (1990) *Foetus into Man*, **Cambridge, MA: Harvard University Press.**
This is an excellent resource for those interested in the details of growth and development of humans. The chapters on 'Puberty' and 'Heredity and the Environment' are particularly useful.

Additional resources for this unit are available on the companion website:
www.routledge.com/textbooks/9780415478724

COGNITIVE DEVELOPMENT

Judy Ireson and Tony Turner

INTRODUCTION

During the secondary school years, pupils develop their knowledge and understanding of a wide range of subjects and also their ability to perceive, reason and solve problems. All of these skills are aspects of cognition (literally 'knowing'). A key feature of cognition is that it involves us as learners in making sense of the world around us. As such it is unlike more basic forms of learning such as memorising a song or rote learning multiplication tables. It includes skills that involve understanding, such as map reading, following instructions to make something and solving problems. Making sense, knowing, understanding, thinking and reasoning develop into adulthood and so cognitive development is an important feature of pupils' mental growth during the secondary school years.

Logical reasoning is one important aspect of cognitive development, along with others such as problem solving, developing expertise and creative thinking. Many school subjects require us to think and reason logically, e.g. when handling evidence, making judgements, understanding when and how to apply rules, untangling moral dilemmas or applying theories. Most Western societies in their schooling of children privilege logical, mathematical and linguistic abilities over other ways of knowing about the world. The tests of ability used by some schools to select new entrants or to allocate pupils to teaching groups are often problem-solving exercises involving pattern seeking, pattern recognition and pattern using and the capacity to think logically and quickly.

We consider some of the ways in which pupils' cognitive abilities develop and are identified, particularly logical reasoning, and discuss briefly some ideas about intelligence, including that there may be a number of discrete intelligences. We also illustrate some of the cognitive demands made by activities in different curriculum subjects. This unit is a continuation of Unit 4.2, which considered physical development. Unit 5.1 addresses in more detail theories of how children learn and develop and can be read in conjunction with this unit.

OBJECTIVES

At the end of this unit you should be able to:

- understand some features of cognitive development and the cognitive demands made by curriculum subjects
- explain some ideas about the nature of intelligence
- identify types of thinking and relate them to learning activities
- begin to use tasks as a way of finding out about pupils' cognitive level
- evaluate the idea of 'matching' the curriculum to pupils' learning needs.

Check the requirements for your course to see which relate to this unit.

DIFFERENCES BETWEEN PUPILS

Differences between children are apparent from an early age. Even before they start school, some children pass developmental milestones such as walking and talking more quickly than others. Children may start reading and counting before they begin school, or become very confident in their physical skills. Those who acquire good language and communication skills tend to be seen as more advanced and may be labelled as brighter than others. At this age it is hard to know whether they have particular linguistic abilities or are more interested in the kinds of activities that encourage this aspect of development.

Motivation

When pupils start primary school, you soon notice that some are better than others at school tasks. One of the reasons that children may be more advanced is that they are interested in the kinds of learning valued in school. It can be argued that school work is a game that children have not chosen to play, but which others, teachers and society, have chosen for them. If this assumption is correct, it is likely that some pupils are not highly motivated by the content and focus of lessons and therefore these pupils may be less successful in school. An alternative view is that we all have an intrinsic motivation to acquire competence and a tendency to protect our sense of self-worth, so those pupils who fall behind in their learning and feel they are not competent become demotivated and act in ways to protect their self-image.

Unfortunately this reaction often involves maladaptive activity such as procrastination, denying interest or playing the class joker. This is challenging for teachers who often resort to extrinsic forms of motivation such as threats or praise. Such encouragement may be effective in the short term but in the long run you are likely to find that it is more beneficial to develop learners' intrinsic motivation, i.e. encourage pupils to see the point of their work and to emphasise their growing competence. It is well documented that learners work best at activities they themselves identify as worthwhile (see Unit 3.2).

Pupils who do well in school subjects are sometimes thought to be 'more intelligent' than others, or more accurately, display more intelligent behaviour. Some people are of the view that intelligence is a fixed capacity or 'entity', which sets a limit on what an individual can achieve. Other people see intelligence as 'incremental', in other words it can grow with learning. An incremental view of intelligence carries with it the potential for change through teaching, whereas an entity view suggests that the effect of teaching is much more limited.

Before reading on, complete Task 4.3.1.

Task 4.3.1 INTELLIGENT PEOPLE

Think of two people who you would say are intelligent – they could be adults or children. In what ways are they similar and how do they differ? Share your ideas with other student teachers in your group and make a note of the characteristics.

Thinking in different curriculum subjects

At this point we turn briefly to some types of thinking and intelligent behaviour called for in different curriculum subjects. A closer look at a few subjects suggests that there are differences but also some overlapping demands. In general, linguistic and logical reasoning seems to be privileged in Western school systems.

It is generally thought that the demands of science and arts subjects are rather different. In a recent discussion about the teaching of art and design in secondary school cognition is described in terms of the acquisition, assimilation and application of knowledge (Addison and Burgess, 2007: 24). These cognitive processes are based on:

■ perception, observation based on experience
■ intuition and reason, both the unconscious and conscious making sense of experience.

These processes require the use of imagination and thinking skills, which are used to transform observations and experiences into material representations. The explicit inclusion of intuition and unconscious making sense of experience is emphasised in art and design in contrast with many other subjects; in the school curriculum knowledge-based, analytical processes are valued in the appreciation and criticism of art and other creative activities (Addison and Burgess, 2000: 26–29; 2007: 24).

By contrast, mathematics and science are usually thought to involve logical and mathematical thinking. Indeed for some people mathematics and art lie towards opposite ends of a spectrum or are 'different cultures', as described by Snow (1960).

Intelligent behaviour thus takes different forms in these areas. Other curriculum subjects seem to overlap to a greater extent, e.g. the work of artists and designers involves similar skills to those of scientists, requiring the manipulation of materials and use of practical techniques. It may be that artists do more of their thinking in the process of manipulating materials, whereas scientists do more thinking and planning before carrying out their practical work. There is no doubt that new ideas and ways of thinking in science and mathematics involve imagination. For example, Archimedes' insight into floating and Watson and Crick's double helix model for the structure of DNA both required imaginative thinking. Similarly, although the use of imagination and creative thinking is most strongly associated with the arts, creativity may be encouraged in many subjects (Qualifications and Curriculum Authority (QCA), 2008c).

A question of intelligence?

Intelligence is most often linked to a pupil's capacity to exercise linguistic and logical mathematical reasoning. This is what is measured by most tests of intelligence. There is considerable evidence to support classical theories of intelligence, which suggest that there is a general factor underlying our performance in a wide range of activities in school and work. However, Gardner criticises these theories for being concerned with only a very narrow range of human ability, namely language and mathematics. He argues that they fail to take account of many other aspects that are important in the real world. In his 'theory of multiple intelligences' he proposes that there are a number of relatively autonomous intelligences. He describes intelligence as 'the ability to solve problems or fashion products that are of consequence in a particular culture, setting or community' (Gardner, 1993a: 15). He has identified many intelligences, some added later as he developed his theory. They include:

Linguistic: use and understanding of language, including speech sounds, grammar, meaning and the use of language in various settings.
Musical: allows people to create, communicate and understand meanings made with sound.
Logico-mathematical: use and understanding of abstract relationships.

Spatial: perceive visual or spatial information, to be able to transform and modify this information, to re-create visual images even when the visual stimulus is absent.

Bodily kinaesthetic: use all or part of one's body to solve problems or fashion products.

Intrapersonal: knowledge of self and personal feelings. This knowledge enables personal decision making.

Interpersonal: awareness of feelings, intentions and beliefs of others.

Naturalistic: the kind of skill at recognising flora and fauna that one associates with biologists like Darwin.

(Gardner *et al.*, 1996: 203)

Existential intelligence: concerned with 'big questions about one's place in the cosmos, the significance of life and death, the experience of personal love and of artistic experience'.

(Gardner, quoted in White, 2005: 8)

Gardner proposes that these relatively autonomous intelligences can be exerted alone or combined in different contexts at different times. A number of intelligences may be needed in order to carry out some tasks, e.g. in the case of art and design both spatial intelligence and bodily kinaesthetic intelligence might contribute to learning.

Look back at the list of characteristics of intelligent people you identified in Task 4.3.1. How well do they map on to Gardner's set of intelligences?

Gardner's 'intrapersonal' and interpersonal' intelligences capture much of what has recently been called 'emotional intelligence' (Salovey and Mayer, 1990). This is the ability to recognise, express and reflect on our own emotional states and those of other people and also to manage these emotions. It is worth noting that for most people learning is an emotional experience which may involve confusion, disappointment, apprehension, fascination, absorption, exhilaration and relief. These emotions can disrupt or facilitate learning and it is easy to see that learners benefit from being able accurately to recognise and manage them effectively. Emotional barriers to learning are acknowledged in the NC for England by requiring teachers to recognise that stress can place hurdles to learning in front of children (QCA, 2008d). For further reading on emotional factors and learning see Goleman (1996); Cherniss (2000); and within mathematics learning, Goulding (2005: 56–58).

Some of the ideas behind the theory of multiple intelligences have received critical reviews (White, 1998, 2005). It is not yet clear just how autonomous these intelligences are and a common view is that a likely model of intelligence is one that operates through a general underlying intelligence backed up by a small number of special abilities. Thus:

we are bound to look critically at evidence and the evidence of the existence for abilities in different intellectual areas, which are quite independent of each other, is not good . . . intelligence is not a monolithic unidimensional ability which allows us with one IQ number to define an individual fully. All measures of different aspects of intellectual ability correlate with one another.

(Anderson, 1992)

Any intellectual behaviour is then a product of a general processing ability and a number of special abilities (Adey, 2000).

The use of the word ability above, rather than intelligence, may echo Gardner's early comments:

nothing much turns on the particular use of this term [intelligences] and I would be satisfied to substitute such phrases as 'intellectual competence', 'thought processes', 'cognitive capacities', cognitive skills', 'forms of knowledge'.

(Gardner, 1983: 284)

In other words it is not clear whether Gardner is describing an innate faculty, a learned process or a structure of knowledge.

However, many teachers and advisers find Gardner's ideas very helpful as they alert us to a variety of ways in which we might recognise and develop intelligent behaviour. An influential report on creativity in education refers specifically to the theory of multiple intelligences in its advocacy of a broader approach to education than has existed under the National Curriculum for England and Wales (Robinson, 1999: 34–37).

Thirty years ago, HM Inspectorate (HMI) in England and Wales reviewed the state of the secondary curriculum and published a forward-looking document outlining a different way for educators to think about what pupils might learn in school (HMI, 1977: 6). They identified eight 'areas of experience' to which pupils should be exposed (see Table 4.3.1).

This publication prompted much debate and about ten years of discussion about the school curriculum in England and Wales. But the decisive 1988 Education Act in England and Wales did not reflect any of that thinking, instead consolidating a curriculum of traditional subjects (for further comparison and discussion see Aldrich (1998: 48)). You might like to consider the correlation between the newly emerging theory of multiple intelligences referred to above and the near forgotten 'areas of experience'. An opportunity to explore further multiple intelligence theory is provided in Task 4.3.2

■ **Table 4.3.1** Areas of Experience

Aesthetic and creative	Ethical	Physical
Social and political	Linguistic	Mathematical
Spiritual	Scientific	

Source: HMS (1976: 4)

Task 4.3.2 MULTIPLE INTELLIGENCES AND DESIGN AND TECHNOLOGY (D AND T)

Locate online the D and T curriculum for England (QCA, 2008e). Select teaching D and T at KS 3 and 4, download the Programme of Study and read the sections 'Key Concepts', 'Key Processes' and 'Attainment Target'.

Using these documents identify the skills and intelligences demanded by D and T at one or more Key Stages. These questions may help you.

1   In which ways do the demands of D and T link to the importance attached in school to linguistic and logico-mathematical aptitude?
2   Using Gardner's 'Theory of Multiple Intelligences', discuss the teaching and learning of D and T as the utilisation and development of different intelligences.

For discussion on the place of D and T in the school curriculum see Owen-Jackson (2008, ch. 1).

Summarise your findings to discuss with your tutor or the D and T teachers in your school. File the final document in your professional development portfolio (PDP).

COGNITIVE DEMANDS ON PUPILS

It is worth noting that all subjects in the NC for England are expected to address the following thinking skills. You should bear this in mind when we turn below to some of the cognitive demands made on pupils through the curriculum.

Information processing skills. These enable pupils to locate and collect relevant information, to sort, classify, sequence, compare and contrast, and to analyse part/whole relationships.

Reasoning skills. These enable pupils to give reasons for opinions and actions, to draw inferences and make deductions, to use precise language to explain what they think, and to make judgements and decisions informed by reasons or evidence.

Enquiry skills. These enable pupils to ask relevant questions, to pose and define problems, to plan what to do and how to research, to predict outcomes and anticipate consequences, and to test conclusions and improve ideas.

Creative thinking skills. These enable pupils to generate and extend ideas, to suggest hypotheses, to apply imagination, and to look for alternative innovative outcomes.

Evaluation skills. These enable pupils to evaluate information, to judge the value of what they read, hear and do, to develop criteria for judging the value of their own and others' work or ideas, and to have confidence in their judgements. (QCA, 2008f, 2008g).

The first pupil related task we have selected (see Task 4.3.3) is adapted from a quiz book (Brandreth, 1981: 118). The problems in these types of books are often abstract, lack a real context but demand reasoning skills, perhaps not too far from the situation commonly found in school.

Task 4.3.3 **A LOGIC PROBLEM**

Try out the following problem on your own; then compare your answer with other student teachers and share how you set about solving the problem.

When Amy, Bill and Clare eat out, each orders *either* chicken or fish, according to these rules:

a   If Amy orders chicken, Bill orders fish.
b   Either Amy or Clare orders chicken, but not both.
c   Bill and Clare do not both order fish.

Who could have ordered chicken yesterday and fish today? (For the solution see Appendix 4.3.1).

The problem in Task 4.3.3 is essentially about handling information according to rules of the type 'If A, then B', commonly found in intelligence tests. In this example the rules are arbitrary and it is not a real-life problem because people don't behave in this way. The problem cannot be solved by resort to practical activity; it is a logico-mathematical task requiring abstract thinking. The puzzle can be done 'in the head', but many people need to devise a way of recording their thinking as they develop their answer and check solutions.

Another kind of reasoning task is a game used in *Thinking through Geography* (Leat, 1998). Pupils are given sets of words and asked to find the odd one out. See Task 4.3.4.

Task 4.3.4 **ODD ONE OUT**

The following sets of words relate to traffic in urban areas. Which is the odd one out in each set? Try this yourself and then compare your answers with others in your group. What went through your mind as you thought about each set?

1	Park and ride	Shopping trips	Bus passes	Ring road
2	Wheel clamp	Tailbacks	Speed cameras	Sleeping policeman

We turn next to an exercise commonly given to pupils in science lessons during Key Stage 3 or early Key Stage 4. Pupils are set a problem-solving task in which they are invited to identify the factors which affect the rate at which a pendulum swings (or 'time period'), see Task 4.3.5. The task is practically based and has real-life connections as pendulums are used to control timepieces, such as a longcase clock which contains a rod (pendulum) with a heavy weight at one end. The length of the pendulum is adjusted to control the accuracy of the timepiece.

The activity concerns *understanding* and *how understanding is gained* rather than knowing and recall. The pupil's task involves planning an investigation, identifying patterns in data and making deductions, in other words, inquiry skills and thinking skills. The exercise illustrates, too, the ways in which pupils respond to data. The analysis of the data requires abstract thought and the ability to handle a complex situation in which several factors (variables) have to be considered.

Listening to pupils as they try to solve the problem can be very valuable as it provides clues about their thinking skills (cognitive processes).

There are some interesting features of this task. Intuitively pupils expect the size of the weight and the 'push' to have an effect on the time period, or rate of swing. They tend to expect heavy weights to 'do more' than lighter weights. The results are contrary to common sense and pupils often think they are wrong. This conflict with everyday conceptions is not an uncommon experience even for adults. Evidence that the magnitude of the weight at the end of the pendulum has no effect on the time period is often rejected intuitively or put down to error. Common-sense notions can be in powerful opposition to evidence.

Some pupils also do not accept that if two variables are changed at the same time, then it is not possible on that evidence alone to make a deduction. In this situation, some pupils may then bring in evidence external to the investigation to support their argument, instead of using the data they have.

When pupils are faced with the need to get evidence for themselves they frequently choose trial and error methods rather than logically constructed enquiries. Trial and error methods often lead to data which do not provide clear-cut answers to questions; this can lead pupils to make unwarranted inferences from the data in an attempt to get an answer.

As the data are not always clear-cut, your judgement may have to be withheld. This may cause mental conflict because there is a powerful expectation that experiments yield positive information. Saying 'this enquiry tells us nothing about the question' is often not an acceptable answer, especially if you have set up the inquiry. Such feelings mean that attitudes of persistence and honesty are critical for the generation of real understanding. For further information about the time period of a pendulum; see Appendix 4.3.2.

This investigation, and others like it, suggests that enquiries that involve handling together several variables (here, the weight, length and push) can be difficult for many pupils. Teaching is

Task 4.3.5 INQUIRY AND UNDERSTANDING: WHICH FACTORS (VARIABLES) AFFECT THE SWING OF A PENDULUM?

Background information

A pendulum is essentially a rod pivoted vertically at one end and free to swing from side to side. A simple example of a pendulum is a piece of string suspended at one end with a weight at the other (see Figure 4.3.1). Pupils are sometimes expected to use experimental data to deduce factors that influence the rate of swing or 'rules of the pendulum'. They may be given the data, or derive the data for themselves. The task is not to learn the rules, but to understand how the rules derive from observation. The exercise for you and for pupils is to work out what can, or cannot, be deduced from the set of data.

The pupil's task

Two pupils were given a task to find out which factors affected the time period, or rate at which a pendulum swings. They were not told exactly what to do but the teacher had suggested investigating the effect of length, weight and push on the time period. They decided to measure the number of swings made by a pendulum in half a minute. They changed variables of the pendulum each time, by varying the:

■ length of the pendulum; they had one short pendulum and one long pendulum (Figure 4.3.1a and b)
■ size of the weight on the end of the pendulum, a heavy weight and a light weight (Figure 4.3.1b and c)
■ height it was raised to set it going – the push. One 'push' was high up, the other 'push' low down (Figure 4.3.1c and d).

In their investigation, pupils sometimes changed one variable at a time but occasionally changed more than one. They collected some readings (Table 4.3.2) and then tried to sort out what the readings meant.

Your task

From the evidence *alone* in Table 4.3.2 what do you think the data tell you about the effect of *length*, of *weight* and of *position of release* on the number of swings per half minute of the pendulum? See Appendix 4.3.2 for further information.

■ **Table 4.3.2** Data on different pendulums obtained by pupils (for Task 4.3.5)

Experiment	Length of the pendulum	Size of weight on the end	Push at start	No. swings in ½ minute
1	long	heavy	large	17
2	short	heavy	large	21
3	long	light	small	17
4	short	light	large	21

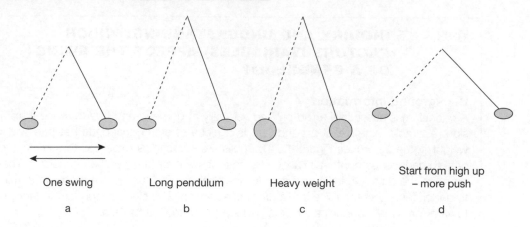

One swing
a

Long pendulum
b

Heavy weight
c

Start from high up
– more push
d

■ **Figure 4.3.1** The pendulum: three variables, length, weight and 'high up'

needed to improve achievement in these skills, as has been shown by the Cognitive Acceleration in Science Project (CASE) (Adey and Shayer, 1994: chs. 5 and 8). Further details can be found in Whylam and Shayer (1978) and Adey *et al.* (1989).

Common-sense beliefs and misconceptions

Everyday beliefs and misconceptions are common in science, as shown in aspects of the pendulum task above, but they also surface in other subjects. Gardner (1991) gives many examples, including a classic case reported by I. A. Richards many years ago. Richards asked undergraduates at Cambridge University to read pairs of poems and then to offer their interpretations and evaluations. He found that the students were heavily influenced by the form of the poem, in other words whether it rhymed, had a regular metre or rhythm, and avoided words that were too common or arcane. Many of them failed to understand the meaning of the poems.

Gardner (1991) suggests that that many of our everyday understandings take the form of 'scripts' and 'stereotypes' that tend to simplify the world around us and make it more manageable. Unfortunately pupils use these to interpret information presented in school subjects. For example, intuitive interpretations of historical events tend to be quite simplistic and stereotypical, there is often a good versus evil narrative, with evil leaders taking on great importance and the good usually winning in the end. Even when pupils have learned that events such as the Second World War have complex causes and that war is seldom due to the behaviour of a single evil leader they may slip back into simplistic ways of thinking.

It is worth getting to know about common misconceptions in your own subject as this can help you see why pupils have difficulty with some new ideas and ways of thinking. You may be able to plan activities to challenge specific misconceptions. We return to this aspect of learning in Task 4.3.6

Understanding percentages

The next task illustrates the demands made by problems involving percentages. Understanding percentage is an important part of everyday economic life, in retail, mortgages, investment and cost of living generally. It is an area of understanding where considerable confusion reigns for both pupils and adults (see Figure 4.3.2).

A supermarket offered olive oil for sale labelled '50% free'. The bottle contained 750 ml and was priced £2.99, the same price as a regular bottle of oil containing 500 ml. A group of adult customers were arguing that this offer was wrong because the price was not cheaper, i.e. £1.50. Despite having the nature of the offer explained to them several times by another customer, most of the group refused to buy the item because it did not cost less than the regular item.

■ **Figure 4.3.2** Percentages in the supermarket

Another example of a common confusion is financial inflation: many adults expect the cost of living to come down when the inflation rate is reduced from, say 3 per cent to 2 per cent. Similarly, some adults have difficulty calculating real costs when sales advertising offers percentage discounts (see question c in Table 4.3.3). When pupils are faced with problems involving percentages, it appears that context is as important as the numbers themselves. In addition, the understanding of what constitutes a right answer is confused with 'what answer is good enough, given the context'. We might sympathise with this last point, e.g. when preparing a dish for four people and faced with a recipe which requires two-thirds of a pint of cream for six people we might estimate rather than calculate exactly. The following example illustrates how context might influence pupils' responses to questions.

Pupils were given three questions on percentages, together with an introduction which explained the meaning of the symbol '%'. The questions (a, b and c) and the number of pupils getting the right answer (the success rate) for each of three year groups is shown in Table 4.3.3.

We suggest you read through the questions in Table 4.3.3 and check your own answers to the questions. Then consider possible reasons for the variation in the success rate shown by these pupils and the different contexts in which the mathematics is set. If ability is being is judged by responses to questions like these then, clearly the context in which the mathematics is set matters. The author describes how pupils arrive at their different answers, the different strategies they use and implications for teaching (Hart, 1981: 96–97).

■ **Table 4.3.3** Pupils' performance on questions involving percentage

a Six per cent (6%) of pupils in school have free dinners. There are 250 pupils in the school. How many pupils have free dinners? The symbol '%' means per cent or per 100, so 3% is 3 out of every 100.

Age/years	13	14	15
Success rate %	36	45	57

b The newspaper says that 24 out of 800 Avenger cars have a faulty engine. What percentage of cars is this?

Age/years	13	14	15
Success rate %	32	40	58

c The price of a coat is £20. In a sale it is reduced by 5%. How much does it cost now?

Age/years	13	14	15
Success rate %	20	27	35

Source: Hart (1981: 96) For further studies on the performance of pupils see Keys *et al.* (1996)

Evidence from this investigation and others like it show that most of us have several strategies we might use when attempting to solve problems such as these. Even young children may have three different strategies for simple addition problems. Students who are able to use more efficient strategies for the percentage problem above, such as multiplying by a fraction, are generally more successful but they might not always use the most advanced strategy. This might be because they are not very confident, or it takes time and effort or the simpler strategy suffices.

Sometimes it is getting the right answer that is the important factor, rather than the understanding of how a right answer can be obtained. Some of these difficulties for pupils may arise from differences in the use of mathematics in and out of school. In everyday life, approximations are often good enough, as in the cooking example above. For further discussion of the cognitive dimension of learning mathematics see Goulding (2005: 58–63).

It is worth noting that adults are often quite happy to admit that they 'are not good at maths', whereas they are more reluctant to admit to literacy difficulties. The drive by government agencies to raise standards of literacy and numeracy is a recognition of the importance of these skills. The Training and Development Agency for Schools (TDA), the body responsible in England for the school workforce and qualifications, requires student teachers to be tested in aspects of numeracy as a condition of gaining qualified teacher status. The TDA has provided online support for student teachers to meet this requirement: TDA Skills Test. Available at http://www.tda.gov.uk/skillstests.aspx, accessed 19.01.09.

DEVELOPING COGNITIVE ABILITIES

Nature and nurture

In general, pupils' cognitive abilities increase with age, as suggested by the percentage example in Table 4.3.2. Older pupils are more capable of abstract, symbolic thinking than younger pupils, who tend to use more concrete representations when solving problems (see the discussion of Piaget's theory of cognitive development in Unit 5.1). It is also clear that in any year group there is considerable diversity, with some pupils able to cope with more demanding work, while others have great difficulty. Several different explanations have been suggested for this diversity and for the general pattern of development during the school years. In the remainder of this section we briefly outline some of these.

So what might influence the development of cognitive abilities? Most answers to this question emphasise biological or environmental factors, or 'nature' and 'nurture'. 'Nature' refers to our inborn, genetic or inherited characteristics and how these unfold during maturation, while 'nurture' refers to the environment in which we grow and develop. Genetic factors undoubtedly affect development, but the extent of this influence is not entirely clear, as it is difficult to disentangle genetic and environmental determinants. The most influential studies compare twins raised together with twins separated at an early age and raised in different families. These studies show that inherited characteristics do have a strong effect but they do not completely determine measured intelligence; the child's environment also makes an important contribution. Criticisms of the research include the view that the measures of intelligence used in these studies give a particular and quite restricted meaning to the notion of intelligence itself (Gardner *et al.*, 1996).

A recent development in the nature–nurture debate is the proposal of an interactive process between nature and nurture, that is viewing development as an interactive, dynamic process between the total environment of the genes and the genes themselves, instead of discussing the effects of genes and environment on development each as separate entities; see, for example, Ridley (2003).

Turning to the environment, children's attainment on entry to secondary school is influenced by their experiences in the home and in primary school. Evidence has accumulated to suggest that a lack of stimulation in early childhood limits the capacity of children to benefit from school and other learning situations. A home life that forms a firm basis for later development is one in which children are well cared for and are encouraged to play, explore and talk and later to enjoy books and learn about the world around them. Children from a good home learning environment have a head start when they enter primary school, as they have started to develop many of the cognitive, communicative and social skills needed in school. Later on, their parents may support them directly and indirectly, e.g. taking them to the local library, explaining homework or enabling them to participate in a wide range of activities. While the majority of parents are able to provide a good home learning environment, some find this difficult, perhaps for economic, social or health reasons. Some parents find a good nursery and primary education for their children, while others find this a problem. The recent government strategy in England, called *Every Child Matters* (DfES, 2003a) is designed to address some of these inequalities. The provisions of the Child Care Act 2006 includes the SURESTART programme and is designed to provide free or affordable childcare for working parents; and in addition to provide an integrated education programme for this age group to address some of the inequalities described above. Further details can be found at the *Every Child Matters* website.

A pupil's performance in attainment tests is thus a reflection of their innate abilities and the influence of the environment. Attainment tests tell you what a pupil knows, understands and can do at a particular point in time, however they are not necessarily reliable predictors of future attainment. Pupils who have had less support for their learning in the past, or who do not speak English as their first language, may be more capable than suggested by these tests. For this reason it is important to take care not to label pupils in a deterministic way as high, middle or low ability on the basis of attainment tests when they enter the school.

Knowledge base

Another important factor that is easily overlooked is the amount of knowledge an individual has acquired. It is sometimes assumed that the more advanced thinking of adults is due to biological maturation, when in fact it may be due to increased knowledge. Comparing children who are very knowledgeable about a subject with adults who are not, e.g. children who are very good at chess and adults who are beginners, illustrates the power of knowledge nicely. When shown a chess board with pieces laid out as they might be during a game children are better than the adults at remembering the positions. If the board is taken away, children are able to place more pieces than adults in the correct position on another, empty board. However, when the pieces are arranged randomly on the board, there is no difference between the adults and the pupils. This indicates that pupils who are good at chess build up knowledge of patterns and configurations as they play and this knowledge helps them remember (see Bransford *et al.*, 1999).

People who are experts are much better at recognising patterns in information and using principles to solve problems. This applies in domains as diverse as medicine, music, physics, computer programming and teaching. Experts build up a strong knowledge base and use it effectively to solve problems. Reading X-ray photographs is one such example (Abercrombie, 1985). For further information see Howe (1998). For pupils, the development of a strong knowledge base in school subjects is important for remembering and thinking. It also makes learning more enjoyable and satisfying. Good teaching and active learning (see Unit 5.2) should help pupils to build up a well-structured knowledge base, but this may be undermined by pressure to cover the curriculum, leaving insufficient time for understanding and consolidation (Biggs and Moore, 1993; Howe, 1998). The growth of testing and its effect on learning may further undermine teaching for understanding (Marshall, 2005: 69–74).

Language and development

Language is the tool of cognitive development; without language it would be difficult to construct ideas, convey ideas, to challenge ideas and so effect change. When we think we do so in symbols, both of language and image; thus language is important for both community and personal development. This is why so much importance is placed on young children learning to talk, then to read and write. Whereas there may be arguments about when children should learn to read and write, no doubt exists that at some stage they should.

If language is crucial to development then the first language or mother tongue is very important for pupils if they are to realise their full potential. For native English speakers it is the full development of the linguistic abilities that becomes important. For pupils for whom English is a second language (E2L) another important issue arises. Whereas, for obvious reasons, it is important that such children learn English (the host language) at the same time, their first language is likely to be the one in which they think. The later a child comes to learning in a second language the more likely is it that thinking skills use the mother tongue. Thus playing down or stopping the development of mother tongue learning not only diminishes its cultural significance but crucially may inhibit their cognitive development and so reduce their opportunity to achieve (Lu, 1998)

The first language skills they learn act also as a bridge to learning the second language; and children are able to use the visual, linguistic and cognitive strategies used in their mother tongue in learning to read and write English. Without this strong bridge their chances of achieving academic success may be reduced and so limit their life chances.

Curriculum and development

Several ideas have now been introduced to explain pupils' performance in school. It has been suggested that performance is a product of inheritance and environment. Also that, as they grow, develop and learn, pupils become more able to handle complex problems and situations and abstract thinking. You are not in a position to influence inheritance or the home learning environment, but you can structure learning experiences so that they are aligned with pupils' current knowledge and understanding. To do this requires a careful analysis of the demands made by a piece of work and the current performance of the pupil or group of pupils with reference to relevant knowledge, skills and understanding needed to do the task. The importance of 'assessment for learning' is emphasised by the recent Office for Standards in Education (Ofsted) report where describing good practice in secondary schools, the Chief Inspector writes: 'Accurate assessment is used to identify focused objectives for learning and is the basis for choosing suitably challenging tasks and resources' (Ofsted, 2007b: para. 73)

Thus with evidence from assessment, future learning activities can then be planned so as better to guarantee some success while offering a comfortable level of challenge (see also Unit 6.1). Curriculum development then becomes a process of matching the curriculum to the needs of the pupils. This strategy depends on you being able to analyse the cognitive demands of the topic and associated learning activities and the relevant capabilities of pupils in order to achieve a match.

Some questions arise from proposing such an approach.

■   Does the cognitive demand made by the teaching material depend on the way the material is presented? Can most concepts be taught to most pupils if suitably packaged and presented?

■   If matching 'curriculum to pupil' is the goal of your teaching how do you build in development, going beyond the current level of performance? How would pupils progress and is there a danger that reducing the chance of failure may remove challenge?

■ Records of pupil performance and development are needed in order to match material to pupil. How do you keep such records?

■ If each pupil has different prior knowledge and understanding, how can you cope with a whole class? See Unit 4.1.

■ If pupils cannot cope with certain concepts because they are not yet ready for them, does this mean that some areas of the curriculum cannot be taught? Is rote learning an acceptable way to overcome this problem? See Unit 4.1 on differentiated learning, Unit 5.1 on theories of learning and Unit 5.2 on active learning.

You may wish to add your own ideas to this list.

An alternative to this approach is to think in terms of 'constructive alignment' (Biggs and Moore, 1993). A basic assumption here is that the learner constructs meaning through relevant learning activities, hence 'constructive'. The 'alignment' part refers to what you do, which is to set up a learning environment that supports the activities appropriate to achieving the desired learning outcomes. The key is that teaching methods and assessment tasks are aligned to the learning activities. Less emphasis is placed on obtaining detailed information about each learner. Such an approach perhaps acknowledges that it is very difficult for you, given the number of pupils in a class and the variety of differences between pupils, to match work on an individual basis. The key point is that you should provide a variety of learning activities carefully designed to enable pupils to achieve the desired outcomes.

The Cognitive Acceleration programme offers a somewhat different approach (Shayer and Adey, 2002) which incorporates several key components. The emphasis is on helping pupils to understand the context, to 'learn how to learn' and begin to construct meaning for themselves. The first component is 'concrete preparation' which involves you setting activities to ensure that pupils are familiar with the context of a problem and any technical vocabulary. You then give pupils problems to discuss in small groups, designed in such a way that all students are able to contribute to the discussion. An important feature of the programme is that pupils are encouraged to think about their own thinking (metacognition). Both you and your pupils are encouraged to think about links between their thinking in other aspects of the curriculum.

Results suggest that pupils can be taught to think in generic ways that help them to learn better across a range of subjects and context (Adey, 2000). Cognitive acceleration started in science but is now being developed in a variety of curriculum subjects (Shayer and Adey, 2002).

CREATIVE PROBLEM SOLVING

In this unit we have taken cognition to encompass acquisition, assimilation and application of knowledge and also problem solving and thinking. Earlier in the unit we used an example from design and technology as fostering capability which is 'the integration of thought and skills into a holistic exercise, rather than a piecemeal exercise of isolated skills'.

A different example of this type of problem solving is shown in a collection of children's responses to problems set by adults. The collection shows the spontaneous work by pupils (upper primary, lower secondary) in response to problems related to everyday events (de Bono, 1972). The tasks include:

■ 'how would you stop a cat and a dog fighting?'
■ 'design a machine to weigh an elephant'
■ 'invent a sleep machine'
■ 'how would you build a house quickly?'

- ■   'how would you improve the human body?'
- ■   'design a bicycle for postmen'.

In this work pupils need knowledge of the context from which the problem is drawn and also use knowledge and skills from both inside and outside the classroom. The work of pupils is occasionally unusual and the solutions sometimes impractical. The responses show much imagination and insight into their everyday world. The book is out of print but used copies are available on the web.

MEASURING COGNITIVE DEVELOPMENT AND INTELLIGENCE TESTS

Much work has been carried out to help us understand pupils' (and adults') responses to problem situations. For example, Piaget devised many tasks that have been used extensively and adapted by others (see Donaldson, 1978, 1992; Child, 2007, ch. 4; Adey and Shayer, 1994). These tasks provide a window on the type and sequence of thought process adopted by learners and reveals much about how pupil's thinking develops. Hopefully, if you have tried some of the tasks in this unit you are starting to see how you might use problems to help you learn more about your pupils' cognitive processes.

The main purpose of many tests and examinations is to assess, rank, select and make predictions about progress (see Unit 6.1). IQ tests were first developed by Binet to identify pupils who may be in need of special education (Gould, 1981: 148 *et seq.*). Later, tests of intelligence were designed to assist educational and occupational selection on a fair, meritocratic basis. Most people are familiar with the notion of 'Intelligence Quotient' (IQ) which is reported as a single number. This is derived from test scores and shows the extent to which the pupil is below or above an average score based on a large sample of pupils. The test is norm-referenced and the average score is given an arbitrary value of 100. As indicated above, intelligence tests generally tap a range of cognitive abilities involving language, numerical and non-verbal thinking and reasoning.

Intelligence testing assumed great importance in the UK after the Second World War due to the 1944 Education Act. Pupils were selected for grammar, technical or 'modern' schools by means of the 11+ examination, a type of intelligence test. Despite being developed into a reliable sophisticated tool, the examination failed to take account of late developers or the effects of pupils' social background. The tests also favoured pupils with good linguistic skills, those who had a good vocabulary and were familiar with middle-class culture, and girls. Girls on average mature faster than boys and the 11+ entry had to be modified to ensure equal access to grammar school by boys and girls. It was shown, too, that performance on the tests could be improved by training, which suggested that in part at least, learned skills were being tested rather than purely innate intelligence. Confidence in the whole issue of selection was undermined by research showing that it had negative impacts on pupils from less advantaged background. In due course many local authorities abolished grammar schools and moved to a fully comprehensive secondary system on grounds of equity, although some retain selection to this day, e.g. Kent and Buckinghamshire (Crook *et al.*, 1999: 1–3; Ireson and Hallam, 2001). Recently, concerns about the impact of home background on pupils' achievement in national examinations have led to a call for the use of cognitive abilities tests in selecting young people for gifted and talented programmes and entrance to university.

Standardised testing is a skilled process and many commercially available tests must only be used by approved persons. IQ testing is sometimes used by educational psychologists to assess pupils who, in various ways, find school difficult. This procedure may be necessary for the identification of special educational needs such as dyslexia (see Unit 4.6). Other reasons for standardised assessment may be for research purposes as part of monitoring a population. See Unit 6.1 for further information on assessment.

We have described some challenges that arise in teaching aspects of a few subjects and drawn attention to the role of cognition in teaching of art and design. Task 4.3.6 asks you to assess the cognitive challenge in your teaching subject.

Task 4.3.6 COGNITIVE CHALLENGE IN YOUR TEACHING SUBJECT

Select a coherent piece of work, say a topic in a unit you teach and identify the concepts pupils usually get wrong, stumble over or misunderstand or you find hardest to teach. As well as your day-to-day classroom experience, use the results of end-of-unit tests or GCSE examination. What is it that makes the concept(s) difficult? Here are some reasons.

- The subject matter is too abstract.
- Pupils do not have the necessary vocabulary.
- The subject matter lacks relevance to everyday life.
- Pupils cannot access sufficient background knowledge.
- Pupils are asked to apply existing knowledge to new situations.
- Pupils are expected to make connections between different areas of knowledge.
- Pupils need to consider multiple factors in a situation.
- The task requires pupils to look for patterns in information.
- Pupils are expected to abstract generalisations from a situation.
- The task asks pupils to make judgements or criticism about propositions or generalisations.

As part of your classroom experience select a topic covered or a task you set and ask pupils to talk about how they completed it and to identify any aspects of the work they found easy and those they found difficult. If the class is not used to doing this, some pupils may find it threatening to admit that something was difficult so you must be careful to ensure that you create an environment in which difficulties are an accepted part of learning. You could ask pupils to work in pairs or small groups before sharing with the rest of the class.

It is likely that more than one factor may contribute to the difficulty. If possible repeat the activity with a different year group where the concepts are revisited and developed further and contrast the results. Keep a record of your conversations and classroom discussions.

Discuss the findings with your tutor; once agreeing on possible causes of the difficulty move on to planning a course of action to ameliorate the problem. See Chapter 5 for further suggestions.

SUMMARY AND KEY POINTS

Cognitive development is described as a process through which pupils develop their knowledge, understanding, reasoning, problem solving and creative thinking, that is *their thinking*. It is influenced by maturation and the learning environment provided at home and in school. We have drawn attention to the priority given by schools in Western society to linguistic skills and logical reasoning especially used in the sciences and mathematics. We compared these with the knowledge, skills and attitudes needed by arts and design and technology courses and asked you to think about the demands of your own subject. The nature of intelligence is subject to debate, and here we have done no more than briefly outline some prevailing views. One is that there is a general component and a number of specialised abilities. Another is that there are multiple relatively autonomous intelligences. Everyday beliefs about intelligence include both the idea that intelligence is a fixed 'entity' and that intelligence can be developed, an 'incremental model'; the latter model implies a potential for change through teaching, unlike the former.

Measures of intelligence or general ability emphasise language and logical-mathematical thinking. Practical tasks in this unit are offered to encourage you to analyse the cognitive abilities involved in completing them. The tasks also illustrate some ways you might go about finding out more about your pupils' cognitive development. Your classes contain pupils with a wide range of cognitive levels, even if your school has ability grouping. By listening to your pupils and finding out about their perspective on learning activities you learn more about their understanding and thinking. This process helps you to design activities to assist their development. We refer you to Unit 5.2, Unit 3.2 and Unit 5.1 for further discussion of related topics.

Check which requirements for your course have been addressed through this unit.

FURTHER READING

Child, D. (2007) *Psychology and the Teacher*, 8th edn, London: Continuum.

A classic text which provides a useful review of cognitive development, theories of learning and intelligence; includes research into classrooms, practice, management and special needs. A useful source of references.

Donaldson, M. (1978) *Children's Minds*, Glasgow: Collins/Fontana.

Essential reading for anyone interested in developmental psychology and Piaget's work on cognitive growth, which is reviewed. See also Donaldson, *Human Minds: An Exploration* (1992) in which the author expands on the ideas expressed in *Children's Minds*, and presents a study of the development of the mind from birth to maturity.

Gardner, H. (1999) *Intelligence Reframed: Multiple Intelligences for the 21st Century*, New York: Basic Books.

Gardner provides a state of the art report on the theory first propounded in *Frames of Mind* (Gardner, 1983). He describes how multiple intelligence theory has evolved and been revised. He introduces two new intelligences, and argues that the concept of intelligence should be broadened, but not so much that it includes every human faculty and value. In addition, he offers practical guidance on the educational uses of the theory, and responds to the critiques levelled against it.

Howe, M. (1998) *Principles of Human Abilities and Learning*, Hove: Psychology Press.

A readable introduction to abilities and how they develop. Explores the nature of abilities and challenges some common-sense notions with interesting examples. Also discusses intelligence and motivation.

APPENDIX 4.3.1 ANSWERS TO TASK 4.3.3

From rules a and b, if Amy orders chicken, Bill orders fish and Clare orders fish. This contradicts rule c. So Amy orders only fish. Then from rule b Clare can order fish or chicken. Clare orders chicken, then Bill orders fish or chicken. If Clare orders fish then Bill can order only chicken. So Clare could have chicken yesterday and fish today.

APPENDIX 4.3.2 COMMENTARY ON TASK 4.3.5

The time period of a pendulum depends on the length of the pendulum; neither the magnitude of the weight nor the position from which the pendulum starts swinging (high or low) affects the time period.

From the pupil readings shown in Table 4.3.2, it follows that investigations 1 and 2 tell you the length has an effect on the time period, since only length was changed, weight and position of release being held the same.

Investigations 2 and 4 tell you that weight has no effect on the time period, since length and position of release are held the same.

Finally, from 1 and 3, *if weight has no effect*, then since the length is constant in both experiments, the position of release has no effect on the number of swings per half minute.

Additional resources for this unit are available on the companion website:
www.routledge.com/textbooks/9780115478724

UNIT 4.4

RESPONDING TO DIVERSITY

Andrew Noyes and Tony Turner

INTRODUCTION

Teaching is a complex process, made interesting by the vast range of educational, social and political contexts and purposes that shape it. Your classroom is likely to contain pupils from several ethnic origins with a range of cultural expectations and belief systems, be a mix of boys and girls and have pupils from family backgrounds with differing views on the value and purpose of education. In addition, there will be differences in the financial status of your pupils' families, and for some families, unemployment. Other pupils may not be living with their parents but looked after in other ways. This unit helps you to understand better the complex make-up of modern UK society, particularly how social diversity, educational opportunity and attainment are interrelated. As well as considering a big picture perspective you are expected to take account of such diversity in your classroom practice and encouraged to examine critically the ways in which you respond to diversity in your teaching.

The standards which student teachers across the UK have to meet make it very clear that the ability to respond appropriately to pupils' diverse needs and circumstances is of primary importance for teachers. The General Teaching Council's (England) (GTCE) 'Code of Professional Values and Practice' states that: 'teachers challenge stereotypes and oppose prejudice to safeguard equality of opportunity, respecting individuals regardless of gender, marital status, religion, colour, race, ethnicity, class, sexual orientation, disability and age' (GTCE, 2006). This is easy to write but much more difficult to implement. It is important to recognise that you are part of that diversity and that your experiences of education and your perspectives on the educational, social and political contexts of the day have a significant influence on the way in which you respond to diversity. In this unit we focus on issues around gender, ethnicity and social class and, although they are examined separately herein, they clearly overlap and intertwine.

OBJECTIVES

At the end of this unit you should be able to:

■ access evidence about the relative academic performance of pupils in relation to gender, ethnicity and class

■ discuss issues of discrimination and bias in relation to gender, ethnicity and class

■ consider school policies and critique classroom procedures in order to promote better opportunities for learning for all pupils.

Check the requirements for your course to see which relate to this unit.

A HISTORY OF DIVERSITY

The presence of people originating from other cultures, faiths and backgrounds has been a feature of British society for many centuries. The notion of 'other' suggests that British society is easily described but 'Britishness' is neither a clearly defined nor a fixed concept. Immigration into the UK has come in waves and even now is a significant political issue both here and across the European Union. A few examples draw attention to the changing ethnic mix that contributes to present day society. The Roman conquest of England at the dawn of the Christian era lasted some six centuries at the end of which came invasion by Germanic peoples. About two centuries later the Viking invasion brought Danes and Norwegians to the mix of peoples. In 1066, the Normans took over much of England and many French people stayed and were assimilated. In the sixteenth century many people fled mainland European countries to escape, for example, religious persecution.

The ethnic mix was added to by Jewish immigration following their persecution at the end of the nineteenth century. Two world wars saw a further migration of people in mainland Europe away from conflict. As many former colonial countries have gained independence, immigration from Africa and Asia altered the ethnic mix in the UK. The important difference associated with this latter wave of immigration was the visibility of newcomers. Prior to 1948 there was only a trickle of black immigrants; the 1950s in the UK saw the active recruitment of Black and Asian families to fill the gaps in the work force (Briggs, 1983: 310–312).

Throughout the last century the population of England and Wales grew steadily, from 35.5 million in 1901 to nearly 49 million in 1991. The 2001 census showed that the population of Great Britain was almost 57.1 million. It would appear that steady population growth has been supplemented by immigration. The minority ethnic population is expected to grow to some 10 per cent of the population by 2020. The freedom of movement of EU citizens and the widening of EU membership has added further to the cultural mix.

Up to the early 1950s immigration was largely white. Many immigrants prior to that date were assimilated into the host culture, although a few immigrant groups maintained their distinctive lifestyle, e.g. the Jewish community. At first, the host community adopted the same attitude towards the Black and Asian immigrants and expected them to adopt the values and lifestyle of the host nation. That this policy did not work may be attributed to the influence of racism. Black people were not accepted in the same way as white immigrants of the past. This attitude towards immigrants and asylum seekers remains today among a minority and far-right political parties such as the British National Party continue to find support in some parts of the UK.

EQUAL OPPORTUNITIES AND EDUCATIONAL EQUITY

For many people it is self-evident that the implementation of equal opportunities policies is a reflection of basic human rights, but how does the notion of equality of opportunity relate to a teacher's response to diversity? Should the same curriculum, teaching styles, etc., be used consistently, or does such 'equality' in fact perpetuate inequality? Sociologists of education have argued for many years that if you assume that all pupils come to school equally prepared, and treat them accordingly, then in reality you advantage those who have been better prepared in their social milieu to succeed at school (Bernstein, 1977; Bourdieu, 1974: 32–56; Willis, 1977). These authors were writing over a quarter of a century ago but their writing remains pertinent for twenty-first-century Britain.

Concerns with equity have shifted between different groups over the years. In the 1980s and earlier there was concern about the underachievement of West Indian pupils (Short, 1986). Now there is growing concern about the low performance of many white working-class boys and pupils from some minority ethnic groups (Office for Standards in Education (Ofsted), 2003a). Although much progress has been made in the last three decades as regards equal opportunities for men and women in the workplace and boys and girls at school, there remain substantial differences in the roles of men and women in society (Equal Opportunities Commission (EOC), 2006). These issues around ethnicity, gender and class comprise the foci of the remainder of the unit.

In many cases, explanations of pupil underachievement have focused on the short-comings of the pupil or their families, the 'deficit model' explanation. More recently the focus has shifted to addressing the educational system as one of the factors contributing to underachievement. This focus is not only at the level of government and school policy but also in the classroom where such policies are interpreted and implemented by teachers (see *TES* (2008) for brief reviews of the relationships between achievement and gender, ethnicity and class).

GENDER

The EOC, which is now part of the Equality and Human Rights Commission (http://www.equalityhumanrights.com), has regularly produced overviews of social differences in our society. These are available online. These overviews include performance in national examinations, take-up and achievement post-16 and in further and higher education, as well as employment patterns and rates of pay in England. There remains much difference in the ways in which boys and girls/men and women exist and move through our society. Focusing on secondary schooling, both boys and girls have been steadily improving their performance in school examinations over the past 20 years but there is now a clear gender gap, particularly in some subject areas. Table 4.4.1 compares the performance of boys and girls at General Certificate of Secondary Education (GCSE) gaining grades A*–C in many of the popular GCSE subjects in 2006.

■ **Table 4.4.1** GCSE attempts and achievements in selected subjects in schools in 2006

Subject	Boys		Girls	
	Entries (thousands)	% passes at grade C or above	Entries (thousands)	% passes at grade C or above
Maths	310	55	305	57
English	307	55	305	69
English Literature	255	61	255	73
Science (any)	298	54	294	55
French	94	58	116	69
German	40	63	45	74
History	106	63	102	69
Geography	104	63	83	69
Art and Design	77	59	112	78
Design and Technology (any)	180	52	159	67
Physical Education	91	60	61	61
Religious Studies	63	63	82	75

Source: Department for Children, Schools and Families (DCSF) (2007a)

■ **Table 4.4.2** GCE A level examination, 2006, take-up by gender

Subject	Entries (thousands)	
	Women	Men
English	53.9	24.2
Mathematics	19.1	30.6
Biological Sciences	27.3	19.3
Physics	5	18.7
Chemistry	16.8	17.7
French	8.3	3.9
Psychology	36.2	12.4
Sociology	18.6	5.7
Art and Design	25.8	11.6
Business Studies	12.1	17.8
History	20.1	20.5
Geography	12.8	15.5

Source: DCSF (2007a)

Given the different performance at GCSE level in England, Table 4.4.2 shows examination entries for General Certificate of Education (GCE) A level courses in 2006. There is clear gender delineation in some areas between what might be considered masculine and feminine subjects (e.g. physics, English, psychology).

The EOC reported that often pupils have very stereotyped ideas about the roles of men and women, and how these stereotypes are themselves linked to what might be termed class (EOC, 2001). There is much research on the different ways in which boys and girls are positioned by, and respond to, moving to a new school (Noyes 2003; Jackson and Warin, 2000), to teachers (Younger *et al.*, 1999), to learning and the curriculum (Paechter, 2000) and to the use of physical space, and so on. These different modes of being are not simply genetic but are socially constructed identities that gradually influence, and are shaped by, boys' and girls' responses to the opportunities and challenges of school (see also Task 4.4.2). Rudduck (2004) explores the challenges of developing, and possibilities for, gender policies in secondary schools in some detail.

So far in this section we have been looking at the relationship between performance and biological differences between pupils; however, the notion of gender is more nuanced, including different masculinities and femininities. Traditional stereotypes are being broken down and at the same time are more fragmented. Such differences are increasingly celebrated and this includes pupils' sexuality. As we said at the outset, teachers' responses to this level of diversity will be influenced by their personal views and experiences. The EOC (2006) reports that there are an estimated 2.3–3.2 million gay, lesbian and bisexual adults in the Great Britain earlier in the decade. Whatever your views, many issues in this area require a thoughtful, professional approach, e.g. homophobic bullying (Intercom Trust, 2008).

ETHNICITY

A report in the mid-1990s showed that considerable progress had been made by many, but not all, groups of minority ethnic pupils (Gillborn and Gipps, 1996). That situation remains today. Two reports at the turn of the century identified the serious underachievement of Black Caribbean pupils and that these pupils have not shared in the rise in standards and achievement shown by many of other pupils, especially white pupils (Fitzgerald *et al.*, 2000; Gillborn and Mirza, 2000). For example, many pupils of Pakistani background failed to keep up with the improvements of most pupils.

These researchers point out that governments, both central and local, have failed to address adequately the issues of underachievement of particular groups of pupils. As a consequence the gap between the high and low achievers widens with time. The under-performance of many white working-class boys serves to emphasise the importance of taking into account how class and gender effects are embedded within the data showing attainments of ethnic groups (see Gillborn and Mirza, 2000).

When data from the Youth Cohort Study (Department for Education and Employment (DfEE), 1999b) are combined with other data and analysed to look at the effect of gender on performance in each ethnic group, some trends emerge. These need to be treated with caution but suggest that:

■ girls are more likely to achieve a higher number of grades than boys in all ethnic groups
■ underachievement is linked with ethnicity for both girls and boys
■ girls from Indian and white ethnic groups are more likely to achieve five GCSE grade A*–C than any other group of pupils
■ Black Caribbean boys show a small decline in an already low relative achievement over the period 1991–1995 (Gillborn and Mirza, 2000: 23); and their achievement is marginally lower than in 1988.

The 2007 Report of the Youth Cohort Study shows that girls in all ethnic groups continue to score higher than boys in GCSE examinations, although both groups of pupils have improved year on year. Black Caribbean pupils improved their performance in GCSE; about one-third of pupils gained 5 grade A–Cs at GCSE in 2003, rising to about one half in 2006. However, relative to most other ethnic groups, black Caribbean pupils remain the lowest achieving ethnic group (see Table 4.4.3 and DCFS, 2008h: 18).

The evidence accumulated by the Swann Inquiry over twenty years ago suggested that many minority ethnic pupils were underachieving (Department of Education and Science (DES), 1985). The research described above, and by Gillborn and Gipps (1996), shows that although achievement has been raised for most pupils, underachievement persists and is very much linked to class and ethnicity.

In eight out of ten local authorities (LAs) that took part in the Youth Cohort Study, and monitored the performance of all their pupils by ethnicity, it was shown that, as a group, Indian pupils attain higher outcomes than their white counterparts. This finding indicates that having English as a second language may not hinder academic achievement. One study has shown that 88 per cent of Indians, 92 per cent of Pakistanis and 97 per cent of Bangladeshis speak a second language (Gillborn and Mirza, 2000: 10). Although some groups of pupils appear to underperform, the research showed that each ethnic group outperforms all other ethnic groups in at least one LA responding to the Youth Cohort Study (Gillborn and Mirza, 2000: 9).

For teachers working with pupils from a range of ethnic backgrounds, this issue of supporting those for whom English is an additional language is important. The issue of language goes far wider than classroom oracy. Many classroom resources are heavily text based and their use of language

■ Table 4.4.3 Achievements at GCSE/GNVQ in 2006, by ethnicity and gender

Ethnicity	% of pupils gaining five or more A*–C		
	Boys	Girls	Total
White	**53.0**	**62.3**	**57.5**
White British	52.9	62.2	57.5
Irish	57.2	65.0	61.3
Traveller of Irish Heritage	14.0	23.2	19.0
Gypsy/Roma	9.3	11.8	10.4
Any other White background	55.0	65.3	60.1
Mixed	**50.7**	**61.2**	**56.1**
White and Black Caribbean	39.9	54.2	47.3
White and Black African	51.8	61.5	56.8
White and Asian	65.6	72.1	68.9
Any other mixed background	54.0	63.3	58.7
Asian	**55.5**	**66.9**	**61.0**
Indian	67.1	76.6	71.7
Pakistani	45.4	57.9	51.4
Bangladeshi	50.9	62.2	56.6
Any other Asian background	57.8	72.4	64.6

Source: DCSF (2007b)

requires not only a level of literacy but also cultural awareness. One area impacted by pupils' non-English home languages is homework and coursework tasks. As is pointed out above, the increased use of home-learning resources, and indeed homework, relies to some extent upon the knowledge and support of parents or other family members. If language provides an extra hurdle only for some pupils then homework has a differential effect. Teachers should consider carefully the language requirements that are made by textbooks, classroom talk and homework tasks.

The above discussion reminds us that the performance of pupils is related to many factors including gender, ethnicity and class. The variables used here, gender, ethnicity and class, are not causes; these variables hide causative factors which contribute to underachievement. Although Table 4.4.3 does not refer to class, there remains considerable variation in the performance of pupil from various ethnic groups in 2006 GCSE examinations. It is also evident that across ethnic groups, gender is a factor in achievement.

Underachievement in school may be a factor in the higher unemployment rate of many minority ethnic adults. In 2005, 9.1 per cent of females and 10.7 per cent of males from minority ethnic groups were unemployed in Great Britain compared to 3.7 per cent of female and 4.6 per cent of male white adults (EOC, 2006). This relatively high level of unemployment has an impact upon the future economic status of those groups and therefore, in a complex way, the chances of their children in school. This is not to say that such inequality of attainment cannot be addressed, or is somehow cyclical, but rather to highlight the limitations of schools' ability to affect social change. Recent government initiatives in England have attempted to improve the employment and access to higher education (DirectGov, 2008).

CLASS

Ethnicity, although considerably more complex than gender, allows us to analyse the attainments of ethnic categories and see which groups are performing above or below what might be expected from national datasets. The theoretical notion of class is far more complex and contested but the effect that pupils' economic circumstances have upon their education is very real. Connolly (2006) makes a convincing argument from statistical analyses of performance data that pupils' social class and ethnicity have a far greater effect on GCSE performance than gender. Understanding social class goes beyond simply looking at economic capital (e.g. measures like free school meals) but relates to other 'capitals' that pupils' families possess. This might be social capital or the cultural capital that includes having well-educated parents and ready access to books, computer media or other learning and cultural experiences. There is much evidence to signal the relationship between culturally rich homes and pupil attainment. This unit has repeatedly referred to the way in which gender, ethnicity and class effects are overlapping and intertwined. In that sense Table 4.4.3, which shows the performance of different ethnic groups, might have as much to do with class as it has to do with gender. Table 4.4.4 uses the rather crude measure of free school meals to show how socio-economic status relates to GCSE performance. The relative underperformance of these 80,000 pupils is striking.

In the same way that our response to gender issues needs to become more nuanced, so here we should include the increasing number of looked-after children and the needs of refugee and asylum seekers children.

The challenge for teachers is of course that we cannot necessarily see who these pupils are but they might well require different kinds of support. The use of school uniforms is argued by some to 'level the playing field' by removing the stigma of poverty but perhaps this in fact disadvantages those most in need of targeted help in school because they are now hidden. Class works in many subtle ways to disadvantage those already disadvantaged. Through language, manners, cultural awareness, etc. the middle classes have a better sense of the 'rules of the game' and so can capitalise better on their educational opportunities. This has recently been seen in the ongoing arguments about Oxbridge recruitment with one Oxford don (Ryan, 2007) explaining that whatever attempt is made to level the playing field for entry to such elite institutions, the elite of society will respond in whatever ways necessary to continue to secure an advantage for their children.

Another aspect of this class discussion is poverty, the effect of which has been described graphically by Davies (2000: 3–22) and can be seen in the GCSE attainments data in Table 4.4.5. It is worth noticing in the table the difference in proportional decrease of 5 A*–C grades when including English and mathematics from high to low deprivation areas. The second-term Labour government (2001) had a commitment to eradicating child poverty by 2020, but many commentators suggest that

■ **Table 4.4.4** Achievements at GCSE/GNVQ in 2006, by free school meals and gender

	No. of 15-year-olds		% Achieving five or more grades A*–C at GCSE	
	Boys	Girls	Boys	Girls
Free school meals	39,498	38,589	28.7	37.4
Non-free school meals	261,971	252,545	56.2	66

Source: DCSF (2007b)

■ **Table 4.4.5** GCSE and equivalent attempts at the end of Key Stage 4 in 2006 by Indices of Multiple Deprivation band

	Percentage of pupils at the end of Key Stage 4 achieving GCSE and equivalents	
	5+A*–C grades	5+A*–C grades inc. English & mathematics
ENGLAND		
0–10 most deprived areas	47.6	29.2
10–20	50.1	34.3
20–30	52.0	37.4
30–40	52.8	39.3
40–50	56.3	42.6
50–60	57.5	45.0
60–70	60.1	47.9
70–80	61.6	49.1
80–90	64.8	53.1
90–100 least deprived areas	68.1	57.6

Source: DCSF (2007a)

the gap between rich and poor is increasing (Gold, 2003: 22; Woodward, 2003:8). This division is happening partly as a result of legislation aimed at the marketing of education by Conservative and Labour governments. Typically more well-educated, middle-class parents have a better understanding of the school system (Power *et al.*, 2003), where to get information from (Hatcher, 1998), how to best support their children's schooling and so generally stand a better chance of maximising the opportunities afforded by the new educational markets (Ball, 2003). Class and ethnicity remain the main factors in educational disadvantage and yet class is possibly the most difficult factor to define, identify and respond to in a way that can precipitate meaningful and long-term change.

Deprivation can be measured by identifying a number of indices such as Income Deprivation, Employment Deprivation, Health Deprivation and Disability, Educational Deprivation (quality of) Living Environment and (incidence of) Crime. Areas, such as LAs, are assessed for these indicators, and ranked according to the prevalence of these indicators. The more indicators, the lower the rank and the more deprived the area. Thus in Table 4.4.5 the data 0–10 per cent identifies the area as in the bottom 10 per cent of the ranking, i.e. most deprived.

SCHOOL POLICY AND CLASSROOM PRACTICE

In this section we begin to examine both school policies and your classroom practice, in the light of the earlier discussion. If schools play a role in the ongoing structuring of inequality in society, albeit not deliberately, then teachers need to reflect critically upon their practice. This involves examining how your own position and action in a diverse society both help and hinder you from challenging unequal treatment in the classroom. See Task 4.4.1.

No matter how concerned the school is to promote equity through good policies, implementing them in the classroom is not an easy matter. It is instructive to use lesson observation time to look

Task 4.4.1 POLICIES TOWARDS EQUAL OPPORTUNITIES (EO)

1   Obtain a copy of the equal opportunities (EO) policy in your school. Read it and try to identify:

   ■ who wrote the policy and were, e.g. parents or pupils involved?
   ■ how old is it?
   ■ are there any later documents, e.g. working party reports?
   ■ what areas of school life does it cover ? Is it the curriculum, playground behaviour, assembly or other aspects of school life? Are any areas of school life omitted from its brief?
   ■ what is the focus of the policy? Is it gender, ethnicity, social class or disabilities?

2   Who knows about the policy? Devise a way of sampling knowledge, understanding and opinion of pupils and staff about the policy. For example:

   ■ are copies of the policy displayed in the school?
   ■ how many staff know about it; have read it?
   ■ how many pupils know about it; have read it?
   ■ who is responsible for EO in the school?

3   Is the policy treated seriously in the school? For example:

   ■ has any in-service programme been devoted to EO issues?
   ■ does the school EO policy influence departmental policy or classroom practice?

Summarise your findings to discuss with your tutor.

Task 4.4.2 RESPONSES TO GENDER: CLASSROOM OBSERVATION

Ask a class teacher if you can observe their lesson. Explain the purpose of your inquiry, which is to find out which pupils participate in the lesson more than others; and which pupils the teacher invites to answer questions or volunteer information. Be prepared to share your findings with the teacher.

Keep a tally of the frequency of attention to, and the time given to, boys and girls. Devise a recording sheet to collect information about one or two of the following behaviours:

■ who puts their hand up to answer a question?
■ who does the teacher select?
■ in class activities, how much time is spent by the teacher with boys, with girls?
■ if the teacher responds to a pupil with praise, criticism or further questioning?
■ when pupils are reprimanded, is there any difference in:
■ the nature of the misdemeanour? That is, what is tolerated, or not by the teacher?
■ the action taken by the teacher?

Consider whether different messages are conveyed to boys, or girls, by the teacher's response to classroom interactions.

Draft notes for discussion with other student teachers in your school experience school. Afterwards, identify and record the implications for your teaching.

Task 4.4.3 **RESPONSES TO ETHNICITY: CLASSROOM OBSERVATION**

Redesign the record sheet you used in Task 4.4.2 to collect data about teachers' response to pupils of different ethnicity. Use similar questions as listed under Task 4.4.2 and observe the same protocols about observing classrooms. You may wish to add to the list of questions:

■ did racist behaviour occur in the class?
■ what was the nature of the racism? Was it name-calling, inappropriate language or metaphors, stereotyping or other actions?
■ how was the racist incident dealt with?

Draft notes for discussion with other student teachers in your school experience school.

at specific aspects of teaching and then to report back to your tutor or tutor group (see Unit 2.1). Some ideas are listed in Tasks 4.4.2 and 4.4.3.

When you first start teaching, your concern is to promote learning through well-ordered lessons. When you feel more confident ask a colleague to observe one of your lessons, focusing on the questions in Tasks 4.4.2 and 4.4.3. As you develop as a teacher you can consider wider issues of inclusion and diversity. These issues might be regarding language or the textual materials used by pupils who, like us, are influenced by words and pictures, particularly moving ones. Access to the World Wide Web has opened up all sorts of material to pupils, and not all of this is helpful for their academic or social development. Task 4.4.4 asks you to review teaching materials for their implicit messages

Now complete Task 4.4.5.

Task 4.4.4 **BIAS AND STEREOTYPING IN TEACHING RESOURCES**

Select a resource in general use in your school, e.g. a book or DVD, and interrogate it for bias and stereotyping. Some questions you could use to address this issue include:

■ are women and girls shown in non-traditional roles?
■ are men shown in caring roles?
■ who is shown in a position of authority? Who is the employer, the decision maker, the technologist?
■ are people and jobs stereotyped, e.g. black athletes, male scientists, female social workers, male cricketers?
■ what assumptions, if any, are made concerning minority ethnic citizens in the UK?
■ how accurate are the images shown of people and of places?
■ how are people in the developing world depicted? Is it to illustrate malnutrition, or their living conditions or the technology employed? Are the images positive or negative?
■ what assumptions, if any, are made concerning underdevelopment in the developing world?

Identify some issues for discussion with your tutor.

> ### Task 4.4.5 WHO IS RECRUITED TO POST-16 COURSES?
>
> If your school has a sixth form, compare the number of pupils in the first year academic and vocational courses, by gender and ethnicity and then compare those numbers with the numbers in the previous year's Year 11 cohort. Identify the subject preferences.
>
> How many pupils left school to carry on education in another institution and what are their gender and ethnic characteristics?
>
> Identify questions arising from the exercise for discussion with your tutor.

RESPONDING TO DIVERSITY IN THE CLASSROOM

Your immediate concerns are focused on the classroom but much of what goes on in the classroom has its origins outside the classroom. These origins include the cultural background of pupils; the teachers' expectations of pupils; the externally imposed curriculum and the school's ethos realised through its policies and practices.

Expectations of academic performance are often built upon both evidence of what a pupil has done in the past and their social position: male/female; white/black; Irish/black Caribbean; working/middle class; stable/unstable family background. A perceived social position is sometimes, if unconsciously, used by teachers to anticipate pupils' progress and their capacity to overcome difficulties (Noyes, 2003). For example, 'Jimmy is always near the bottom of the class, but what do you expect with his family background?' Or 'The trouble with Verma is her attitude, she often seems to have a chip on her shoulder and doesn't respond well to discipline even when she is in the wrong. She never gives herself a chance; I'm always having a go at her.' You might have found yourself thinking these things, or even making assumptions about pupils from the time you first saw their names on the register. What might you expect of a Chantelle, Saima or Harriet, or of Wayne, Edward or Manbwe?

The interaction of the teacher with pupils in the classroom is often revealing. Some teachers may subconsciously favour asking boys, rather than girls, to answer questions. Once established, the reasons for this behaviour can be explored. Similar questions can arise about the way teachers respond to pupils' answers. Whereas one pupil might make a modest and partly correct response to a question to which the teacher's response is praise and support, to another pupil, offering the same level of response, a more critical attitude may be adopted by the teacher.

Are these different responses justified? Is the pupil who received praise gaining support and encouragement from praise; or is the pupil being sent a message that low-level performance is good enough? It is teacher expectations that direct and control such responses. If, as has been documented in the past about the performance of girls, the praise is implicitly saying 'you have done as well as can be expected because you are a girl' and the critical response is implying 'come on now, you're a boy; you can do better than this', then there is cause for concern.

Such interpretations depend very much on the context. A comparison of teacher behaviours in different lessons might reveal the influences on teaching and learning of the subject and the gender, age and social and cultural background of teachers and pupils (Pearce, 2005). Tasks 4.4.2 and 4.4.3 addressed this suggestion.

Now complete Task 4.4.6.

Task 4.4.6 ETHNICITY AND LEARNING

Some schools are predominantly or entirely white, whereas others are more diverse. However, the issues of race and ethnicity are relevant for *all* schools and pupils.

Question: What has been the impact of legislation and initiatives on the achievement of pupils in your placement school?

1   Read the executive summary of the report *Aiming High: Raising the Achievement of Minority Ethnic pupils,* (DfES, 2003g).

2   Read in full the following paper: Tomlinson, S. (2005) 'Race, ethnicity and education under New Labour', *Oxford Review of Education*, 31 (1): 153–171.

3   Find out from staff, school documents and inspection reports (Ofsted or LA) information about the following features of your placement school:

- the ethnic mix of pupils
- the different languages spoken by pupils
- the number of pupils needing and receiving support because English is not their first language
- the cultural and religious background of pupils
- the academic performance of pupils, e.g. in National Assessment Tasks, GCSE and GCE A level.

Are the academic data analysed by ethnicity?

4   Obtain copies of the following documents in your placement school:

- the equal opportunities policy
- the behaviour policy, including how any advice or policy on racist behaviour.

5   Find out to what extent your subject department policy and practice relates to the whole school ethnicity policy. Focus particularly on the curriculum, for example:

- Is a commitment to multiculturalism explicit at subject level? If so, identify how this policy influences teaching and learning.
- Examine teaching and learning resources to see if they reflect this policy.
- Are there different strategies for supporting pupils from different backgrounds?

6   Is there evidence of positive approaches to ethnicity and anti-racism in the broader life of the school? For example, in display material, in assemblies, in the recognition of different cultural and religious practices at different times of the year, in extra-curricular activities. You may find other examples.

Summarise your findings and analyse the information using, for example, the reports cited above. Draw out some key issues for discussion with your tutors as a draft document. Later revise the report in the light of discussions and file in your PDP.

SUMMARY AND KEY POINTS

In order to promote equity in educational contexts teachers need to 'understand how pupil and young people develop and that the progress and well-being of learners are affected by a range of developmental, social, religious, ethnic, cultural and linguistic influences' (Training and Development Agency for Schools (TDA), 2007c, Q18).

But this is only the first step! Beyond 'understanding' needs to come action and this involves noticing, critiquing and changing your own practices if necessary, in order to create learning environments in which all pupils can thrive and succeed, and where prejudice is rooted out. Some of the developments that need to take place might involve changing resources, grouping procedures, developing other teaching styles. Through such actions teachers can begin to challenge some of the inequalities in our society. Alternatively teachers can simply maintain the status quo and although their discourses might welcome diversity, their practices might be maintaining inequity.

Sometimes you hear a teacher say 'I didn't notice their colour, I treat them all the same'. Learning opportunities are enhanced by not being 'gender blind' or 'colour blind' or 'class blind'. We suggest that not recognising pupil differences, including culture, is just as inadequate a response to teaching demands as is the stereotyping of pupils. Pupils learn in different ways and a key part of the differentiated approach to learning is to recognise those differences without placing limits on what can be achieved (see Unit 4.1).

This brings us back to expectations, preconceptions and even prejudices (however unintended) of the teacher. If you expect most Asian girls to be quiet and passive and good at written work, then that is not only what they do, but also perhaps all they do. Individuals respond in different ways to teachers; you should try to treat each person as an individual and respond to what they do and say, making positive use of your knowledge of the pupils' culture and background. This unit has addressed inclusion in relation to gender, ethnicity and class; another dimension of inclusion is addressed in Unit 4.6 on inclusion and special educational needs.

Check which requirements for your course you have addressed through this unit.

FURTHER READING

Gillborn, D. and Mirza, H. S. (2000) *Educational Inequality: Mapping Race, Class and Gender; A Synthesis of Research Evidence*, London: Ofsted.
This report says that black pupils failed to share in the dramatic rise in attainment at GCSE, which took place in the 1990s, to the same degree as their white peers. Black and ethnic minority youngsters are disadvantaged in the classroom by an education system, which perpetuates existing inequalities. Differences in the achievements of boys and girls and pupil of professional and working-class parents are compared and contrasted.

RELEVANT WEBSITES

Department for Children, Schools and Families (DCSF): http://www.dcsf.gov.uk
The Resources and Tables and Statistics sections of the DCSF website contain extensive information that helps you to consider the performance and characteristics of pupils from different backgrounds. The Youth Cohort Study offers longitudinal data over a number of years; their annual reports can be accessed online (see DCFS, 2008h).

Equal and Human Rights Commission (EHRC): http://www.equalityhumanrights.com
This site brings together three commissions: the Commision for Racial Equality (CRE), the Disability Rights Commission (DRC) and the Equal Opportunities Commission (EOC). The site is a wealth of information and includes many easily downloadable statistical summaries of social life in Britain. From these reports you can examine the changing nature of the society in which you live and teach. Many reports are repeated on an annual basis.

TES (*Times Educational Supplement*) (2008) Gender/Race/Class: TES Series challenging education stereotypes, London: TES: http://www.multiverse.ac.uk/ViewArticle2.aspx?anchorId=247&selectedId=298&contentId=14485
How pupils perform at school is influenced by their gender, race and social class. The gaps between the performances of different groups have led to public concern and a multitude of national and local initiatives. In this four-part series, the TES examines each factor in turn to see how it affects teachers and pupils.

Additional resources for this unit are available on the companion website:
www.routledge.com/textbooks/9780415478724

VALUES, EDUCATION AND MORAL JUDGEMENT

Ruth Heilbronn and
Tony Turner

INTRODUCTION

In England all tcachers are expected to uphold the values expressed in the Statement of Professional Values and Practice for Teachers of the General Teaching Council for England (GTCE, 2006). Many other countries have similar requirements; check with your tutor the requirements that apply in your context. As a student teacher you arc expected to have high expectations of your pupils and a commitment to ensuring that they achieve their full educational potential. You are expected, too, to develop a positive relationship with your pupils by being fair, respectful and supportive; and to demonstrate the same qualities that you expect of your pupils. For student teachers in England these qualities are described in more detail by the Training and Development Agency (TDA, 2007a: 7). Professional standards for teachers that apply at all stages of a teacher's career underpin this fundamental aspect of teaching as a professional practice.

In guidelines to understanding the standards to qualify as a teacher in England the Training and Development Agency for Schools (TDA) states that

> education is part of the process through which people acquire values and learn to apply those values in the attitudes they adopt and the ways they behave . . . Children and young people are more likely to thrive if they feel that they are valued and are confident that their teachers and their peers will support them. They are more likely to behave in a positive and constructive manner, and adopt appropriate values and attitudes, when they encounter such behaviours, values and attitudes in their teachers.
>
> (TDA, 2007b: 3)

Writers on this area of education often use the terms 'values education' and 'moral education' interchangeably. It might be argued that it would be better to see 'moral education' as only one part of values education. Some educators suggest that:

> morality is the area of values which affects how we treat each other and hencc the area of values which in an important sense are not optional. It is about what we owe to each other, what we may blame others for not living up to.
>
> (Haydon and Hayward, 2004: 165)

Talking about morality involves wider notions of 'goodness' and the nature of right and wrong than talk about 'values'. In this unit the term 'values' is used to mean 'that which we think is desirable' and is related to notions of 'worth'. Our values determine how we choose a course of action or a belief in situations where different views are held. 'Value judgements' on controversial issues differ from judgements of fact, for which evidence can provide a foundation. The National Curriculum (NC) in England for Citizenship is one vehicle through which 'moral development' can be introduced; we discuss citizenship later in the unit.

Any adult in a relationship with a child is in some sense teaching about values, and such teaching takes place in various situations, such as in one-to-one conversations, classroom teaching, parent or carer interactions. Teaching can be considered a practice in which the good teacher exemplifies both the skills and the values of the practice (Dunne, 2003: 353–371) and we expect teachers to be exemplary figures in the course of their work (McLaughlin, 2004: 339–353). When teachers stop a class because they have overheard a racist remark, which they tackle sensitively, knowledgeably, successfully, while upholding the value of tolerance, we would say they dealt competently with the situation, drawing on their own values and their experience, in a 'deliberate exercise of principled judgement in the light of rational knowledge and understanding' (Carr, 1993: 253–271).

Schools as well as individual teachers uphold values and express these in their policies and practices, which often makes for considerable differences in ethos between different schools, as evidenced in the behaviour, attitudes and priorities of the pupils and staff. Research shows that these differences are important to the individual teachers' ability to thrive and develop in school (Heilbronn *et al.*, 2002: 371–389) and to the pupils' expressions of respect for one another and their socially responsible behaviour in the classroom (Hansen, 1995: 59–74). The ethos of the school is created largely through the school leadership and the way in which it develops and manages a vision of the kind of school it wishes to maintain. Some tangible expressions of a school's ethos might be found in the way parents are welcomed into the school; the relationships between adults and young people outside as well as inside the classroom; staff sensitivity to cultural and faith differences in the school, or how the school celebrates the success of its pupils. The ethos of a school contributes as much to values education as does the prescribed curriculum.

The focus of much of this unit is on the opportunities in your daily work with pupils to foster and develop values and moral judgement and to help pupils understand the need for a set of common values and what those values might be. The Further Reading section at the end of this unit offers an opportunity to pursue some of the issues in more depth, e.g. Langford (1995). See also Unit 5.1 which introduces you to theories of how learning happens.

OBJECTIVES

By the end of this unit you should be able to:

■ describe the legal responsibility of the school towards pupils in the area of moral, spiritual, social and cultural development

■ begin to identify opportunities in school to promote these aims

■ try out some methods of teaching towards these aims and evaluate them

■ place moral and values education in a subject and school context, including citizenship education

■ relate these aspects of a teacher's role to the standards/competences expected of a newly qualified teacher.

Check the requirements for your course to see which relate to this unit.

VALUES EDUCATION

Values underlie all aspects of school life. In some schools there is a common understanding about where many of the values come from, as they are based on definable moral codes. So in a faith school there is a clear understanding about what is 'right' and 'wrong', based on tradition and 'scripture'. In a secular school this understanding is not so clear-cut. Nevertheless, even in a faith school, there are many areas where value judgements are made, which are not clearly indicated by the underlying faith ethic, such as judgements about what is 'fair' in a particular situation, which may involve judgement in choosing one claim above another. Teachers frequently have to make such judgements in particular situations. School policies may provide guidelines, for example when dealing with racist or bullying incidents. In these policies 'respect' is usually a core value and many of the school rules and procedures may derive from this stated value. However, what cannot be prescribed in advance of any situation is what any individual person *ought* to say and to do in response to any particular situation. As a teacher, it is a matter of judgement how to mediate the policy. So in acting and being in the classroom a teacher stands as a moral example. An example of such a situation might be that a teacher might punish a pupil who steals from another pupil without much discussion or might decide to talk to the pupil, to get them to understand that what they had done was wrong. Choosing how and when to have this discussion is also a matter of judgement.

Talking with a pupil about the rights and wrongs of an aspect of their behaviour is important, as it acknowledges and respects the pupil's ability to develop an understanding of the consequences of their actions and behaviour. Such talk encourages a sense of responsibility and agency and so helps the pupil to develop moral understanding. The policies of the school are in some sense a formal interface between the individual child and teacher. So the teacher in the example could relate the discussion about why stealing is wrong, to the school's formal statement or rule relating to respect. In fact, 'everyday classroom life is saturated with moral meaning. Even the most routine aspects of teaching convey moral messages to students' (Hansen, 1995).

It is suggested that pupils judge their teachers on the way they respond to unacceptable behaviour and it is significant that pupils tend to have good relationships with teachers who are good at engaging in reasoned discussion and who are able to articulate the values which underlie the rules about what is acceptable and non-acceptable behaviour (Lickona, 1983). These teachers are more likely than those who merely assert the rules to trusting and respectful relationships. In a research review article Nucci sums up this research to report that

> students rated highest those teachers who responded to moral transgressions with statements focusing on the effects of acts ('Joe, that really hurt Mike'). Rated lower were teachers who responded with statements of school rules or normative expectations ('That's not the way for a Hawthorne student to act'). Rated lowest were teachers who used simple commands ('Stop it!' or 'Don't hit').

> (Nucci, 1987: 86–92)

It is important to attend to any emerging issues of values in any particular classroom situation. The pupils experience a lesson as individual young people and not as disembodied 'learners' of a particular curriculum area. As participants in the classroom, their experiences need to be attended to. Van Manen has written of 'the tact of teaching' and how good, trustful relationships rely on 'pedagogical thoughtfulness' on the part of the teacher (van Manen, 1991). It may not be possible to take all the time necessary to deal in depth with particular issues as they arise, but this too can be acknowledged to your pupils. What is important is to reflect back to your pupils the example of the kind of behaviour that is acceptable and the reasons behind the values expressed, rather than ignoring or suppressing what is not acceptable.

Diversity and common values

A review of research shows that quite young children can distinguish moral judgements related to justice and fairness, from judgements about social conventions, such as how to address people properly or follow dress codes (Nucci, 1987: 86–92). It is important in a pluralistic society to be able to distinguish a number of fundamental common values that do not rely on any specific cultural or religious foundations. In a 1999 review of the first NC a forum was set up, entitled the National Forum for Values in Education and the Community, with representatives from all the faith groups and non-faith members. This forum developed guidance for teachers in the form of a 'Statement of Values' that set out to identify common and enduring values in society about which most people would agree, irrespective of culture or belief, and incorporated into the NC for England. The published statement is shown in Figure 4.5.1.

The National Forum moved from the broad general statement of values above to the translation of those values into practice. For example, under the heading of 'the Self' the National Forum team listed the following goals for pupils:

- develop an understanding of our own characters, strengths and weaknesses
- develop self-respect and self-discipline
- clarify the meaning and purpose in our lives and decide, on the basis of this, how our lives should be lived
- make responsible use of our talents, rights and opportunities
- strive, throughout life, for knowledge, wisdom and understanding
- take responsibility, within our capabilities, for our own lives (QCA, 2007g).

Similar detailed statements of values for 'relationships', 'society' and the 'environment' are given in the current NC for England (see Figure 4.5.2).

A statement of values

1 The self
We value ourselves as unique human beings capable of spiritual, moral, intellectual and physical growth and development.

2 Relationships
We value others for themselves, not only for what they have or what they can do for us. We value relationships as fundamental to the development and fulfilment of ourselves and others, and to the good of the community.

3 Society
We value truth, freedom, justice, human rights, the rule of law and collective effort for the common good. In particular, we value families as sources of love and support for all their members, and as the basis of a society in which people care for others.

4 The environment
We value the environment, both natural and shaped by humanity, as the basis of life and a source of wonder and inspiration.

■ **Figure 4.5.1** A set of values

Source: Qualifications and Curriculum Authority (QCA) (2007f)

The National Curriculum in England states that education should reflect the enduring values that contribute to personal development and equality of opportunity for all, a healthy and just democracy, a productive economy, and sustainable development. These include values relating to:

- *the self*, recognising that we are unique human beings capable of spiritual, moral, intellectual and physical growth and development.
- *relationships* as fundamental to the development and fulfilment of ourselves and others, and to the good of the community. We value others for themselves, not only for what they have or what they can do for us.
- *the diversity in our society*, where truth, freedom, justice, human rights, the rule of law and collective effort are valued for the common good. We value families, including families of different kinds, as sources of love and support for all their members, and as the basis of a society in which people care for others. We also value the contributions made to our society by a diverse range of people, cultures and heritages.
- *the environment*, both natural and shaped by humanity, as the basis of life and a source of wonder and inspiration which needs to be protected.

■ **Figure 4.5.2** Values underpinning the National Curriculum
Source: QCA (2007b)

In England, the current NC has clearly stated underlying aims, to promote 'the spiritual, moral, cultural, mental and physical development of learners at the school and within society'. The NC should enable 'all young people to become responsible citizens' (QCA, 2008i). Schools therefore have a responsibility to promote values education in some form, for example through a programme of Personal, Social and Health Education (PSHE), including Citizenship. For further discussion of curriculum for Wales, Scotland Northern Ireland and England see Chapter 7.

Successive Education Acts since 1988 have required the Office for Standards in Education (Ofsted) to inspect the contributions which schools make to pupils' spiritual, moral, social and cultural education and how well pupils' attitudes, values and other personal qualities are developed in relation to the aims underlying the NC, including the statement of values. Ofsted have published guidance entitled *Promoting and Evaluating Pupils' Spiritual, Moral, Social and Cultural Development,* which is available online (Ofsted, 2004c).

Settings and opportunities for values education and moral development

A young infant follows rules for behaviour and an adult reasons. How does the ability to reason in situations of dilemmas and conflicting choices develop? How does a child develop a sense of moral judgement? This question is a source of continuing interest and research. Building on Piaget's foundational work (Piaget, 1932), Kohlberg developed his own detailed theory of the development of moral reasoning. Both writers stressed maturation factors, linking the development of moral judgement with cognitive capability and arguing that mature moral judgement is dependent on a capacity to reason logically: it develops as children's reasoning ability develops (Kohlberg, 1985: 27–87). In discussing the development of children's ability to use increasingly nuanced moral reasoning, the term 'moral development' is frequently used, particularly in policy and curriculum documents.

For further study, the following articles provide an introduction to the work of Piaget and Kohlberg, with hyperlinks and further suggestions for reading: Atherton (2005); Crain (1985: 118–136); Huitt and Hummel (2003) and Nucci (2007).

Arguably children and young people's understanding of the school's values, and more widely of right and wrong, is 'caught', rather than 'taught'. Many argue that the only way for values and moral reasoning to develop is through experience, and more specifically through the reflection which arises out of experienced situations, usually in discussion with others, adults or peers. Research has shown that there *are* common features of schools that seem to have a positive impact on the development of student values, such as participation in the communal life of the school and the classroom; encouragement to behave responsibly; provision of an orderly school environment, and clear rules that are fairly enforced (Battisch *et al.*, 1998). A school's explicitness about its values and the extent to which teachers actually practised shared values also has an important influence on the pupils' own development of moral understanding and responsibility. There is also agreement that parental influence in values formation is far more significant than that of the school. This reinforces the importance of a partnership approach between schools and their local communities (Nucci, 1987: 86–92).

The NC for Citizenship is one of the vehicles through which 'moral development' can be channelled, since citizenship education, because of its subject matter, is inherently related to values. Political and social issues concern the question of how we should collectively live, which is at root an extension of the basic moral question 'how should I live my life?' (Haydon and Hayward, 2004: 161–175). Although many areas of public life can be taught about in a fairly descriptive manner they remain value-laden in nature and may often call for sensitive handling. The NC also outlines a strong, though not statutory programme of PSHE and this curriculum area, together with Citizenship, is intended to foster 'moral development' and values education. These subjects are the primary vehicles for engaging in values education in England and Wales.

There have also been curriculum initiatives which promote the development of moral reasoning through specific learning activities. Examples may be found in various programmes of 'character education', which create contexts, topics and examples giving structured opportunities to discuss values (see for example, 'Good Character.com': http://www.goodcharacter.com/).

The question remains open as to whether values can be *taught* or only *caught*? Kohlberg *did* comment on the artificiality of creating a specific curriculum area purely to foster moral development. In an evaluation of one of his own early projects he stated

While the intervention operation was a success, the patient died. When we went back a year later, we found not a single teacher had continued to engage in moral discussion after the commitment to the research had ended.

(Kohlberg, 1985: 80).

This underlines the importance of personal experience and commitment to engaging with discussion about values. When we talk to a pupil about something we have both experienced, we engage in a process of talking about values in a direct and practical way.

Thus:

The use of discussion acknowledges that social growth is not simply a process of learning society's rules and values, but a gradual process in which students actively transform their understanding of morality and social convention through reflection and construction. That is, students' growth is a function of meaning-making rather than mere compliance with externally imposed values.

(Nucci, 1987: 86–92)

Pupils can also learn from each other by exchanging information, sharing experiences and debating them. The new understandings which are 'constructed' in peer exchange can be powerful in developing pupils' moral understandings (Berkowitz *et al.*, 1980: 341–357).

Engaging in discussion also trains children in processes involved in moral reasoning, although these discussions need to be appropriately related to the pupils' own experience or imaginative capacities. Beliefs and conventions change as we follow and contribute to debates about behaviour or about decisions which impact on human life. An example might be the case of separation of conjoined twins, where one only could live, which was resolved legally in the face of conflicting moral viewpoints in the English courts in 2000. Other areas where beliefs and conventions are debated and change relate to the environment and the various choices which individuals may make in their own lives about such matters as endangered species or global warming. Young people need the skills and confidence to engage with such debates and these skills develop through practice.

In this unit we identify some areas where values education can take place, focusing on issues of a generic nature rather than those arising from curriculum subject such as citizenship or religious education. Our approach is to make suggestions for tasks from which you select those appropriate to your school and your needs. It is not intended that all tasks be undertaken.

Aims of schools

Aims are discussed also in Units 7.1 and 7.2; see Task 7.1.1.

Task 4.5.1 (pp. 224–5) focuses on the aims of a school and how these are interpreted.

Subject teaching

Now turn to Task 4.5.2, p. 225.

Class management

We have drawn attention earlier to the importance of teachers setting an example for pupils. If values are preached but not practised, pupils attach little significance to them much less uphold them. Task 4.5.3 (p. 226) illustrates the problem of acting fairly. See also Unit 3.3.

Assessing pupils

Cheating in tests, for example, is rare but when spotted often involves either collusion (with other pupils) or answers being hidden on the person. This situation may be recognised by direct observation during the test, by comparing pupil scripts or by recognising a familiar phrase likely to have been copied.

One cause of cheating is fear. The fear may be of the teachers' wrath or punishment in response to poor marks. Some teachers may, for example give a roll call of marks when the scripts have been marked; or make public comment on individual performances. If mistakes are treated with anger, with punishment or by humiliation, pupils are not encouraged to think freely or attempt solutions. Cheating often means the pupil gets the task correct, instead of wrong. This denies the pupil access to feedback from the teacher about the source of error and it denies them the opportunity to learn from mistakes.

Another cause of pupil anxiety may arise through trying to maintain personal esteem. This situation can occur when the pupil tries to 'keep their end up' at home. Some parents regularly ask their children how they got on and a pupil does not like admitting to low marks. On the other hand,

Task 4.5.1 AIMS OF THE SCHOOL AND HOW THEY ARE INTERPRETED

This task contains several sub-tasks which focus on the aims of the school. Carry out one or more sub-tasks, either alone or in pairs and discuss your findings with other student teachers or your tutors.

1   Collect together the aims of your placement school and, if possible that of another school. Compare and contrast the aims of each school.

■   Does the school in its aims express a responsibility for the moral development of pupils?

■   How are these aims expressed and in what ways do the school statements differ?

■   How do your schools' aims meet the aims of the NC, described above?

2   Consider these statements of broad aims for a school curriculum. The curriculum should:

■   allow pupils to make decisions and act on them

■   expose pupils to situations in which their contribution is necessary for the success of a project

■   foster pupils' self-image, or self-esteem

■   help pupils develop independently but without losing contact with their peers

■   enable pupils to interact with teachers as young adults with adults.

Respond to the statements above under the following two headings.

■   In what ways do the statements contribute to the moral and values education of pupils?

■   How might the aims of this school curriculum be translated into opportunities for the pupil to achieve them?

3   In what ways does the PSHE programme contribute to the moral and values education of your pupils? Select a topic, for example:

■   acceptable and unacceptable food, and the different practices for its preparation

■   the celebration of festivals and holidays across cultures and nations

■   preparation for careers, job application and interviews

■   identifying bias and stereotyping and developing strategies to deal with it

■   care of the environment

■   practices and attitudes towards marriage.

4   How does school assembly contribute to the moral and values education of pupils? You could:

■   attend assembly; keep a record of its purpose, what was said and the way ideas were presented. After an assembly interview some pupils and compare your perception of the session with theirs. Did they understand the message? Or believe the message? In what way is the message relevant to their life and their family?

■   help plan an assembly with a teacher and with the support of other student teachers.

You may find these questions helpful in planning your inquiry.

■ Does the school have a programme for assemblies? How is the content and approach agreed within your school?
■ Does assembly have a Christian bias, a requirement of the 1988 Education Act?
■ How does the content and messages of assemblies relate to the cultural mix of your school?
■ Does assembly develop a sense of community or is it merely authority enhancing?
■ How is success celebrated and whose success is mostly recognised?

5   Is any part of in-service education and training (INSET) for staff devoted to moral, ethical and values education? How is such work focused and by whom? Your school tutor can direct you to the staff responsible for INSET.

Task 4.5.2 **THE PLACE OF SUBJECT WORK IN PROMOTING MORAL DEVELOPMENT AND VALUES EDUCATION**

1   Identify a social, ethical or moral issue that forms part of teaching your subject. Review the issue as a teaching task in preparation for discussing it with other student teachers. Your review could include:

■ a statement of the subject matter and its place in the curriculum
■ the moral focus for pupils, e.g. the decriminalisation of drugs; use of animals for research purposes; road building through the green belt
■ a sample of teaching material
■ an outline teaching strategy, e.g. a draft lesson plan
■ any problems that you anticipate teaching it
■ any questions that other student teachers might help you resolve.

2   In what ways do the curriculum resources of your department encourage a moral dimension to the teaching of your subject curriculum? Explore and discuss.

This task could be tackled alone or with other student teachers on a subject basis, later pooling and comparing information. For example choose a topic for which you have to prepare lesson plans and consider what might be the moral dimensions of that topic. What visual aids are there to support that topic in this way? Are there any useful teacher networks and websites to support your planning?

Make a summary of your response to this task and use it to check the standards/ competences for your course.

Task 4.5.3 **CLASS CONTROL**

Pupils dislike being punished for offences which they did not commit or the misdemeanours of others. Under stress, it is easy to respond inappropriately to a misdemeanour.

A class is reading from a set text; from time to time there is a pause in the reading in order to discuss a point. As the lesson progresses, some children get bored and seek distraction by flicking pellets at other pupils seated towards the front. The teacher knows it is happening but is unable to identify the culprit. One or two pupils take offence at being hit by pellets and object noisily. The teacher asks for the action to stop but it continues; noise rises and teacher threatens to keep the person in when they are identified. The lesson is stopped and the culprit asked to own up. No one volunteers; the class is given 5 minutes to sort this out; no one owns up and the whole class is kept in after school for 15 minutes.

Some pupils object to detention on the grounds that they did not offend. Some walk out of the detention and refuse to stay.

How might the teacher:

■ deal with this situation for those left in detention?
■ deal with those who walked out?
■ deal with any complaint of unfair practice?
■ have responded differently to the whole incident?

How have your responses to this task helped you meet the requirements for your course?

the source of anxiety could be about being moved to a lower set. Where streaming is the method of grouping pupils, failure may lead to demotion, losing friends and having to make new ones; or to joining the class of an unpopular teacher.

Cheating may be a sign that the pupil cares. The misdemeanour may not entirely be their fault; children who don't care are less likely to cheat than those who do. The 'don't care' attitude is a different issue with which to contend. In Task 4.5.4 we invite you to share with others your responses to pupils caught cheating in examinations.

Task 4.5.4 **YOUR RESPONSE TO CHEATING IN CLASS?**

How might teachers respond to a pupil caught cheating? It is not possible to condone the action but at the same time pupils are not mature adults.

Draw up some school guidelines to

■ prevent cheating
■ publicly deal with cheating
■ help the pupil concerned.

Check if there are guidelines in your school about cheating in school, work or public examinations. How do these guidelines relate to your views? In what ways does Task 4.5.4 help to develop an understanding of values for the pupil? How have your responses to this task helped you meet the standards/competences for your course?

Critical incidents

Many incidents which arise in school have at their heart moral judgements. We mentioned one above, about cheating. How we respond to them taxes our notions of right and wrong; our decisions may guide pupils towards establishing their own moral position. Two examples of incidents are presented in Task 4.5.5; your experience in school will furnish you with others. The incidents attempt to show different aspects of moral judgement that need to be exercised by the teacher and the pupil. The first problem concerns personal morality; the individual has to wrestle with their own conscience and decide what to do. In the second example the teacher is put into a position where their decision is likely to be widely known. The teacher is faced by a situation in which there is a conflict of responsibilities towards the parents on one hand and the pupils on the other.

Task 4.5.5 CRITICAL INCIDENTS

Read through the following extract and respond to the questions below. You may wish to prepare your response to these situations prior to talking with other student teachers and your tutor. This procedure may help you get a focus on the legal responsibilities you face as a teacher. See also Unit 8.3.

1 Wayne is in school during break in order to keep an appointment with a teacher. On the way, he has to walk past his own form room. As he does so he sees a pupil from his class, whom he recognises, going through the desk of a classmate. The pupil is not facing Wayne but it seems obvious to him that items are being removed from the desk and pocketed, but he can't identify the items. Wayne walks away unseen by the other pupil. Wayne decides to tell his class teacher.

2 Serena is 15 years old, the daughter of practising Muslim parents, and attends the local girls' comprehensive school. Her parents do not approve of her mixing, at her age, with other pupils outside school hours and expect her to return home promptly after school. Nevertheless, in the company of other girls, Serena sometimes walks home part of the way with boys from the adjacent boys' school. Under pressure from peers, she agrees to meet one boy after school. She arranges for an alibi with another girl. On the proposed day of the meeting, the girl friend panics at the prospect of having to lie to Serena's parents and backs out of the agreement. That afternoon the two girls argue in class to the extent that they are kept back by the teacher after school. The teacher becomes unwittingly privy to the arrangements.

For each case:

■ Identify the issues.
■ Describe the advice would you give in each situation to the teacher and the pupils.
■ How have your responses to this task helped you meet the requirements for your course?

Valuing diversity

Some schools draw their pupils from families of widely differing backgrounds and ethnicity. This rich mix of pupils can be an opportunity for a school through its curriculum and its policies to help pupils understand the differences between people. In this way, schools can help pupils make judgements about other people based, for example on knowledge rather than hearsay, on how they behave rather than on ethnicity. The importance of recognising the differences between pupils in preparing lessons has been recognised in the NC for England (see also Units 4.4, 4.6 and 7.3).

In some schools the ethnic mix of pupils is more narrow but is the aim of valuing diversity altered because of this difference? Commenting on the experiences of minority ethnic teachers placed in a largely all white school, researchers wrote:

> We live in a hugely diverse society yet, for reasons of where people choose to live, children may come into contact with only a restricted range of the wider society of which they are a part. The experiences related to us by ethnic minority trainees who had been placed in mainly all white schools were a striking testimony to the important part minority teachers can play in such settings.
>
> (Carrington and Tomlin, 2000: 155)

One of the teachers interviewed by Carrington and Tomlin said of her first day in her school experience school:

> The first time I went into that school I saw all these white faces. And it really scared me, because it was the first time I was ever in a situation like that. Because I have always lived in a multicultural place, I find it really hard to go somewhere that isn't multicultural and ... difficult to deal with.
>
> ('Jill' in Carrington and Tomlin, 2000: 151)

Jill did not describe her own ethnicity, classifying herself as 'Other'; it is not possible to say whether she declined to be categorised from a list of ethnic groups, or simply did not recognise herself from the categories on offer (Carrington and Tomlin, 2000: 142).

Now complete Task 4.5.6.

Task 4.5.6 **MAINLY WHITE SCHOOLS AND NATIONAL ETHNIC DIVERSITY**

Respond to the evidence offered by Carrington and Tomlin above using the following questions. We suggest that first you tackle this on your own and then bring your responses to a seminar, either with other student teachers or with your tutor.

1   How would you respond to and support a teacher who has the experience that Jill describes above?
2   Would schools with mainly white pupils find It harder to prepare pupils for living in an ethnically diverse society than others with a more diverse population?
3   How and in what ways do the aims of your school experience school address the challenge of preparing pupils for living in an ethnically diverse society?
4   What evidence do you have that your school experience school adopts a positive approach to ethnic diversity?

Task 4.5.7 asks you to explore more deeply the identification of values, where they come from and to consider how pupils develop their sense of moral values.

Task 4.5.7 TEACHING RIGHT AND WRONG

Consider the following questions and draft a response to them before reading the extracts from Smith and Standish (1997). In the light of your reading review your responses.

1   In your opinion can values be explicitly 'taught' or only 'caught' by example and through experience? Draft a response to this question before reading the text: see Smith and Standish (1997: 75–91).
2   'If we are supposed to instil values in the young, then whose values are we supposed to instil?' Where do values come from?
3   What do you think about the statement of values of the National Forum as a basis for education in schools? See Figure 4.5.1 in this unit and Smith and Standish (1997: 1–14).
4   Investigate the Kohlberg stages of moral development, e.g. through the paper by Nucci (2007). What is your view of these stages? Do they tally with your views on how a moral sense develops? Do people pass through various stages of moral development as they grow up? See Smith and Standish (1997: 93–104).

Respond to these three issues around the requirement of schools and teachers to provide a set of moral values through which to help the growth and development of their pupils. Summarise your thoughts and discuss them with other student teachers before finalising your response. File the summary in your professional development portfolio (PDP).

SUMMARY AND KEY POINTS

Teachers have a responsibility in law in England, with similar statements in other countries of the UK, for the promotion of 'the spiritual, moral, cultural, mental and physical development of pupils at the school'. Moral judgement and moral behaviour derive from personal, social, cultural, religious and political viewpoints and conventions. Societies in which the norms of some groups of people differ from the main group have the potential for enrichment or friction. The statement of values is a step towards developing a consensus about values which may support social cohesion as outlined in the NC for England.

   In school there are many opportunities for developing these qualities in both subject work, including the citizenship programme and pastoral work. We have identified some situations in which your judgement is called into play and may call for the resolution of dilemmas of a moral nature, such as dealing with tale telling. On occasions you may find it difficult to know how to respond; in your teacher education course you should ask experienced and qualified staff for help.

   There are occasions when moral and ethical issues provide the explicit aim of the lesson. The use of simulations is a good way of introducing such issues as are organised discussions. We give advice on handling discussion and using simulations in Appendices 4.5.1 and 4.5.2 respectively. All these issues are likely to remain on your agenda during your induction period as a newly qualified teacher (NQT). It is important that you know the school policy about the issue that confronts you.

   The section 'Further Reading' gives more detailed advice and background about the development of values education.

   Check which requirements for your course have been addressed through this unit.

FURTHER READING

Haydon, G. (2006a) *Values in Education*, London: Continuum.

Addresses the issues of 'what are the fundamental aims and values underlying education'? And how can education promote values in a world of value pluralism? The text also discusses morality and should schools teach it. In a secular society, how should schools treat the links between morality and religion? This is an updated version of an earlier text 'Teaching about values' (1997).

Haydon, G. (2006b) *Education, Philosophy and the Ethical Environment*, London: Faber and Faber.

This book offers a critical analysis of some of the fundamental questions about the nature and purpose of education, using the concept of 'ethical environment'. It addresses many ideas about values education including the contrasting ideas of relativism and universal values, indoctrination, the relationship between values and sense of identity and the demands of pluralism.

Langford, P. (1995) *Approaches to the Development of Moral Reasoning*, Hove: Erlbaum.

A survey of approaches to the development of moral reasoning, including the contributions of Freud, Kohlberg and Piaget. The text provides a critical review of stage theory of moral development and alternative approaches are described.

Smith, R., and Standish, P. (eds) (1997) *Teaching Right and Wrong: Moral Education in the Balance*, Stoke on Trent: Trentham.

This discusses the work of the National Forum on Values Education and also argues for alternative approaches to values education.

RELEVANT WEBSITES

There are now many resources for developing a values education programme in schools, or for finding ideas and materials for work with pupils, where values education appears in a specific curriculum area, such as PSHE, RE or Citizenship. The following websites are helpful:

Belief, Culture and Learning Information Gateway: http://www.becal.net

An online resource database and very useful gateway to international societies and associations working in values education. It gives access to theory and research as well as resources.

Values Education Council of the United Kingdom (VECUK): http://www.ve1.valueseducation.org.uk

British Humanist Association: http://www.humanism.org.uk

Citizenship Foundation: http://www.citizenshipfoundation.org.uk

CitizED subject resource bank: http://www.citized.info

APPENDIX 4.5.1 HANDLING DISCUSSION WITH CLASSES

A common technique for raising moral and ethical issues is through discussion. Discussion activities have great potential as a vehicle for moral development through sharing ideas, developing awareness of the opinion of others, promoting appropriate social procedures and profitable debate. It can be a vehicle for learning. On the other hand, discussion can be the mere pooling of ignorance, the confirmation of prejudices and a stage for showing off.

Young people can be taught the protocols of discussion. It is necessary to help them realise that discussion may not end in a clear decision or agreement, but can lead to a deeper understanding of the issues and the position of others. For the teacher, discussion is one of the more difficult strategies to implement. Perhaps that is why it is not often used. See also Unit 3.1 on communicating with pupils. Some worked-through discussion of examples of controversial issues can be found in Levinson (2005b: 258–268): see Appendix 4.5.2.

The following notes may help you develop your strategies for conducting discussion. They adopt the neutral chairperson approach. (There are other ways of running discussions and for further help on running discussions and other ways of introducing moral issues see for example QCA materials relating to *Citizenship at Key Stage 3 (Year 7): What Are Ground Rules for Discussion?* (QCA, 2008h).

Four factors need to be considered. They are:

■ rules and procedures for discussion

■ provision of evidence – information

■ neutral chairperson

■ outcomes expected.

1 Rules and procedures

You need to consider:

■ choice of subject and length of discussion (young pupils without experience may not sustain lengthy discussion)

■ physical seating; room size; arrangement of furniture so that most pupils have eye contact

■ protocols for discourse; taking turns; length of contribution; abusive language

■ procedures for violation of protocols, e.g. racist or sexist behaviour

■ how to protect the sensitivity of individuals; pupils may reveal unexpected personal information in the course of a discussion

■ stance of the chairperson.

2 Provision of evidence

In order to stimulate discussion and provide a clear basis for argument, you need to:

■ know the age, ability and mix of abilities of the pupils

■ know what information is needed

■ know sources of information

■ decide at what point the information is introduced (before, during).

3 Neutral chairperson

A neutral stance may be essential because:

■ authority of the opinions of the chair should not influence the outcome

■ the opinions of pupils are to be exposed, not those of the teachers

■ the chairperson can be free to influence the quality of understanding, the rigour of debate and appropriate exploration of the issues

■ pupils will understand the teacher's stance if it is made clear at the start.

4 Possible outcomes

The strategy is discussion not instruction. Pupils should:

■ learn by sharing and understand the opinion of others

■ be exposed to the nature and role of evidence

■ realise that objective evidence is often an inadequate basis for decision making

■ come to know that decisions often rely on subjective value judgements

■ realise that many decisions are compromises.

Action

Try out these rules by setting up a discussion with other student teachers on the topic of: 'Equal opportunities for girls enable them to join the power structure rather than challenge it'.

APPENDIX 4.5.2 SIMULATIONS AND ROLE PLAY

General and specific subject advice may be found in some of our companion subject books in the series, e.g.:

Geography: Lambert. D. and Balderstone, D. (2000): 271-88.

History: Haydn, T., Arthur, J. and Hunt, M. (2001): 82 and 145–146.

Modern Foreign Languages: Pachler, N. and Field, K. (2002): 97, 117, 219

Science: Levinson, 2005: 258–268.

A fuller description of simulations is given in:

Turner, S. (1995) 'Simulations' in J. Frost (1995) *Teaching Science*, London: Woburn Press.

> **Additional resources for this unit are available on the companion website:**
> www.routledge.com/textbooks/9780415478724

AN INTRODUCTION TO INCLUSION, SPECIAL EDUCATIONAL NEEDS AND DISABILITY

Nick Peacey

INTRODUCTION

In England, the National Curriculum established a commitment to inclusion for pupils with special educational needs (SEN) and disabled pupils, within the duty to 'include' all those vulnerable to exclusion (Department for Education and Employment and Qualifications and Curriculum Authority (DfEE/QCA), 1999a). The recent revision of the secondary curriculum in England (QCA, 2007a) has given additional possibilities to teachers as they seek to meet that commitment.

The term 'inclusion' is used to mean every pupil's entitlement to feel they belong and can achieve in school; it underpins the expectation that the mainstream school is the place in which increasingly all pupils will be educated. Understanding approaches to the diverse needs of all their pupils is essential to all teachers' professional development (see also Unit 4.4).

OBJECTIVES

At the end of this unit you should be able to:

- explain how the terms inclusion, SEN and disability are used
- understand recent legislation and regulation in this area, in particular teachers' responsibilities at what are known as School Action and School Action Plus of the revised *Special Educational Needs Code of Practice for the Identification and Assessment of SEN* (DfEE, 2001a) and the Disability Discrimination Act (DfEE, 1995, revised 2005)
- recognise that teacher attitudes may limit pupils' achievement of their potential
- start developing your knowledge of teaching strategies which may be used with pupils with different SEN within the whole-class inclusion approach.

Check the requirements for your course to see which relate to this unit.

DEFINITIONS

The publication *Index for Inclusion,* which was sent with government support to every school in England, describes inclusion in these terms:

> Inclusion in education involves the processes of increasing the participation of students in, and reducing their exclusion from, the cultures, curricula and communities of local schools. Inclusion is concerned with the learning participation of all students vulnerable to exclusionary pressures, not only those with impairments or categorised as having special educational needs. Inclusion is concerned with improving schools for staff as well as for students.
>
> (Centre for Studies on Inclusion in Education (CSIE), 2000)

The authors of *Index for Inclusion* note that schools cannot remove all barriers to inclusion. There are limits, for example, to how much schools can do about the barriers created by poverty. (They are, of course, under a duty, as are all other services for children and young people, to support pupils' achievement of economic well-being, as set out in *Every Child Matters* outcomes (Department for Education and Skills (DFES), 2004f).

DEFINING SEN AND DISABILITY

SEN is essentially a *relative* term: pupils with SEN are said to require something 'additional to' or 'different from' that offered to other pupils. The definition of disability is rather different: the Disability Discrimination Act (DDA) (DfEE, 1995) describes a disabled person as having 'a physical or mental disability which has an effect on their ability to carry out normal day-to-day activities'. That effect must be substantial (i.e. more than minor or trivial), adverse and long term (has lasted or is likely to last at least a year or for the rest of the life of the person affected). These criteria do not apply to all conditions: the DDA 2005 expanded the definition to include 'automatically' people with cancer (whether or not it is in remission), severe disfigurement, HIV and multiple sclerosis (DfES, 2005).

The DDA applies to a wider group of pupils than those defined as having SEN. For example, it covers those with medical or physical impairments for whom there are no educational barriers to learning. Some authors feel we are outgrowing the notion of SEN. They feel it places too much emphasis on locating difficulty within the individual, the 'medical model'. This model prevents us seeing diversity as a resource rather than a problem and discourages us from giving appropriate consideration to the possibilities of removing barriers to the learning of minorities by changing the culture, policies and practice of schools.

But much government statute and regulation (including that related to funding) is still set in an SEN model, so the term will be with us for some years. In this unit, we use the terminology adopted in the revised SEN Code of Practice (DfEE, 2001a) which, while not claiming that there are hard and fast categories of special educational need, groups pupils' special educational needs under four main categories:

■ communication and interaction
■ cognition and learning
■ behaviour, emotional and social development
■ sensory and/or physical.

The Code suggests that:

> children will have needs and requirements which may fall into at least one of four areas, although many children will have inter-related needs which encompass more than one of these areas. The impact of these combinations on the child's ability to function, learn and succeed should be taken into account (in planning learning).
>
> (DfEE, 2001a: 7:62)

We return to these categories later in the unit and consider some of the classroom approaches involved. By the end of your course you are expected meet certain standards with respect to your understanding of SEN issues. The example that follows is related to the requirements in England.

REPORTS AND LEGISLATION

The National Curriculum (NC) for England 1999 and the revised secondary curriculum 2007

The National Curriculum for England is built on aims, values and principles which support the celebration of diversity (the recognition of everything that diversity of learners brings to a school community) and the rights of all pupils to participate and be partners in their own learning (see Unit 5.3).

The rights of all pupils within the National Curriculum were set out in a set of principles that have become known as the 'general inclusion statement'. This statutory statement has two strands:

1   a duty on teachers not to ignore the three principles of inclusion below in their planning and teaching
2   substantial flexibility to allow teaching to be matched to the needs of all pupils. For example, when necessary, secondary teachers should draw on material from Key Stage 2 programmes of study. The recent revision of the secondary national curriculum has given teachers even greater flexibility.

The inclusion statement is based on three principles.

1   The need to set suitable learning challenges. This principle encourages high expectations for all within a teaching and learning model which matches the approach used to pupils' abilities.
2   Responding to pupils' diverse learning needs. This principle emphasises the need for thought about learning environments. This includes:

   ■   the way adults structure communication with pupils
   ■   the layout of the classroom space
   ■   making sure that all pupils can hear.

   It stresses the need for all pupils to feel safe and secure and that views, particularly stereotypical views, about disability and impairment must always be challenged.
3   The overcoming of potential barriers to learning and assessment. This principle includes the statement 'Teachers must address the needs of the minority of pupils who have particular learning and assessment requirements which, if not addressed, could create barriers to learning.'

Some background: the developing legislative framework

1971 Legislation in England and Wales brings those considered 'ineducable' into education.

1981 The 1981 Education Act included specific duties on local education authorities (LEAs) and school governors to make provision for SEN, defining responsibilities and procedures for SEN and the establishment of parents' participation in special educational assessments, along with a right of appeal.

1988 The 1988 Education Act introduced the national curriculum and reinforced the duty to consider special educational needs, though many felt that proper attention was not given to the area.

1994 The *Code of Practice on the Identification and Assessment of SEN* (Department for Education (DfE), 1994b) came into effect on 1 September. This statutory code of practice (approved by both Houses of Parliament) cannot be ignored by anyone to whom it applies.

2000 The revised national curriculum for England (containing the 'inclusion statement') became law.

2001 A revised Office for Standards in Education (Ofsted) framework for inspection included specific provisions for the inspection of inclusion in schools. The current framework can be consulted on the Ofsted website (http://www.ofsted.gov.uk).

2001 The Disability Discrimination Act 1995 was revised by the addition of a Part 4. This, for the first time, places duties on schools and other educational institutions not to treat disabled pupils less favourably and to make 'reasonable adjustments' to ensure that disabled pupils are not put at a substantial disadvantage. Disabled school staff have rights under Part 2 of the Act, 2001. The revised SEN Code of Practice re-emphasises the importance of whole-school approaches. 'The effective school will identify common strategies and responses across the secondary curriculum for all pupils designed to raise pupils' learning outcomes, expectations and experiences.' The curriculum available for all pupils will directly affect the need to intervene at an individual level (DfEE, 2001a).

2004 *Removing Barriers to Achievement: The Government's SEN Strategy (DfES 2004g)*. This document gives a detailed outline of the government's planning for SEN over the next ten years and provides a useful summary of many areas of development.

2005 The Disability Discrimination Act 2005 puts a duty on public bodies to 'promote disability equality'. This brought disability equality the same legislative status as gender equality and race equality. All schools should have a Disability Equality Scheme.

The SEN Code of Practice: a little more detail

The SEN Code identifies some detailed approaches to inclusion (DfEE, 2001a). These include:

▪ All teaching and non-teaching staff should be involved in the development of the schools' SEN policy and be fully aware of the school's procedures for identifying, assessing and making provision for pupils with SEN.

▪ Practice encourages high expectations by emphasising that:

1 All pupils should know what is expected of them.
2 Secondary schools' general marking policies should be consistent in all subjects.
3 Schools should be similarly consistent in other areas, making clear, for example, how they expect all pupils to behave and to present their work.
4 Emphasis on literacy across the curriculum will help to achieve consistency in handwriting, spelling, punctuation and presentation.

5   Thus for all subject areas and for all pupils including those with SEN, there will be a common set of expectations across the school which are known to everyone, and a further commitment to support those pupils who have difficulty meeting those expectations.

■   Schools should not assume that pupils' learning difficulties always result solely, or even mainly, from problems within the young person. Pupils' rates of progress can sometimes depend on what or how they are taught. A school's own practices make a difference, for good or ill.

■   Planning for SEN should consider the views and wishes of pupils 'from the earliest possible age' as well as those of parents and carers; and that 'Partnership between schools, pupils, parents, LAs, health services, social services, voluntary organisations and other agencies is essential in order to meet effectively the needs of pupils with SEN.'

Task 4.6.1 asks you to investigate the provision for SEN in your school experience school.

Task 4.6.1 **THE CODE OF PRACTICE ON THE IDENTIFICATION AND ASSESSMENT OF SPECIAL EDUCATIONAL NEEDS AND THE SCHOOL'S POLICY**

Read the school policy on SEN provision in your school experience school and the extracts from the revised SEN code of practice above. Refer also to the revised code of practice online (DfES, 2001a). Discuss the implementation of the code of practice and school policy with your tutor and the SEN co-ordinator in your school experience school. Useful questions include: how does your school ensure that the requirements of the SEN code of practice are met? How are all the teachers in school involved in implementing the code of practice? Discuss your findings with those of another student teacher who has undertaken the same exercise in another school.

Make a summary of the similarities that arise and try to account for the differences.

THE THREE-STAGE APPROACH TO MEETING NEEDS

At the heart of this approach is 'a model of action and intervention designed to help pupils towards independent learning'. The Code outlines a three-stage approach to match the needs of pupils with the special education provision available to them.

School Action

The first stage, known as School Action, comes into play when it is clear that 'the pupil requires help over and above that which is normally available within the particular class or subject'. According to the Code, the school's SEN co-ordinator should facilitate assessment of the pupil's particular strengths and weaknesses, and plan for support and monitoring of progress.

There is an expectation that parents, carers and pupils will take a part in this process: their advice and ideas are invaluable in making sure any planning is effective.

The pupil's subject teachers remain responsible for working with the pupil on a daily basis and for planning and delivering a programme which takes individual needs into account. Suggestions on how to do this are often written in an individual education plan.

INDIVIDUAL EDUCATION PLAN (IEP)

An IEP typically records that which is *additional to* or *different from* curriculum provision for all pupils. Most IEPs include information about:

■ the short-term targets set for the pupil (these are typically reviewed two or three times a year)
■ the teaching strategies to be used
■ the provision to be put in place (see next section)
■ when the plan is to be reviewed
■ the outcome of the action taken.

Good IEPs are also likely to include some details of the pupil's strengths and interests.

SUPPORT AND PROVISION ON AN INDIVIDUAL PLAN

This requirement may include (particularly where concerns go beyond School Action) the deployment of extra staff to enable extra help and sometimes one-to-one tuition for the pupil. The plan may also prescribe:

■ different learning materials or special equipment
■ group support
■ cover for teacher and specialist time for planning support
■ staff development and training on effective strategies
■ support with classroom review and monitoring.

Policy on IEPs has changed in recent years. Following the introduction of personalised learning (see Units 4.1 and 5.1) and so individual planning for all pupils becomes more commonplace, there is less need for plans for pupils with SEN to be special. So you may find that you work in a school where there are few IEPs or where the reviews of IEPs are incorporated into the ordinary review systems for all pupils.

ADDITIONAL STAFF IN THE CLASSROOM

Where other adults are supporting pupils in your classroom, you need to liaise closely with them to ensure the pupil gains maximum benefit from this support. The support staff should have the lesson plans well in advance. You can check with them that the materials you are providing are appropriate for a particular pupil. See Task 4.6.2.

Task 4.6.2 WORKING WITH SUPPORT TEACHERS IN THE CLASSROOM

For a pupil with SEN to be fully supported in the classroom, the classroom teacher and the support teacher must develop an effective working relationship. Ask some support teachers for advice about how you can best work together. Observe support teachers working in classrooms and consider what has to be done to ensure the pupils make maximum progress.

Identify the implications for lesson planning for your classes; file your notes in your professional development portfolio (PDP).

Teaching staff should be clear that the learning of all pupils in lessons is their responsibility. There are concerns (see e.g. Ofsted, 2006b) that in some schools pupils with SEN and/or disabilities are substantially taught by teaching assistants and that this could damage their attainment.

School Action Plus

If School Action is not helping a pupil, schools may explore the next stage, School Action Plus. Schools should consult specialists when planning for a pupil at School Action Plus. The Code states:

> At *School Action Plus* external support services, both those provided by the LA and by outside agencies, should advise subject and pastoral staff on new IEPs and targets, provide more specialist assessments, give advice on the use of new or specialist strategies or materials, and in some cases provide support for particular activities. The kinds of advice and support available to schools will vary according to local policies.
>
> (DfEE, 2001a: 5:55)

There is of course no reason why specialist advice cannot be taken at any other point.

This IEP sets out revised strategies to promote the pupil's success in the school setting. The Code insists that the delivery of the plan is largely the responsibility of subject teachers, supported by their departments or faculties.

RECORDS FROM TRANSFER BETWEEN SCHOOLS

When working with pupils at School Action or School Action Plus you can consult the school records created at transfer from primary to secondary school. These records can include:

■ detailed background information collated by the primary school SEN co-ordinator (SENCO)
■ copies of IEPs prepared in support of intervention through primary School Action or School Action Plus.

You can contact the SENCO or the pupil's form tutor if you feel such information would help your planning. Task 4.6.3 allows you to find out more about the assessment of pupils with SEN.

Task 4.6.3 **PUPIL ASSESSMENT ON ENTRY TO SCHOOL**

Ask what, if any, assessments are carried out on pupils on entry to the school and if you can see examples of them. You should also ask what use is made of the information collected by these assessments. Select the assessments made on one, or two, pupils with SEN and describe how the information is used to plan their learning. File the information in your PDP.

The statement of SEN: the third stage of the Code of Practice for SEN

You may hear colleagues describe a pupil as 'statemented' or 'having a statement'. If School Action Plus is not succeeding, schools and parents or carers consider a multi-disciplinary assessment to examine the possibility of a pupil's having a statement of SEN. Because the process is expensive – assessment alone can cost several thousand pounds – local authorities (LAs) have the duty to consider carefully whether to embark on it. Many LAs, supported by the government, have reduced the number of statements they issue and give the funds released directly to schools for SEN provision. (The government's SEN strategy in England has encouraged delegation of funding to schools, rather than distribution through the 'statementing' process (DfES, 2004g). This process may support more speedy and effective use of the funds. If you are interested in the arguments, see Pinney, 2004.)

But the statement is popular with parents and schools because it guarantees the resources written into its clauses. Failure to provide can be challenged in the tribunal for SEN and disabilities (SENDIST) and, if necessary, in the courts. So the voluntary bodies that support parents and carers of pupils with SEN are unlikely to allow the present system to be abolished at any time in the near future: they do not yet feel that the current checks and balances guarantee that schools will put the appropriate provision in place.

Assessing for a statement and (if it results) a statement's implementation and review are 'high stakes' operations. So, unlike School Action and School Action Plus, where schools have very substantial freedom in how they implement the Code, everything to do with the statement is embedded in a network of statutory rules: the earlier stages. Teachers need to know about pupils with statements of SEN in their class. You should discuss with your head of department or faculty how you can plan for pupils in your classes who have statements of SEN. Find out more about statemented pupils in your school experience school through Task 4.6.4.

Task 4.6.4 **HOW DOES A STATEMENT COME ABOUT IN YOUR SCHOOL?**

Arrange a convenient time with the SEN co-ordinator to discuss the procedure by which a statement of special educational need is drawn up for a pupil and how it is reviewed. Has the LA/schools practice been changed by new funding approaches? Summarise your findings and arrange to have it checked by your tutor or the SENCO.

THE RESPONSE OF SCHOOLS

English as an additional language and SEN

The SEN Code of Practice notes 'The identification and assessment of the special educational needs of young people whose first language is not English requires particular care' and 'Lack of competence in English must not be equated with learning difficulties as understood in this Code' (DfEE, 2001a: 5.15 and 5.16).

The main points to be aware of are:

■ pupils learning English go through well-researched stages: they may for instance say little or nothing for some time, but are learning nonetheless
■ pupils learning English benefit from high-quality learning environments: they do not as a rule need individual programmes; see for example Unit 3.2 which discusses motivation
■ if the English learning stages are not proceeding as they should, the possibility of a learning difficulty may be considered. This is the time for teachers to consult specialist help, such as the school SEN co-ordinator.

Those wishing to take this area further should explore the ideas discussed in Hall (1996).

Special schools, units and resource bases

The proportion of pupils with statements educated outside the mainstream has remained almost the same between 1999 and 2007, though the number of local authority special schools has gone down. Mainstream schools often have special units or resource bases attached to them.

A resource base provides support within a school to pupils who are normally registered with 'ordinary' classes. A special unit is one step more separate: its pupils will be on the roll of the unit and normally based there. Inspection Reports identify increased interest in resourced mainstream provision by rating it highly compared to that in other primary, secondary and special schools. Overall, pupils were as likely to make good progress with their academic, personal and social development in primary, secondary or special schools. But pupils had the best chance of making good progress in all three areas in resourced mainstream schools. A greater proportion of this provision was outstanding and it was seldom inadequate. High quality specialist teachers and a commitment by leaders to create opportunities to include all pupils were the keys to success.

(Ofsted, 2006b, para. 280–283)

Access arrangements for examinations and assessments

All awarding bodies for public examinations, such as General Certificate of Secondary Education (GCSE) and General Certificate of Education (GCE) Advanced (A) level make access arrangements for pupils with EAL or learning difficulties. Access arrangements are also available for other pupils with special education needs and/or disabilities. If you feel that a pupil you are teaching needs such arrangements for an examination or national test (National Curriculum Test CNCT) and such arrangements are not in place, you should contact the SENCO or the schools examination co-ordinator to find out what can be done. It is important that the arrangements are in place well before the assessment: their use benefits from practice, like so many other things.

Concerns about a pupil not recognised by the school as having SEN, i.e. not at one of the stages of the SEN Code of Practice

You may feel a concern about a pupil's progress and wonder if they have unrecognised SEN. It is always worth comparing notes with another teacher to see how the pupil responds in their class. While sensory impairments of hearing or vision are sometimes not picked up until secondary age, the most frequently unrecognised SEN/disabilities are delays or impairments in speech, language or communication (see Communication and Interaction below).

The SEN Code of Practice notes that: 'Some young people may also raise concerns about their own progress and their views should be treated seriously' (DfEE, 2001a, 6:13). Looking back to your time in secondary school you may reflect that this will frequently require some courage from the pupil. The issues brought to you can be the tip of the iceberg. Once again, discussion with a colleague can help to get the perspective clear.

You should be alert to signs of eating disorders, particularly anorexia, whether they emerge through observation, written work or conversation; such instances should always be reported for urgent consideration.

HELPING PUPILS WITH SEN TO LEARN

This section provides some guidance on working with pupils with SEN in mainstream classrooms. Overall, you need to remember that the evidence from research supporting the effectiveness of differentiated pedagogies (teaching strategies; see Unit 5.3) for different groups of SEN is slender. Such evidence as there is suggests that 'good normal pedagogy' is the key to success (National Association for Special Educational Needs (NASEN), 2000).

The principles of the inclusion statement of the National Curriculum for England (QCA, 2007c) can guide planning for success for all pupils. Part of this planning should involve (at a tactful moment) asking pupils with SEN how they like to be taught. A good example of how to do this is provided by the British Stammering Association which has produced an excellent video on consulting youngsters worried about a speaking up in class. This resource can guide teachers into making successful approaches to pupils (British Stammering Association, n.d.).

We suggest that you develop your teaching approaches, i.e. concentrate on good normal pedagogy. Seek advice from your tutor. You should:

■ concentrate on developing your overall teaching style and the best possible learning environments in classrooms you use

■ use the education plan, the knowledge of others in the school and books and websites (e.g. Training and Development Agency for Schools (TDA), 2008b) to build your awareness of the issues in learning for a pupil; do not assume that you need to teach that pupil differently or separately

■ remember that motivation is a key part of learning, whatever the SEN: special treatment can 'turn off ' adolescents (see Unit 3.2)

■ remember that the timing and intensity of interventions often makes the difference rather than the pedagogy

■ have high expectations of homework and classwork. Clarity in setting homework and care in checking it has been recorded properly helps those likely to find the work difficult.

The SENCO forum on the Virtual Teacher Centre website provides a space for teachers to discuss general SEN issues; see website addresses at the end of this unit. Using information and communications technology (ICT) is an important way of helping some pupils with SEN: see Unit 5.5. and Leask and Pachler (2005).

We turn now to discuss briefly the particular of groups of pupils with specific educational needs. The headings are derived from the revised Code of Practice (DfEE, 2001a: 7:52 *et seq.*).

Communication and interaction

This section of the unit covers pupils with speech and language delay, impairments or disorders, specific learning difficulties, such as dyslexia and dyspraxia, hearing impairment and those who demonstrate features within the autistic spectrum. These difficulties may apply also to some pupils with moderate, severe or profound learning difficulties.

PUPILS WITH LANGUAGE IMPAIRMENT

Pupils in this group may have receptive (i.e. limitations in comprehending what is said to them) or expressive (i.e. they find it hard to put their thoughts into words) language impairments. Obviously, the first impairment is the harder to identify. You need to be aware that:

- emotional and relationship difficulties often go with language and less often with hearing impairments
- language impairments need specialist attention, normally from a speech and language therapist.

When teaching these pupils you should be sure:

- to check understanding
- to use visual aids and cues to the topics being discussed
- that the pupil is appropriately placed to hear and see
- that you explain something several different ways if you have not been understood the first time
- like a good chairperson, you repeat what pupils say in discussion or question and answer sessions (in any case, others in the class may not have heard)
- that you allow time for pupils to respond in question and answer sessions and, if necessary, ensure they are pre-prepared for responding, perhaps by a teaching assistant. See Unit 6.1, the section on questioning.

For further information see the website of the charity 'I Can' (ICAN) and the materials available within the Inclusion Development Programme (Department for Children, Schools and Families (DCSF), 2008i).

PUPILS WITH AUTISM SPECTRUM CONDITIONS (SOMETIMES KNOWN AS AUTISM SPECTRUM DISORDERS (ASD))

Those with autism spectrum conditions typically lack 'mentalisation', i.e. the ability to picture what another person is thinking. Any social context can create barriers for them. You may also hear of Asperger's Syndrome. Autistic pupils are typically described as having little or no language as well as social impairments and pupils with Asperger's as having excellent language but lacking social awareness.

Autistic pupils' absorbing interests (such as train timetables) and lack of social focus mean that co-ordinated planning is essential. You need to be aware that:

- many researchers feel autistic pupils' behaviour is a form of stress management for their hyper-sensitivity to many different stimuli, including touch and sounds
- pupils with conditions on the autistic spectrum can learn 'intellectually' how to act socially, e.g. in the matter of eye contact
- suggestions on approaches to autism vary widely; you need to be absolutely clear what agreed strategies are being used at school and home and work within them.

Further help may be obtained from the National Autistic Society (see useful websites listed at the end of this unit; see also DCSF, 2008i).

Cognition and learning

Pupils who demonstrate features of moderate, severe or profound learning difficulties or specific learning difficulties, such as dyslexia (reading difficulty) or dyspraxia (developmental co-ordination disorder), require specific programmes to aid progress in cognition and learning. Such requirements may also apply to some extent to pupils with physical and sensory impairments and those on the autistic spectrum. Some of these pupils may have associated sensory, physical and behavioural difficulties that compound their needs.

LEARNING DIFFICULTIES

Pupils identified in this group are likely to be attaining well below others of their age in a range of areas. Fletcher-Campbell argues that such pupils do not need different teaching approaches from those used with typically developing pupils, but benefit from planning that sets appropriate objectives at appropriate levels (Fletcher-Campbell, 2004). Pupils with severe learning difficulties are described as 'unlikely to reach level 2 of the National Curriculum by the age of 16' (DfEE/QCA, 2001). The term 'learning difficulties' thus covers a huge range of need so you need to check exactly what the pupil's IEP advises and seek specialist advice.

DYSLEXIA

You should be aware that:

■ the term dyslexia covers a wide range of needs
■ the current emphasis is now on examining the individual's skills, such as phonological awareness, and working on them as a way forward.

Phonological awareness is the conscious sensitivity to the sound structure of language. It includes the ability to auditorily distinguish units of speech, such as a word's syllables and a syllable's individual phonemes (the smallest functional unit of sound such as 'c' in *cat* or 'th' in *that*).

Most schools and all local authorities have staff with specialist knowledge in this area. Help can also be sought from the major voluntary organisations working in the field. These include the British Dyslexia Association; Dyslexia Action; the Dyslexia Forum and the Professional Association of Teachers of Students with Specific learning difficulties (PATOSS); see website addresses at the end of this unit. The Inclusion Development Programme has materials to support you in this area (DCSF, 2008i).

DYSPRAXIA

Dyspraxia often shows as clumsiness and or a lack of co-ordination. It may be defined as difficulty in planning and carrying out skilled, non-habitual motor acts in the correct sequence. Pupils with dyspraxia need the support of whole-school systems, particularly in terms of ensuring that they are taking enough exercise. If you are concerned about a pupil with dyspraxia, it may well be worthwhile talking to the physical education staff as well as to the SENCO.

Difficulties with handwriting are a specific co-ordination issue. The publications and conferences of the National Handwriting Association and Dyspraxia Foundation are helpful; see useful websites at the end of this unit.

Behaviour, emotional and social development

The revised Code of Practice indicates that there are:

Children and young people who demonstrate features of behavioural and emotional difficulties, who are withdrawn or isolated, disruptive and disturbing, hyperactive and lack concentration;

those with immature social skills; and those presenting challenging behaviours arising from other complex special needs.

(DfES, 2001a: 7:60)

BEHAVIOURAL, EMOTIONAL AND SOCIAL DIFFICULTIES (BESD)

Behavioural problems are discussed in Units 3.1, 3.2, 3.3 and 4.1 and tasks are set there to develop your skills in this area. The differentiation case studies in Unit 4.1 are relevant here.

The SEN Code of Practice states: 'Where a pupil with identified SEN is at serious risk of disaffection or exclusion the IEP should reflect appropriate strategies to meet their need.' This brief sentence only scratches at the surface of a critical issue in secondary school organisation. A pupil presenting consistent problems of behaviour may, for instance:

■ have special needs: a language impairment or an autistic spectrum disorder
■ have mental health needs: may be depressed, for instance
■ be responding to an unsatisfactory school environment: bullying in the playground, or an irrelevant curriculum.

The definition 'emotional behavioural difficulties' is often applied to pupils whose behaviour is consistently poor and not obviously related to the circumstances and environment in which pupils find themselves. Pupils who are withdrawn also fit into this category. Traditionally, secondary schools have three management approaches to concerns about pupil behaviour. These approaches are: (1) the pastoral teams (form tutor, year/house head, etc.); (2) SEN teams; (3) subject department or faculty.

The inter-relationship of these management teams is critical to a school staff's success with emotional behavioural difficulties. Important points to bear in mind in your response to pupils with BESD include:

■ understanding that learning, not counselling, is the teacher's contribution to resolving BESD: many heads of BESD schools talk of the 'therapy of achievement'
■ that many pupils identified as having BESD have speech, language and communication needs and benefit from teaching that recognises those needs
■ being aware that 'scaffolding' success for such pupils can be demanding and you should expect support with your planning (and lessons learnt) from experienced colleagues. Scaffolding is discussed in Unit 5.1.

Your lesson planning should address:

■ knowing the strengths and interests of any pupils with BESD
■ knowing the levels of language and literacy of pupils with BESD
■ considering alternatives within lesson plans if one learning project is not succeeding.

Among pupils identified as having BESD you may come across some pupils known as having:

■ attention deficit disorder (ADD)
■ attention deficit hyperactivity disorder (ADHD).

These terms are medical diagnoses. Medication by means of stimulants is a widely used treatment. The diagnoses in themselves tell you nothing about teaching such pupils. Further help and guidance may be found in Cooper and Ideus (1996) and the websites at the end of this unit.

Sensory and/or physical needs

The SEN Code of Practice advises that:

> There is a wide spectrum of sensory, multi-sensory and physical difficulties. The sensory range extends from profound and permanent deafness or visual impairment through to lesser levels of loss, which may only be temporary. Physical impairments may arise from physical, neurological or metabolic causes that require no more than appropriate access to educational facilities and equipment; other impairments produce more complex learning and social needs; a few children have multi-sensory difficulties. For some children the inability to take part fully in school life causes significant emotional stress or physical fatigue.
>
> (DfEE, 2001a: 7:62)

MEDICAL CONDITIONS

The SEN Code of Practice notes that:

> a medical diagnosis does not necessarily imply special educational needs. It may not be necessary for a child or young person with any particular diagnosis or medical condition to have a statement, or to need any form of additional provision at any phase of education. It is the child's educational needs rather than a medical diagnosis that must be considered.
>
> (DfEE, 2001a: 7:64)

A medical condition can affect a child's learning and behaviour. The effect may also be indirect: time in education can be disrupted, there may be unwanted effects of treatment; and through the psychological effects which serious or chronic illness or disability can have on a child and their family. Schools and the pupil's carers and the medical services should collaborate so that pupils are not unnecessarily excluded from any part of the curriculum or school activity because of anxiety about their care and treatment.

DEAFNESS AND HEARING IMPAIRMENT

If you have a deaf pupil in your class, you should be aware that:

■ specialist advice is usually not difficult to come by, you should ask the SENCO for direction in the first instance
■ many of the teaching checkpoints set out under language impairment apply
■ pupils have individual communication needs: you need to know what they are.

Further advice can be obtained from the Royal National Institute for the Deaf (RNID) and the National Deaf Children's' Society; see list of useful websites at the end of this unit. The RNID has produced a valuable series of booklets on teaching deaf and hearing-impaired pupils, including material on subject teaching.

VISUAL IMPAIRMENT

If you have a pupil with visual impairment in your class, specialist advice is again usually not difficult to come by. The SENCO in your school and the individual education plan for the pupil concerned can help in the first instance. Less obvious visual problems may affect pupils' reading. Specialist optometrists sometimes prescribe overlays or tinted spectacles to help with this problem.

There is substantial online support to teachers from the Royal National Institute for the Blind; see list of useful websites at the end of this unit. Further information may be found in Miller (1996). We suggest you investigate in depth some pupils with special educational needs in your classes; see Task 4.6.5.

Task 4.6.5 ANALYSING THE NEEDS OF INDIVIDUAL PUPILS

Identify, with the help of your tutor, pupils with whom you have contact who have SEN of different types. Draw up a table similar to the one here in your diary and complete it, drawing on the expertise of different staff as necessary. Use fictional names to preserve confidentiality and compare practice in your school with student teachers from different schools.

Description of pupil's needs	Paul has severe hearing loss. He can lip read reasonably well and his speech is fairly clear.
Pupil's strengths and interests	
Adaptation to classroom work in my subject	
General adaptation to classroom work	
The role of the school/SEN co-ordinator	
The role of outside agencies (local authority services, educational psychologists, educational welfare officers)	

For one or more of your pupils identify:

■ the implications of your findings for lesson planning
■ how you can work with the support teacher.

Share your ideas with your tutor and/or the SENCO.

SUMMARY AND KEY POINTS

This unit has introduced you to a range of emotional, behavioural and learning difficulties of pupils with special needs and set those needs in the context of the *Code of Practice* for SEN in England (DfEE, 2001a).

Understanding the requirements of that Code is important for your practice. We have not addressed the special needs of the gifted pupil (see Unit 4.1).

Every child is special. Every child has individual educational needs. A major problem experienced by pupils with SEN is the attitudes of others to them. For example, pupils who have obvious physical disabilities such as cerebral palsy often find they are treated as though their mental abilities match their physical abilities when this is not the case. How will pupils with SEN find you as their teacher?

Teachers need to ensure that all their pupils learn to the best of their abilities and that pupils with SEN are not further disabled by the lack of appropriate resources to support their learning, including software (Leask and Pachler, 2005).

In your work as a student teacher, we expect you to develop your understanding of the teacher's responsibilities for pupils' SEN so that, when you are in your first post, you are sufficiently aware of your responsibilities and that you ensure that your pupils' SEN are met. The basic rule for you to remember is that you cannot expect to solve all pupils' learning problems on your own. You must seek advice from experienced staff.

Check which requirements for your course have been addressed through this unit.

FURTHER READING

Cheminais, R. (2000) *Special Educational Needs for Newly Qualified and Student Teachers: A Practical Guide*, Abingdon: Routledge.

DfES (Department for Education and Skills) (2005a) *Implementing the Disability Discrimination Act in Schools and Early Years Settings.* Available online at: http://www.teachernet.gov.uk/wholeschool/sen/disabilityandthedda.

Farrell, M. (2009) *The Special Education Handbook: An A-Z Guide for Students and Professionals*, 4th edition, Abingdon: Routledge.

Garner, P. (2009) *Special Educational Needs: The Key Concepts*, Abingdon: Routledge.

Lewis, A. and Norwich, B. (2004) *Special Teaching for Special Children? Pedagogies for Inclusion*, Maidenhead: Open University Press.

Ofsted (Office for Standards in Education) (2006b) *Inclusion: Does It Matter Where Pupils Are Taught? An Ofsted report on the provision and outcomes in different settings for pupils with learning difficulties and disabilities.* Available online at: http://sen.ttrb.ac.uk/viewarticle2.aspx?contentId=12629.

QCA (Qualifications and Curriculum Authority) (2001a) *Planning, Teaching and Assessing Pupils with Learning Difficulties*, Sudbury: QCA Publications. Available online at: http://www.qca.org.uk/qca_11583.aspx.

RELEVANT WEBSITES

Attention Deficit Disorder (ADD) Information Service: http://www.addiss.co.uk

British Dyslexia Association: http://www.bdadyslexia.org.uk/

Dyslexia e-mail forum, run by Becta: http://lists.becta.org.uk/mailman/listinfo
This and many other forums on other areas are online at this web address.

Dyspraxia Foundation: http://www.dyspraxiafoundation.org.uk/

I-Can: http://www.ican.org.uk
This is a charity for people with speech and learning difficulties.

NASEN (National Association for Special Educational Needs): http://www.nasen.org.uk/

National Autistic Society: http://www.nas.org.uk

National Deaf Children's Society: http://www.ndcs.org.uk

National Handwriting Association: www.nha-handwriting.org.uk

The Professional Association of Teachers of Students with Specific Learning Difficulties (PATOSS): http://www.patoss-dyslexia.org

Removing Barriers to Achievement: the Government's SEN Strategy: http://www.teachernet.gov.uk/wholeschool/sen/senstrategy/

Royal National Institute of the Blind: http://www.rnib.org.uk

Royal National Institute for the Deaf (RNID): http://www.rnid.org.uk/

See also Unit 1.4 Relevant Websites for more web addresses.

Additional resources for this unit are available on the companion website:
www.routledge.com/textbooks/9780415478724

5 HELPING PUPILS LEARN

This chapter is about teaching and learning. As you work through these units we hope that your knowledge about teaching and learning increases and that you feel confident to try out and evaluate different approaches.

Unit 5.1 introduces you to a number of theories of learning. Theories about teaching and learning provide frameworks for the analysis of learning situations and a language to describe the learning taking place. As you become more experienced, you develop your own theories of how the pupils you teach learn and you can place theories in the wider context.

In Unit 5.2 teaching methods which promote learning are examined. How you use these methods reveals something of your personal theory of how pupils learn. At this point in your development we suggest that you gain experience with a range of teaching methods so that you are easily able to select the method most appropriate to the material being taught.

Unit 5.3 provides you with details about teaching styles. Again we suggest that as you gain confidence with basic classroom management skills you try out different styles so that you develop a repertoire of teaching styles from which you can select as appropriate.

We have talked at various points in this book about the characteristics of effective teaching. Unit 5.4 is designed to provide you with information about methods for finding out about the quality of your own teaching and that of others – through the use of reflection using action research techniques and drawing on the evidence base underpinning educational practice. During your initial teacher education course you are using action research skills in a simple way when you observe classes. In this unit we explain the key aspects of action research and reflective practice.

Units 5.5 and 5.6 are new in this edition of the book and they reflect significant changes in expectations of teachers. Unit 5.5 focuses on the concept of personalising learning. Information and communications technologies (ICTs) have made it possible to monitor pupils' actual attainment in tests against their expected grades across all subjects and this means that teachers can be sent alerts automatically where pupils are not achieving what is expected. This allows for teachers to intervene more quickly to help pupils falling behind and provide a more 'personalised' learning experience. The concept of 'personalised learning' has many other features and these are discussed in this unit.

Unit 5.6 introduces you to exciting new research in neuroscience which is yielding new knowledge about brain function and how teachers can maximise pupil learning. This new area of work is possible because activity of the brain can be monitored more easily than before using the latest MRI scanners and other techniques.

Readings for Learning to Teach in the Secondary School: A Companion to M Level Study
brings together essential readings to support you in your critical engagement with key issues raised
in this textbook.

WAYS PUPILS LEARN

Diana Burton

INTRODUCTION

The primary aim of a teacher's work is, fairly obviously, that pupils should learn *how, what, whether* and so on in relation to the subject of study; teaching is therefore a means to an end, not an end in itself. However, the interaction between the activities of teaching and the outcomes of learning is critical. In order to develop presentation and communication techniques that facilitate effective learning a teacher must have some understanding of how pupils learn. Course lectures and school experience add to and refine the ideas you already have about learning and reveal differences in how individuals learn. Psychological research is concerned with this individuality of *cognition* – knowing, understanding, remembering and problem solving. Research reveals information about human behaviour, motivation, achievement, personality and self-esteem, all of which impact on the activity of learning.

Several theoretical perspectives contribute to our understanding of how learning happens, e.g.:

- behaviourism, which emphasises external stimuli for learning
- gestalt theory, which expounds principles of perception predicated on the brain's search for 'wholeness'
- personality theories which are located in psycho-analytic, psycho-metric and humanist research traditions.

Discussion of such work can be accessed quite easily in libraries and journals (see Child, 2007, for a very good overview) so this unit is organised around theories that have been influential amongst educators in terms of pedagogic strategy. These theories include:

- Piaget's cognitive-developmental theory of maturation
- metacognition, the way in which learners understand and control their learning strategies
- social constructivist theories which emphasise social interaction and scaffolded support for learning
- constructivist theories about the strength of pupils' existing conceptions
- information processing approach to concept development and retrieval
- learning style theories, which question qualitative distinctions between ways of learning, stressing instead the matching of learning tasks to a preferred processing style.

In Unit 4.3 you met multiple intelligence theory which suggests a multi-dimensional rather than a singular intelligence (Gardner, 1983, 1993a, 1993b). Many educators consider this to have application when considering how individuals learn. Similarly, emotional intelligence theory, addressed in Unit 4.3 and emphasising the potency of the learner's emotional state, is also of interest to teachers (Goleman, 1995; Mayer and Salovey, 1997; Salovey and Mayer, 1990).

OBJECTIVES

At the end of this unit you should be able to:

■ appreciate the interaction between ideas about learning and pedagogic strategies
■ begin to explain and differentiate between some psychological perspectives on learning
■ appreciate more about your own approach to learning.

Check the requirements for your course to see which relate to this unit.

HOW IDEAS ABOUT TEACHING AND LEARNING INTERACT

Decisions you make about how to approach a particular lesson with a particular pupil or group of pupils depend on the interplay between your subject knowledge and pedagogic knowledge, your knowledge of the pupils and your understanding about how learning happens. Let us imagine that you have spent 10 minutes carefully introducing a new concept to your class but a few pupils unexpectedly fail to grasp it.

You test the reasons for their lack of understanding against what you know about the pupils' and their:

■ prior knowledge of the topic
■ levels of attention
■ interest and motivation
■ physical and emotional state of readiness to learn, etc.

You consider also factors relating to the topic and the way you explained it, for example:

■ the relevance of the new material to the pupils
■ how well the new concept fits into the structure of the topic
■ the level of difficulty of the concept
■ your clarity of speech and explanation
■ the accessibility of any new terminology
■ the questioning and summaries you gave at intervals during your explanation.

Finally you draw on explanations from educational psychology which have informed your understanding of how pupils learn and provided you with questions such as:

■ does the mode of presentation suit these pupils' learning styles?
■ has sufficient time been allowed for pupils to process the new information?

■   does the structure of the explanation reflect the inherent conceptual structure of the topic?

■   do these pupils need to talk to each other to help them understand the new concept?

A teacher's ideas about learning usually derive from a number of theories rather than from one specific theory. This is fine because it allows for a continual revision of your ideas as you gain more experience. This process is known as 'reflecting on theory in practice'. Some of the theories from which your ideas may be drawn are reviewed in this unit. In order to contextualise that review it may be helpful to think about the types of learning activities you engage your pupils in.

Gagné (1977) identified five factors which contribute to effective learning (see Task 5.1.1). Each of these factors is used in teaching all subjects in the curriculum, interacting in complex ways. It is helpful to consider them separately so that a particular pupil's learning progress and needs can be monitored and so that you can plan lessons that foster all types of learning. See also Unit 2.2. Stones (1992) discussed the need to employ different pedagogical strategies depending on the type of learning planned (see also Unit 5.). Task 5.1.1 invites you to analyse the demands on pupils in a lesson using the factors identified by Gagné.

Task 5.1.1 ANALYSING LEARNING ACTIVITIES FOR DEMANDS ON PUPILS

Select a learning activity from a lesson you have taught or observed. Describe the types of learning involved using the table below for each of the five factors. An example from science has been provided to guide you.

Activity	Intellectual skills	Verbal skills	Cognitive strategy	Attitudes	Motor skills
Science: a group activity using particle theory of matter	Discussing how to set up the activity to test an hypothesis	Defining solids, liquids and gases	Recalling previous knowledge about particles	Listening to and sharing ideas	Manipulating equipment

Share your findings with your tutor or class teacher. Identify their implications for planning future lessons.

The theory of multiple intelligences provides a different, and possibly more attractive, framework than Gagné's typology for categorising the different ways pupils learn (Gardner, 1983). Gardner proposes that there are a number of relatively autonomous intelligences, such as 'linguistic intelligence' and 'musical intelligence' and includes nine or more different 'intelligences'. These intelligences may act alone or be combined in different ways in different contexts. Gardner was not concerned about the use of the word 'intelligence' as such, so that the words 'gift' or 'skill' would be equally appropriate This theory is discussed more fully in Unit 4.3. Noble argues that it helps teachers make sense of their observations that pupils have different strengths and learn in different ways (Noble, 2004: 194). Since each intelligence has a different developmental trajectory, pupils may engage in higher order thinking in areas of intellectual strength and engage in lower order, less abstract thinking in areas of weaker 'intelligence'. Noble relates this to Bloom's revised taxonomy of educational objectives which orders cognitive processes from simple remembering to higher order creative and critical thinking (Anderson and Krathwohl, 2001). From this viewpoint, Noble developed a matrix with Multiple Intelligence (MI) dimensions on the horizontal axis and the Revised Bloom Taxonomy of educational objectives (RBT) on the vertical axis. Teachers used this matrix to design learning outcomes and activities so that pupils could demonstrate what they had learned at different levels of complexity across intellectual domains.

This approach has clear application in terms of differentiation in the classroom (see Unit 4.1 for more about differentiation). For instance, in a science project about the processes involved in the formation of volcanoes, pupils were given the opportunity to explain the process in a number of ways, such as drawing a flow chart (spatial intelligence) or setting the words to the tune of a well-known song (musical intelligence). Thus a pupil understands better how volcanoes are formed by using, for example a flow chart rather than by writing a paragraph. In this way the types of learning pupils engaged in were broadened so that each pupil had greater access to the curriculum and could tackle a task at the relevant level of complexity depending on the strength of a particular intelligence.

PSYCHOLOGICAL PERSPECTIVES ON LEARNING

A determining factor in lesson preparation is the knowledge that learners already possess. Unfortunately, identifying this knowledge is not as simple as recalling what you taught the pupils last time for knowledge is, by definition, individualised. Each pupil's experiences of, attitudes towards, and methods of processing, prior knowledge are distinct. Psychologists are interested in how learners actively construct this individualised knowledge or 'meaning' and different theories offer different notions about what constitutes knowledge. *Cognitive developmental theory* depicts knowledge as being generated through the learners' active exploration of their world (Piaget, 1932, 1954). More complex ways of thinking about things are developed as the individual matures. *Social constructivism* explains knowledge acquisition by suggesting that learners actively construct their individual meanings (or knowledge) as their experiences and interactions with others help develop the theories they hold (Brown, 1994; Rogoff, 1990). The constructivist approach draws attention to pupils' prior knowledge which interacts with new knowledge offered through the curriculum. The interaction may support, or conflict with, the teacher's objectives. *Information processing theories* view knowledge as pieces of information which the learner's brain processes systematically and which are stored as abstractions of experiences (Anderson *et al.*, 1996). The relatively new theory of *situated cognition* does not recognise knowledge as existing outside of situations, but rather as 'collective knowledge', i.e. the shared, ongoing, evolving interaction between people (Lave and Wenger, 1991; Davis and Sumara, 1997).

The first three of these theories have been influential in shaping pedagogy. They are outlined briefly below.

Cognitive developmental theory

PIAGET

Jean Piaget was a Swiss psychologist who applied Gessell's (1925) concept of maturation (genetically programmed sequential pattern of changing physical characteristics) to cognitive growth. He saw intellectual and moral development as sequential with the child moving through stages of thinking driven by an internal need to understand the world. His theory implied an investigative, experiential approach to learning.

According to Piaget the stages through which the child's thinking moves and develops is linked to age. Piaget thus provides a criterion-referenced rather than norm-referenced explanation of cognitive development (Shayer, 2008). From birth to about 2 years children understand the world through feeling, seeing, tasting and so on. Piaget called this the *sensory-motor* stage. As children grow older and mature, from about 2 to 6 years, they begin to understand that others have a viewpoint. They are increasingly able to classify objects into groups and to use symbols. Piaget termed this the *preoperational* stage.

The third stage, from 6 to about 12 years, identified by Piaget is the *concrete operational* stage. Children are still tied to specific experience but can do mental manipulations as well as physical ones. Powerful new internal mental operations become available, such as addition, subtraction and class inclusion. Logical thinking develops.

The final stage of development sees children manipulating ideas in their heads and able to consider events that haven't yet happened or think about things never seen. Children can now organise ideas and events systematically and can examine all possibilities. Deductive thinking becomes possible. Piaget suggested that this final stage, called the *formal operational* stage, begins at about 12 years of age.

The learner's stage of thinking interacts with experience of the world in a process called *adaptation*. The term *operations* is used to describe the strategies, skills and mental activities which children use in interacting with new experience. Thus adding 2 and 2 together, whether mentally or on paper, is an operation. It is thought that discoveries are made sequentially so, for example, adding and subtracting cannot be learned until objects are seen to be constant. Progress through the sequence of discoveries occurs slowly and at any one age children have a particular general view of the world, a logic or structure that dominates the way they explore the world. The logic changes as events are encountered which do not fit with their *schemata* (sets of ideas about objects or events). When major shifts in the structure of children's thinking occur, a new stage is reached. Central to Piaget's theory are the concepts of *assimilation*, taking in and adapting experience or objects to existing strategies or concepts, and *accommodation*, modifying and adjusting strategies or concepts as a result of new experiences or information (see Bee and Boyd, 2006, for a full description of Piaget's theory).

THE INFLUENCE OF STAGE THEORY

Piaget's work influenced the way in which some other psychologists developed their views. Kohlberg (1976) saw links between children's cognitive development and their moral reasoning, proposing a stage model of moral development (see also Unit 4.5). Selman (1980) was interested in the way children make relationships, describing a set of stages or levels they go through in forming friendships. Stage models of development were also posited in relation to personality growth, quite independently of Piaget's work. Influenced by Freud's (1901) psychoanalytic approach, Erikson's (1980) stages of psychosocial development explain the way in which an individual's self-concept develops, providing important insights into adolescent identity issues, such as role confusion. Harter's research (1985) indicated the significance of teacher–pupil relationships for pupils' feelings of self-worth in relation to learning competence. You may be aware of the fragility of an adolescent's self-concept, of the fundamental but often volatile nature of adolescent friendships and of the increased interest adolescents show in ethical issues. It is essential to consider these factors when planning for learning since they impact so heavily on pupils' motivation for, and capacity to engage with, lesson content.

Bruner also developed a stage model of the way people think about the world. He described three stages in learning, the enactive, iconic and symbolic. Unlike Piaget's stages, learners do not pass through and beyond Bruner's different stages of thinking. Instead, the stage or type of representation used depends on the type of thinking required of the situation. It is expected, however, that as pupils grow up they make progressively greater use of symbolic representation (Bruner, 1966).

The stress on the idea of 'stages' in Piaget's theory was far-reaching, especially the implication that pupils need to be at a particular developmental level in order to cope with certain learning tasks. However, research studies have substantially refuted such limitations on a pupil's thinking (Donaldson, 1978). Thus the construct of a staged model of cognitive maturation probably has less currency than the features of development described within the various stages. Flavell (1982), a former student of Piaget, has argued that, while stage notions of development are unhelpful, Piaget's ideas about the sequences learners go through are still valid. They can help teachers to examine the level of difficulty of topics and curriculum material as a way of deciding how appropriate they are for particular age groups and ability levels (see e.g. Unit 4.3 which discusses with examples aspects of cognitive development). The key task for teachers is to examine the progress of individuals in order to determine when to increase the intellectual demand on them. Bruner (1966) argued that difficult ideas should be seen as a challenge that, if properly presented, can be learned by most pupils.

More recent work by Adey and colleagues has revealed that learning potential is increased if pupils are metacognitively aware, i.e. if they understand and can control their own learning strategies. These strategies include techniques for remembering, ways of presenting information when thinking, approaches to problems and so on. Shayer and Adey developed a system of cognitive acceleration in science education (CASE; see Unit 4.3), which challenges pupils to examine the processes they use to solve problems (Adey, 1992; Shayer and Adey, 2002). In doing so it is argued that pupils are enhancing their thinking processes. In Jordan, researchers have identified statistically significant differences in seventh grade pupils' achievement in science when using a metacognitive teaching strategy (Namrouti and Alshannag, 2004). Kramarski and Mevarech (1997) showed that metacognitive training helped 12- to 14-year-old pupils draw better graphs. Other researchers reported that transfer of learning between tasks is enhanced where the teacher cues learners into the specific skill being learned and encourages them to reflect on its potential for transfer (Anderson et al., 1996). It seems, then, that the higher-order cognitive skills of Piaget's stage of formal operations can be promoted and encouraged through a focus on metacognition. Learners need, above all, to learn to learn, and teachers' endeavours would be more profitable if they concentrated on teaching pupils how to learn (Claxton, 2002). Task 5.1.2 asks you to plan a lesson in which metacognitive strategies are introduced. The project 'Learning How to Learn', part of the 'Teaching and Learning Research Programme' (TLRP and ESRC, 2007), confirms that emphasis should be placed on practices that have potential to promote autonomy in learning (Black et al., 2006). The greater availability of information, particularly on the Internet, requires pupils to have the confidence, curiosity and skills to utilise it.

Task 5.1.2 ANALYSING LEARNING ACTIVITIES: LEARNING HOW TO LEARN

Consider how you can give pupils at least one opportunity during each lesson to think about the strategies they are using in their learning. Ask them, for instance, how they reached a particular conclusion, how they tackled the drawing of a 3D shape or how they would undertake a journey from point A to point B. Summarise your findings and file in your professional development portfolio (PDP).

Social constructivist theory

The ideas of Russian psychologist Lev Vygotsky and of Jerome Bruner in the USA have increasingly influenced educators in recent years. Vygotsky's work dates from between the 1920s and 1930s (he died in 1934) but was suppressed in Soviet Russia. His most influential work *Thought and Language* was not available in English until 1962 (Vygotsky, 1962).

VYGOTSKY

Juxtaposing Piaget's and Vygotsky's theories shows their 'common denominators as a child centred approach, an emphasis on action in the formation of thought, and a systematic understanding of psychological functioning'; and their biggest difference as their understanding of psychological activity (Kozulin, 1998: 34). For Vygotsky (1978), psychological activity has socio-cultural characteristics from the very beginning of development. Concepts can be generated from a range of different stimuli, implying a problem-solving approach for pupils and a facilitator role for the teacher. Whereas Piaget considered language a tool of thought in the child's developing mind, for Vygotsky language was generated from the need to communicate and was central to the development of thinking. He emphasised the functional value of egocentric speech to verbal reasoning and self-regulation, and the importance of socio-cultural factors in its development. In communicative talk, too, the development is not just in the language contrived to formulate the sentence but also through the process of combining the words to shape the sentence because this shapes the thought itself.

Thus Vygotsky's work highlights the importance of talk as a learning tool and reports in the UK have reinforced the status of talk in the classroom, highlighting its centrality to learning (Bullock Report, 1975; Norman, 1992). Vygotsky believed that such communicative instruction could reduce pupils' 'zone of proximal development', i.e. the gap between their current level of learning and the level they could be functioning at with adult or peer support: 'What a child can do today in cooperation, tomorrow he will be able to do on his own' (Vygotsky, 1962: 67). Myhill (2006) conducted a two-and-a-half-year study which investigated the nature and quality of classroom discourse, how teachers use questions, how they capitalise on pupils' prior knowledge and how they help pupils become independent learners. By analysing how teachers used talk to develop and build on pupils' learning within episodes of whole class teaching Myhill found that teacher discourse can variously support or impede pupil learning and that cognitive or conceptual connections in pupils' learning are often ignored. Task 5.1.3 invites you to investigate pupil talk in your lessons.

Task 5.1.3 **ANALYSING LEARNING ACTIVITIES: PUPIL TALK**

Consider how often you make opportunities for your pupils to talk in lessons, to each other, to you, to a digital or video recorder or via a CD-ROM. Listen to how the talk develops the thinking of each member of a small group of pupils. Notice how well-timed and focused intervention from you moves the thinking on. Write up examples of showing the advantages of promoting pupil talk with your pupils; file the report in your PDP.

Other research in this area has indicated that reciprocal peer and cross-age tutoring can also promote learning (Smith *et al.*, 2003; Telecsan *et al.*, 1999; Topping, 1992). See also Unit 5.3. Crucially however, pupils need to be given specific preparation and guidance by the teacher in order to work effectively. Homogeneous pupil grouping and pairing, such as setting provides, have advantages in promoting argument and sharing complex ideas (Rogoff, 1990). However, in pair work a slight difference in the intellectual functioning of partners was best because it promoted cognitive conflict (Doise, 1990). The idea of a learning community has been developed, where group work and seminars provide the main vehicle for learning (Brown, 1994). The concept of learning communities has become popular as a professional development tool for teachers as well as a means of promoting pupil learning. In both Australia and England networked learning communities have been set up in which teachers can work together to share and develop their practices (Burton and Bartlett, 2004; Clayton *et al.*, 2008; O'Brien *et al.*, 2006). Unit 5.2 provides discussion and examples of active learning strategies which often involve pupil–teacher and pupil–pupil dialogue, both examples of strategies illustrating a Vygotskian approach to learning.

BRUNER

Bruner also placed an emphasis on structured intervention within communicative learning models. He formulated a theory of instruction, central to which is the notion of systematic, structured pupil experience via a spiral curriculum where the learner returns to address increasingly complex components of a topic. Current and past knowledge is deployed as the pupil constructs new ideas or concepts. Thus the problem of fractions in Year 5 will be tackled via many more concrete examples than when it is returned to in Year 7. Learning involves the active restructuring of knowledge through experience; the pupil selects and transforms information, constructs hypotheses and makes decisions, relying on a developing cognitive structure to do so.

Teachers should try to encourage pupils to discover principles by themselves through active dialogue with the teacher. Thus, for Bruner, the teacher's job is to guide this discovery through structured support, e.g. by asking focused questions or providing appropriate materials; he called this process 'scaffolding' (Bruner, 1983). Maybin *et al.* (1992) have used 'scaffolding' in relation to classroom talk. The ideas of pupils emerging through their talk are scaffolded or framed by the teacher putting in 'steps' or questions at appropriate junctures. For example, a group of pupils might be discussing how to solve the problem of building a paper bridge between two desks. When hearing an idea emerge the teacher can intervene by asking a question that requires the pupils to address that idea explicitly, which may help pupils find the solution. Bruner argued that the scaffolding provided by the teacher should decrease in direct correspondence to the progress of the learner. Wood (1988) has developed Bruner's ideas, describing five levels of support which become increasingly specific and supportive in relation to the help needed by the pupil:

- general verbal encouragement
- specific verbal instruction
- assistance with pupil's choice of material or strategies
- preparation of material for pupil assembly
- demonstration of task.

Having established the task the pupils are to complete, a teacher might give general verbal encouragement to the whole class, follow this up with specific verbal instruction to groups who need it, perhaps targeting individuals with guidance on strategies for approaching the task. Some pupils need physical help in performing the task and yet others need to be shown exactly what to do, probably in small stages. Thus the different ways in which teachers support pupils is a further example of differentiation (see Unit 4.1). Scaffolding is also discussed in Unit 3.1

The five levels of support above can help you think systematically about the nature of the support you should prepare for particular pupils, and to keep a check on whether the level is becoming more or less supportive. This has obvious relevance for differentiation in your classroom, workshop or gym. Task 5.1.4 addresses scaffolding and differentiation.

Task 5.1.4 **ANALYSING LEARNING ACTIVITIES: DIFFERENTIATION**

For a topic you are planning to teach, use the levels to prepare the type of support you think might be needed during the first lesson. Once you start teaching the topic choose two particular pupils who need different levels of support. Plan the support for each one, lesson by lesson, noting whether they are requiring less or more support as the topic progresses. What does this tell you about the way you are teaching and the way the pupils are learning?

Choose an example of scaffolding you have used and write a report to discuss with your class teacher or tutor. File the report and comments in your PDP.

Constructivism

Another theory of learning, called 'constructivism', is that described by Driver and Bell (1986). Whilst sharing a Brunerian and Vygotskyan emphasis on the social construction of meaning, constructivism places much more importance on learners' individual conceptions and gives them responsibility for directing their own learning experiences. Thus pupils construct meaning for themselves from their experiences and do not learn ready-made knowledge; they already have explanations for events and phenomena in the world from their own experience. These explanations may conflict with accepted views thus raising the importance of knowing about your pupil's prior knowledge. The pupil's prior understanding has to accommodate new experiences, such as that provided by the school curriculum. The pupil then has to accommodate the new experiences. Development of knowledge and understanding thus becomes a dynamic process constantly changing in response to new experiences. This process creates a demand on the pupil's willingness to accommodate new knowledge, thus motivation becomes a challenge. One useful teaching strategy is 'cognitive conflict', moving the pupil from a comfortable state of understanding to an uncertain state by providing experiences that challenge that current understanding. Constructivism implies that understanding is the goal, not just recall of knowledge which, unless placed in an explanatory framework, may soon be forgotten. It is this framework that the pupil has to construct and develop. It is in the science subject area that constructivist ideas have been most influential (Levinson, 2005a: 95–100). Naylor and Keogh (1999), for instance, have devised concept cartoons, which enable learners to examine the very core of their conceptions.

Information processing theory (IP theory)

This approach to learning originated within explanations of perception and memory processes and was influenced by the growth of computer technology. The basic idea is that the brain attends to sensory information as it is experienced, analyses it within the short-term memory (STM) and stores it with other related concepts in the long-term memory (LTM).

Psychologists saw the functioning of computers as replicating the behaviour of the brain in relation to the processing of information. Information is analysed in the STM and stored with existing related information in the LTM. This process is more efficient if material can be stored as abstractions of experience rather than as verbatim events. If I ask you to tell me the six times table, you recite 'one six is six, two sixes are twelve' and so on. You do not explain to me the mathematical principle of multiplication by a factor of six. On the other hand, if I ask you the meaning of the term 'economic enterprise', it is unlikely that you will recite a verbatim answer. Instead, your STM searches your LTM for your 'schema' or idea of enterprise. You then articulate your abstracted understanding of the term, which may well be different from that of the person next to you. In terms of intellectual challenge, articulating the second answer requires greater mental effort, although knowing one's multiplication tables is a very useful tool. Thus teachers need to be absolutely clear that there is a good reason for requiring pupils to learn something by rote because rote learning is not inherently meaningful so cannot be stored in LTM with other related information. Rather, it must be stored in its full form, taking up a lot of 'disk' space in the memory. Information stored in this way is analysed only superficially in STM, the pupil does not have to think hard to make connections with other pieces of information. For further information about information processing theory see Unit 5.3.

It can often be difficult in secondary schools for teachers to determine precisely the prior knowledge of their pupils. You observe that good teachers cue learners in to their prior knowledge by asking questions about what was learned the lesson before or giving a brief resume of the point reached in a topic. Such strategies are very important because, if previous learning has been effective, information is stored by pupils in their LTMs and needs to be retrieved. Psychologists suggest that learning is more effective, i.e. more likely to be understood and retained, if material is introduced to pupils according to the inherent conceptual structure of the topic (Ausubel, 1968; Gagné, 1977; Stones, 1992). By this is meant identifying the basic concepts that need to be taught first, then broadening the concepts to widen their application, thus introducing new concepts. To understand, for example, momentum, you need to understand mass and velocity (which in turn requires an understanding of speed and direction), and so on. Concept maps can reflect the conceptual structure of a topic. Information which is stored using a logical structure is easier to recall because the brain can process it more easily in the first instance, linking the new ideas to ones which already exist in the memory. As a teacher, this requires you to have thought through the structure in advance, and to know how the concepts fit together; hence the importance of spending time on schemes of work even where these are already produced for you. See Task 5.1.5 and concept maps in Unit 5.2.

This emphasis on structure and sequence can be found perhaps most readily in the teaching of modern foreign languages (MFL) and mathematics. Mitchell (1994) found that MFL teachers use a 'bottom–up' language learning theory, encouraging recognition and acquisition of vocabulary first, followed by the construction of spoken and written sentences. This approach might be described as moving from the general to the specific but in most subjects the process is the other way around. It might be argued that in mathematics we start with the general concepts of addition, subtraction and multiplication, moving to more specific computations like calculating area, solving equations or estimating probabilities.

Concept development

It is helpful to consider briefly how information processing relates to concept development. The material held in the LTM is stored as sets of ideas known as 'schemata' ('schema' is the singular). A schema is a mental structure abstracted from experience. It consists of a set of expectations with which to categorise and understand new stimuli. For example, our schema of 'school' consists of expectations of pupils, teachers, classrooms, etc. As teachers, our school schema has been refined and developed as more and more information has been added and categorised. Thus it includes expectations about hierarchies, pupil culture, staffroom behaviour and so on. It is probable that our school schema is different from, and more complex than, the school schema held by a parent, simply because of our involvement in schools.

When children are young their schemata do not allow them to differentiate between pieces of information in the way that those of older pupils do. A one-year-old's schema of dog might

Task 5.1.5 **STRUCTURING A TOPIC FOR EFFECTIVE LEARNING**

Choose a topic from your subject area, one covering some 4–6 lessons and one you are expecting to teach. For a couple of minutes think freely about some of the ideas contained within it, jotting them down haphazardly on paper. Now think about how those ideas fit together and whether, in teaching the topic, you would start with the general overarching ideas and then move to the specific ones or vice versa.

You can organise your topic by drawing up a conceptual hierarchy of it like the one started below for a Personal and Social Education (PSE) topic. This hierarchy moves from the general to the specific.

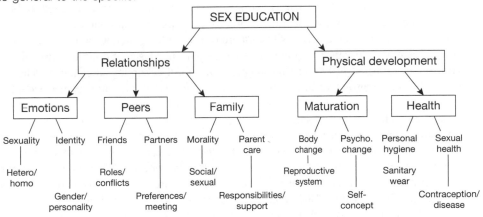

Give your completed structure to other student teachers in your subject for comment. Use their responses to finalise your hierarchy, adding notes for clarification as needed.

From the school's Scheme of Work (SoW) identify the topic and the lessons that are related to it. Sketch out the sequence of ideas in the SoW and identify how the teaching is organised, from general to specific or vice versa. Compare the sequencing of ideas in your framework and that in the SoW.

Try to account for the differences and discuss them with your subject tutor. How are the differences explained? Are they to do with an understanding of the topic, or related to the ways pupil respond to the topic or a practical issue to do with resources?

Write an account of the study, including implications for the way you approach teaching this topic. File in your PDP.

include expectations about cats too because they have insufficient experience of the two animals to know them apart. As they experience cats as furry, dogs as hairy, cats meowing, dogs barking, etc., greater differentiation is possible. Since the object of school learning is to promote pupils' concept differentiation in a range of different subjects, teachers should encourage comparison between objects or ideas and introduce new ideas by reference to concrete examples. Even as adults, whilst we can think abstractly, we find new ideas easier to grasp if we can be given concrete examples of them. Teaching in the context of IP theories stresses the application of knowledge and skills to new situations. The teacher's role is to help pupils find new ways of recalling previous knowledge, solving problems, formulating hypotheses and so on. Montgomery (1996) advocates the use of games and simulations because they facilitate critical thinking and encourage connections to be made between areas of subject knowledge or experience.

We have discussed cognitive development theories, social constructivist theories and information processing theories. In each of these theories about how learning occurs, there has been an emphasis on the individual and the differences between them. Looking at what is known about how learners' styles and strategies differ equips us further to understand individual differences.

LEARNING STYLES, STRATEGIES AND APPROACHES

You may encounter teachers who try to identify whether learners are predominantly visual, auditory or kinaesthetic learners (VAK) (Dryden and Vos, 2001). However, there are a number of such classifications based on different constructs and using a range of measurement tools. Some of these ideas, such as the VAK construct and preferred environmental conditions or stimuli for learning, have been incorporated, along with a range of other ideas, such as neuro-linguistic programming, into accelerated learning programmes such as that of Alastair Smith (Smith and Call, 2002); see Burton and Bartlett (2006) or Burton (2007) for a critical appraisal of these ideas.

It can be attractive to feel that we can categorise pupils into fixed learning approaches or styles and then teach to a formula that such a categorisation suggests. However we know that learning is complex and context-dependent with pupils employing different approaches in different settings. It is important that you take an eclectic approach to learning style classification and measurement lest you too readily pigeonhole pupils and fail to provide a range of approaches to learning, including choice of activities and resources to maximise access for all to the curriculum. In this section you are introduced to categorisations that have been researched over a considerable number of years but for a full overview of the range of research in this area consult Riding and Rayner (1998).

There is often confusion about what constitutes learning style as distinct from learning strategy. The *cognitive* or *learning style* may be a fairly fixed characteristic of an individual, is static and a relatively in-built feature of them (Riding and Cheema, 1991). In contrast *learning strategies* are the ways learners cope with situations and tasks. Strategies may vary from time to time and may be learned and developed.

Learning style

Understanding how in-built features of learners affect the way they process information is important for you as a student teacher. Learning styles may be grouped into two principal cognitive styles. The first, the *Wholist-Analytic Style,* identifes whether an individual tends to process information in wholes (wholist) or in parts (analytics). The second, known as *Verbal-Imagery Style,* describes whether an individual is inclined to represent information during thinking verbally (verbalist) or in mental pictures (imagers) (Riding and Rayner, 1998).

The two styles operate as dimensions so a person may be at either end of the dimension or somewhere along it. Think about what your own learning style might be. Do you:

- approach essay writing incrementally, step by step, piecing together the various parts or do you like to have a broad idea of the whole essay before you start writing?
- experience lots of imagery when you are thinking about something or do you find yourself thinking in words?

Discuss your style with other student teachers. In doing so you are developing your metacognitive knowledge, that is, about your own way of learning.

These styles are involuntary so it is important to be aware that your classes contain pupils whose habitual learning styles vary (Riding and Rayner, 1998). You need to ensure that you provide a variety of ways in which pupils can work and be assessed. It would not be sensible to present information only in written form; if illustrations are added, this allows both 'Verbalisers' and 'Imagers' easier access to it. Similarly, 'Wholist' pupils are assisted by having an overview of the topic before starting whilst 'Analytics' benefit from summaries after they have been working on information.

This is not to suggest that you must determine the style of each pupil, but that there must be opportunities for all pupils to work in the way that is most profitable for them. Unlike the way in which intelligence quotient (IQ) is used, the higher the IQ score the better the performance is expected, the determination of learning style does not imply that one way of processing is better than the other. The key to success is in allowing learners to use their natural processing style. It is important for you to be aware of your own style because teachers have been found to promote the use of approaches that fit most easily with their own styles.

Learning strategy

Learning strategy describes the ways in which learners cope with tasks or situations. These strategies develop and change as the pupil becomes more experienced. Kolb's work describing two dimensions of learning strategy is the most widely known in the area of strategy theory (Kolb, 1976, 1985). Kolb envisaged a cyclical sequence through four 'stages' of learning (concrete experience, reflective observation, abstract conceptualisation, active experimentation) arising from the interaction of two dimensions (see Figure 5.1.1). The east–west dimension represents 'processing' (how we approach a task) and the north–south dimension 'perceiving' (our emotional response, or how we think or feel about it). Kolb argued that these two dimensions interact and that, although learners use preferred strategies, they could be trained to develop aspects of other strategies through experiential learning.

Thus Kolb suggests that learners need, at the concrete experience stage, to immerse themselves in new experiences. Learners reflect on these experiences from as many perspectives as possible at what Kolb calls the reflective observation stage. This reflection enables the learner to create concepts which integrate their observations into logically sound theories at the abstract conceptualisation stage, which are then used to make decisions and solve problems at the active experimentation stage (Fielding, 1996). Learners have a predilection for one of the stages. It can

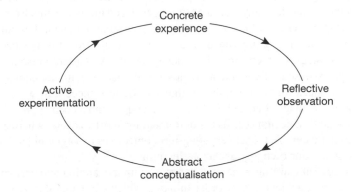

■ **Figure 5.1.1** Figure 5.1.1 Kolb's experiential learning cycle
Source: Adapted from Kolb (1985: 21)

be argued that learners should be provided with experiences that ensure that they use each of the stages in the cycle at some time, in addition to their preferred one (McCarthy, 1987). Indeed, without involving pupils in each of the stages, deep learning (higher order learning) may not emerge (c.f. Bloom's Taxonomy of Educational Objectives, in Anderson and Krathwohl, 2001).

We suggest now you try Task 5.1.6.

Task 5.1.6 WHAT IS YOUR PREFERRED STYLE OF LEARNING?

Try to work out which stage of Kolb's cycle describes the way you learn yourself most of the time. See Figure 5.1.1.

Think about how you process information, e.g.

■ are you more comfortable reflecting on ideas where you have had lots of concrete experience?

■ do you prefer to draw from abstract theory and experiment with it to solve problems?

Do you think Kolb's ideas could help you to process information differently? Would it be helpful to 'practise' different ways? What are the implications for your work with pupils?

Learning approaches

Other researchers are interested in the motivations and attitudes pupils and students bring to their learning, described as 'approaches to learning'. Researchers have investigated learners' approaches to study and how learning approaches interact with learning strategies (Biggs, 1978, 1987, 1993; Entwistle, 1981, 1993).

Entwistle described different orientations to learning, such as being oriented towards discovering the meaning of a topic or being oriented simply to scratch the surface. Combinations of these orientations with extrinsic factors, such as the need to pass examinations or the love of a subject, were thought to lead to learning strategies which characterised certain approaches to study, from 'deep' to 'surface' levels of thinking.

A pupil's approach is a function of both motive and strategy and motives influence learning strategies (Biggs, 1993). Thus, a pupil with an instrumental (surface) motive is likely to adopt reproducing or rote-learning (surface) strategies. Deep motive results from an intrinsic desire to learn and can inspire the use of deep strategies, emphasising understanding and meaning. An achieving motive might be an egotistical need to pass examinations; from this perspective the learner can derive achieving strategies which stress time management, well-ordered resources and efficiency.

Pupils whose motives and strategies are compatible with the demands made by learning tasks are likely to perform well. Pupils are likely to be less successful where motives and strategy are incompatible with task demand. For example, a pupil with a deep approach to learning is constrained by superficial task design such as a requirement for short answers, whereas a pupil with an achieving motive may be deterred if set very long-term objectives or vague objectives. Task 5.1.7 invites you to investigate your own approaches to learning.

Successful learning, if defined in terms of understanding and permanence, is linked with deep approaches that achieve higher order thinking which can be taught. A number of researchers have stressed the impact of social and cultural environment on learning approaches; Phan and Deo (2007) further confirmed this finding in two large-scale studies in the South Pacific region. The achievement-

Task 5.1.7 **YOUR OWN APPROACHES TO LEARNING**

Refer to the paragraphs above on learning approaches. Do you recognise any of these learning approaches in relation to your own learning? Do your motives and associated learning strategies stay the same over time or do they depend on the task and the reason you are completing it? Prepare some notes on your own ways of learning for discussion with your tutor.

driven context within which secondary school pupils in England currently learn, for instance, could militate against the possibility of teaching deep approaches because of time constraints. An important development in this regard is the UK government's recent emphasis on personalised learning and the use of interactive technology to this end (Leadbeater, 2004). The use of individualised learning, facilitated by technology, may encourage the use of deep approaches to learning if pupils have the time and autonomy to pursue a topic in depth. Currently there are worries that the use of the World Wide Web develops a surface approach, with learners simply cutting and pasting from unregulated information sites.

Increasingly sophisticated online resources will be available to teachers; in addition you should endeavour to maintain variety in the learning experiences you design for pupils, in the ways you present information, in the resources pupils use and the tasks they undertake and in the ways you assess their progress.

SUMMARY AND KEY POINTS

The importance of developing your own models of learning and refining these as you gain more experience of pupils and learning contexts has been established. You continually draw on your knowledge of the pupils, of the subject and of how learning happens in the teaching process. The symbiosis between theoretical positions and pedagogic practices has been emphasised. For instance, concept development is enhanced where pupils are introduced to new ideas via concrete examples and retention is aided if topics are taught according to their inherent conceptual structure.

The key features of cognitive development, social constructivism and information processing theories, as they apply to pupil learning, have been outlined and your attention has been drawn to the implications of learning style and strategy for teaching techniques. The benefits of adopting a facilitative, interventionist approach and of aiming for a variety of approaches in presentation, resource, task and assessment can be extrapolated from all the theories that have been discussed. Learning is likely to be most effective where pupils are actively involved with the material through critical thinking, discussion with others and metacognitive awareness of their own learning strategies.

Check which requirements for your course have been addressed through this unit.

FURTHER READING

Bartlett, S. J. and Burton, D. M. (2007) *Introduction to Education Studies*, 2nd edn, London: Sage/Paul Chapman Publishing.
   Chapters 6 and 7 provide a full outline of major psychological theories and contemporary psychological research into how pupils learn.

Bee, H. and Boyd, D. (2006) *The Developing Child*, 11th edn, London: Allyn and Bacon; Child, D. (2007) *Psychology and the Teacher*, 8th edn, London: Continuum.

These two books are excellent for taking you further into theories of learning and child development. They are both very well researched and provide comprehensive coverage of the issues, delivered in a contemporary style with practical illustration to aid understanding.

Riding, R. (2002) *School Learning and Cognitive Styles*, London: David Fulton.

Riding incorporates recent psychological developments on individual learning differences with practical classroom applications. He presents new approaches in three key areas: processing capacity, cognitive style and understanding the structure of knowledge. These are central to the understanding of pupil differences and they affect our perception of how pupils can be helped to learn; why pupils find some aspects of their school work difficult; and why pupils behave as they do. See also Riding and Rayner (1998).

Shayer, M. and Adey, P. (eds) (2002) *Learning Intelligence: Cognitive Acceleration across the Curriculum from 5 to 15 Years*, Buckingham and Philadelphia: Open University Press.

This edited collection describes how children's general ability to process information – their 'intelligence' – can be improved by appropriate cognitive acceleration methods. Through examples of cognitive acceleration in a variety of contexts, from Year 1 to Year 9 and in science, mathematics and arts subjects, each chapter draws on research or development experience to describe effects of cognitive acceleration programmes. The book also discusses at the psychological theory that underlies cognitive acceleration.

Additional resources for this unit are available on the companion website:
www.routledge.com/textbooks/9780415478724

ACTIVE LEARNING

Françoise Allen and
Alexis Taylor

INTRODUCTION

A key element of the educational process is the development of concepts. The definition of 'concept' in this unit follows Vygotsky's use of the term to mean 'word meanings':

> a concept is more than the sum of certain associative bonds formed by memory, more than a mere mental habit; it is a complex and genuine act of thought that cannot be taught by drilling but can be accomplished only when the child's mental development itself has reached the requisite level.
>
> Practical experience also shows that direct teaching of concepts is impossible and fruitless. A teacher who tries to do this usually accomplishes nothing but empty verbalisation, a parrot like repetition of words by the child, simulating knowledge of the corresponding concepts but actually covering up a vacuum.
>
> (Vygotsky, 1986: 149–150)

The extract above suggests that you cannot do the learning for the pupil and that in order for understanding to occur the pupil has to be active in the learning process. Active learning is then meaningful learning, in which something of interest and value to the learner has been accomplished and understood. Some writers use the term 'deep learning' instead of meaningful learning, in contrast with 'shallow learning', or surface learning, which is learning with little understanding.

This unit addresses ways in which you can help pupils to engage in meaningful learning. Active learning supports meaningful learning because it enables the learners to develop not only their knowledge of the subject taught but also their skills for learning and their ability to reflect on the processes involved in that learning.

OBJECTIVES

At the end of this unit you should be able to:

■   explain the term 'active learning' and discuss the advantages of active learning to the teacher and the learner
■   be aware of ways of promoting active learning
■   consider the use of resources to support active learning in your lessons.

Check the requirements for your course to see which relate to this unit.

CASE STUDIES: TWO PUPILS LEARNING MATHEMATICS

Manjeet and Robert are two pupils in Year 8. They are taught the same syllabus at the same school and complete the same activities and assignments. They had similar scores on entry from primary school. Both Manjeet and Robert have worked hard. Here's what they have to say about one of the subjects they take.

Manjeet: I enjoy mathematics because I can think, which is not something I always do in other subjects. I try to see the mathematics that is being taught in other areas as well. For example, I use it at home to help my younger brother and also in the supermarket when we go shopping, and also when I go to town with my friends. I also watch programmes on television – you know, those consumer programmes – and link what we've been doing in class. I never used to like mathematics in my primary school, but I do now. I enjoy doing the homework, and ask the teacher for some additional work. I also try to make up sums and mathematics tasks of my own. This works because I passed the latest mathematics tests. The way I work is to try to understand the information first, then try the activity. If I don't understand I will ask the teacher, or a friend, or my dad – he's good with mathematics! I think that someone who is not so good at mathematics would struggle a bit. You know, to get interested in it. To find different ways to understand it. In our mathematics class we do a lot of group discussions and pair work and the teacher really makes us think by asking some very difficult questions. She also lets us work on our own by getting us to do investigations and surveys and looking on the Internet.

Robert: The way I do mathematics is to keep trying the same activity until I get it right . . . you know, trying different techniques to solve the same problem. I know I have to get the mathematics right because my parents say it will be useful for me in the future – to get a good job. I usually try to write down formulas and learn them by heart. I always do the classwork and homework the teacher has given me (I'd get a detention if I didn't!), and also so I can get a good enough mark . . . but I only spend a short time on it as I like to go out with my friend Sam and play football. The lessons are usually the same. The teacher explains something to us, then shows us some examples on the board. Then we usually have to do some exercises from a text book or worksheet. We usually work on our own. Sometimes we do past test papers as practices, but I didn't do well on the last one!

Task 5.2.1 helps you to reflect further on Manjeet's and Robert's different perceptions of their learning.

Task 5.2.1 **TWO PUPILS' EXPERIENCES OF MATHEMATICS**

Discuss with other student teachers the observations made by these pupils from the standpoint of (1) their response to the tasks set and possible reasons for these responses; (2) the way the tasks may have been set by the teacher.

Record your findings in your professional development portfolio (PDP).

The responses of Manjeet and Robert suggest two important things. The first of these is that they experience mathematics differently. Manjeet and Robert appear to approach their learning in two qualitatively different ways. Manjeet appears to take a deep approach to his work, while Robert appears to adopt a shallow (surface) approach. Nevertheless, Robert's willingness to try different ways of solving problems is a positive feature of his attitude to work. Those who take a deep approach tend to have an intrinsic interest in the topic and the tasks and aim to understand and seek meaning. They adopt strategies that help them to satisfy their curiosity and to look for patterns and connections in other areas. They think about the task. Those who adopt a surface approach see tasks as work given to them by others. They are pragmatically motivated and seek to meet the demands of the task with the minimum of effort (Entwistle, 1990; Prosser and Trigwell, 1999). Deep and surface learning are discussed further in Units 3.2 and 5.3.

We suggest that Manjeet is engaged mainly in active learning and Robert is engaged more in passive learning. This may mean that Manjeet is able to think abstractly and is actively involved in the process of learning. This may involve, for example, learning through doing, trying things out, getting it wrong and knowing why. It is not just the learner's attitude but the way the task is set. Active learning has to be encouraged. Perhaps the way Robert engages with the subject is as much to do with the way the lessons are taught and the learning structured as it is with Robert's own approach to learning. Finally it is important to remember that we are not discussing the ability of the pupils but their response to the subject and the way tasks are set.

WHAT IS ACTIVE LEARNING?

Active learning occurs when a pupil has some responsibility for the development of the activity. Supporters of this approach recognise that a sense of ownership and personal involvement is the key to successful learning. Unless the work that pupils do is seen to be important to them and to have purpose and unless their ideas, contributions and findings are valued, little of benefit is learned. Active learning can also be defined as purposeful interaction with ideas, concepts and phenomena and can involve reading, writing, listening, talking or working with tools, equipment and materials, such as paint, wood, chemicals, etc. In a simple sense it is learning by doing, by contrast with being told.

Active learning may be linked to experiential learning. Experiential learning is also learning by doing but with the additional feature of reflection upon both action and the results of action; only where pupils are 'engaged actively and purposively in their own learning is the term experiential appropriate' (Addison and Burgess, 2007: 35–36). Both active and experiential learning contribute to meaningful learning. See also the CASE project (Unit 4.3).

Active learning strategies benefit both teachers and pupils. As a teacher they enable you to spend more time with groups or individuals, which allows better-quality formative assessment and feedback to take place (both of these are key features of 'assessment for learning' – see Unit 6.1). Active learning can also enhance your support of pupils with special educational needs. Activity methods encourage autonomous learning and problem-solving skills, important to both academic and vocationally based work. There is, of course, an extra demand on you in the planning and preparation of lessons.

The advantages of active learning to pupils include greater personal satisfaction, more interaction with peers, promotion of shared activity and team work, greater opportunities to work with a range of pupils and opportunities for all members of the class to contribute and respond. It can encourage mutual respect and appreciation of the viewpoint of others.

It is important to realise that learning by doing, by itself, is not enough to ensure learning. The proverb 'I hear and I forget; I see and I remember; I do and I understand' was reformulated by a prominent educationalist as 'I do and I am even more confused' (Driver, 1983: 9). The essential step to learning and understanding is reflection through discussion with others, especially the teacher; such discussions involve 'thinking' as well as recalling, i.e. experiential learning.

ACTIVE LEARNING AND MOTIVATION

You support pupil learning by identifying clearly defined tasks which have purpose and relevance. Relevance may arise because of personal interest, i.e. it is intrinsic; or the motivation may be extrinsic, e.g. to please the teacher. Outside interests become increasingly important as the pupils get older. If the school task links with some future occupation, employment training or higher education, motivation is increased and engagement promoted. Motivation is considered in greater detail in Unit 3.2.

The development of thinking skills is an important part of the Secondary National Strategy (Department for Education and Skills (DfES), 2005d). This is reaffirmed in the Qualifications and Curriculum Authority's (QCA) Framework for Personal, Learning and Thinking Skills (PLTS) (QCA, 2008g), which plays a significant role in guiding schools to meet the aims of the new secondary curriculum. The six areas of the PLTS framework all involve aspects of active learning, for example: 'analyse and evaluate information, judging its relevance and value' (Area 1: independent enquirers); 'propose practical ways forward, breaking these down into manageable steps' (Area 2: effective participants); 'assess themselves and others, identifying opportunities and achievements' (Area 3: reflective learners); 'collaborate with others to work towards common goals' (Area 4: team workers); 'organise time and resources, prioritising actions' (Area 5: self managers); 'try out alternatives and new solutions and follow ideas through' (Area 6: creative thinkers).

We shall return to teaching to develop higher level intellectual skills later in the unit, but first we need to consider how tasks can engage pupils with their need to know. If the task fails to do this, then learning is on sufferance, leading to problems. Such problems may include poor recall of anything learned or rejection of learning tasks, which in turn may lead to behaviour problems.

LEARNING HABITS

Learning how to learn is a feature of active learning. By promoting activities which engage pupils and require them to participate in the task from the outset, you foster an approach to learning which is both skill based and attitude based. Active learning methods promote habits of learning which it is hoped are valuable in the workplace, the home and generally enhance pupils' capacity to cope with everyday life.

School can be a place where pupils learn to do things well and in certain ways. Some skills are developed which are used throughout life. For example, pupils learn to consult a dictionary or a thesaurus in book form or through a word-processing package, in order to find meanings or to counteract poor spelling. These skills become habits, capable of reinforcement and development. Reinforcement leads to improved performance. Many of our actions are of this sort, like dressing and eating. Important attitudes are developed, such as the confidence to question statements or to believe that problems can be solved and not to be put off by difficulties.

Many professional people depend, in part, on habits and routines for their livelihood; these include the concert pianist who may well practise time and time again a piece of music already well established in her repertoire. Practice commits the sequence of notes and pauses to memory, leaving time to concentrate on expression. Actors, too, depend on learned routines for skilful performance.

ACTIVE LEARNING, DISCOVERY LEARNING AND ROTE LEARNING

Active learning is sometimes associated with discovery learning, which attempts to motivate pupils by helping them to learn things for themselves. By contrast, rote learning suggests that rote methods do not require the learner to understand what is learned. Ways in which rote learning can be made into an active process are also discussed below.

Discovery learning

Discovery learning at its simplest occurs when pupils are left to discover things for themselves but it is difficult to imagine when such learning happens in the organised classrooms of secondary schools (see also Mosston and Ashworth, 2002, in Unit 5.3). Much more common is the use of a structured framework in which learning can occur, sometimes called 'guided discovery'. But is the intention of guided discovery that pupils come to some predetermined conclusion or is it that learning should take place but the outcomes vary from pupil to pupil? You need to be clear about your reason for adopting guided discovery methods. If the intention in discovery learning is to move pupils to a specific end point, then as discovery it could be challenged. This approach might preclude, for example, considering other knowledge that surfaced in the enquiry. If discovery learning focuses on how the knowledge was gained, then the activity is concerned with processes, i.e. how to discover and how to learn. The question is one of means and ends. Are discovery methods then concerned with:

■ discovery as 'process' – learning how to learn? or
■ discovery as 'motivation' – a better way to learn predetermined knowledge and skills and to support personalised learning with pupil and teacher working together to set the direction of open-ended work? or
■ discovery as 'change in self' – learning something about yourself as a teacher and using this knowledge to deal with pupils' learning differently?

Rote learning

It is a fallacy, we suggest, to assume that pupils can learn everything for themselves by discovery methods. Teachers are specialists in their fields of study; they usually know more than pupils and one, but not all, of their functions is to tell pupils things they otherwise might not know but need to know. You need to consider what you want to achieve and match the method with the purpose.

Pupils need to be told when they are right; their work needs supporting. On occasions you need to tell pupils when they are wrong and how to correct their error. How this is done is important but teachers should not shirk telling pupils when they underperform or make mistakes. See Unit 3.2 on motivating pupils, which includes a section on giving feedback.

Rote learning may occur when pupils are required to listen to the teacher. There are occasions when you need to talk directly to pupils, e.g. to give facts about language, of spelling or grammar; about formulae in science; or health matters such as facts about drug abuse or safe practice in the gymnasium. Other facts necessary for successful learning in school include recalling multiplication tables; or remembering vocabulary; or learning the reactivity series of metals; or recalling a piece of prose or poetry. Some facts need to be learned by heart, by rote methods. There is nothing wrong with you requiring pupils to do this from time to time, provided that all their learning is not like that. Such facts are necessary for advanced work; they contribute to the sub-routines which allow us to function at a higher level. Habits of spelling, of adding up, of recalling the alphabet are vital to our ability to function in all areas of the curriculum and in daily life.

Sometimes pupils need to use a routine as part of a more important task but which they may not fully understand. You may decide that the end justifies the means and that through the experience of using the routine in different contexts understanding of that routine develops. Many of us learn that way.

Learning facts by heart usually involves a coding process. For example, recalling telephone numbers is easier if it is broken into blocks such as 0271 612 6780 and not as 02716126780. Another strategy is the use of a mnemonic to aid recall, such as recalling the musical scale E G B D F by the phrase 'Every Good Boy Deserves Fun'. Do you know any other mnemonics? What other ways are there of helping pupils to learn by rote?

Use Task 5.2.2 as a starting point to evaluate the different types of learning your teaching provides.

Task 5.2.2 **IDENTIFYING TYPES OF LEARNING**

Review a selection of lessons that you have taught and identify where you have used active learning, guided discovery and rote learning. Critically reflect on the reasons for your choices and whether these were appropriate. If you have not given the pupils opportunities to learn in these ways try to work out some lesson plans where you build them in.

Record your reflections in your PDP.

AIDS TO RECALL AND UNDERSTANDING

Sometimes information cannot easily be committed to memory unless a structure is developed around it to help recall. That structure may involve other information which allows you to build a picture. In other words, recall is constructed. Structures may include other words, but often tables, diagrams, flow charts or other visual models are used. Other ways of helping learners to remember ideas or facts are to construct summaries in various forms. Both the act of compiling the summary and the product contribute to remembering and learning.

Figure 5.2.1 shows a model of learning based on the idea that personal development proceeds best by reflection on your own actions. Reflection in the model is incorporated in the terms 'review' and 'learn'. This model is presented as a cyclic flow diagram which enables you to keep in mind the essential steps:

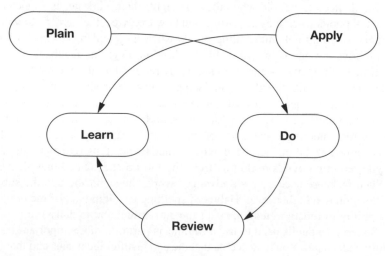

■ **Figure 5.2.1** An active learning model: plan, do, review, learn, apply
Source: Watkins *et al.* (2007: 77)

This model can be applied to pupils' learning as well as your own. To develop your teaching, the key ideas in the model could be interpreted as follows:

■ Do: plan and teach a lesson to defined objectives.
■ Review: identify where learning took place and what contributed to successful learning. Which ideas or processes gave difficulties? Try to identify factors contributing to success and difficulties.
■ Learn: clarify what has been learned about your teaching and what questions remain. Seek help about difficulties, from mentors, tutors, books and colleagues. Identify problems you can solve and those that are more deep rooted. Look for evidence about what works.
■ Apply: how can I use the new knowledge to improve the next lesson? Planning and further action.

See also Figure 6.1.4 'The planning/teaching/assessment/planning loop' (Unit 6.1), where the model is applied to lesson planning.

Spider diagrams and concept maps

It is often helpful to pupils and teachers to 'brainstorm' as a way of exploring their understanding of an idea. One way to record that event is by a spider diagram (or burr diagram) or mind map in which the 'legs' identify the ideas related to the topic. Figure 5.2.2 shows a spider diagram constructed by a pupil of some ideas associated with 'fruit'.

Concept maps are developments from spider diagrams and are used to display important ideas or concepts which are involved in a topic or unit of work and, by annotation, show the links between them. An example is shown in Figure 5.2.3. Concept maps can be made by pupils as a way of summarising their knowledge of a unit of work. The individual map reveals some of the pupil's understanding and misunderstandings of the topic. An example of a Year 10 pupil's concept map can be found in Lambert and Balderstone (2000: 203). Making a concept map at the start of a unit

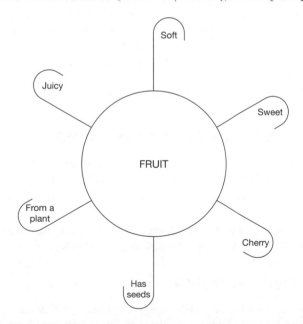

■ **Figure 5.2.2** A pupil's meaning of fruit – or a pupil's understanding of the concept of fruit

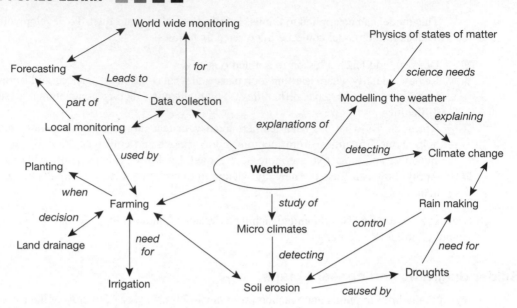

■ Figure 5.2.3 Concept map: weather

helps to probe pupils' prior knowledge of the subject. In either case, you may need to provide a list of ideas with which pupils can work and to which they can add their own ideas, as illustrated in Figures 5.2.2 and 5.2.3.

Concept mapping is useful as part of your lesson preparation, particularly when beginning a new unit of work. See also Task 5.1.5 in Unit 5.1. Concept mapping enables you to gain an overview of the unit, to consolidate links between several ideas and may reveal weaknesses or gaps in your own understanding. Concept mapping (sometimes called 'mind maps') is a useful way of linking topics in the curriculum so as to promote continuity and breadth of understanding in your teaching (see, for instance, concept maps in science in Frost and Turner, 2004, chapter 5.2). Concept maps are difficult to construct but the process of drafting one is a valuable exercise. For further discussion see the Teacher Training Resource Bank (TTRB) (http//:www.ttrb.ac.uk) on writing mind maps; and White and Gunstone (1992).

Task 5.2.3 asks you to prepare a concept map.

Task 5.2.3 **USING CONCEPT MAPPING**

Prepare a concept map for a unit of work that you are going to teach. Discuss this map with your tutor and reflect on its value to enhance active learning in the lesson.

We turn now to a particular group of activities designed to aid recall and learning, familiar to teachers as Directed Activities Related to Texts (DART).

DART AND ACTIVE LEARNING

DART are ways of engaging pupils in active reading, writing and listening in order to foster their understanding of a text and their ability to reflect. They have been in existence for decades (see, for instance, Davies and Greene, 1984; Gilham, 1986: 164) and continue to be widely used (e.g. Davison and Dowson, 2003: 69; 79; 254). DART can involve the use of textbooks, but take in ways of using a variety of written and other visual materials including resources downloaded from the web. DART are devised to ensure that pupils interact with a text. Interaction includes, for example, underlining certain types of word; listing important words; drawing diagrams or reformulating a labelled diagram into continuous prose. The level of demand, i.e. differentiation, is adjusted by you to meet the needs of the pupils. A listening activity may be designed to help pupils understand instructions given by the teacher. Some examples of DART are discussed below.

Giving instructions

This includes activities as diverse as making bread, carrying out a traffic survey or gathering information on the effects of the Black Death. A common complaint by teachers is that pupils do not read instructions or, if they do, are unable to comprehend them. Sometimes the language level is too high; or pupils may understand each step but not the whole or just lack confidence to act. Sometimes it is because pupils do not have any investment in the project; it is not theirs. Ways of alleviating such problems depend on the ability and attentiveness of your pupils but can include:

■ Making a list of the instructions on a sheet and given to each pupil, then reading out loud to check understanding. As the instructional steps are completed pupils could be asked to tick off that step.
■ Writing instructions on numbered cards. A set of cards can be given to a group who are instructed to put the cards into a working order. Discrepancies are discussed and the order checked against the purpose in order to agree the acceptable sequence, which can then be written or pasted in pupils' books.
■ Matching instructions to sketches of events: ask pupils to read instructions and select the matching sketch and so build a sequence.
■ Discussing the task first and then asking pupils to draft their own set of instructions. After checking by the teacher the pupils can begin.

The same approach can be applied to how to do something or explain a process, for example, helping pupils explain how:

■ ice erodes rock
■ a newspaper is put together
■ to interrogate a database
■ to use a thesaurus.

Listening to the teacher

Sometimes you want pupils to listen and enjoy what is being said to them. There are other occasions when you want pupils to listen and interact with the material and keep some sort of record. It may be to:

■ explain a phenomenon, e.g. a riot
■ describe an event, e.g. a bore in a river
■ describe a process, e.g. making pastry
■ demonstrate a process, e.g. distillation

■ design an artefact, e.g. a desk lamp
■ give an account, e.g. of work experience or a visit to a gallery.

There are a number of ways you can help pupils in their work. For example:

■ identifying key words and ideas as you proceed, signalling to pupils when you expect them to record them
■ identifying key words and ideas in advance on a worksheet and asking pupils to note them, tick, underline or highlight as they are discussed. These words can be written on the board or transparency for reference
■ adopting the above strategy, but develop as a game. Who can identify these ideas? Call out when you hear them
■ using a diagram which pupils annotate as the lesson proceeds. This diagram might be used to: label parts; describe functions; identify where things happen. Pupils could keep their own notes and then be asked to make a summary and presentation to the class. Some pupils may need a word list to help them.

A possible way to support pupils in preparing a summary could be to give them a depleted summary and ask them to complete it. The degree of help is a matter of judgement. For example:

■ Give the summary with some key words missing and ask pupils to add the missing words.
■ Give the depleted summary with an additional list of words. Pupils select words and put them in the appropriate place. The selection could include surplus words.
■ Vary the focus of the omitted words. It could be on key words, or concepts, or focused on meanings of non-technical words, e.g. on connecting words or verbs, etc. (Sutton, 1981: 119; Frost and Turner, 2004: 181–184).

Another possibility could be to provide pupils with a writing frame. A writing frame consists of 'visual guidance on the construction of each paragraph or section of a piece of writing, which includes all or part of a topic sentence and bullet points identifying items which pupils should include' (see Moss in Davison and Dowson, 2003: 151). So, in the example given above of writing an account of a visit to a gallery, a possible frame might look like this:

■ give details of the journey to the gallery (e.g. times of departure and arrival, mode of transport, route)
■ describe the gallery (e.g. size, age of the building)
■ list three artists whose work is exhibited there and give some information about what you have learnt about them
■ say whether you would recommend this gallery to friends and give at least two reasons.

Characterising events

You may wish to help pupils associate certain ideas, events or properties with a phenomenon, for example, what were the features of the colonisation of the West Indies; or what are the characteristics of a Mediterranean-type climate? As well as reading and making notes:

■ List ideas on separate cards, some of which are relevant to the topic and others not relevant. Ask pupils to sort the cards into two piles, those events relevant to the phenomenon and those not directly related. Pupils compare sorting and justify their choice to each other.

■   Mix up cards describing criteria related to two phenomena. Ask pupils to select those criteria appropriate to each event. A more complex task would be to compare, for example, the characteristics of the Industrial Revolutions of the eighteenth and twentieth centuries.

Interrogating books or websites or reading for meaning

Learners often feel that, if they read a book, they are learning and don't always appreciate that they have to work to gain understanding. Learners need to do something with the material in order to understand it. There are a number of ways of interrogating the material in order to assist with learning and understanding. There are some general points to be considered. It is important that pupils:

■   are asked to read selectively – the length of the reading should be appropriate
■   understand why they are reading and what they are expected to get out of it
■   know what they are supposed to be doing while they read, what to focus on, what to write down or record
■   know what they are going to do with the results of their reading; e.g. write, draw, summarise, reformulate, précis, tell others, tell the teacher, carry out an investigation.

Homework which says simply 'read through this chapter tonight and I will give you a quick test on it tomorrow' is not helpful to learning because the focus is unclear. The following notes suggest ways of helping pupils read texts, worksheets, posters, etc. with purpose.

Getting an overview

Using photocopies of written material is helpful; pupils annotate or mark the text to aid understanding. Pupils read the entire text quickly, to get an overview and to identify any words they cannot understand and to get help from an adult or a dictionary. Ask pupils to read it again, this time with a purpose, such as to list or underline, or group key words or ideas.

Reformulating ideas

To develop understanding further, pupils need to do something with what they have read. They could:

■   make a list of key words or ideas
■   collect similar ideas together, creating patterns of broader concepts
■   summarise the text to a given length
■   turn prose into a diagram, sketch or chart
■   make a spider diagram
■   design a flow chart, identifying sequence of events, ideas, etc.
■   construct a diagram, e.g. of a process or of equipment with labels
■   turn a diagram into prose, by telling a story or interpreting meanings
■   summarise using tables, e.g. relating structure to function (e.g. organs of the body), historical figures' contribution to society (e.g. emancipation of women), form to origin (e.g. landscapes and erosion).

Where appropriate, pupils could be given a skeleton flow chart, spider diagram, etc., with the starting idea provided and asked to come up with further ideas.

Reporting back

A productive way of gaining interest and involvement is to ask pupils to report their findings, summaries or interpretation of the text to the class.

The summary could take one of the forms mentioned above. In addition, of course, a pupil could use the board, overhead projector, poster or a computer assisted presentation such as PowerPoint. Pupils need to be prepared for this task and need to be given sufficient time.

Public reporting is demanding on pupils and it is helpful if a group of pupils draft the presentation and support the reporter. The presentation could be narrative, a poem, a simulation, diagram, play, etc. A suggested sequence of events is shown in Figure 5.2.4.

The feedback loop, from 'Class discussion' to 'Task' in Figure 5.2.4, can be introduced depending on the time available and the interest and ability of the class.

DART and related types of learning activity emphasise the importance of language in learning and in assessing learning. For further discussion on the role of language in learning, see, for instance, Burgess (2004).

LESSON PLANNING FOR ACTIVE LEARNING

It is important to distinguish between 'activities' and 'active learning'. It is relatively easy to fill a lesson with a series of activities which keep pupils busy and apparently enjoying it, yet these may provide an insufficient learning challenge. Such work may be well within the pupils' grasp and so they do not have to think much about what they are doing or why. Many pupils take seriously copying from the board, book or worksheet but such activities are often superficial and should be used very sparingly, even though they can keep a noisy class quiet.

Some lesson planning in the early stages of learning to teach may be to ensure that all your learning outcomes are addressed, or that your discipline is effective; this requirement may lead to a lesson that is teacher dominated. For example, you may have explained orally the lesson and its purpose and have asked some bridging questions; you may have then used a video and asked the pupils to complete a worksheet in response to watching the video; finally, you may have given out homework based on the class work. You, as the teacher, have been very active in the lesson and may well feel exhausted at the end of it! From the pupils' point of view, however, the lesson may have been quite passive because they were told what to do at every step, with little or no input into what they were to do and learn. In these circumstances many pupils may not fully engage in learning except in a superficial sense.

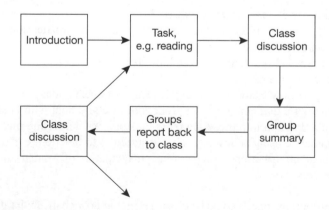

■ **Figure 5.2.4** Reporting back

Task 5.2.4 and Table 5.2.1 provide a springboard to explore these issues.

Task 5.2.4 **ACTIVE CLASSROOM OR ACTIVE LEARNING?**

You may find this task best discussed in a tutor group.

A structured sequence for the learning of new vocabulary in a modern foreign language lesson is presented in Table 5.2.1.

The pupils seem to be busy but to what extent do you think the pupils are actively engaged in learning?

Identify any changes you would make to the lesson and explain why.

After you have carried out this task compare your response to that of an experienced teacher.

■ **Table 5.2.1** An example of a structured sequence for the oral introduction of new vocabulary for beginners in a modern foreign languages lesson

Activity	Pupils' physical involvement	Pupils' mental involvement
Teacher presents new vocabulary items with visuals (e.g. downloaded pictures, flash-cards, etc.)	Pupils repeat each word at a time, in chorus and individually at random	Pupils try to work out the meaning of the word. The repetition helps them to start to commit the new vocabulary to their short-term memory
Teacher numbers visuals on the board, then calls out the words at random (or this could be done with the use of a list on tape)	Pupils write the number which corresponds to the word they have heard	Pupils draw upon their short-term memory to identify the correct item.
Teacher asks question: 'what is this' in the target language	Pupils put their hands up, selected individuals speak	Pupils draw upon their short-term memory to produce the words
Teacher asks pupils to work in pairs and ask each other what the new words are. Teacher encourages the class to answer in short sentences, i.e.: this is a . . .	All pupils involved in speaking	Pupils recall the words as well as the question. They begin to formulate a simple sentence.

This short lesson sequence illustrates how to scaffold the learning at the initial stage (see references to scaffolding in Units 3.1 and 5.1). It provides ample opportunities for pupils to take part, but the cognitive demands remain mostly at word or phrasal level and the activities are tightly controlled by the teacher. For pupils to progress further, they would need to be led gradually towards the challenge of identifying patterns, generating sentences of their own, integrating this vocabulary into new structures and contexts. Furthermore, the memorisation techniques illustrated here may well be helpful in the short term but are unlikely to be sufficient to secure deep learning by way of retaining the newly acquired words in the longer term.

Planning for learning requires you to focus on what the pupil is doing as well as what you are doing. To encourage pupils to participate fully in the lesson and hence promote meaningful learning there are some important features to bear in mind when planning your lesson. As well as planning strategies to include input from pupils into the development of the lesson you should:

- share the learning outcomes with your pupils and give an example of what your pupils' finished work should look like, i.e. what counts as a successful piece of work
- focus some learning outcomes on process rather than content, i.e. on pupil action and contribution to their own learning
- illustrate your criteria for assessment
- link the lesson to the pupils' prior knowledge and include your strategies for eliciting it
- prepare contingency plans for differentiating for both faster and slower pupils
- think about ways to help pupils in difficulties, i.e. give support (see Units 4.1 and 5.1).

Task 5.2.5 helps you to consider strategies to activate pupils' prior knowledge.

Task 5.2.5 **ACTIVATING PUPILS' PRIOR KNOWLEDGE**

Select a strategy for probing pupils' prior knowledge of a topic and use it in a lesson in which you are being observed. Evaluate the effectiveness of the activity yourself and also ask your tutor or class teacher to give you feedback on its effectiveness. Strategies you could use include a question and answer session, brainstorming in small groups (i.e. eliciting spontaneous recall of relevant information or randomly listing ideas/suggestions in relation to a particular topic or question) followed by a plenary session, asking pupils to prepare a spider diagram summarising what they know and using them to plan the next lesson.

File feedback in your PDP.

Active learning goes hand in hand with an approach to teaching which encourages pupils to develop and progress as individuals and not merely to receive information from the teacher. Active learning, therefore, is a process that is:

- structured and organised, a purposeful activity through which pupils can achieve the intended learning outcome as you have planned it
- transformational, enables pupils to consider alternatives, to think differently and develops attitudes and values
- communicative, involves engagement with others within and beyond the classroom and develops higher order skills such as analysis, communication, investigation, listening
- generative, pupils are engaged in the process of their own learning and generates deeper understanding by challenging pupils' understanding
- supportive of meaningful learning.

If this description is correct then a lesson must invite pupils to participate in the work, contribute to its development and, consequently, begin to shape their own learning. The demands on the pupil in such a situation move learning to higher-order skills to which we now turn. See above under 'What is active learning?' and 'Active learning and motivation'; and Unit 3.1.

Developing pupils' higher-order thinking skills

It is now generally acknowledged that the explicit development of thinking skills needs to take place alongside teaching of factual content and that the emphasis in learning is not just on the outcomes, but also on the processes. Teaching, therefore, needs to be designed to enable pupils to:

- develop logical reasoning in order to apply it to new contexts (formal thinking approach)
- deconstruct problems in order to find solutions to them (heuristic approach);

■ reflect on, and evaluate their own learning (metacognitive approach) (Muijs and Reynolds, 2005).

These three approaches are at the heart of active learning because they promote the learners' engagement with the task and encourage pupils to make sense of their learning.

Bloom identified six levels of thinking of gradually increasing complexity, which make increasing demands on the cognitive processes of learners (Bloom, 1956; see also Unit 3.1). These six levels are listed in column one of Table 3.1.2. While there is the potential for active learning at every level, the last three levels can be linked to the higher-order thinking skills mentioned above. (For examples of the use of Bloom's taxonomy in teaching, please refer to the DfES Standards Site, http://www.standards.dfes.gov.uk/secondary/keystage3/casestudies.)

Providing suitable challenges is fundamental to active learning. One strategy used extensively by teachers is questioning (see also Unit 3.1). The importance of questioning as a teaching and learning strategy is long established and well documented in educational research, e.g. Wragg and Brown (2001), Kerry (2004). Many studies show how questions can take various forms and how they can be adapted to serve a variety of purposes to promote active learning, such as:

■ capturing pupils' attention and interest
■ recalling and checking on prior knowledge
■ focusing pupils' attention on a specific issue or concept
■ checking and probing pupils' understanding
■ developing pupils' thinking and reasoning
■ differentiating learning
■ extending pupils' power of analysis and evaluation
■ helping pupils to reflect on how they learn.

To the student teacher, questions asked by experienced teachers may appear intuitive and instinctive whereas in reality good questioning develops by reflection on experience. Questions should not just be 'off the cuff' but prepared in advance and related to the learning outcomes, so that pupils' learning is structured. A useful observation schedule identifying good principles for effective questioning is available (Good and Brophy, 2000: 412).

Experienced teachers have the skill of asking questions beyond those planned in response to pupils' replies. Experience here depends largely on knowing your subject and how to use this knowledge, that is, how to put across your knowledge of the subject to the pupils in a way that enables them to learn; and on knowing your pupils and how they respond to your subject and your teaching.

Effective questioning is central to the teaching and learning process. How Bloom's taxonomy relates to the use of questions by the teacher is shown in Table 3.1.2. This table identifies the purposes of questioning at the various levels and gives examples of the sort of question that may be asked. For further discussion of questioning techniques, see *Pedagogy and Practice: Teaching and Learning in Secondary Schools Unit 7 Questioning* (DfES, 2004c) but as a starting point complete Task 5.2.6.

Task 5.2.6 **DEVELOPING HIGHER-ORDER THINKING SKILLS THROUGH THE USE OF QUESTIONING**

Observe a lesson and script the questions used by the teacher to promote learning. Try to classify against Bloom's (1956) taxonomy of educational objectives. How does the type of question used impact upon active learning? Discuss your observations with the teacher or your tutor and record them in your PDP folder.

COMMUNICATING WITH PUPILS: VISUAL AIDS

A multisensory classroom experience can also contribute towards active learning. Visual aids can generate interest, bring 'reality' into the classroom and enable an interactive approach to be adopted. Visual aids may take the form of a simple prompt, e.g. a picture, a poster, an object, an overhead projector transparency, a PowerPoint presentation or an interactive whiteboard.

Visual aids are a powerful tool for focusing attention, stimulating memorisation and conveying meaning, all of which contribute to active learning. They enable the teacher not only to present information but also to clarify concepts and meanings and build ideas, models, diagrams, sequences with the class. For example, you can hide definitions that the pupils then complete. Pupils can hypothesise from partly revealed screens, share their ideas and complete some creative writing from these ideas as a class activity. Equally importantly, the pupils can use these aids to display their own understanding.

The interactive whiteboard is a touch-sensitive projection screen which is connected to a computer and a projector. By simply pointing at active elements on the board with a finger or with an appropriate electronic 'pen', you or your pupils can highlight or move what is displayed on the board. Text can be written on the whiteboard, for instance a whole-class correction of an exercise, and then saved for further use at a later date. You and your class can also interface with downloaded interactive materials from the Internet, or simply use downloaded images, graphs, texts, etc. These added advantages make the interactive whiteboard a powerful tool for engaging pupils in learning. Although it requires a fair amount of preparation time, it helps to make smooth transitions between activities and there is an increasing range of time-saving resources being developed for use in the classroom. (For suggestions on use, please refer to British Educational Communications and Technology Agency (BECTA) g, 2005).

Other equipment, e.g. the overhead projector (OHP) if the interactive whiteboard is not accessible can be equally effective for some activities. Such facilities are easy to use, but it is important that their use is effective and not cosmetic.

Whatever visual aids you choose to use, it is worth bearing in mind that they require management. There are practical implications for maximising the impact of visual aids, including their clarity, lack of ambiguity and appropriateness of the language level to your class. Another practical implication is to make sure all pupils can see and, if appropriate, read your visual prompts. It is useful, for example, when using visual aids to practise where to stand so as not to obstruct pupils' vision.

Task 5.2.7 invites you to consider the use of visual aids in one of your units of work.

Task 5.2.7 **USING VISUAL AIDS**

In one of your units of work, plan how you can involve pupils in using visual aids themselves. For example, ask a group of pupils to put together a PowerPoint presentation or, if using an overhead projector, complete their own transparencies to report back on a group discussion. Teach the lessons incorporating this use of visual aids and then evaluate the effectiveness of this approach and how you might improve their use in future lessons.

Finally, complete Task 5.2.8.

Task 5.2.8 **REFLECTING ON YOUR USE OF ACTIVE LEARNING**

Read the article: Powell, E (2005) 'Conceptualising and facilitating active learning: teachers' video-stimulated reflective dialogues', *Reflective Practice*, 6: 3, 407–418.

Make notes in your PDP. Arrange to video-tape a lesson, or a part of a lesson which aims to promote active learning. If possible, discuss the video extract with your tutor, using Moyle's reflective framework (see Appendix 1(b) of Powell's article). Write a critical analysis of your teaching, drawing upon the research literature that you have read.

SUMMARY AND KEY POINTS

Teaching is an enabling process; teachers can guide pupils' learning but cannot do the learning for them. Pupils need to engage mentally with a task if learning is to take place; thus you need to enthuse and motivate pupils, give purpose to their learning tasks and to provide active learning experiences. Some of these activities, e.g. DART, introduced study skills, important for pupils preparing for public examinations.

This unit has used examples of reading, writing, listening and talking activities designed to improve learning and learning skills. There is hands-on activity in practically based subjects, such as art and design, home economics, science and technology. Working with your hands does not guarantee that learning takes place; both hand and brain need to be involved. Pupils need to have a say in the design, execution and evaluation of practical work in the same way as we have stressed the need for their active involvement in reading and listening.

The advantages claimed for the active learning include, too, an emphasis on co-operative learning which can provide opportunities for pupils to take some responsibility for their own learning by, for example, active participation in the development of the task. This approach requires pupil self-discipline and may contribute to that wider goal of education. For the teacher, it opens up a wider range of teaching methods to develop personalised learning, allows the growth of resource-based learning and provides space for monitoring pupil progress and giving formative feedback. Aims can be widened; as well as encouraging acquisition of knowledge and understanding, active learning can be used to promote process skills and higher order skills.

The key to good teaching is preparation. This is particularly important if you select active learning strategies. These strategies are a major part of your teaching repertoire and as such contribute significantly to the professional standards required of an NQT. Further advice and guidance on active learning is available in the Further Reading section.

Check which requirements for your course have been addressed through this unit.

FURTHER READING

Muijs, D. and Reynolds, D. (2005) *Effective Teaching: Evidence and Practice*, 2nd edn, London: Paul Chapman.

This book provides a comprehensive introduction to what are considered to be key elements of effective teaching, as evidenced by recent research on practice. The chapters on interactive teaching, collaborative small group work, constructivism and problem-solving and higher-order thinking skills will be particularly useful to deepen your understanding of what active learning involves and how you can bring it about in your own teaching.

Watkins, C., Carnell, E. and Lodge, C. (2007) *Effective Learning in Classrooms*, London: Paul Chapman.

This book focuses on learning, what makes it effective and how to promote it in classrooms. The authors identify active learning as a core process for promoting effective learning in classrooms. Drawing upon international research as well as case studies involving practising teachers, they provide you with useful ideas and frameworks for developing your own conception of active learning.

RELEVANT WEBSITES

The British Education Research Association (BERA): http://www.bera.ac.uk

A useful starting point to find out about recent research projects on active learning and thinking skills.

Pedagogy and Practice: Teaching and Learning in Secondary Schools: http://nationalstrategies. standards.dcfs.gov.uk/node/97131?uc=force_deep

This site contains 20 self-study units for school improvement. Unit 7 Questioning and Unit 11 Active Engagement Techniques are particularly relevant to this chapter.

Qualifications and Curriculum Authority (QCA) (2007): http://curriculum.qca.org.uk/

This site gives you information on the new secondary curriculum. Click on Key Stages 3 and 4 and go to Skills, then Personal, Learning and Thinking Skills to see the framework for PLTS. You can also explore how your subject relates to PLTS (go to Subjects and follow the link).

Teachernet: http://www.teachernet.gov.uk

A large section of this site is dedicated to teaching and learning, including information about official documents and teaching strategies. It contains examples of case studies and links to resources, e.g. lesson plans on various topics and themes, and guidance is regularly updated.

> **Additional resources for this unit are available on the companion website:**
> www.routledge.com/textbooks/9780415478724

ACKNOWLEDGEMENT

The authors would like to acknowledge the input of Tony Turner into the first four editions of the unit.

TEACHING STYLES

John McCormick and
Marilyn Leask

INTRODUCTION

This unit is concerned mainly with *individual teaching styles*. Although individual teaching styles are very personal, analysis of teaching has identified common approaches in teaching episodes and these provide a framework in which personal styles can develop. The unit introduces some of these approaches and provides opportunities for you to use theories encountered in other units in this book to widen your range of teaching styles within them.

OBJECTIVES

At the end of this unit you should be able to:

- ■ choose teaching styles to match the different learning strategies of pupils
- ■ analyse and outline different teaching styles in sequences of your lesson plans
- ■ map your developing style to the requirements for your course and use your mapping to set targets for further development
- ■ appreciate the difference between personal teaching styles and theoretical models of teaching styles
- ■ appreciate the importance of *mobility ability* (Mosston and Ashworth, 2002), which refers to the capability good teachers have for switching their style to meet different needs of learners.

Check the requirements for your course to see which relate to this unit.

The appendix to the unit provides an example of peer tutoring as a personalised learning teaching strategy.

TEACHING STYLES AND STRATEGIES

We define your *teaching style* as resulting from the combination of your underlying behaviours (personality plus characteristics), your microbehaviours (ways you communicate through speech and body language) and the teaching strategies you choose (where we use the term strategy, other writers may use terms such as 'teaching approach' or 'teaching method'). Thus, we define *teaching strategy* as an approach or method chosen for a particular teaching episode.

In choosing the *teaching strategies* for a lesson you make decisions about the best ways to achieve the desired learning outcomes through

■ how you present the material
■ activities and tasks
■ whether pupils are passive or active. You will for example decide on the kind of talk you want pupils to engage in: questioning, debating, discussing or working in silence.
■ grouping pupils
■ feedback – when, how, why.

The teaching strategies we introduce in this unit include the following:

■ closed, framed and negotiated strategies (see Table 5.3.1)
■ command, practice/task, reciprocal, self-check, inclusion, guided discovery, convergent discovery, divergent discovery, learner initiated, self-teaching strategies (see Table 5.3.2)
■ peer tutoring (see Table 5.3.3).

We use the term *learning style* to refer to the preferences a learner may have for learning by seeing (visual), hearing (auditory) doing (kinaesthetic). Using teaching strategies which stimulate several senses is considered to deepen learning. The learning strategies pupils use, i.e. techniques to help them learn, change according to the material to be learned.

Task 5.3.1 provides an example of how you may need to adapt your teaching style when moving from one school to another where pupils are used to using different learning styles.

Unit 5.1 introduced the notions of pupils' learning styles (preferences), which tend to be fixed and individual, and learning strategies, by which learners 'cope with situations and tasks'. Sources differ, e.g. Department for Education and Skills (DfES),(2003c; 2004a) Sousa (1997, cited by Tileston, 2000), but whereas about 35 per cent of people tend to prefer visual learning and 40–45 per cent prefer kinaesthetic, only about 25 per cent prefer auditory learning. One possibility in the example in Task 5.3.1 is that, irrespective of their preferences, the learning strategies of the pupils in the first school had developed to cope with auditory delivery and individual work. However, in the second school teachers had taken account of preferred learning styles, with the result that learning strategies of the pupils centred on visual and kinaesthetic presentation of work, rather than oral questions and instructions. The student teacher had developed a *teaching style* to match the learning strategies of the first pupils, and the mismatch in the second school was enough to render learning, and teaching, ineffective.

It was not until the student teacher's strategies changed, resulting in a new and successful style of teaching, that successful learning was achieved.

As you progress through your school experience you should refine your practice to develop teaching styles which suit different circumstances and achieve diverse learning outcomes. Usually you will need to adopt a range of styles within a single lesson in order to ensure all of the pupils achieve the learning outcomes.

However, teachers do not always develop or use the most suitable teaching style for their pupils. Research (DFES, 2003c) suggests that mismatch between pupils' preferences and teachers' styles arises from a number of reasons, including a belief on the part of individual teachers that

---

Task 5.3.1 **TEACHING AND LEARNING STYLES**

Consider the following situation.

A student teacher moves to a second school after a highly satisfactory first school experience.

Knowing the pupils in both schools are of a similar background the student teacher decides to play safe for his first solo lesson. His starter activity worked well in the first school and he is comfortable with it. A stimulating picture is projected onto a whiteboard and pupils are asked questions, which they are to answer by finding and using information in the picture.

However, the activity does not work. The pupils repeatedly ask for explanations and clarifications, and they interrupt the student teacher so much that he is unable to complete the starter. He has to abandon it and move to the body of the lesson, where he tries to tell the pupils how to carry out an activity. Again, he has trouble in making his instructions clear and he continues to be interrupted by requests for clarification.

For the next lesson, after advice, the student teacher projects a picture onto a whiteboard but, as they enter, pupils are handed a worksheet with 6 questions and instructions to work in pairs to decide answers. After a few minutes adjacent pairs are asked to spend two minutes deciding the best answers and to choose one member from each of these larger groups to feed back one of their answers to the class. Following this the pupils continue in pairs, using a Directed Activity Related to Text (DART; see Unit 5.2) to decipher the instructions for the next part of their lesson

The lesson starts well.

What has caused the difference?

---

what helps them to learn also helps pupils. The effect is that teachers use a style which is actually best suited to themselves and not necessarily to their pupils. The position may be even worse for student teachers; Calderhead and Shorrock (1997) suggest teachers, especially student teachers, are initially more comfortable with structures and styles they experienced as pupils than with new ideas and Evans (2004) reports 41 per cent of a sample of student teachers felt they taught in the way they had been taught even though many of them had not enjoyed that way. Therefore, it is important that when you critically reflect on your own teaching that you take active steps to incorporate effective practice, including ideas gained from discussion, observation of other teachers and reading, to become comfortable and confident with a wide repertoire of strategies.

Task 5.3.2 illustrates how reflective reading can produce questions which help you improve your practice.

The OFSTED report referenced in Task 5.3.2 also noted that, even in good schools, achieving consistency across the curriculum remains a challenge and lessons are not always good. It is important that you bear this in mind. Evans (2004) reported many of the student teachers she questioned were unsure of their own preferred styles; and you can only find your own style through exploration and reflection. Some strategies you use will not work well, or possibly you will use a strategy without appreciating its shortcomings, but you should not be dissuaded in exploring teaching strategies by the occasional apparent failure to engage pupils.

Task 5.3.2 REFLECTION ON EFFECTIVE TEACHING STYLES

Think about these questions, derived from Ofsted (Office for Standards in Education) observations of successful schools (Ofsted, 2002), in the context of your own teaching.

■ Does my behaviour instil confidence in high attainers? Is it open and encouraging to the less able?

■ Is my style of presentation formal or informal and what effect does this have on my ability to communicate enthusiasm?

■ Is my style of teaching personal to every pupil?

■ Do I build a variety of activities into my lesson plans with a style that incorporates a range of teaching strategies?

■ Does my style of delivery make my objectives clear at the start and allow me to refer to objectives at key points in the lessons?

■ When I mark and assess work is my style of giving feedback personal and informative?

Make notes about how you wish to improve your current practice and return to this list periodically to check that you are making progress.

Teaching strategies and meaningful learning

Meaningful, or deep, learning (Entwistle, 1990) requires pupils to engage in an active reconstruction of information, to make new links and test old ones, to resolve contradictions and to identify underlying principles. The knowledge and understanding they gain by these activities tend to be retained and can be applied to new situations. On the other hand, shallow, or surface, learning occurs when pupils do not really understand principles, or when they hold misconceptions. For example, they can often use formulae but do not really understand their derivation and cannot apply them to new contexts; or they hold two conflicting views at the same time. Research and evidence from practice (Mercer, 2000; Wegerif and Dawes, 2004) is showing that actively teaching pupils how to learn, how to explain themselves, how to ask probing questions, sharing the lesson objectives and learning outcomes with them and teaching them how objectives link with activities and assessment tasks leads to raised achievement and motivation.

The distinction between deep and shallow learning is important. As a student teacher, your teaching strategies may be focused on immediate needs (such as an impending test) and might not always promote meaningful learning. In contrast, expert teachers are able to enhance both types of learning, enabling pupils to better understand (Hattie, 2003). When you reflect on your teaching think about the degree to which you are teaching 'to the test', and use your reflections to develop a style in which your use of assessment promotes meaningful learning.

Meaningful learning happens best where social interaction, particularly between a learner and more knowledgeable others, is encouraged, and where there is a co-operative and supportive ethos. Teaching styles therefore need to take account of the need for discussion, both between pupils and, following constructivist ideas (see Unit 5.1) between pupils and teacher. Your teaching style should make allowances for the learning process and should not only promote discussion but encourage pupils to challenge their own and others' ideas and to go back and forth in a non-linear way. To do this you may need to move from a *closed* strategy (Barnes *et al.*, 1987) where the teacher tightly controls the content of the lesson, the learning environment and the outcomes of the lesson, through a *framed* strategy, where the teacher controls the topics and has clear expectations of outcomes but allows the pupils to propose and test alternatives, to a *negotiated* strategy, where the pupils have much more freedom in determining the area of investigation and the way in which work is reported, with, possibly, variable outcomes. Table 5.3.1 describes the main differences between these strategies.

Using SACK to develop a style

In choosing which of these broad styles to use you need to consider the learning objectives of the lesson. Not all objectives require deep learning. You may be familiar with the acronym SACK (or CASK), which is intended to remind you to check whether the balance of your lesson objectives between Skills, Attitudes, Concepts and Knowledge is appropriate. Developing process skills by, for example, training pupils to perform a simple action might best be achieved through the use of a *closed* style, whereas developing an attitude or coming to understand a concept, which does require meaningful learning to take place, might best be achieved through the use of a *negotiated* style. When one of the objectives for the lesson concerns the development of knowledge it can be more difficult to decide on the best style to use and you might need to refer to other theories, such as information processing models, to help you.

■ **Table 5.3.1** Teaching strategies: closed, framed, negotiated

	The participation dimension		
	Closed	**Framed**	**Negotiated**
Content	Tightly controlled by the teacher. Not negotiable	Teacher controls the topic, frames of reference and tasks; criteria made explicit	Discussed at each point; joint decisions
Focus	Authoritative knowledge and skills; simplified monolithic	Stress on empirical testing processes chosen by teacher; some legitimation of pupils' ideas	Search for justifications and principles; strong legitimation of pupils' ideas
Pupils' role	Acceptance; routine performance; little access to principles	Join in teacher's thinking; make hypothesis, set up tests	Discuss goals and methods critically; share responsibility for frame and criteria
Key concepts	'Authority': the proper procedures and the right answers	'Access'; to skills, processes criteria	'Relevance': critical discussion of pupils' priorities
Methods	Exposition; worksheets (closed); note giving; individual exercises; routine practical work. Teacher evaluates	Exposition; with discussions eliciting suggestions; individual/group problem solving; lists of tasks given; discussion of outcomes, but teacher adjudicates	Group and class discussion and decision making about goals and criteria. Pupils plan and carry out work, make presentations, evaluate success

Source: Adapted from Barnes *et al.* (1987: 25)

Information processing models and teaching style

As well as constructivist models of learning (see Unit 5.1) there are models which represent learners as information processing systems. These can run parallel to constructivist views and do not necessarily contradict them; pupils with poor information processing capabilities, who needed a lot of help and progress slowly, would be those who had small Zones of Proximal Development. (Vygotsky, 1986)[1]

Information-processing models can illustrate why your teaching style may need to change depending on whether your lesson's learning objectives relate to skills, knowledge or attitudes or concepts.

Learners can only take in a certain amount of information in a set time. There is limitation of capacity and bottlenecks can occur when a lot of information is transmitted. When bottlenecks occur, not all of the transmitted information is received in the memory. Therefore your teaching style must not deliver too much information too quickly or introduce digressions. Reflect on whether a closed or framed style enables you to get your key points across with most effect. At the start of the lesson, when objectives and activity are outlined or, for example, when demonstrating a technique or ensuring safety, a very focused, closed, instructional style might be required.

After information is received, short- and long-term memory work together dynamically. Information is processed in short-term memory, where fresh information and retrieved existing knowledge are used together to make meaning of new situations. This meaningful learning can then be stored in long-term memory. A teaching style which presents information in different ways and provides a variety of perspectives matches a range of preferred learning styles. More pupils are able both to take in the information and to link it to existing knowledge, increasing the probability that it is processed rather than disregarded. In this part of the lesson a closed style is not as effective in promoting meaningful learning, even if that learning concerns knowledge rather than understanding of concepts, because it is the links which are important and closed styles tend not to allow pupils to develop these as well as other styles.

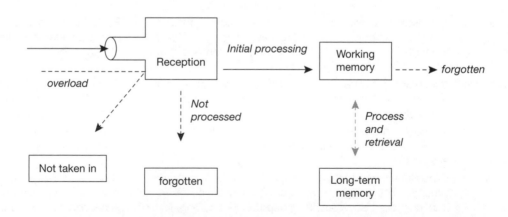

■ **Figure 5.3.1** A flowchart for an information processing model. The flow chart shows the typical pattern of information flow and processing suggested by a multi-store model of memory proposed by Atkinson and Shiffrin (1968)

Source: Adapted from Gross (2001: 244–248)

Once you have considered the teaching styles most likely to support the achievement of your lesson objectives consider your strategies. For example, according to sources cited earlier (DFES, 2003c; Sousa, 1997) the majority of pupils are not likely to be auditory learners. Therefore you should consider an alternative *strategy* to talk, or use talk in conjunction with, for example, a PowerPoint reinforcement of key themes. You might also think about your grouping strategy. Kutnick *et al.* (2006) have carried out an extensive literature review and suggest no one form of grouping benefits all pupils. Grouping by ability can benefit the most able but demotivate the less able; friendship groups can sometimes provide a non-threatening environment which promotes learning, at other times it can reinforce cultural stereotypes which prevent progress. Be prepared, therefore, to seek advice about the organisation strategies you employ within the classroom and how best to exploit the different opportunities they present.

Task 5.5.3 asks you to reflect on the application of the closed, framed and negotiated strategies.

Task 5.3.3 **INFORMATION PROCESSING THEORY AND CLOSED, FRAMED AND NEGOTIATED TEACHING STRATEGIES**

Reflect on the pros and cons of closed, framed and negotiated strategies in:

■ ensuring you gain pupils' attention, so that they receive information which they are then able to process. Is a closed style the best way of achieving this end?

■ identifying and stressing key information so it is processed. Does a framed strategy allow pupils to develop a personal perspective and thus make links which gives them a better chance of processing it?

■ making links to known features clear. Alternatively, does a closed strategy identify the links, with the result that pupils learn more effectively than with a framed strategy? Or could a negotiated strategy allow pupils to identify links which you are not aware of?

■ using patterns and mnemonics as memory aids to help pupils remember them. Is it better to provide mnemonics (closed) or allow pupils to make their own (framed)? If they are making their own are groups based on friendship a good strategy for allowing them to explore and experiment without fear of ridicule?

Use your reflections when you develop your teaching. Begin with your scheme of work. Can you identify opportunities for different strategies? Then develop your ideas in individual lessons. Finally, when you analyse your lessons consider the effectiveness of the styles you used.

Experiment and consistency

Early school experience should have helped you to realise that routines in the classroom are important. So although you need to experiment in developing a personal style you also need to think about consistency. Consistency can be helped by developing assertive respectful behaviour in which you share your strategies with the pupils. Involving learners more actively improves their learning (see, e.g. Black *et al.*, 2003: 78). Task 5.3.4 asks you to consider what teaching styles are appropriate for particular activities.

Task 5.3.4 **CASE STUDY – ELECTRICITY BILLS**

This case study is taken from Key Stage 4 science, on the theme of calculating electricity bills. Electrical use is calculated in Kilowatt-hours (1 KWh = 1 Unit) and the cost of running an electrical appliance depends on how quickly an appliance uses a unit. A 3 Kw kettle might use one unit of electricity in 20 minutes. On the other hand, a modern DVD player in standby mode might only use one unit of electricity in six months.

What teaching styles (e.g. closed, framed, negotiated) do you think are attempted and achieved in each stage of the lesson outlined below?

Preparation

Pupils are given a sheet to take home and complete. This requires them to fill in details from the electrical rating plates of some domestic equipment, including a kettle, an iron and their personal audio, video or computer equipment.

Starter activity

As the pupils arrive the first three to enter are given a small script to rehearse. The script centres on parental response to a high electricity bill which is blamed on the use of personal CD and DVD equipment by their pupils. The script is read after the last member of the class arrives.

Introduction

The teacher introduces the lesson by saying she is going to give the pupils the power to refute parental arguments that high bills are their fault. Lesson objectives are given as

■ to understand the concept of Kilowatt-hours and units of electricity
■ to be able to calculate electricity bills.

Exposition

The teacher discusses and, using a whiteboard, outlines what KWh are and how they are used in calculating bills. A text book with relevant formula is used to support this – pupils read the page silently and answer questions.

Activity

Pupils work in groups to pool and use their information to work out the cost of running each appliance for 1 hour. The teacher circulates and helps as appropriate, telling pupils what to do.

Plenary

One member from each group outlines which is the cheapest and most expensive appliance to run and how they could explain this to their parents.

Analyse the activity and decide how much control the teacher has maintained over the teaching agenda. How might the need to maintain control have affected her style?

Analysis of the activity in Task 5.3.4 might suggest the teacher controls the teaching agenda quite carefully. The styles in use in the lesson tend to be closed. However, in practice, analysis of style is not always simple. Therefore, discussion of observations with teachers during school experience should include teaching styles. As well as clarifying actual and intended styles this discussion might help you to appreciate that boundaries between styles are not hard and fast. Styles form a continuum and you need to be able to move along the continuum and vary your style to suit the needs of your pupils. Continue Task 5.3.4 to explore the possibilities this episode provides.

Task 5.3.4 **CONTINUED CASE STUDY – ELECTRICITY BILLS**

Review the electricity case study and think of alternative styles for teaching each activity in the lesson. For example, the starter results from a fairly closed style – pupils are chosen and read the script. A framed style would introduce the idea of the play but might allow the pupils to write and perform the script themselves. A more negotiated style might allow the pupils to vary the delivery, so they might produce a news broadcast instead of a play. Which is the most suitable style, if there is one, and how might it be influenced by teacher behaviour or teaching strategy? Apply the same procedure to some of your own lessons.

Think about the styles you intend to use and the styles you actually do use in your teaching.

Does your need to control the teaching agenda compromise your style? For example, do you plan a lesson with, say, a negotiated style for part of the work but find in your post-lesson analysis that you have not allowed the pupils enough latitude?

You might find you move between closed and framed styles during a lesson but use a negotiated style less often. Negotiated styles are often associated with project or extended work. However, you should not neglect them. There are times in individual lessons when they can be used particularly to challenge the more able pupils.

Mosston's 'Spectrum of teaching styles'

Mosston uses the term 'style' where we would use 'strategy'.

In analysing your lessons it is helpful to refer to more sophisticated models than those presented by Barnes *et al.* (1987). One such model is characterised by a focus on the decision-making process which is central to planning and teaching. In your teaching you make decisions about what you and your pupils will and will not do. Some of these decisions are made before teaching, some during teaching and some after teaching. Underpinning theory holds that these decisions 'are the pivotal element in the chain of events that form the teaching-learning relationship' (Mosston and Ashworth, 2002: xx) and gives rise to what Mosston describes as the *spectrum of teaching styles* (Mosston, 1972). This spectrum is independent of individual or personal style but it can enable teachers to fully understand the effects of their style(s) on pupils' learning. Mosston, and later Mosston and Ashworth, argue that the spectrum is applicable to all subjects and can serve as a guide to selecting the best approach for a particular purpose. Therefore knowledge about the 'styles' contained in the spectrum, and about the decision-making processes within them, allows you to develop a teaching repertoire which better promotes learning and help pupils to become independent learners.

Mosston's spectrum is similar to the closed, framed and negotiated strategies mentioned above in that it proposes a continuum of teaching styles in which pupils are able to exercise a greater and greater amount of intellectual freedom, creativity and inquiry but it has more features. Over the years there have been a number of models of the spectrum. Typically at one end of the spectrum is *command*, where the teacher makes every decision, before, during and after the teaching episode. Further along the continuum are, for example, *practice* or *task*, where pupils have a limited amount of freedom to decide how they complete a set activity, and *reciprocal*, where pupils work together and are involved in the evaluation of each other's performance. *Inclusion* takes account of differences in learners and allows them to decide on the level at which they perform. As the continuum proceeds there are increasing amounts of freedom and a transfer of responsibility. With *discovery* the teacher begins to hand responsibility for all decisions to the learner, who decides what to study, for how long, with what resources, how findings will be reported, and so on. This results in a *learner-initiated* style and finally, but not in classrooms, a *self-teaching* style. Table 5.3.2 illustrates the differences between these styles.

Although different styles have different implications for independent learning, no single style is more or less important than another; what is important is that the teacher appreciates their potential and can move between different styles as circumstances demand. This is described by Goldberger (2002) as *mobility ability* – 'the ability to shift comfortably from one teaching style to another in order to meet learner objectives' (Mosston and Ashworth, 2002: xi). As a student teacher, you might approach this from a different perspective. Where are your style(s) on the spectrum and can you widen your range so you can change styles to meet different demands? Task 5.3.5 asks you to look at your own styles of teaching. It may be helpful to carry out this activity with the help of a mentor who observes lessons you have chosen to provide a variety of teaching activities.

Task 5.3.5 **ANALYSING YOUR TEACHING STYLE**

Analyse a number of your own lessons in terms of the decisions you make about the strategies to use. What decisions do you take and which decisions are made by pupils? Do the lessons have common features which give you a guide to your prime location(s) on the spectrum?

If you make all of the decisions about the lesson, for example, what will be done, what specific actions will be completed, how big groups will be, what answers will and will not be accepted, how feedback will be provided, this corresponds to the *command* style.

Giving some latitude, by, for example, allowing pupils to make their own decisions about how they carry out a task you have defined corresponds to the *practice* style.

Giving more latitude, by, for example, allowing pupils to work in groups of their own choice to solve a problem which you have defined but where there is a predetermined solution corresponds to the *guided discovery* style. Note here that there is a set problem with a predetermined answer – these decisions have been made by the teacher.

Giving pupils a problem with a number of possible outcomes and allowing them to make their own decisions about groups, method and reporting back, followed by neutral feedback which values all of the responses, corresponds to the *divergent discovery* style.

Now use the following table when evaluating some of your lessons to help you analyse your teaching style and to develop it.

Stage of lesson	Usual strategy	Overall style	Assessment of this style on the quality of pupil learning. What can I do to widen my range of styles?
Introduction			
Activity 1			
Activity 2			
Activity 3			
Plenary			

■ **Table 5.3.2** Mosston's 'Continuum of teaching styles' (2002 version)

The command style
This style is often described as autocratic or teacher centred. It is appropriate in certain contexts, e.g. teaching safe use of equipment, learning particular routines in dance.

The practice style
Whilst similar to the command style, there is a shift in decision making to pupils and there is more scope with this style for the teacher to work with individuals whilst the group are occupied with practice tasks such as writing for a purpose in English or practising skills in mathematics.

The reciprocal style
The pupils work in pairs evaluating each other's performance. Each partner is actively involved – one as the 'doer' and one observing, as the 'teacher partner'. The teacher works with the 'teacher partner' to improve their evaluative and feedback skills. This style provides increased possibilities for partners to improve their evaluative and feedback skills. This style provides increased possibilities for 'interaction and communication among students' and can be applied when pupils are learning a foreign language or learning routines in gymnastics. Pupils learn to judge performance against criteria.

The self-check style
This style is designed to develop the learner's ability to evaluate their own performance. The teacher sets the tasks and the pupils evaluate their performance against criteria and set new goals in collaboration with the teacher. Pupils start at the same level and move up when the teacher deems them ready.

The inclusion style
In this style, differentiated tasks are included to ensure that all pupils gain some feeling of success and so develop positive self-concepts, e.g. if an angled bar is provided for high jump practice, all pupils can succeed as they choose the height over which to jump. They decide at what level to start.

Guided discovery
Mosston sees this as one of the most difficult styles. The teacher plans the pupil's learning programme on the basis of the level of cognition development of the learner. The teacher then guides the pupil to find the answer – reframing the question and task if necessary but always controlling the teaching agenda. Pupils with special educational needs are often taught in small groups and this approach might be used by the teacher to develop an individualised learning programme for each pupil.

Convergent discovery style
There is a single desired outcome to the learning episode but the learners have autonomy over processes and presentation. The teacher provides feedback and clues (if necessary) to help them reach the correct outcome.

Divergent discovery style
Learners are encouraged to find alternative solutions to a question, e.g. in approaching a design problem in art. Multiple solutions are possible and the learners assess their validity, with support from the teacher if necessary.

The learner designed individual programme
A pupil designs and carries out a programme of work within a framework agreed and monitored by the teacher. Pupils carrying out open-ended investigations to answer a particular question in science provide an example of this style. The knowledge and skills needed to participate in this method of learning depend on the building up of skills and self-knowledge in earlier learning experiences.

Learner-initiated style
At this point on the continuum, the stimulus for learning comes primarily from the pupil, who provides the question to investigate as well as the method of investigation. Thus the pupil actively initiates the learning experience and the teacher provides support. Giving homework which allows pupils freedom to work on their own areas of interest in their own way might fall into this category. However, Mosston and Ashworth make the point that this kind of learning arises 'only when an individual approaches the teacher (authority figure) and initiates a request to design his/her own learning experiences' and 'when teachers ask learners to *do a project* it cannot be construed to be an example of this style'. (Mosston and Ashworth, 2002: 284–285).

Self-teaching style
This style describes independent learning without external support. For example, it is the type of learning that adults undergo as they learn from their own experiences.

Source: Adapted from Mosston and Ashworth, 2002

Personalised learning and independent learners

Personalised learning requires that you 'shape teaching and learning around the different ways children learn' (after Barnes and Harris, 2006: 1). In some circumstances, such as using resistant materials workshops, science labs or working with javelins, your choice of actions is quite constrained. Nevertheless, it is unlikely that you adopt a complete command style during school experience and allow pupils no discretion over their work. It is even less likely that you have the confidence or desire to adopt a discovery style, and hand over the majority of the decisions to the pupils. However, it is important that you do encourage learners' independence in order to further the personalisation of their learning. Mosston and Ashworth (2002) argue that there is a 'discovery threshold' and independent learning is not possible if teachers do not use discovery or later styles. Therefore your repertoire should incorporate at least discovery styles and provide pupils with challenging opportunities for independent work both within and outside the classroom. Rudduck *et al.* (2006) have advice to offer here. Their project used discussions with pupils to investigate personalised learning. Two of the key strategies identified by the pupils as important were oral praise and feedback, which provides immediate support and helps clarify misunderstandings, and the use of negotiated and realistic targets. Combining these contributes to a style which encourages all pupils to develop the skills of independent learning. This is considered in more depth, in the context of the 14–19 curriculum, by the Qualifications and Curriculum Authority (QCA). Task 5.3.6 asks you to consider how you might develop independent learners.

Task 5.3.6 DEVELOPING INDEPENDENT LEARNERS: ADVICE FROM QCA

QCA provides advice for the 14–19 phase (at http://www.qca.org.uk/14-19/6th-form-schools/index_s3-3-learning-teaching.htm). Reflect on this advice and use it when considering the following:

▪ When you set and clarify learning objectives, expectations and boundaries how do you share these with pupils? Do you instruct or allow pupils to construct their own understanding?

▪ How do you help pupils to acquire knowledge, skills and understanding? Do you tell them or do you ask them open-ended questions? Do you accept different answers as being of equal value?

▪ How structured are the opportunities you provide for pupils to demonstrate, practise and apply what they have learned? Who decides the format for demonstrating learning?

▪ How do you support learners in becoming independent? Is it by helping them to reflect and build on their existing learning through open-ended questions or allowing trial and error? Alternatively, do you have a 'this is how-to-do it' approach?

The questions above are open ended. They do not all have hard and fast answers. If your answers are of the 'usually I would . . . but sometimes' or 'when I started I would . . . but now I' you are beginning to adjust your style in response to your experience and to develop as a teacher. You are beginning to widen your repertoire and developing *mobility ability* in order to develop independent learners.

Unit 5.5 provides further advice on personalising learning.

SUMMARY AND KEY POINTS

You should appreciate that although any teaching style is individual it tends to be identifiable within a continuum of styles. In this unit we have tried to identify individual factors, behaviours and strategies which affect teaching and have asked you to analyse them so you can begin to place your own styles on the continuum and then, with support, widen your range. Developing a repertoire of styles to match different needs through different behaviours and the use of a wider range of strategies is important to your development as an effective teacher. Check which requirements for your course you have addressed through this unit.

FURTHER READING

Gipps, C. and Stobart, G. (1997) *Assessment: A Teacher's Guide to the Issues*, London: Hodder and Stoughton.
Chapter 2 provides a useful discussion of deep and shallow learning.

Jensen, E. (2006) *Super Teaching,* 3rd edn, San Diego CA: The Brain Store.
This is an easy-to-read book which contains many ideas which will help the development of teaching strategies and styles.

Joyce, B., Weil, M. and Cahoun, E. (2009) *Models of Teaching*, 8th edn, Boston, MA: Allyn and Bacon.
The authors identify models of teaching and group them into four 'families' which represent different philosophies about how humans learn. This is a comprehensive text designed for those who wish to deepen their knowledge of teaching and learning issues.

Street, P. (2004) 'Those who can teach – deconstructing the teacher's personal presence and impact in the classroom', *Triangle Journals*.
This paper on the impact of personal attributes on teaching was presented as a paper at a Vocational Education conference in 2004.

University of Wollongong's teaching and learning newsletter, *UniTeaching UOW*, 2 (3) (November 2001).
Provides a personal account of teaching styles.

For the 'Thinking Together' project see texts such as:

Mercer, N. (2000) *Words and Minds: How We Use Language to Think Together*, London: Routledge.
Wegerif, R. and Dawes, L. (2004) *Thinking and Learning with ICT: Raising Achievement in Primary Classrooms*, London: Routledge.

RELEVANT WEBSITES

Geoff Barton's website: www.geoffbarton.co.uk
Geoff Barton has been involved in teacher education and consultancy for the DFES, QCA and TDA. His website is well worth a visit, in particular the PowerPoint presentation 'What do we know about learning'. What are its implications for your teaching style? See his comments on teaching styles at www.geoffbarton.co.uk/files/longman/Longman%2012%20-%20Teaching%20Styles.doc

APPENDIX 5.3.1 PEER TUTORING AS A TEACHING STRATEGY

We acknowledge the contribution of Dr Lyn Dawes and Professor Neil Mercer to this section.

Peer tutoring is using one another's minds to do better together than you would do separately.

Peer tutoring is a well-established teaching strategy when setting paired and group work. Pupils of all ages have long used peer tutoring as an examination revision strategy – working together and testing each other. The advantages of peer tutoring include the active engagement of pupils with their own learning and that of others, and increased opportunities for true collaboration. Pupils learn well if supported by, and supporting, other pupils who are working at a different level of challenge. The challenges for teachers implementing peer tutoring in classroom contexts include teaching the pupils how to work most effectively in this way, organising pupils into the most suitable pairs or groups, setting tasks which offer the right level of challenge, ensuring that discussions remain focused on the learning task (scaffolding), and helping pupils to recognise the value of having access to a range of different points of view.

Research and evidence from peer-tutoring practice in schools indicates that where systematic and carefully constructed approaches to peer tutoring are applied across the pupils' school experience, there are marked gains in pupils' achievement across a number of measures. The proforma at the end of this appendix illustrates the rules governing talk which pupils are taught by teachers using the 'Thinking Together' teaching strategy. The Thinking Together website contains further examples.

As a teaching strategy, peer tutoring avoids the situation which can easily occur, where teaching is focused on the middle range of ability leaving both weak and bright pupils unengaged.

The 'Thinking Together' teaching strategy, for example, has been adopted by increasing numbers of local authorities (LAs) and schools. Pupils are given carefully designed lessons in how to talk and work together, so as to help each other to think and learn. Pupils are given guidance about how to explain their ideas clearly, how to listen, to question and to work together constructively. Teaching and learning activities across the curriculum then utilise the new skills the pupils have gained.

Lyn Dawes, who worked on the project as a teacher-researcher developing ideas and materials for ten years, comments:

> The children are learning a lot more collaboratively, and listening to each other rather than just hearing each other and they make sure that everyone in the group is involved. They feel more empowered.

There is evidence from ten years of pupil data that the 'Thinking Together' teaching strategy raises pupils' achievement. Pupils in schools which have adopted the approach have achieved significantly higher Standard Assessment Tasks (SATs) scores in maths and science than pupils in matched control schools who have not had the 'Thinking Together' training. They also achieved significantly higher scores in a test of non-verbal reasoning, indicating that the experience has helped the development of their 'thinking skills'.

The Ofsted inspection report of Eggbuckland Community College (2003) also provides examples of the success of peer tutoring. Pupils are taught skills in questioning, listening, etc. which make peer tutoring more than a conversation between two pupils. Ofsted described the strategy at Eggbuckland as one where pupils:

■ research topics individually and in groups

■ teach one another topics which they have struggled to master and freely question one another about these

■ divide up tasks in order to achieve more in a given time

■ readily share the fruits of their researches or other ideas.

All pupils have their own wireless laptops and are being taught in rooms with wireless connections. Teachers across all subject areas agreed to adopt very interactive teaching styles which use peer tutoring. Newly qualified teachers mentor student teachers.

See Table 5.3.3 for an example of a framework for supporting peer tutoring.

■ **Table 5.3.3** Thinking Together – a framework for supporting peer tutoring

Working with a partner when one of you is teaching the other

Our names are_____

The activity we are working on is_____

Learning objectives_____

Success criteria _____

How to teach	How to learn
• Ask what you can help with • Try not to take over – give time and space • Give a clue or a suggestion rather than the answer • Be encouraging and helpful • Be patient – wait and listen • Think how you can show or explain things a different way. • Can you draw a diagram, make it into a story, use an example, find a picture? • Remember that by explaining, you are learning too!	• Ask lots of questions • Explain what you do understand and say what you don't • Accept advice and suggestions in the friendly way they are offered • Be prepared to try something new • Say what you have learned • Think of ways you can show or use what you have learned • Say thank you!

• Listen attentively to each other
• Try not to get distracted
• Use your time together well

Source: DFES (2004i)

Additional resources for this unit are available on the companion website:
www.routledge.com/textbooks/9780415478724

NOTE

1   'The discrepancy between a child's actual mental age and the level he reaches in solving problems with assistance indicates the zone of his proximal development' (Vygotsky, 1986: 187).

IMPROVING YOUR TEACHING

An introduction to practitioner research, reflective practice and evidence-informed practice

Steve Bartlett and Marilyn Leask

INTRODUCTION

In order to improve pupils' learning experiences you are expected to reflect regularly upon your lessons. You consider the content, how it was taught, the involvement of individual pupils and the class as a whole and, ultimately the most important consideration, what the pupils learned. Whilst continuing to cast a critical eye over what you do, it is important that you also share and discuss your 'findings' with fellow professionals. It is by doing this that you can refine your teaching methods, discover new approaches and compare how others have tackled similar situations. Thus, evaluation and reflection is central to the development of good teaching.

Other aspects of school life, such as the pastoral care system, work experience, community service and a host of extra-curricular activities, also play a significant part in pupil learning. It is important that these aspects are also reflected upon to ensure that they meet the needs of pupils in creating an orderly and vibrant learning community. Therefore, reviewing and evaluation takes place in all areas of school life in order to make the whole experience more fulfilling for pupils and also teachers.

'How do you know your lesson went well?' is a question you can expect to be asked from time to time and you need to be able to provide answers. The fact that pupils are quiet, busy, happy or good and look as if they are working industriously is no guarantee that the learning you have intended is taking place. In this unit we introduce you to simple techniques that may help you find answers to questions about your teaching and other school activities. In carrying out the tasks throughout this book you are engaging in *reflective practice*. Rather than relying on your own opinions or superficial anecdotal observations, you should gather evidence that can be examined critically. In this way your evaluation becomes more rigorous and can be regarded as being practical research into teaching. This kind of research undertaken in collaboration with university staff can lead to M level accreditation. See the website for this book for further information (www. routledge.com/textbooks/9780415478724). The work in this unit provides a brief introduction to this area and we suggest that once you gain qualified teacher status you extend your knowledge and understanding of the tools of *practitioner research* including techniques for reflective thinking and reviewing evidence of effective practice as part of your continuing professional development (CPD). Increasingly teachers are able to access research and evidence to inform their decisions. This approach gives rise to the phrase 'evidence-informed' practice.

OBJECTIVES

At the end of this unit you should be able to:

■ demonstrate an understanding of practitioner research
■ identify different forms of evidence on which you can draw to enable you to make an informed decision concerning an aspect of practice
■ apply research strategies to evaluate and improve aspects of your teaching
■ develop your ability to reflect on practice based on evidence from research to acquire higher levels of professional knowledge and judgement.

Check the requirements for your course to see which relate to this unit.

THE PROCESS OF PRACTITIONER RESEARCH

Schön (1983) used the phrase 'reflective practitioners' to explain how enlightened professionals work in modern society. Dewey (1933) introduced the notion of 'reflective thinking'. These concepts signify how professionals, such as teachers, are able to analyse the effectiveness of their actions and to develop different ways of working as a result. Thus professionals are constantly learning about what they do and so improving their practice. Drawing upon the concepts advanced by Schön and Dewey, in addition to those proposed by Stenhouse (1975, 1983) of the 'teacher as researcher' and Hoyle and John's (1995) distinctions between 'extended and restricted' professionals, Zwozdiak-Myers (2009) identifies nine dimensions of reflective practice. You may find it helpful to keep these in mind as you work through this unit and consider how you might collect data on your own practice to provide a foundation for improvement. Figure 5.4.1 sets out these dimensions.

Investigation into their own practice by professionals themselves came to be known as 'action research', 'practitioner research' or 'teacher enquiry'. It is this link between action and research that Hopkins (2002) suggests has a powerful appeal for teachers. Thus McNiff and Whitehead say that:

> Action research involves learning in and through action and reflection . . . Because action research is always to do with learning and learning is to do with education and growth, many people regard it as a form of educational research.

> (McNiff and Whitehead, 2002: 15)

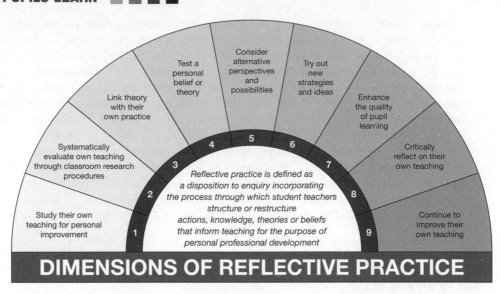

The dimensions of reflective practice shown:

Test a personal belief or theory

Consider alternative perspectives and possibilities

Try out new strategies and ideas

Link theory with their own practice

Enhance the quality of pupil learning

Systematically evaluate own teaching through classroom research procedures

Critically reflect on their own teaching

Study their own teaching for personal improvement

Continue to improve their own teaching

4 5 6 7 8 9 3 2 1

Reflective practice is defined as a disposition to enquiry incorporating the process through which student teachers structure or restructure actions, knowledge, theories or beliefs that inform teaching for the purpose of personal professional development

DIMENSIONS OF REFLECTIVE PRACTICE

■ **Figure 5.4.1** Dimensions of reflective practice, Zwozdiak-Myers (2009)

Action research is often presented as a cycle that involves the identification of a problem or issue, information or data is collected on the problem/issue, action is then planned and implemented on the basis of this improved knowledge, resulting changes are monitored and evaluated. This final monitoring and evaluation stage is likely to lead to plans for even further action and so the cycle becomes presented as a continuous process or spiral of development (see the action research models of Kemmis and McTaggart, 1988: 14; Elliott, 1991: 71). Criticisms of action research include the implication that teachers can only conduct research if they have a particular issue that needs to be solved and that the research is not built on existing evidence of effective practice. It may be, however, that a teacher uses research purely as a way to develop a wider understanding of their practice. The General Teaching Council for England (GTCE) 'Research of The Month' initiative, Training and Development Agency for Schools (TDA) Teacher Training Resource Bank (TTRB) the Department for Children, Schools and Families (DCSF) TRIPS database, and the EPPI Centre's Systematic Reviews have all been developed to produce access to evidence to underpin practitioner research. Schools have access to benchmarking data and pupil data e.g. from Centre for Evaluation and Management (Durham University) pupil tests (e.g. ALIS and YELLIS, MidYS, SOSCA as described on http://www.cemcentre.org/) tests or the GL Assessment Cognitive Abilities Tests (previously called National Foundation for Educational Research (NFER) CATS tests).

Practitioner research may be research by individuals into their own teaching and classrooms, groups of teachers researching into an issue they have noticed such as pupil achievement in a particular subject, or the school monitoring process may have identified areas such as pupil behaviour at lunchtime, truancy or pupil preparation for assessment at Key Stage 3 (KS3) as being in need of investigation. Table 5.4.1 outlines points to consider in planning a practitioner research project. This type of research, if undertaken with the guidance of university teacher educators can lead to masters-level accreditation.

Very often when new initiatives are introduced into schools, such as the development of personalised and flexible approaches to learning they are evaluated through the techniques of practitioner research and enquiry.

■ **Table 5.4.1** Planning a practitioner research project

Research focus

You need to be clear as to the focus of the research. It is a useful exercise to write a paragraph at the outset explaining what is going to be investigated and why this is seen as worth doing. When the focus is decided a number of research questions need to be devised in order to identify precisely what it is you wish to find out. These are the central questions of the research and are important as they provide you with a clear trail to follow.

Data sources

It is important to consider current research findings and what has already been written about your focus (see the further reading and websites at the end of this unit). This will add to your professional knowledge and inform your research. You then need to decide what data you need to collect and where that data might be held. The most likely sources of data for teacher researchers include pupils, teachers from the same or different schools, parents, other adults working with children, documents such as pupil reports, school records and written policies, inspection reports, government or similar publications.

Research methods

After identifying the sources you can now decide upon appropriate methods of data collection. Designing the 'tools' to gather information may at first seem daunting but this should not be the case. In your teaching you are continually using data collection skills through questioning pupils, scanning your classroom, marking pupils' work, and analysing curriculum documents. Carrying out practitioner research enables you to use and further develop these skills. Basic tools for collecting data are through interviews, questionnaires, observation and through analysis of documents.

Timeline

Now that you have a clear research focus and research questions, the sources of data have been identified and the methods have been decided upon, you can develop a timeline for the collection and analysis of data. A clear plan helps ensure that nothing important has been missed out and gives you more control over the process by fitting the research into your existing work commitments. It is particularly important to have a clear timeline when evaluating initiatives where data has to be collected at specific points. For instance, the evaluation of a curriculum initiative on the teaching of a history module, using drama techniques, to a Year 7 class, would have to be carefully planned beforehand as data could only be collected during the teaching and immediately after.

Before embarking on practitioner research you, along with other colleagues involved, need to have thought through and discussed the purpose of the enquiry. This enables you to construct a plan that identifies what is going to be investigated, how the research will be carried out and what will be the expected outcomes in terms of data and analysis. These factors can then be put in a time frame that offers targets to work towards. Whilst it is important to have a clear plan, as a researcher you must be prepared to adapt and change according to altering circumstances. In this way planning and conducting classroom research mirrors the teaching process itself. The research encompasses some of the methods you have already been using, such as observation, keeping a diary, obtaining the perspective of different interested parties (pupils, staff, parents) by the use of interview and questionnaire, and examining documents. Table 5.4.1 provides you with guidance for planning a practitioner research project. Evaluating evidence from practice and research is a professional skill. Task 5.4.1 is designed to give you practice in this.

Task 5.4.1 **EVALUATING RESEARCH AND EVIDENCE**

You need to understand data collection methods not just so that you can conduct research yourself but also to evaluate evidence from other larger scale research projects that may be used to make informed judgements on your own practice. In this way teaching becomes an evidence-based profession. Choose a research article from a major education research journal such as the *British Educational Research Journal* or from the Teacher Training Resource Bank or a systematic review from the EPPI Centre on an area of teaching that particularly interests you. How were the data in the article collected? What are the key findings and do they add to your knowledge or understanding in any way?

RESEARCH TECHNIQUES FOR USE IN THE CLASSROOM

Perhaps the most familiar methods of collecting data are interviews, questionnaires, observations, diaries and analysis of documents. The texts in the further reading provide more detail. When used these need to be carefully designed to ensure the information you want is obtained while also considering the feelings of those from whom it is being collected.

Interviewing

Interviews can take many forms depending upon whom you are interviewing, where the interview is being held and what the focus is. Interviews can be very formal, as is the case when a candidate is interviewed for a teaching post, or they can be very informal and part of an everyday conversation, as when a teacher asks a pupil from their tutor group how they are getting on with their General Certificate of Secondary Education (GCSE) coursework. Different types of interview are a normal part of a school day and teachers can become very skilled at gathering information from pupils by such methods. Consider, for instance, how you might 'interview' a pupil about an incident on the corridor between lessons, or a pupil who is finding work in a particular subject difficult. These are both instances where you employ your professional judgement and skills to find out what has been going on and what the issues are. You are then in a position to act appropriately. Teachers are also used to 'interviewing' or talking with parents when working together to aid the progress of their children. It is also helpful to ask small groups of pupils about particular issues in what can be termed 'focus group interviews'.

When conducting interviews as part of practitioner research, it is important to consider beforehand what questions need to be asked, how they will be asked and how the data will be recorded. You may wish to tape record them though it is often easier to make brief notes under key headings during or immediately after the interviews.

Questionnaires

A questionnaire is useful for surveying pupil or parental views. They can enable the collection of information from a large number of people comparatively quickly and anonymously if appropriate. In a questionnaire the wording and layout of the questions is very significant. They need to be framed so that those being asked, Year 7 pupils for instance, can understand and answer appropriately. How the completed questionnaires are collected in also needs to be considered. If they are given to whole groups of pupils it is possible to explain the purpose of the questionnaire, read out the questions and then collect them all in at the end of a lesson. Always test out the questions with a small group to check the questions are understood and that the answers are likely to be relevant to the topics being researched.

Observation schedules

An observation schedule provides a structured framework for recording classroom behaviours. Observations should be carefully planned so as to cause the least disruption to the lesson. Unit 2.1 provides information about observation schedules as does Unit 5.3. You should by now have used forms of observation schedules to observe classroom routines. It is not possible to record everything that happens in a classroom so you need to focus on, for example, a particular group or pupil or aspect of the teacher's work and record behaviour over time. It is important that you devise your own observation schedules to suit your particular purpose. Video recordings provide an additional way of recording data about classroom activities.

Paired observation

This is a streamlined procedure which enables you to obtain feedback on aspects of your work which are difficult for you to monitor. The example in Unit 5.3, of two student teachers working together with one providing feedback on the topic chosen by the other, is an example of paired observation in practice. Paired observation works in the following way: Two colleagues pair up with the purpose of observing one lesson each and then giving feedback about particular aspects of the lesson or the teaching of the person observed. The person giving the lesson decides the focus of the observation. The three stages of a paired observation are:

Step 1: You both agree the focus of the observation and what notes, if any, are to be made.
Step 2: You each observe one lesson given by the other. Your observations and notes are restricted to the area requested.
Step 3: You give each other feedback on the issue under consideration.

The cycle can be repeated as often as you wish.

Research diaries and other documents

Diaries can provide valuable data and useful records over a period of time. They may be kept by the researcher and or those involved in the research such as pupils or teaching colleagues. They can be designed in different ways, for instance, a decision may be made to write under specific headings giving short relevant pieces of information such as the subjects that give homework each night, the particular tasks set and the length of time spent on each. Alternatively the diarist may be allowed to express themselves more freely, for example explaining how they feel each day's lessons have gone and the reasons why they think this. The structure chosen depends upon the nature of the research project and is a decision to be made by the researcher often in conjunction with the participants. All sorts of documents can provide useful information to the practitioner researcher, e.g. school policies, pupil work and curriculum documents. For instance, analysis of pupil work, asking pupils to write about or draw images of a topic/issue being researched, biographical accounts written by participants, all provide useful insights into life in school and the learning that is taking place.

You need to be aware that there are in fact many ways of collecting information and it is important to be creative as well as adaptable in considering data collection. Some of the texts in the recommended reading give more detailed advice on how to design and use the methods noted above.

Task 5.4.2 is designed to give you practice in planning the kind of research project which student teachers are likely to undertake.

Task 5.4.2 **PLANNING YOUR RESEARCH PROJECT**

Many ITE courses include a small research project as an assignment. Check your course requirements with your tutor. The process is as follows.

Identify an issue or problem associated with your teaching for further investigation (e.g. challenging the more able pupils or the development of active learning techniques).

▪ Outline the focus of the research and explain why this is an important area for you to investigate.
▪ Write a number of key research questions (about four) that identify what you need to find out.
▪ Check what has been done before (use the websites at the end).
▪ List the likely sources of data.
▪ Identify methods to collect the data.
▪ Collect and analyse the data.

We suggest you discuss your findings with your tutor and other student teachers.

EVIDENCE ABOUT TEACHING AND LEARNING

The evidence available for drawing conclusions about teaching and learning can take different forms. This evidence includes:

▪ Quantitative data in numerical form that can be collected from a range of sources e.g. statistical returns, questionnaires, school management information systems. League tables of school performance are a good example of how quantitative data can be presented and used. This type of data is useful for measurement and comparison on a large scale but often lacks explanation for individual differences and can feel very impersonal.
▪ Qualitative data which are more descriptive and often include detailed personal explanations. These data are 'richer' and give a feel for particular cases but are sometimes harder to analyse and do not lend themselves easily to measurement. Such data are collected through observation, interview, analysis of documents, diaries, video, photographs, discussions, focus groups brainstorming.

Task 5.4.3 is designed to give you an overview of the performance data available in your school.

As a practitioner-researcher you need to develop a strategy for data collection that is most appropriate for the chosen research focus. This strategy often involves the use of both quantitative and qualitative data. For instance, if you were investigating a problem of pupil truancy at your school, you would be likely to require quantitative data showing the extent of truancy and any relationship this has to factors such as age, gender, pupil performance, the time of day or particular

> ## Task 5.4.3 REVIEWING PERFORMANCE DATA AT YOUR SCHOOL
>
> Schools have developed information systems that provide data showing how they are performing. Subject departments collect such data as part of the annual reviewing process and to inform future development plans. Find out about the different types of evidence and the process used in your school to evaluate the effectiveness of teaching, pupil's learning and the monitoring of individual pupil progress. Consider how this data informs the setting of future targets for pupils, teachers, subject departments and the school.

lessons. You may also want qualitative data that gives a more in-depth and personal explanation of truancy from the perspectives of truants, their teachers and parents. Using both types of data helps to give a fuller understanding of the issues.

As part of any previous school experience you may have been set assignments that involved the collection of a range of information to support your analysis. These tasks will have started with a clear focus or a question to answer, such as what routines does the teacher use in managing the work of the class? You may have collected evidence from various literature sources to help you to answer that question initially. In addition, you may have observed and made notes about what the pupils and the teacher actually did during a variety of lessons; you may have looked at the pupils' work and the teacher's lesson plans; you may have cross-checked your perceptions with those of the teacher as a way of eliminating bias, improving accuracy and identifying alternative explanations. In doing such assignments you have been involved in basic classroom research. In such research, data are gathered from different sources, checked for alternative perceptions/explanations, as exemplified in the fifth dimension of Zwozdiak-Myers' (2008) model of reflective practice, and conclusions are drawn from this information so as to develop teaching in the future. This process, whereby you approach the topic of the research from as many different angles and perspectives as possible in order to gain a greater understanding, is called triangulation. Miles and Huberman (1994) have suggested that the constant checking and double-checking involved means that triangulation becomes a way of life in such research. Burton *et al.* (2008) provide a more detailed discussion of how to improve the validity of your research project.

ETHICAL ISSUES

Teachers and other professionals working in classrooms have a duty of care for their pupils. Within this professional way of working there is a need to respect others, share information as appropriate, yet also maintain trust and individual confidentiality. There are ethical considerations to be taken into account when you are collecting data from pupils and teachers. You must be open about the purposes of your research and obtain agreement from those who are in a position to give it; your tutor may advise you to get the permission of the headteacher and others involved. You should consider the role of the pupils in your research and how much to involve them. Pupils are invariably as interested as teachers in educational improvements and the positive developments in school life that can result from such research. You need to take your responsibility in the area of ethics seriously. It is worth consulting the British Educational Research Association ethical guidelines (http://www.bera.ac.uk) for a more detailed consideration of ethical issues. Table 5.4.2 outlines the key areas to consider.

■ **Table 5.4.2** An ethical approach to practitioner research

You must take responsibility for the ethical use of any data collected and for maintaining confidentiality. Before starting, check the ethical requirements at your institution. We suggest that you should as a matter of course:

1 Ask a senior member of staff as well as the teachers directly involved with your classes for permission to carry out your project.

2 Before you start provide staff involved with a copy of the outline of your project which should include:
- the area you are investigating
- how you are going to collect any evidence
- from whom you intend to collect evidence
- what you intend to do with the data collected, e.g. whether it is confidential, whether it will be written up anonymously or not
- who the audience for your report will be
- any other factors relevant to the particular situation.

3 Consider if you need to ask for pupil and or parental consent.

4 Think about how you want the pupils to be involved in the research.

5 Finally, check whether staff expect to be given a copy of your work.

6 If you store data electronically, then you should check that you conform to the requirements of the Data Protection Act. For example, you should not store personal data on computers without the explicit authorisation of the individual.

SUMMARY AND KEY POINTS

Developing your teaching skills is one important aspect of your professional development. But other important attributes of the effective teacher that we stress in this book are the quality and extent of your professional knowledge and judgement. Skills can be acquired and checked relatively easily. Building your professional knowledge and judgement are longer-term goals which are developed through reflection and further professional development. You may find that the work required for M level work, or Doctorate in Education work provides a structure for reflection and research in your work context. The website for this book provides further information.

The World Wide Web provides easy access to educational ideas for parents and pupils. Increasingly as a professional you can expect to be asked to provide evidence to underpin your approach to education.

In this unit, we have opened a door on a treasure trove of strategies which you can use to reflect on the quality of your teaching. We suggest that you come back to this work during the year and again later in your career. The application of practitioner research to your work at that later stage opens your eyes to factors influencing your teaching and learning which you may not have known existed. Critical reflection aided by practitioner research, by individuals or by teams, provides the means by which the quality of teaching and learning in the classroom can be evaluated as a prelude to improvement.

You should now have ideas about practitioner research and how to evaluate the quality of your teaching through using a continuous cycle of critical reflection so that you can plan improvement based on evidence. If you intend to develop your research skills, then we suggest that you read several of the texts in the further reading and consult with experienced colleagues as, in this unit, we have provided simply a brief introduction into an important area of professional practice and accountability.

Check which requirements for your course have been addressed through this unit.

FURTHER READING

Baumfield, V., Hall, E. and Wall, K. (2008) *Action Research in the Classroom*, London: Sage.
This book is a good introduction to practical classroom research. It explains the process in a clear and straightforward manner and as such is very useful for the novice teacher researcher.

Burton, D. and Bartlett, S. (2004) *Practitioner Research for Teachers*, London: Paul Chapman.
This text discusses the nature of practitioner research and its importance for your professional development as a teacher. It also provides useful practical guidance on the designing of research projects and data collection.

Burton, N., Brundrett, M. and Jones, M. (2008) *Doing Your Education Research Project,* London: Sage.
The authors take you through the process of designing and conducting a research project. It is a useful text for studying research methods as part of an academic qualification or if you are seeking to conduct research in your professional situation.

Hopkins, D. (2002) *A Teacher's Guide to Classroom Research*, 3rd edn, Milton Keynes: Open University Press.
This book is concerned with the development of teachers as researchers as a means to improving classroom practice. It gives clear advice on all aspects of the research process.

Koshy, V. (2005) *Action Research for Improving Practice – A Practical Guide*. London: Paul Chapman.
This book provides practical guidance for undertaking action research in your classroom.

RELEVANT WEBSITES

British Educational Research Association: http://www.bera.ac.uk

DCSF The Research Informed Practice Site: http://www.standards.dfes.gov.uk/research/

ESCALATE: http://www.escalate.ac.uk

EPPI Centre: http://www.eppi.ioe.ac.uk

GTCE Research of the Month: http://www.gtce.org.uk/research/romtopics

Teacher Training Resource Bank: http://www.ttrb.ac.uk

Additional resources for this unit are available on the companion website:
www.routledge.com/textbooks/9780415478724

PERSONALISED LEARNING

Carrie Winstanley

INTRODUCTION

Personalised learning occurs when learning is tailored to individual needs, interests and aptitudes. The aim is to ensure that each pupil achieves the highest standards possible for their abilities, regardless of their personal background and circumstances. A further emphasis is on equipping pupils for more autonomy in their learning which is considered vital to cope in the rapidly changing world of work. The notion of personalising learning has become popular in many countries and contexts where there have been drives to raise standards and meet complex pupil needs. This is particularly apparent in England where the government (initially the Department for Education and Skills (DfES) and now the Department for Children, Schools and Families (DCSF)) has introduced a personalised learning initiative. It links closely with the *Every Child Matters* agenda (DfES, 2003a) and has been presented as a way of tailoring education to individuals, helping all pupils fulfil their educational potential.

This unit looks at what is generally understood by the concept of personalised learning. It then deals with practical considerations of personalising learning, pragmatic aspects of classroom realisation, and meeting the needs of all pupils. Personalised learning includes aspects of good practice in learning and teaching, such as the importance of knowing pupils' interests, motivations, strengths and problems and helping pupils become independent learners. The role of the teacher in ensuring consistent best practice, meeting pupils' needs and facilitating acceptable levels of achievement is therefore very important.

As well as these practical considerations however, if you are learning to teach in England it is vital to have a clear sense of the nature of education policy related to personalised learning and to have some understanding about how and why it has been introduced. This is particularly important as the personalised learning initiative has been criticised for being little more than an emphasis on existing good practice; basically a case of dressing-up old ideas in new clothes. To this end, after explaining the idea of personalised learning, the policy and some of the criticisms are introduced.

OBJECTIVES

At the end of this unit you should be able to:

- understand what is meant by personalised learning
- recognise instances of good practice in personalised learning
- be better equipped to embed personalised learning in your own teaching.

Check the requirements for your course to see which relate to this unit.

WHAT IS PERSONALISED LEARNING?

There is no single widely agreed definition of personalised learning and so there is some confusion about what is meant by the term. Thus, it is perhaps easier to say what it is not, rather than describing it in the affirmative. It is not the same as 'individualised learning' where pupils learn alone, following a solo agenda, essentially working in isolation. Likewise, it is not the development of an entirely personal programme of learning for each and every subject.

Personalised learning incorporates whole-class work and, in particular, interactive group work with interventions for pupils who need additional support. Personalising learning means shaping teaching and learning around the different ways pupils learn. The onus is on schools and teachers to ensure that what they provide in terms of the teaching, curriculum and school organisation are designed to reach as many pupils as possible by providing a wide range of learning experiences for diverse pupil needs across a unit of work. Thus, every pupil's learning is maximised in the context of the current curriculum; it does not mean that each pupil has a personalised curriculum.

The focus on personalised learning reinforces some current teaching methods in schools and classrooms as well as requiring modifications and even, in some cases, the creation of new practices. Personalising learning therefore impacts on classroom practice.

The principles and components of personalised learning

Personalised learning influences your work in the classroom and also affects pedagogy throughout the education system. The principles, a reflection of existing good practice, are:

■ Pupils' individual needs must be assessed and these needs must be a focus of attention in the classroom as well as in the broader whole-school, home and local communities.
■ Teachers should have high expectations and focus on the achievement of all pupils, providing support for pupils to achieve.
■ Schools, through suitable systems and organisation, must develop the kind of ethos and culture that supports pupils, encouraging high expectations and achievement for all.
■ Parents and carers should develop active partnerships with schools in order to help them value and become involved in their children's education.
■ Local Authorities (LAs) and government agencies (including DCSF) should provide a supportive and flexible environment to help schools achieve personalised learning aims.

Five key components of these underlying principles have been identified in the personalised learning strategy adopted in England. The first three listed are called the 'inner core' and focus on classroom practice. The last two concern 'whole-school' and 'beyond the school' issues and consequently are not the focus of the practical aspects of this unit. The key components of personalised learning are as follows:

1   Assessment for learning
    ■   setting personal targets
    ■   effective feedback
    ■   use of data to plan learning
    ■   improved transition and transfer
    ■   peer and self-assessment.

2   Effective teaching and learning
    ■   lessons in learning

- mentoring strategies
- wider teaching repertoire
- interactive, inclusive teaching programmes
- ICT across the curriculum.

3   Curriculum entitlement and choice
- pupil choice for study and learning
- models and materials for catch-up and extension
- creating time for tailored curriculum
- flexibility leading to relevant qualifications for all.

4   Organising the school
- leadership and management focus on learning and teaching
- workforce organised appropriately
- buildings facilitate personalised learning
- clear behaviour and attendance policies.

5   Beyond the classroom
- parental involvement
- learning in community context
- co-ordinated services in/out of schools to support the whole child
- network and collaborations.

Task 5.5.1 asks you to consider the impact of personalised learning for you as a teacher.

Task 5.5.1 **DEMANDS OF PERSONALISED LEARNING**

Identify some of the demands personalising learning makes of you in the context of your teaching. Use the components of personalised learning to frame your response.

Record your responses in your professional development portfolio (PDP) and discuss with other student teachers.

PRACTICAL CONSIDERATIONS

What does personalised learning look like in the school context and how have schools been adapting to personalised learning? Some ways to personalise learning include:

- preventing pupils falling behind through earlier interventions
- focusing assessment on its value for learning rather than merely for measuring
- making use of assessments and the associated data in order to plan pupils' targets
- pupils taught in smaller groups to meet their needs better (accomplished by using teaching assistants and by extending the school day)
- focus on developing learning strategies and independent learning skills
- ICT used more often and with a creative approach
- providing additional support for pupils when needed to facilitate achievement
- making better use of a virtual learning environment that is pupil-friendly and accessible off-site

- a focus on safe school environments where hindrances to learning and teaching (e.g. bullying, poor behaviour) are tackled promptly and decisively
- teachers celebrating achievement and keeping expectations of pupil performance high
- ensuring that all pupils have increased opportunities to study safely, particularly pupils from disadvantaged backgrounds (achieved through opening libraries in the evening, running supervised homework sessions etc.)
- augmenting or replacing parents' evenings with 'pupil review days' that allow personal meetings for pupils to discuss targets and progress with teachers
- shifting the way that parents and carers are involved in their child's education so they have a more active role rather than merely receiving reports of progress or problems
- reorganising school staffing to create different approaches to leadership and teaching
- dynamic approaches to grouping pupils
- relaxing the timetable to increase choice and opportunities.

The benefits for pupils

Some of the main potential advantages for pupils in terms of the skills developed through personalised learning are listed below (assembled from a range of teaching and government web sources). Pupils should:

- become more independent workers, requiring less supervision as they build project management techniques
- develop team work skills through varied small group work
- learn about the importance of being a reliable, consistent team member
- become increasingly confident in their approach to problems, building resilience and demonstrating persistence in resolving difficulties through expanded opportunities to control their own work
- improve their written and oral communication skills due to increased small group work and more contact with teachers in deeper-level conversations (rather than short answers in a class context)
- increase the responsibility for their own actions by planning their own work, not just responding to short in-class tasks and homework
- respond to resources with an evaluative and critical mindset, questioning new ideas effectively
- become more motivated, engaged and excited by their work, encouraging creative responses and deeper learning.

As a student teacher, your major concern is what this means you might do in the classroom. We now turn to some commonly recommended approaches closely associated with personalised learning that you can adopt in your classroom, with short explanations and some suggestions for further developments. Since personalised learning is such a broad concept, many of these approaches are already discussed in other units; you should therefore also refer to these other units. (See for example, Units 2.3, 3.1, 4.1, 4.3, 5.3, 6.1.)

CLASSROOM APPROACHES

We consider the following pedagogies and organisational issues here:

1 Grouping pupils and group work
2 The importance of questioning techniques

3   Task setting, problem-solving and investigations
4   Cognitive issues: accelerated learning, learning styles and metacognition
5   Goal setting and assessment for learning
6   Emotional intelligence and motivation
7   Differentiation.

1 Grouping pupils and group work (see also Unit 2.3)

Collaborative group work is central to personalised learning and so it is worth spending some time considering different options in grouping pupils. Flexibility is the key to ensuring pupils have a range of experiences and you must be willing to move groups around for different tasks and sometimes even just for a refreshing change. The main methods for grouping pupils require you to have a good knowledge of your classes, but sometimes random grouping for shorter tasks can be very effective and allow pupils to show propensities and skills that had not had a chance to shine through. (Be warned: these discoveries can be negative as well as positive!) Some suggestions for different groupings are as follows:

■   prior knowledge, established skill level, previous experience or expertise
■   preference for particular styles of learning
■   ability (determined through cognitive tests, pupil/peer referral)
■   attitude and potential effort
■   motivation and personal interest
■   friendships
■   mixed ages.

2 The importance of questioning techniques (see also Unit 3.1)

The quality of questions both to pupils and from pupils is important and the 1956 work of Bloom is a still much-used (often adapted) way of thinking about higher order questioning. Bloom's taxonomy has been criticised as well as lauded, but most agree that it is a useful guide and by ensuring that Bloom's higher level thinking is included in all lessons, more appropriate learning is likely to result. The taxonomy is generally expressed as shown below, with 1–3 described as 'lower order' and 4–6 as 'higher order':

1   Knowledge: retrieving and recalling information, naming, sequencing, recognising, listing and describing, finding and sequencing information
2   Comprehension: explaining ideas, interpreting, précis, cataloguing, organising, classifying, rewording
3   Application: adapting ideas to different contexts, implementing ideas, operating, directing concepts and knowledge to various problems and tasks
4   Analysis
5   Evaluation
6   Synthesis.

However, there are many versions of the taxonomy, some that use different language (e.g. DfES 2006b, web version) but sometimes with the order of the top two components switched so that the 'highest' skills come first. The question is whether it is a higher order skill to create or 'synthesise', or whether 'evaluation' is more complex. In practice, the priority issue is to ensure that pupils are challenged appropriately and the value of the taxonomy is measured in how it helps teachers match tasks to pupil need. The higher order questions, therefore, involve the following kinds of tasks:

Analysing:

■ classification and organisation
■ conducting a comparison
■ identifying the values, aims or motives of something
■ creating a model of an abstract idea
■ supplying evidence for an idea or view.

Evaluating:

■ prioritising evidence
■ making value judgements
■ assessing different points of view
■ creating criteria for making decisions
■ settling controversial issues.

Synthesis (creating):

■ generating ideas (measured for their originality and fluidity)
■ creating some kind of new and original outcome (could be writing a poem, designing something on paper; building an object; choreographing a dance, etc.)
■ a new combination of old ideas to create something new
■ predicting responses and making inferences
■ examining 'What if?' scenarios.

3 Task setting, problem-solving and investigations (see also Unit 5.2)

Real-life problems, interesting investigations and meaty projects in which pupils can be involved in depth are preferable to isolated, disconnected tasks and activities. These more intricate, often open-ended tasks do need time limits (particularly for homework) as pupils can end up expending too much energy on one activity. Some people have warned that real-life problem-solving scenarios could reduce the amount of content covered by pupils. Other researchers have found that this is not the case and, in fact, investigations and problems introduce greater complexity to tasks and work well with the teacher as facilitator or 'navigator' (Gallagher and Stephen, 1996).

One useful technique for helping pupils plan and execute projects is 'Thinking Actively in a Social Context' (TASC) (Wallace, 2000), a model that can be used by pupils to plan long-term project work or short-term activities. It is a flexible tool for thinking and discussing ideas and tasks. The elements are below. There is no prescribed order; the model is usually presented in a circle with no specified start or end point and there are verbs to explain what is being done with a linked trigger question.

Gather/Organise – What do I know about this?
Identify – What is the task?
Generate – How many ideas can I think of?
Decide – Which is the best idea?
Implement – Let's do it!

Evaluate – How well did I do?
Communicate – Let's tell someone!
Learn from experience – What have I learned?

The model works well for a variety of ages and there is plenty of guidance on its use on the National Association for Able Children in Education website (http://www.nace.co.uk/).

4 Cognitive issues: accelerated learning, learning styles and metacognition (see also Units 4.3, 5.1)

Many contemporary ideas about how children's cognitive processes should be harnessed have been controversial. There is little empirical evidence to support the notion of 'accelerated learning', but some teachers are devoted to using the techniques to help pupils learn and their anecdotal evidence is of great success. Similarly, there are many who feel that the best way to meet needs is to have a clear understanding about pupils' learning styles, for example to know whether visual, auditory or kinaesthetic approaches would help understanding (see Unit 5.1). This is related to the work of Gardner (1999) on multiple intelligences. Also, the idea of metacognition is to focus on how we learn and how we can improve our learning. Useful metacognitive techniques aim to help pupils analyse their own strengths and weakness and develop their self-understanding in relation to their work (see Units 4.3, 5.1 for more on multiple intelligences and metacognition).

5 Goal setting and assessment for learning (see also Units 3.2, 6.1)

Teachers should help pupils ensure they develop strategies for reviewing their own performance. Once these have been cultivated, pupils can be more involved in setting their own learning goals in collaboration with their teachers. This provides vital clarity and transparency and making the next steps explicit helps keep pupils and teachers on target. Where pupils are progressing particularly well, there is potential to compact curricula, allowing pupils to skip tasks that do not provide them with challenge.

When assessing pupils' work, you should emphasise progress and achievement, rather than failure, and avoid making unhelpful comparisons between pupils. Feedback must be constructive and should nurture learners' strategies to help them manage their work. The aim is to involve both teachers and learners in reflection and dialogue, enabling all to have learning opportunities and ensuring that pupils are recognised for their efforts.

6 Emotional intelligence and motivation

Together with understanding your pupils' metacognition and learning styles, knowledge of their emotional intelligence allows insights to their responses to tasks and assists you with pupil groupings (see Unit 4.3).

Personalised learning should make it easier to keep pupils motivated as they have been involved in choosing much of their learning (see Unit 3.2). McLean (2007) suggests that teachers employ 'energisers' in order to 'whet pupils' appetites for learning'. He highlights the following:

■ Engagement: how teachers show they are interested in and value pupils
■ Structure: provides clear pathways towards learning goals and boundaries that let pupils know what is expected of them
■ Stimulation: comes from a curriculum that highlights the relevance of activities and sets achievable goals

■ Feedback: provides information that lets pupils know how they are doing, guiding them from where they are to where they need to be.

He contrasts these characteristics with 'drainers', the polar opposites of energisers and highlights the need to understand pupils holistically, including their emotional intelligence.

7 Differentiation (see also Unit 4.1)

Central to personalised learning is differentiation in order to meet pupils' individual needs. There are many varied ways to differentiate and you need to choose what is most appropriate in relation to the required learning outcomes. Options include differentiating by adjusting the:

■ content of the task: depth and breadth of work (including its complexity or level of abstraction)
■ pace of the task: perhaps set a one-day project, or set a strict time or word limit or make more time for independent study
■ outcome, or the form of the response (oral, or video, rather than written)
■ roles within group work: change the leader, time-keeper, scribe etc.
■ amount and nature of support or scaffolding you provide
■ access to resources
■ style of the research materials: some more complex than others, some just aural etc.

See Unit 4.1 for more on differentiation.

Now complete Task 5.5.2 which provides you with opportunities to test out personalised strategies in the seven classroom approaches mentioned above.

Task 5.5.2 **PERSONALISING LEARNING IN YOUR CLASSROOM**

This is a group task in which you are asked to work with six other student teachers, each selecting one of the seven classroom approaches above. Investigate the approach further using other units in this book and a variety of other sources. Prepare a short presentation exploring the approach in depth and identifying how what you need to think about in your practical teaching.

Try some of the approaches in your teaching and evaluate their effectiveness. Present the findings to the rest of the group.

THE ROLE OF THE TEACHER AND CONTINUING PROFESSIONAL DEVELOPMENT

Since the personalised learning agenda is about consolidating and embedding good practice, there is no radical shift in the practice of teaching for already effective practitioners; any modifications are minimal; teachers still organise and manage learning, but one shift may be a need to include more of the pupils' voice in how this is realised. The teacher is not a key person delivering knowledge; rather is someone who facilitates and leads learning through more of a mentoring and tutoring role. The challenge here is to be able to monitor pupils successfully and to be confident enough to pass planning and decision-making to pupils and to do this genuinely, providing a guiding role, rather than paying lip-service to choice.

SUMMARY AND KEY POINTS

Personalised learning is not a new *idea*, although in England it became a national initiative in 2004. It involves shaping teaching and learning around the different ways that pupils learn, which has implications for teaching methods, curricula structure and school organisation.

You are likely to find that many of the elements of personalised learning are already part of what you plan to do in the classroom as they mostly emphasise ideas widely recognised as best practice. These include really getting to know your pupils, varying pedagogies in the classroom and encouraging learners to understand their own learning strategies in order to raise their achievement. One of the key challenges for all teachers is to recognise where you are already meeting the requirements of the personalised learning agenda and making these explicit.

Check which requirements for your course have been addressed through this unit.

FURTHER READING

James, A. and Pollard, A. (eds) (2004) *Personalised Learning,* London: TLRP/ESRC.

A very useful and thorough review of the agenda including reviews of related TLRP projects that feed into many of the recommended strategies for meeting pupils' needs through personalisation. A balanced view is presented with real account taken of practical concerns, a good understanding of the research evidence and some useful and illuminating case study material. The work is presented in an accessible fashion.

West-Burnham, J. and Coates, M. (2007) *Personalizing Learning: Transforming Learning for Every Child,* London: Network Educational Press Ltd.

This book is written with a teacher audience in mind and presents a useful combination of theory and practice, taking in ideas such as multiple intelligences and the restructuring of the curriculum. It focuses on the importance of personalised learning for equity, inclusion and effectiveness and includes some case studies of successful schools who have taken the agenda on board.

RELEVANT WEBSITE

TeacherNet: http://www.teachernet.gov.uk/management/newrelationship/personalisedlearning/

This website has links to many of the documents mentioned in this unit (e.g. 'The National Conversation about personalised learning') and is a very useful archive showing how the strategy has developed since its introduction in 2004. The broader TeacherNet site has further links to practical ideas with suggestions for different tasks in National Curriculum subjects at all key stages.

Additional resources for this unit are available on the companion website:
www.routledge.com/textbooks/9780415478724

BRAIN, MIND AND EDUCATION

An emerging science of learning

Jonathan Sharples

INTRODUCTION

> Only by understanding how the brain acquires and lays down information and skills will we be able to reach the limits of its capacity to learn.
>
> (Blakemore and Frith, 2005)

> I have learned that we can all engage with experts and academics about brain-science, understand it and relate it to our knowledge of teaching.
>
> (Baroness Morris, Institute for the Future of the Mind (IFOM), 2007)

The relevance of neuroscience and psychology to education and its potential impact on pedagogy is increasingly recognised by scientists, educationalists and policy makers. Our thoughts, perceptions and actions change the strength of connections between neurons in the brain, such that these physical connections come to represent our understanding and knowledge of the world around us, so called *brain plasticity*. Education, therefore, plays a crucial role in shaping the shifting balance of strengthening and weakening of connections in the brain (Ansari, 2006). It has been argued that teachers are the only professionals required to change brain connectivity in young people on a daily basis! (IFOM, 2007)

At the same time, significant insights are being revealed about developmental stages (e.g. early years, adolescence), the impact of environmental factors on the brain (e.g. nutrition, sleep, exercise) and the processes by which the brain acquires essential learning skills and abilities such as memory, attention, motivation, numeracy and literacy (Blakemore and Frith, 2005; Goswami, 2006; Teaching and Learning Research Project (TRLP) and ERSC, 2006).

BRIDGING THE GAP BETWEEN NEUROSCIENCE, PSYCHOLOGY AND EDUCATION

The growing body of scientific research into the nature of learning and education is matched by the enthusiasm of many teachers to incorporate this understanding into their experience and expertise. In a survey of 71 teachers attending in-service training, 64 teachers (90 per cent) thought that knowledge of the brain was important, or very important, in the design and delivery of teaching (Pickering and Howard-Jones, 2007). Yet despite this enthusiasm, and the immense potential for interdisciplinary collaboration between the brain-sciences and education, there has traditionally been little dialogue between these fields. As result, there are too few insights from neuroscience and psychology research informing and supporting education practice, and insufficient input from teachers into the design of educationally relevant research in the laboratory (Ansari, 2006; TRLP/ERSC, 2006; Goswami, 2006). The challenge facing translation between these fields is described below.

> You can already see where there is understanding in neuroscience blossoming, that is making contact, or could make contact with what we are doing in the classroom. There is enough knowledge in neuroscience to enhance and develop today's teaching. But it can't be done by taking something off the shelf that the neuroscientists have discovered. There has to be some system of transfer.
>
> (Howard-Jones, 2008)

One consequence of this lack of interaction is that there are an increasing number of products and ideas permeating into education predicated on unfounded scientific and educational claims such as 'Left/right brain thinking', 'BrainGym' (Goswami, 2006; IFM, 2007). Without an effective knowledge base on which to draw to appraise such technologies and ideas, many teachers feel vulnerable to being told what works from a position of unchallenged scientific authority (Pickering and Howard-Jones, 2007).

Against this background, there is a growing call to deliver cross-disciplinary initiatives that develop and assess teaching methods generated from translating the latest scientific research to classroom practices (Ansari, 2006; Goswami, 2006). Two types of multidisciplinary approaches have been highlighted as being crucial to progress:

■ laboratory studies that investigate the effect of different educational methods on the functional and structural organisation of the brain

■ classroom-based studies that develop and access teaching methods drawn from existing neuroscience and psychology insights.

So, how to proceed? The only effective way is through a balanced dialogue between scientists and educators, drawing upon the knowledge and expertise with each group. This approach empowers educationalists without devaluing established good practice that is yet to be supported by science. Existing programmes that are combining scientific expertise with teachers' experiential expertise of learning are delivering novel improvements in pedagogy that have both scientific validity and direct relevance for the classroom.

The case studies below highlight some of the exciting advances that are available when teachers engage with accessible, accurate and educationally relevant insights on 'The Learning Brain' and apply them to their practice.

'Science of Learning' professional development project

During 2008 the IFM ran a pilot professional development project with Gloucestershire Advanced Skills Teachers, exploring how engaging practitioners with some scientific perspectives on learning can develop classroom practice. The programme consisted of a series of workshops over six months investigating scientific concepts in areas relevant to learning, such as attention, memory, creativity, emotion, gifted and talented children and development disorders (Sharples *et al.*, 2008).

A key feature of this programme was allowing teachers sufficient time to go away and apply some of the ideas in the classroom and discuss with colleagues and scientists how the science might relate to practice. This very 'hands-on' method of engagement proved crucial in building a deeper understanding and practical engagement of the science. By not being prescriptive about how to apply the science, the teachers were able to employ their new knowledge in extremely diverse ways.

What kind of scientific insights were translated to the classroom? The example of 'Motor Learning' represents just a snapshot of the type of transfer that took place from science to practice; see Table 5.6.1.

Once you have considered the ideas in Table 5.6.1, try Task 5.6.1.

Task 5.6.1 **BUILDING LEARNING THROUGH MOVEMENT IN YOUR LESSONS**

For a series of lessons identify the learning you are planning for your pupils. Choose a concept that lends itself to 'Motor Learning'. Plan an activity that involves motor learning and ask a colleague to observe and make notes about the lesson in which you try this out. If the school considers what you are doing ethical – *you must check this* – teach two classes the same concept using different approaches. Test the pupils' understanding of the concept a month or two later and analyse the impact of the strategies you adopted.

An example of motor learning in action is available on Teachers TV, *'School Matters: Neuroscience, Schools and the Future'*. Available online at: http://www.teachers.tv/video/ 28021.

This work could be developed as an action research project but you will need to discuss this with your tutor (see Unit 5.4).

■ **Table 5.6.1** Motor learning

Have you ever noticed how it is often easier to remember a phone number when you have a phone in front of you to trace the movements of the numbers? What about riding a bike? Isn't it remarkable that you can jump onto a bike after 20 years and still manage to get from A to B, even if somewhat wobbly? Our memory for movement (motor skills) is extremely powerful and offers an often overlooked tool for improving learning in the classroom.

An exciting aspect of the human brain is that our thoughts, perceptions and actions mould the strength of connections between neurons, so that these physical connections come to represent our understanding and knowledge of the world around us. It therefore follows that if you can maximise the strength, and breadth, of connections associated with a particular concept you increase the chance of making that concept stick.

As educationalists, we often focus on using words and visual representations to deliver concepts in a classroom, which use only a fraction of the neural networks available at our disposal. In contrast, the motor systems of the brain occupy a large portion of our brain and include some of the most evolutionarily advanced systems. Thus, incorporating movement into the learning process offers a powerful tool for broadening the neural representation of a particular concept and an opportunity for making it stick.

Of course, the idea of motor learning is not new, however, the Advanced Skills Teachers on the 'Science of Learning' project found that an understanding of the science basis behind motor learning enabled them to explore the idea in the classroom with much more confidence. The group took this idea of motor learning and applied it in some impressively imaginative ways.

Sarah Shaw, a primary teacher from Longlevens Junior School in Gloucester, taught her Year 5 class 'The life and stages of a river' exclusively through a one-hour dance lesson. After providing the class with a pictorial representation of the water cycle, the pupils worked in groups to recreate the cycle using movement, dance and music.

Sarah recalls how the lesson really stuck with the group: 'Six months after the lesson I re-tested the class on their knowledge of the lesson. Many of the children could remember the dance explicitly and were surprised how clearly they could recall the facts. The lesson had certainly been a powerful learning experience.'

During 2004–2005, neuroscientists at the IFM worked with the South Australian Neuroscience Institute (SANI) to set up a postgraduate course for teachers, engaging them with scientific insights on learning – the Graduate Certificate in Neuroscience/Learning. Out of that course came a fantastic example of how an understanding of motor learning could lead to powerful learning outcomes.

A teacher on the course had been struggling to teach the rather abstract concept of 'negative numbers' to her primary school class. Seeing they were in difficulties, the teacher took the class into the playground, lined them up side-by-side and drew out a number line. She explained that as they walked one step forward they were to count up, and as they stepped backwards, count down. Two steps forward, move onto 2, a further step forward, 3.

'Now what happens when we move four steps backwards? Where do we go?' she asked.

Immediately, the class visualised how you could move 'below' zero and grasped the concept of negative numbers. When testing the children at the end of term she notice the class wriggling in their chairs as they tried to mentally walk out the number line once more!

Source: Adapted from Greenfield (2008)

Task 5.6.2 asks you to turn your attention to visualisation and learning.

Task 5.6.2 VISUALISATION AND LEARNING

Not only does movement activate large regions of the brain, but just *imagining* a movement activates similar brain regions as physically doing that activity. Using your own research, find out about how visualisation works in the brain and can be applied to improve learning. Write a 1,000-word article on your findings and use these notes to plan a lesson that incorporates visualisation.

Fostering creative thinking: co-constructed insights from neuroscience and education

The neuroscience and cognitive basis of creativity has been investigated over the last 10 years. Although creativity is sometimes seen as somewhat of a 'black art' that cannot be characterised, researchers in the field are beginning to understand the cognitive processes that produce creativity and the brain-activity that underlies this thinking (Howard-Jones, 2008).

Working with student teachers and teacher education staff, researchers at Bristol University explored how these emerging scientific insights on creativity might relate to creative strategies in the classroom (Howard-Jones, 2008). Sixteen student teachers took part in a short programme of seminars and activity-based workshops exploring scientific concepts about creativity, as part of an action research cycle. The research team monitored and supported the student teachers interpretation of the science to produce a series of co-constructed insights and ideas on how creativity can be encouraged in the class. Some of the ideas they produced included:

- ■ What is creativity? Where in the brain is creativity?
- ■ What do we do when we are being creative?
- ■ How does the classroom environment influence creativity?
- ■ What is a 'creative strategy' for a teacher?
- ■ What can and can't the sciences of the mind tell us about creativity?

The ideas they describe are pushing at the boundaries of creative pedagogy and promise to yield some innovative strategies for generating creativity and innovation in the classroom. The report is available online (Howard-Jones, 2008).

Task 5.6.3 asks you to consider the outcomes of this project in the light of your own teaching.

Task 5.6.3 DEVELOPING CREATIVE STRATEGIES IN THE CLASSROOM

Read the project report described above and using the ideas in the article, design a lesson that will encourage the two patterns of thinking required for creativity, *generative* and *analytical*. Consider factors such as the layout and design of the classroom, the goals and rewards on offer, the level of restraint you encourage, and the context in which you present and discuss material.

'Brainology' – transforming students' motivation to learn

A key idea to emerge from neuroscience research over the last ten years has been an appreciation of the flexibility, or plasticity, of our brains as we continue to learn and experience new things throughout our lives (Ansari, 2006). Whilst it may be clear, therefore, that education plays a crucial role in shaping the configuration of the developing brain, what is far less obvious is how an actual *awareness* of the learning brain can itself impact back on learning and education.

Dweck and colleagues have shown that what students believe about their brains, whether they see their intelligence as something that is fixed or something that can grow and change, has significant impact on their motivation, learning and school achievement (Dweck, 2008). See also Unit 4.3.

The research team have shown that those students who believe that intelligence is fixed ('fixed' mindsets) are less motivated to learn, more afraid of effort, and more likely to quit after an academic setback. In contrast, students who believe that intelligence is potential that can be developed through effort ('growth' mindsets) are much more committed to learning, show determined reactions to setbacks and ultimately show significantly improved academic trajectories (Dweck, 2008; Blackwell *et al.*, 2007).

Using this as a starting point, they have gone on to demonstrate that growth mindsets can be taught directly to children, by cultivating a malleable 'I am what I choose' attitude. Significantly, the key messages of this intervention programme were based on applying an awareness of brain-plasticity to illustrate the flexible nature of intelligence: first, learning changes connectivity within the brain, second, this process takes place throughout our lives and third, you, as an active learner, are in charge of that process. In a rigorous, randomised controlled study, an eight-session intervention workshop developed by Dweck and colleagues was shown to have a broad, catalytic effect on motivation in young adolescents, which resulted in a reversal in academic trajectories for mathematics grades (Blackwell *et al.*, 2007).

Task 5.6.4 asks you to consider learner mindset and prepare teaching strategies to encourage your pupils.

Task 5.6.4 **THE IMPORTANCE OF LEARNER MINDSET FOR CLASSROOM MOTIVATION**

Watch the video of Dweck discussing how a learner's view of intelligence influences their motivation to learn (Dweck, 2007), Then read the summary article by Dweck (2008).

Using these resources, summarise three teaching strategies you can use to encourage your pupils to develop a flexible view of intelligence.

In Task 5.6.5 you are asked to integrate the ideas raised in the tasks above into your lesson plans.

Task 5.6.5 **'BRAINOLOGY' – DEVELOPING YOUR OWN CLASSROOM INTERVENTION**

Read the summary article by Dweck (2008), as in Task 5.6.4.

Now read the original research paper, Blackwell *et al.* (2007). Note that the Appendix section at the back of the paper contains a detailed lesson plan for the intervention programme.

Based on an awareness of the learning brain as described in the resources, develop your own intervention designed to encourage a flexible view of intelligence with your pupils.

SUMMARY AND KEY POINTS

The examples described in this unit illustrate that learning not only changes the brain, but that an understanding of the brain can directly change learning. When scientific ideas are combined with educational expertise in such a way, exciting co-constructed insights can emerge that can improve learning in the classroom. The examples described here are just a taster of some of the activity taking place in the 'Brain, Mind and Education' at present. The 'Further Reading' section provides other areas to explore in this exciting and rapidly evolving field.

Check which requirements for your course have been addressed through this unit.

FURTHER READING

Blakemore, S. and Frith, U (2005) *The Learning Brain: Lessons for Education*, Oxford: Blackwell.

Levitin, D. J. (2006) *This is Your Brain on Music: the Science of a Human Obsession*, New York: Penguin.
> A recent book that covers much of the amazing recent work on how the brains creates and interprets music. The author is a professional musician and producer as well as a neuroscientist.

National Research Council, USA (2000) *How People Learn: Brain, Mind, Experience, and School*, Washington, DC: National Academy Press.
> An excellent resource put together by the Committee on Developments in the Science of Learning under the auspices of the National Academy of Sciences, USA.

Sala, S. D. (1999) *Mind Myths: Exploring Popular Assumptions About the Mind and Brain*, Chichester: Wiley.
> A collection of articles looking at the origins and true explanations of many of the common misconceptions about how our brain works (eg the 'we-only-use-10%-of-our-brain' myth).

Sousa, D. A. (2006) *How the Brain Learns*, 3rd edn, Thousand Oaks, CA: Corwin Press.
> Sousa generally presents neuroscientific information that is up to date and accurate, and assumes no prior knowledge. His focus is clearly on the classroom environment, but he is careful not to over-interpret the science. Sousa also has some more specialised titles (e.g. *How the Brain Learns to Read*, 2005), which are equally well written.

Teachers TV, *Ready to Learn*. Available online at: http://www.futuremind.ox.ac.uk/impact/related_media.html.
> Video clips of Professor Susan Greenfield, Dr Jonathan Sharples (IFM) and Dr. Paul Howard-Jones (Department of Education, Bristol University) discussing future prospects for collaborations between neuroscience and education.

The Child's Mind. Available online at: http://www.futuremind.ox.ac.uk/downloads/gloucestershire/childs_mind.pdf.
> A special edition of Scientific American Mind that contains articles on lots of educationally relevant brain-science issues (development disorders, attention, literacy strategies etc).

Times Educational Supplement (TES) *'Brain and Behaviour'* articles, (2006–2008). Available online at: http://www.tes.co.uk.

Transcripts from the 'All-Party Parliamentary Group on Scientific Research in Learning and Education' seminars. Available online at: http://www.futuremind.ox.ac.uk/impact/parliamentary.html. Access is through clicking on the article links on the page.
> This cross-party group of MPs and peers is considering how insights from cognitive science can be used more effectively to inform education policy.

Wolfe, P. (2001) *Brain Matters: Translating Research into Classroom Practice*, Alexandria, VA: Association for Supervision and Curriculum Development.
A simple, straightforward and generally accurate presentation of how the brain works.

There are many popular science magazines around now that often have good articles on various aspects of neuroscience that may be relevant to learning. First and foremost, in terms of readability at least, is *New Scientist,* which comes out weekly. There is also now a separate edition of Scientific American called *Scientific American Mind,* which is highly recommended.

RELEVANT WEBSITES

The Brain from Top to Bottom: http://thebrain.mcgill.ca/flash/index_d.html

Institute for the Future of the Mind (IFM): http://www.futuremind.ox.ac.uk/impact/

The International Mind, Brain and Education Society: http://www.imbes.org

Additional resources for this unit are available on the companion website:
www.routledge.com/textbooks/9780415478724

6 ASSESSMENT

Assessment and its reporting is a central issue in teaching and learning throughout education. In England, since the 1988 Education Reform Act national testing has taken central stage in monitoring standards in schools.

This chapter addresses the purposes of assessment, their relationships to teaching and learning and recent changes in assessment practice and reporting. The importance of formative assessment in supporting learning and raising achievement is discussed. Formative assessment is contrasted with summative assessment as 'high stakes' vs. 'low stakes'. The concepts of validity and reliability in assessment are addressed, including the tensions between them. Learning objectives which are difficult to assess but educationally important are included.

Assessment provides information about individual pupils' progress, helps you devise appropriate teaching and learning strategies and gives parents helpful information about their child's progress. Assessments are used to compare pupils and schools across the country. This chapter discusses the extent to which a test can provide all this information.

Unit 6.1 gives an overview of the principles of assessment, of formative and summative assessment, diagnostic testing and ideas of validity and reliability. The difference between norm-referenced testing and criterion-referenced testing is introduced and the nationally set tests discussed in the light of these principles. This unit shows how assessment can be used to identify progress and diagnose problems. The management of assessment is addressed. The unit compares what could be assessed with what is assessed, linking the issues to public accountability at school, local authority (LA) and national levels. It touches on the broader issue of what could be assessed, as opposed to what is assessed together with the role of assessment in the public accountability of teachers, school governing bodies and LAs.

Preparing pupils for public examinations is an important feature of a teacher's work as well as preparing them for national tests and so Unit 6.2 considers assessment as exemplified by SATs, GCSE and GCE A level. This unit links national monitoring of standards with your classroom work, raising issues of accountability. Public examinations grade pupils on a nationally recognised scale and exercise control over both entry to jobs and higher education. Unit 6.2 addresses how national standards are maintained and national grades are awarded. Recent national developments in vocational education and the current drive to improve the status of vocational education in relation to academic education is highlighted. Contrasts are drawn between the assessment methods used for vocational courses and academic courses.

Readings for Learning to Teach in the Secondary School: A Companion to M Level Study brings together essential readings to support you in your critical engagement with key issues raised in this textbook.

ASSESSMENT, PUPIL MOTIVATION AND LEARNING

Terry Haydn

INTRODUCTION

Assessment covers all those activities that are undertaken by teachers and others to measure the effectiveness of their teaching and the extent to which the pupils learned what they were trying to teach. Assessment includes not only setting and marking pupils' work, tests and examinations but also the recording and reporting of the results. Assessment is central to the process of all teaching and learning, not something that occurs from time to time at the end of a course of study.

At the end of every lesson you need to consider whether the pupils were successful in achieving the learning outcomes you planned for. This information enables you to identify what was effective, what you might do differently next time, and what to teach next.

Recent research into teachers' assessment practice suggests that intelligent, thoughtful and well-informed assessment can have a major role in ensuring that pupils achieve their educational potential (Assessment Reform Group (ARG), 2002; Black *et al.*, 2003; Clarke, 2005).

It is important to develop a clear understanding of current arrangements for assessing, reporting and recording pupils' work and progress, the purposes to which this assessment is put and the tensions which arise from these differing purposes.

Assessment is a contested area of educational policy and practice. It has a complicated and sprawling agenda making it difficult to gain a comprehensive grasp of assessment issues. This unit provides an introduction to assessment issues and carries suggestions which provide 'a grounding' in assessment. As in other areas of teaching it requires initiative, application and thoughtfulness to use assessment effectively.

OBJECTIVES

At the end of this unit you should be able to:

- identify the differing purposes of assessment and the tensions between them
- explain what is meant by 'assessment for learning'
- use correctly some important assessment terminology
- select strategies for making assessment and evaluation purposeful and manageable
- use assessment data in various ways to help pupils to make progress in their learning.

Check the requirements for your course to see which relate to this unit.

UNDERSTANDING THE VOCABULARY OF ASSESSMENT

Assessment has become a much more technical and complex issue in recent years. Part of acquiring 'a grounding' in assessment issues is understanding the terms commonly used. For example, a distinction is often made between 'formative' and 'summative' assessment: are you clear about what is meant by these terms? The term 'value added' is often used about assessment data, what does this mean? What does it mean to say that an assessment is 'valid' or 'reliable' or that an assessment has been 'criteria referenced' rather than 'norm referenced'? You may also encounter the phrases 'high stakes' and 'low stakes' testing. A substantial section of this unit is devoted to formative assessment, or 'assessment for learning' as it is sometimes termed. For an explanation of the terms used in this paragraph, go to http://www.uea.ac.uk/~m242/historypgce/assess/welcome.htm.

Turning to Figure 6.1.1 you could ask yourself the questions listed there about a piece of work you are going to mark. It is important to be clear about the purpose of your marking because you spend much time and effort assessing pupils' work.

Assessments serve a variety of purposes and sometimes tensions arise because of this. It is difficult to devise 'all-purpose' assessment instruments which serve all these purposes. One common tension is whether the assessment is more useful for supporting pupil learning ('formative' assessment)' or for making judgements ('summative' assessment).

One purpose of assessment is to show pupils that you take an interest in their work; this can help to motivate pupils, but does not necessarily have this effect (see the section in this unit on feedback to pupils). Assessing pupils' work can also give you some idea about how well the pupils are doing, monitoring their progress. It can sometimes ascertain whether pupils have particular difficulties in their learning which might be alleviated with specifically targeted interventions and assistance once diagnosed (see also Unit 4.1 and differentiation). Assessment data on a pupil's prior attainment can also help to detect pupil underachievement, where a pupil may be 'coasting', or going through 'a bad patch' for some reason, which can then be investigated. Under the terms of the 1988 Education Reform Act for England and Wales (ERA, 1988), schools also have a statutory responsibility to report to parents about pupils' progress at least once a year, and more regular communication between the school and parents (for example, through a weekly diary system) can be a valuable way of safeguarding pupils' academic and social well-being. As well as reporting on achievement, many schools also use assessment data to make comparisons between pupils or to 'group' pupils into particular teaching sets. Some pupils are motivated by competition, and try harder to do well and 'come top'; however, some commentators suggest that 'we should spend less time ranking children and more time helping them to identify their natural competences and gifts and cultivate these' (Gardner, 1986). Assessment is not just about gaining information about pupil performance: teachers can use assessment data to reflect on and evaluate *their own* performance; to gauge which elements of learning they taught effectively, and which less so. Assessment data also makes it possible for teachers, schools and local authorities (LAs) to be held to account for their performance. Task 6.1.1 invites you to review the evaluation of some recent lessons.

Why am I marking this?
Why do we have assessment?
Why are there so many different forms of assessment?
Is this assessment activity helping my pupils to learn more effectively?
Is this assessment activity helping me to teach more effectively?

■ **Figure 6.1.1** The purposes of assessment: why do we assess pupils' work?

Task 6.1.1 SETTING TASKS TO ASSESS LEARNING

Part of the challenge of becoming an effective teacher is devising appropriate and helpful learning experiences for pupils. In most lessons, at some stage the pupils are 'doing something' other than simply listening to the teacher. Some benefits to pupils undertaking a lesson task include:

a   enthusing the pupils, making them keen to learn and find out more about the learning problem

b   moving pupils forward in their learning

c   providing information for the teacher on what the pupils have learned, (which bits of the topic they understood and which they didn't fully grasp)

d   providing information to the teacher about *why* the pupils failed to understand aspects of the topic, the mistakes they made or the misconceptions they hold

e   enabling the teacher to know which pupils have understood the topic, and which have not

f   enabling pupils to take a particular move forward in their learning; the activity has made a particular point in a powerful way, changing pupils' ideas about the topic

g   providing information for parents about how their child is progressing, and how well the pupils are likely to do in an external examination.

Select one class you teach and then some tasks you have given pupils to do in recent lessons.

1   Which of the benefits described above have been a feature of the tasks? (You may wish to identify other benefits). Was there a task which delivered all these benefits or was it difficult to devise a task which incorporated all the benefits described?

2   Over a period of several lessons with a class, do your tasks address all the points described above or to they tend to fall into a limited band within the list?

3   Write brief notes summarising your analysis and the implications this may have for planning future tasks for your class.

UNDERSTANDING AND TAKING ACCOUNT OF THE LIMITS OF TEACHING EFFECTIVENESS

A common mistake made by student teachers is to assume that because they have explained certain concepts, and pupils have undertaken activities intended to develop their understanding of these concepts, pupils will have learned what the student teacher was trying to teach. It is quite common for student teachers to write in their lesson evaluations that all the learning objectives for the lesson were achieved. And yet, in how many lessons do all the pupils learn everything that the teacher is trying to teach? What percentage of what is taught is learned in classrooms, in the sense of being understood, retained and, where appropriate, applied after the lesson? As Fontana noted, 'we each of us receive a constant and varied stream of experiences throughout our waking moments, each one of which can potentially give rise to learning, yet most of which apparently vanish without trace from our mental lives' (Fontana, 1993b: 125). The tendency to overestimate the extent to which teaching has been effective is not limited to student teachers. Research has been undertaken to discover experienced teachers' views on what proportion of pupils understood the concepts that they were trying to teach before their lessons and what proportion understood these concepts after the series of lessons (Sadler, 1994). Some outcomes of this research are given in Figures 6.1.2 and 6.1.3.

Before teaching		After teaching	
Teachers' predictions	Actual	Teachers' predictions	Actual
25%	30%	60%	15%

■ **Figure 6.1.2** Percentage of pupils with a basic understanding of the concept of gravity before and after teaching

Before teaching		After teaching	
Teachers' predictions	Actual	Teachers' predictions	Actual
25%	30%	60%	15%

■ **Figure 6.1.3** Percentage of pupils with a basic understanding of the concept of planetary motion before and after teaching

Sadler suggests that teachers often disturb pupils' ideas without decisively changing them, with the result that they are often confused by the lessons, with teachers then moving quickly on, in order to cover an overloaded curriculum. Teachers do sometimes ask the class whether everyone has understood what they have just talked about, but how many pupils are confident enough to put their hand up to say they do not understand something?

It is because of these 'learning deficits' that assessment is so important. It is helpful for you to know which pupils have grasped what you were trying to teach, and which have not; which aspects of the topic pupils did understand, and which aspects passed them by. This information helps to inform both your next lessons with the class, and how you approach planning and teaching the lesson next time round. This idea of a 'feedback loop', whereby your assessment of pupil learning informs your future planning is represented in Figure 6.1.4.

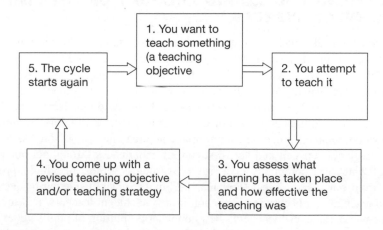

■ **Figure 6.1.4** The planning/teaching/assessment/planning loop

Thus when planning a series of lessons, the learning activities must be backed up by appropriate assessment strategies. However Kyriacou remarks:

> This does not mean that every lesson must have some in-built testing of attainment: rather, a more subtle form of ongoing probing and reviewing should be employed which will be sufficient to enable you to feel confident that the intended learning is occurring. Nevertheless, there is a role here for formal tests from time to time and also the use of homework to explicitly probe the learning covered.
>
> (Kyriacou, 1998: 25)

One of the most difficult tasks which you face on teaching placement is the challenge of writing consistently useful and insightful evaluations of your lessons, in a way that demonstrates over time an increasingly sophisticated understanding of your own learning and that of your pupils. If lesson evaluations do not focus in a genuine and thoughtful way on the full range of factors influencing pupils' learning and your own learning, they can become a formulaic, repetitive and dreary chore. Evaluations can be very general, vague and descriptive, almost like a personal diary of what went on in the lesson reporting, for instance that pupils were 'on task', 'behaved well' and that the lesson 'went OK'. But thoughtfully done, they can provide an insightful analysis of the learning that occurred, possible explanations or hypotheses for learning 'deficits' and demonstrate rigorous and perceptive self-awareness of your own developing strengths and weaknesses as a student teacher. See Task 6.1.2 on the following page.

One of the main purposes of your lesson evaluations should be to assess the extent to which the pupils learned what you were trying to teach, although there is also a place for reflecting on what you learned from your experience of teaching the lesson. Comments might include things such as:

- ■ What do I need to think about for the next lesson with this class in the light of this lesson?
- ■ How appropriate was the material presented to pupils?
- ■ To what extent were learning objectives achieved?
- ■ Were the learning outcomes what I expected?
- ■ What might be changed to improve the lesson?
- ■ Now I know what they've learned, what do I teach next?
- ■ Why did the class (or individual pupils) respond as they did?
- ■ What was it that made it a good/bad lesson?
- ■ Which bits worked well and which bits didn't?
- ■ What have I learned from the experience of teaching this lesson?
- ■ Did I learn anything useful about the learning of particular pupils within the group?
- ■ To what extent did I provide access and challenge to all pupils in the class?
- ■ Which of the requirements for my course have I made progress in/am I struggling with?

Over a series of lesson evaluations, it can be helpful to aim at a balance of reflecting on your own learning, and that of the pupils. Vary the focus of your evaluations. Task 6.1.2 addresses aspects of lesson evaluation.

Assessment for understanding and the importance of knowledge application

The way that you formulate a question for pupils can have a significant influence on the chances of pupils responding with a correct answer. Several examples of this are provided in Unit 3.1. How

> Task 6.1.2 **THE ROLE OF LESSON EVALUATIONS IN THE ASSESSMENT OF LEARNING**
>
> Student teachers often evaluate their lesson before they have marked pupils' work and done 'recap' questioning or 'check-up' testing at the start of the next lesson. With at least one lesson evaluation, make a conscious attempt to refrain from writing it up until you have marked pupils' work arising from the lesson, and given them either an oral or written test on the previous lesson next time you teach them to assess how much knowledge and understanding of the lesson they have retained.
>
> Write a summary of your findings and use it to identify the advantages and disadvantages of delaying your evaluation of the lesson. Discuss your findings with your tutor.

useful is it if pupils only answer the question correctly if you ask them in a particular way? That is to say, if you test for understanding with exactly the same question, and the same context as in the way you explained it to them, rather than asking them in a slightly different way? (Do they in effect, just give you back what you told them, without full understanding?). Do they make connections with contingent aspects of the topic? Which bits have they grasped, and which bits are still unlearned or only partially understood? Only if you have an accurate idea about pupils' current understanding of the topic can you make intelligent decisions about what to teach next. In other words 'If you want to give someone directions, one of the first questions you would ask is "Where are you now?"' (Driver, 1983: 42). It is therefore helpful if teachers probe to test that pupils have fully understood what they are trying to teach. This is not because we are trying to catch them out, it is because we really want them to learn. What is important is not just 'covering content': 'Helping pupils to make sense of what they are learning is the most important thing' (Clough, 2001).

Assessment should be seen by pupils as part of their learning, rather than just a check that they have learned, and that should test real learning rather than simply regurgitation of information (Harlen, 1995: 14). Learning implies understanding which can be demonstrated by using new knowledge; as one writer has put it 'The chasm between knowing X and using X to think about Y' (Wineburg, 1997: 256). This problem is one of knowledge application:

> Schools worldwide are relatively successful at teaching bodies of knowledge, but they are much less good at getting pupils to be able to use and apply this knowledge in new settings or in problem solving . . . This failure to apply acquired skills is evident in all areas of the curriculum. It is a long-standing challenge to educational systems everywhere.
>
> (Desforges, 2002: 4)

Holt gives some helpful pointers of ways of assessing genuine understanding, which might lead to a greater likelihood of pupils being able to apply the knowledge they have gained to solve problems and use in other contexts (see Table 6.1.1).

UNDERSTANDING PROGRESSION IN YOUR SUBJECT AND BEING ABLE TO SEE 'THE BIGGER PICTURE'

To improve your assessment of pupils you need to have a clear grasp of progression in your subject, that is, to know what it means 'to get better at geography' (or whatever). There is more to getting better in school subjects than simply acquiring more factual knowledge. For example in history,

■ **Table 6.1.1** What it means to understand something

Holt writes: 'It may help to have in our minds a picture of what we mean by understanding. I feel I understand something if I can do some, at least, of the following:

•   state it in my own words

•   give examples of it

•   recognise it in various guises and circumstances

•   see connections between it and other facts or ideas

•   make use of it in various ways

•   foresee some of its consequences

•   state its opposite or converse.

The list is only a beginning, but it may help us in the future to find out what our students really know as opposed to what they can give the appearance of knowing, their real learning as opposed to their apparent learning'

Source: Holt (1964: 176)

there is more to progression than simply 'knowing more stuff'. As well as developing the depth and breadth of their historical understanding, pupils should make progress in their understanding of the discipline of history, their ability to make intelligent historical enquiries themselves, and their ability to communicate their understanding orally and in writing. We are not just filling up pupils' 'hard disk' space with facts, we are helping to develop their ability to acquire, process and communicate subject related information effectively (see Table 6.1.1).

How are you going to make intelligent assessments of your pupils' progress if you do not have a sound grasp of what it means to get better in the subject you teach? Understanding progression in your subject is an important prerequisite for being able to use assessment in a focused and appropriate way in your teaching.

Now consider Task 6.1.3.

Task 6.1.3 **DEVELOPING AN UNDERSTANDING OF PROGRESSION IN YOUR SUBJECT**

If you were asked at interview what it means to get better at history, geography or whatever your subject is, how well would you be able to answer this question? If you are not confident that you have a sound understanding of progression in your subject, read the specifications for your subject in the National Curriculum (NC) and in the NC attainment targets for each subject. In addition, identify research and inspection evidence related to progression in your subject, in the form of Office for Standards in Education (Ofsted) and Qualifications and Curriculum Authority (QCA) reports, and journal articles. Give examples of progression from two or more topics you expect to teach or have taught, indicating, for example, development across Key Stages 3 and 4.

You need to be aware of the full breadth of benefits which pupils might derive from the study of your subject and from being in your classroom. All NC subjects provide opportunities to promote pupil learning in a range of areas beyond subject specific learning, e.g. to contribute to the development of pupils' 'key skills'. These are defined in the NC for England as communication, application of number, use of information communications technology (ICT), the ability to work with others, the ability to problem solve, and the ability to improve their own learning performance (QCA, 2008k).

There are opportunities in your lessons to develop other aspects of pupils' education, in areas such as citizenship, aspects of personal, social and health education (PSHE), cultural education and work-related learning. Sometimes student teachers lose sight of these broader learning agendas. Keep in mind the question 'what is the full breadth of benefits which this pupil might derive from the study of my subject and being in my classroom?' and try to ensure that your assessment practices reflect this breadth (see Figure 6.1.5).

In an effort to draw attention to the wider dimensions of learning, the Education Committee of the Inner London Education Authority (abolished in the 1980s) drafted a broad set of aims for schools called Aspects of Achievement (Figure 6.1.6). The aims were published in an attempt to counter the emphasis placed on examination success to the exclusion of other important pupil achievements.

Hargreaves' idea should be a strong reminder to teachers that there should be more to assessment than just measuring the cognitive development of pupils.

In 1999, for the first time, the NC for England included a statement of its overall values, aims and purposes (Department for Education and Employment (DfEE)/QCA, 1999b: 10–13). Although modified since that date the current aims are very similar and include the following statements:

> The school curriculum should develop enjoyment of, and commitment to, learning as a means of encouraging and stimulating the best possible progress and the highest attainment for all pupils. It should build on pupils' strengths, interests and experiences and develop their confidence in their capacity to learn and work independently and collaboratively. It should equip them with the essential learning skills of literacy, numeracy, and information and communication technology, and promote an enquiring mind and capacity to think rationally.
>
> . . . the curriculum should enable pupils to think creatively and critically, to solve problems and to make a difference for the better. It should give them the opportunity to become creative, innovative, enterprising and capable of leadership to equip them for their future lives as workers and citizens.

(QCA, 2008b)

Our experiences have shown that some student teachers have only a limited awareness of this component of the NC. Task 6.1.4 explores further the aims of the English NC.

Task 6.1.4 **TO WHAT EXTENT DOES YOUR ASSESSMENT PRACTICE TAKE INTO ACCOUNT THE VALUES, AIMS AND PURPOSES OF THE NC FOR ENGLAND?**

Read the section of the NC which explains the values, aims and purposes underpinning the NC (see reference above). Consider the extent to which your assessment practice contributes to the delivery of these aims. How might your assessment of pupils be improved in order to take account of these aims? Discuss this aspect of lesson planning with your tutor.

> What will be the effect of assessment on the motivation and attitude of pupils who are very weak in my subject if assessment focuses entirely on pupil attainment in my subject?

■ **Figure 6.1.5** The influence of restricted assessment practice on the motivation of less able pupils

"We offer these four aspects (of achievement) in order to clarify our own and our readers' thinking about achievement and we believe that such a scheme is appropriate to the educational aims of comprehensive schools and secondary school teachers."

Aspect I
That aspect most strongly represented in GCSE: written expression, capacity to retain propositional knowledge and select from it in order to answer questions. Examinations tend to emphasise knowledge more than skills; memory more than problem solving or investigation; writing more than other forms of communication; speed more than reflection; and individual more than group achievement.

Aspect II
The capacity to interpret and apply knowledge. The emphasis here is practical rather than theoretical; oral rather than written; investigation and problem solving rather than recall. This aspect forms one part of public examinations, but which is rarely important.
It is more time consuming and expensive to assess than Aspect I.

Aspect III
Personal and social skills. The capacity to communicate with others face to face; to work co-operatively in the interest of a wider group; initiative, self-reliance and skills of leadership. This is not an element assessed directly by traditional public examinations, including the GCSE.

Aspect IV
Motivation and commitment; the willingness to accept failure without destructive consequences; readiness to persevere and the self-confidence to learn despite the difficulty of the task. This is often seen as a prerequisite of achievement rather than achievement in its own right, and yet motivation can be seen as an achievement in most walks of life. It is in many ways the most important aspect as it can affect outcomes in the other three.

■ **Figure 6.1.6** Aspects of achievement
Source: Hargreaves (1984: 2)

There is a danger of assessing things that are comparatively easy to assess (or which are assessed in public examinations), at the expense of assessing learning experiences that are valuable and worthwhile, but difficult to assess.

Some skills are relatively straightforward to identify, others are harder to describe and identify. Thus recall is often referred to as a lower-order skill and may refer to recall of facts, of knowledge 'of' and knowledge 'how'. Lower-order skills are important and form the foundation on which many higher-order skills function. Being able to *apply knowledge* moves skills up a level but even here we can distinguish applying knowledge to familiar situations, or taught situations, from the context of a new situation which is harder for a learner, a higher skill level (Bloom, 1956).

The analysis of data or of situations and the synthesis of information into explanations of events both require even higher-order skills. The use of imagination, being inventive, thinking laterally are other ways of looking at these higher-order skills, all often difficult to assess.

There is often a tension between validity and reliability when making assessments of learning. It is easier to test for lower-order skills than for higher skills. For example, testing a pupil's ability to do simple addition is comparatively straightforward; a pupil is either correct or incorrect and the test has high reliability and validity. Assessing the ability to develop reasoned judgements is complicated, longer and becomes more subjective in its assessment.

The key skills of the current NC in England are a good example of this tension. Progress in what have been termed the 'hard' key skills, namely application of number, communication and use of information and communications technology (ICT) is comparatively straightforward. There are a range of basic tests which can be devised to measure these skills, such as those used in England to test student teachers' skills in these areas, in order to gain Qualified Teacher Status (QTS). It is much harder to devise simple tests in the area of the 'soft' key skills – problem solving, working as part of a team and 'learning how to learn' (metacognition). It can be difficult to provide individual assessment of pupils where they have been working collaboratively on a group task, and yet co-operative learning has been identified as a very effective way of moving pupils forward in their learning (Kagan, 1994). The danger is that you (and the school system generally) can drift into a tendency to assess only that which is easy to assess, or which is directly tested in public examinations, rather than assessing the full range of educational aims outlined in the values, aims and purposes of the NC in England (QCA, 2008b).

The impact of assessment on pupil motivation

There is evidence that some assessment practice may be harmful to pupils' learning, that testing pupils from the age of five onwards is having a negative effect on some pupils' self-esteem and motivation to learn (Black and Wiliam, 1998). Other evidence suggests that test results can influence friendship groups, with high-achievers rejecting less successful pupils at primary school level (Bloom, 2007). Assessment can be valid and reliable and yet have a profoundly dispiriting influence on pupils' attitude to a subject and to school in general. It has been argued that the ways in which some assessment has been developed in recent years has contributed to disaffection and disengagement in schools (Elliott, 1998; Macdonald, 2000). Others have indicated that there is a danger that 'pupils who see themselves as unable to learn usually cease to take school seriously, many are disruptive within school, others resort to truancy' (Black and Wiliam, 1998: 4). As Holt points out, pupils 'at the bottom of the pile' often do not respond by exhibiting determination to get to the top:

> Most people . . . understand it [education] as being made to go to a place called school, and there being made to learn something that they don't much want to learn, under the threat that bad things will be done to them if they don't. Needless to say, most people don't much like this game and stop playing as soon as they can.
>
> (Holt, 1984: 34)

Although Holt wrote this in 1984, might there still be some truth in this assertion today? You might give a moment's thought to what a day in school is like for a pupil who is weak in all subjects, and the role that assessment plays in this pupil's life.

In how many classes are all the pupils 'good learners' who are doing well in the subject and who have good learning habits and attitudes?

In how many classes are all the pupils trying their best to do well in the subject and working hard to learn everything that the teacher is trying to teach?

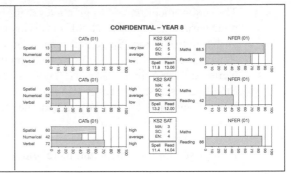

■ **Figure 6.1.7** A day in the life of a pupil (see Figure 6.1.9 for a larger version of the Year 8 table)

If you can, spend a day with a class as they move from subject to subject and consider the question in Figure 6.1.7 in the light of your experience.

ARG found 'strong evidence of the negative impact of testing on pupils' motivation', qualifying this with the caveat that 'this varied in degree with the pupils' characteristics and with the conditions of their learning' (ARG, 2002: 2). Recent research on assessment has focused on how to minimise the potentially harmful effects of assessment on pupil motivation and progress. This research is discussed in the following sections of the unit.

ASSESSMENT FOR ACCOUNTABILITY AND ASSESSMENT FOR LEARNING: RECENT DEVELOPMENTS IN ASSESSMENT PRACTICE IN THE UK

There has been a revolution in assessment policy in the UK over the past two decades. You need to understand the implications of these changes, and in particular the *tensions* between different aspects of assessment. Task 6.1.1 introduced you to the tension between assessment for accountability and assessment for learning. This section provides a context for current emphases in assessment policy. It is important for new teachers to understand *why* there has been a move away from the assessment arrangements which operated prior to the inception of the NC in England in 1991. Assessment has been one of the battlegrounds of education policy over the past two decades, and the comparative importance attached to assessment for accountability and assessment for learning has fluctuated over this period.

Public accountability in the pre-National Curriculum (NC) era in England and elsewhere in the UK was in part through school reports which were limited in depth and detail and often quite vague (see http://www.uea.ac.uk/~m242/historypgce/assess/sixties.htm). This vagueness was one of the factors which led to the introduction of a NC for England and Wales, more rigorous and detailed in specifying and measuring learning outcomes for each school subject.

Policy makers saw the introduction of more precise assessment arrangements as a way of 'driving up' educational standards by making outcomes more transparent and making teachers, schools and LAs more accountable to the public purse. This accountability required extending the scope of standardised methods of measuring educational outcomes right across compulsory education, instead of limiting it principally to external examination results at the ages of 16 and 18.

In England, each NC subject has a series of statements, called Attainment Targets, which set out in general terms the expected performance of pupils at different stages as they progress through the eleven years of compulsory schooling. These stages are termed 'levels' and describe progression.

The requirement now is to give a level for every pupil at the end of each Key Stage, but there are wide variations between schools in the way they report pupils' progress, with some schools reporting at the end of a Key Stage and other schools requiring departments to report more frequently. In England testing takes place in English, mathematics and science at the end of a Key Stage, called National Curriculum Tasks (NCTs) (sometimes called National Assessment Tasks and formerly Standard Assessment Tasks (SATs)). Since 2008 statutory testing at the end of Key Stage 3 has been abandoned but schools retain the option of using National Curriculum Tasks (National Assessment Agency, 2008).

In Task 6.1.5 you are asked to asked to explore the use of levels to report on pupil progress in your school placement and more widely.

Task 6.1.5 **TEACHER ASSESSMENT OF PUPIL LEVELS OF ATTAINMENT**

This task has two parts:

1   The use of 'levels' to describe pupil attainment is an important element of the current NC system of assessment. Teacher assessment of pupil progress in the levels of attainment for each school subject is an important part of current assessment arrangements. Find at least three articles or chapters from the recent literature on assessment which include discussion of 'levels', and make brief notes on their comments about the use of the 'levels' system. Use these notes as a basis for discussion of the levels system with your tutor. What are the views of the teachers you work with about 'best use' of the levels system, and to what extent do these reflect or differ from the views expressed in the documents you have annotated? After discussion, summarise your findings and file the summary in your professional development portfolio (PDP).

2   What are the arrangements for recording and reporting teacher assessment of pupils' levels of attainment across school subjects in your placement school? Identify the advantages and disadvantages of recording and reporting on pupils' progress through level statements only at the end of a Key Stage, rather than on a monthly, half-termly, termly or yearly basis. Write up your findings to discuss with your tutor. After revision, file the document in your PDP.

In England the 'levels of attainment' in English, maths and science, (together with public examination results), are used to make comparisons of pupil performance across the curriculum. These same data are used to compare departments and teachers in school, schools within an LA and LAs nationally. This process is an example of one assessment instrument being used to measure many different educational factors.

Teachers in England are now in possession of much more information about their pupils' progress than was the case in the past and in-school assessment of pupil progress has become a much more precise, detailed and sophisticated process than was the case twenty or thirty years ago.

However, commentators argue that the statistics are flawed, that this form of comparison is 'not a fair contest' between schools, and that the increased emphasis on testing of pupils has had harmful effects on pupils' self-esteem, well-being and attitudes towards school and learning (ARG, 2002; Bawden, 2007; Crace, 2007; Russell, 2007).

Many teachers and educationalists express concern at some of the 'unintended outcomes' which derived from these changes to assessment systems in UK schools. They point out that time spent on testing pupils (and preparing for the tests) means that too much time is spent 'weighing the pig instead of fattening it', taking time away from teaching and learning. There is also the danger that teachers might be tempted to 'teach to the test' and concentrate on what is to be tested and what is easy to test rather than teaching the full breadth of the curriculum (Stobart and Gipps, 1997; Wiliam, 2001; Torrance, 2002). Until 2008 in the UK, only England required all pupils to be tested by a common examination at the end of Key Stage 3 (formerly called Standard Assessment Tasks), accompanied by the publication of those results comparing schools and LAs.

How are you to handle the very different views about current assessment arrangements? Whilst you are entitled to have a constructively critical opinion of the helpfulness of current assessment arrangements, your professional responsibility is to make the best of whatever system operates at the time. Inspectors report that some teachers, departments and schools make better use of assessment than others in order to improve the learning outcomes of their pupils (Ofsted, 2003b): the same could be said of student teachers.

Most teachers and many schools wrestle with the problem of how to make best use of assessment, how to ensure that the time and effort involved in assessing teaching and learning is not wasted, or even counter-productive. Assessment is an aspect of education where research has had a significant influence on policy and practice. As later sections of this unit indicate, there is a body of research evidence available to teachers and student teachers to inform their practice in assessment and to ensure that they make the most of the potential of assessment to improve pupil attainment. This research may help teachers avoid some of the pitfalls and 'collateral damage' that assessment practice can occasion.

Why is assessment for learning important?

There is ongoing concern and debate about some aspects of the current arrangements and emphases in assessment and testing, and there is a consensus that the development of formative assessment, or 'assessment for learning' as it is sometimes termed, is now one of the most important agendas in education policy and practice.

Several research studies have shown that the use of assessment *to develop pupils' future learning* makes a substantial difference, not just to pupils' attainment, but to their attitude to learning, their engagement with school subjects and their motivation to do well in these subjects (Black and Wiliam, 1998; Murphy, 1999; Black *et al.*, 2003).

The purposes of assessment can be categorised as assessment for *accountability*, assessment to provide *certification*, and assessment for *learning* (Black *et al.*, 2003). These authors argue that assessment for learning is not currently used to best effect in many schools in the UK, and that the development of assessment for learning could do more to improve educational outcomes than almost any other investment in education.

Teachers have to spend many hours assessing, recording and reporting pupils' work. You want to make sure that this investment in time spent marking pupils' work and giving feedback on their progress justifies the time and effort you spend on it. It is not a question of how much assessment of pupils' work you do, or even how *much* feedback you provide, but how *intelligently* you use assessment to inform your future teaching, and your feedback to pupils. Teachers who can use assessment to promote pupils' learning 'come to enjoy their work more and to find it more satisfying because it resonates with their professional values. They also see that their students come to enjoy, understand and value their learning more as a result of the innovations' (Black *et al.*, 2003: 3).

You should be aware that there are some tensions between 'assessment for learning' and other functions of assessment, and to be clear about exactly what 'assessment for learning' means.

What does 'assessment for learning' mean?

One definition of assessment for learning is 'any assessment for which the first priority is to serve the purpose of promoting students' learning.' (Black *et al.*, 2003: 2). An assessment activity can help learning 'if it provides information to be used as feedback by teachers, and by their students in assessing themselves and each other, to modify the teaching and learning activities in which they are engaged. Such activity becomes *formative assessment* when the evidence is used to adapt the teaching work to meet learning needs' (Black *et al.,* 2003: 2).

Some features of teaching methods which incorporate assessment for learning are listed in Table 6.1.2 and their characteristics discussed below.

■ **Table 6.1.2** Teaching methods which support 'assessment for learning'

1   The integration of a variety of assessment instruments into day-to-day classroom activities
2   Sharing learning goals
3   Self and peer assessment
4   Questioning and dialogue with pupils
5   Feedback

The integration of a variety of assessment instruments into day-to-day classroom activities

An assessment should be a worthwhile learning experience for pupils rather than simply an administrative task to measure what they have learned. In Gardner's words:

> It is extremely desirable to have assessments occur in the context of students working on problems, projects or products which genuinely engage them, which hold their interest and which motivate them to do well. Such exercises may not be as easy to design . . . but they are far more likely to elicit a student's full repertoire of skills and to yield information that is useful for subsequent advice and placement.
>
> (Gardner, 1999: 103)

Assessment should be deliberately and thoughtfully 'planned into' the series of teaching and learning activities for the topic. This brings us back to the idea explained earlier in the unit of 'the planning loop' (Figure 6.1.2), which encompasses your learning objectives, how to achieve them, teaching, assessing and evaluating, and from which a revised set of learning objectives emerges. When you have taught for a while, this process is informed by your previous experiences of teaching the topic, and you make adjustments to your planning in the light of what seemed to work well or not work well on the previous occasion.

Another advantage of assessment tasks that are seen as routine and low-key classroom activities is that this avoids the negative effects which 'high stakes' testing has on some learners. It has been suggested that pupils are most likely to engage with learning tasks if they are a combination of 'low stress' and high challenge – difficult enough to be interesting and worth exploring, but not accompanied by anxiety of being formally tested, with associations of judgement, possible failure and comparison to other pupils.

You should also employ a wide range of strategies to gain insights into pupils' understanding of their learning. As well as using written tasks you can ask pupils, in small groups, to draw mind maps or spider diagrams (see Unit 5.2) expressing their ideas about the topic; or to give a short oral presentation or present a poster summarising the main points of their task. Employing a wide range of assessment instruments also guards against the danger of over-reliance on one form of testing, such as pupils' written work.

Sharing learning goals

Recent research suggests that it is helpful if teachers are explicit about their learning objectives and share them with pupils in a way that pupils can understand (ARG, 2002; Black *et al.*; 2003; Clarke, 2005).

This approach can include 'modelling' or showing them what a good response might look like and involving them in an active dialogue, talking to them and getting them to talk about their learning. This is not easy: 'How to draw the concept of excellence out of the heads of teachers, give it some external formulation and make it available to the learner is a non-trivial problem' (Sadler, in Gipps, 1997). Some ways forward include:

■ 'model' examples which might help pupils to grasp what it is you are looking for
■ when pupils are doing presentations, give them explicit performance indicators which show them the criteria on which you will judge their performance, such as quality of preparation and research, marshalling of information, structure, delivery, eye contact, answering of questions, etc.
■ discuss with pupils what constitutes a good essay
■ describe effective ways of working in a group.

An assessment instrument for pupils designed to help them think about what is involved in working successfully in groups is shown in Figure 6.1.8.

Grouping pupils can be important; some research has shown that girls-only groups function better than boys-only and mixed groups in terms of collaboration and co-operation, unless boys-only groups are told explicitly that collaboration is one of the key learning objectives (Underwood, 1998). Recent research into pupils' understanding of why they did history at school revealed that many pupils had very little understanding of why history was on the school curriculum (Adey and Biddulph, 2001; QCA, 2005). These finding suggests that you should discuss 'the object of learning' with your pupils, so they have some sense of 'what it is about' or why they are learning something.

Self and peer assessment

Involving pupils in their learning and their understanding of the processes, issues and problems involved in learning is considered to be effective practice. For example you could ask pupils to:

■ make their own judgement on the extent to which they have learned something
■ mark each others' work
■ give 'critical friend' advice to their peers
■ draw up a mark scheme and performance criteria.

Name of pupil:	
People in my group:	
Task:	
Did I. . . ?	
Make sure I knew what we were supposed to do?	
Make a contribution to discussions?	
Listen to other people's ideas?	
Make suggestions to improve other people's ideas?	
Encourage other people to share their ideas?	
Give people good reasons when I did not agree with them?	
Avoid putting other people down?	
Avoid getting cross with other people?	
Talk to someone I don't usually talk to?	
Develop my understanding of the task by working with others?	
As a group, did we. . . ?	
Make sure that everyone had something to do and joined in?	
Keep on task?	
Come to agreements on what to do?	
Make compromises about what to do?	
Finish the task on time?	
Feedback our ideas clearly to the rest of the class?	

■ **Figure 6.1.8** Helping pupils understand what is involved in working successfully in groups

An editable version of Figure 6.1.8 is available on the companion website:
www.routledge.com/textbooks/9780415478724

All these strategies shift assessment responsibilities towards the pupil and help the pupil to take responsibility for their own learning. In particular, you need to convey to pupils the nature of the 'gap' which exists between their current level of performance and what is desirable, and what it takes for them to bridge that gap (Clarke, 2001). You can't do the learning for the pupils; they have to do it for themselves and your job is to show them how to do this. Unless pupils are involved in working things out, learning may become memorisation without real understanding, and be soon forgotten.

Questioning and dialogue with pupils

One of the best (and quickest) ways of checking to see if pupils have understood something is to ask them questions, but traditional modes of questioning in the classroom are often laboured, time consuming and inefficient. Clarke illustrates a common lesson scenario related to questioning:

Most teachers' lessons begin with an almost automatic question and answer recall session of the whole class . . . The typical response is that the same few children continually have their hands up and, in order to elicit the right answers, the teacher usually chooses the right children. If the wrong answer is given (usually children who would give the wrong answer don't have their hands up) the teacher gives a side-stepping response and moves on to someone who will give the correct answer.

(Clarke, 2005: 54)

At best, you only find out if one pupil has understood the answer to the question. Alternative methods of questioning are:

■ allow pupils to discuss the answer to the question in small groups before asking the group to feedback

■ ask pupils to put their hands up at intervals to indicate their degree of confidence about particular points

■ use a 'traffic light' system, whereby pupils hold up a coloured card to indicate whether they feel confident they have understood something (green), they are sure they do not understand it (red), or they are somewhere in between (amber). The teacher can then look for opportunities to talk to pupils one-to-one or in the different colour bands about their work

■ pupils who have raised a green card can be asked to try and explain things to those less certain of the learning problem

■ vary the type of questions you ask pupils, combining quick factual recall questions with more open questions which require some time for thinking (and perhaps collaborative working) before the pupils respond

■ increase the 'wait time', especially for 'open' questions, so that you evince more considered and developed answers from pupils

■ get pupils used to thinking about their own questions in relation to the topic or issue they are learning about. Wiliam stresses the importance of trying to get pupils to think, and work things out for themselves in learning, and getting pupils to devise questions which might be asked of a particular topic (Wiliam, 2007)

■ use voting technology, which enables pupils to provide feedback to questions electronically (and anonymously), and where you have access to such facilities, it can be interesting and helpful, but like many ICT applications, it can be used skilfully or maladroitly. You need to think about the range and type of questions you are asking if you are to avoid the danger of producing stultifying multiple choice ordeals for pupils (see Walsh (2006) for an evaluation of the use of such software).

Clarke (2005) gives a wide range of suggestions for varying the format of questioning in classrooms. Over time, you should develop your skills in asking questions across the full range of Bloom's taxonomy, from simple recall and comprehension to complex evaluative questioning (Bloom, 1956).

The key to success in developing fruitful channels of dialogue with pupils about their learning lies in creating a relaxed and trusting overall climate in the classroom, so that pupils are not afraid to answer questions, contribute to discussions and make public their thinking or volunteer their uncertainties to you and the rest of the class (Clarke, 2005). Your capacity to do this requires sensitivity, knowledge of your pupils, first-rate interpersonal skills as well as perceptiveness in formulating adroitly posed questions. You should consider what proportion of your pupils are willing to admit they don't understand something, or risk giving you an answer which they are not sure is right, and try to cultivate a classroom climate where more and more pupils are prepared to 'open up' and share their thinking about their uncertainties. High quality questioning has been identified as 'the single most important factor in students' achievement of high standards, where questions were used to assess students' knowledge and challenge their thinking' (Ofsted, 1996: 23). It is an important skill to develop; a useful source of information is EBS (Educational Broadcasting Services) (2004).

Feedback

It is important that the information gleaned from an assessment activity is used in some constructive and purposeful way to move pupils forward in their learning.

The term 'consequential validity' has been coined to mean 'does anything happen as a result of this assessment, will anything useful be done with this information?' (Gipps, 1997). The full benefits of diagnostic assessment only accrue if teachers, having found out about pupil misconceptions and gaps in understanding, then go on to use this information in some purposeful way (Black *et al., 2*003). This situation means that you should revisit the same topic focusing on elements of the work where there were gaps in pupil understanding. Pupils could work together to address these gaps. Another element of consequential validity relates to the impact assessment had on the learner. For this to happen feedback is essential so that pupils know how they have done, where they did well, where they made mistakes and what they might do to learn from their errors.

Pupils' *attitude* to learning has a big influence on how successful they are at learning. Feedback influences not just the chances of them getting it right, or doing better next time, but their attitude to learning and the amount of effort and thought that they are willing to put into learning. Thus 'a number of pupils do not aspire to learn as much as possible, but are content "to get by", to get through the period, the day or the year without any major disaster, having made time for activities other than school work' (Perrenoud, 1991: 2).

You need to develop several feedback skills, including:

■ providing constructive, honest feedback to pupils who are not doing well in your subject, without discouraging them
■ ensuring that able pupils who are doing well do not coast and work below their capabilities
■ persuading all the pupils in your care to want to make progress in your subject.

Feedback and marking

Inspectors have been critical of the vague and unfocused nature of some teacher assessment and of the failure to use it to any useful purpose. Thus a recent inspector's report said:

In the most effective schools, teachers and pupils often review learning against precise objectives regularly during lessons. Good quality marking is a common characteristic in these schools. Assessment remains, overall, the weakest aspect of teaching. Where assessment is ineffective, teachers do not routinely check pupils' understanding as the lesson progresses. Many teachers still struggle to use the information from assessment to plan work that is well matched to the pupils' needs, and to involve the pupils in assessing their own work.

(Ofsted, 2007b: 66)

The thrust of the comments echoes the report of Ofsted in 1996; assessment remains a difficult area of teaching. Your assessment of pupils should make clear they are capable of learning and can improve their performance. You should ensure that your feedback to pupils has a positive impact on their subsequent efforts to learn. One study of the impact of feedback on performance found that in two out of five instances, the effects were negative (Kluger and DeNisi, 1996).

The time and effort involved in assessment is worthwhile when it provides clear guidance for pupils on how to improve:

> There's an awful lot of giving smiley faces at the bottom of children's work and very elaborate praise and stars and so on. They are fine for maintaining pupils' motivation and making children feel good about it, but unless it's accompanied by more direct specific advice about what to do to make the piece of work better it's actually of very little help to pupils as a learning activity. It actually helps the pupil to be told directly but kindly, what it is they are not doing very well so that they know how to do it better.
>
> (Gipps, 1997)

The Assessment Reform Group has argued that too much emphasis has been put on 'teaching to the test', 'drilling' pupils for these tests, and giving feedback that foregrounds marks and grades and encourages competition between pupils, with the result that some pupils become disaffected from learning. They suggest ways of using assessment which minimises demotivating pupils (see Table 6.1.3).

The overarching idea behind these suggestions is towards the idea of the teacher acting as a 'coach' and critical advisor to pupils, helping them achieve their 'personal best' in the way that an athletics coach might work with an athlete. Talk to teachers with whom you work about how they handle the challenge of providing feedback to pupils who are trying to do well in the subject, but whose progress is limited by lack of ability rather than lack of effort. How do experienced teachers engage pupils who are reluctant to commit themselves to learning?

Using assessment data

Teachers have access to a wide range of data about their pupils' prior academic achievement. In addition to primary school reports there are, in England, the results of the Key Stage 2 NCT (former SATs) in English, maths and science passed to secondary schools as well as teacher assessment of their level of attainment in all school subjects. In addition, teachers have pupils' reading ages. Some secondary schools also use the National Foundation for Educational Research (NFER) Cognitive Ability Tests (CATs), usually taken by pupils at the start of Year 7. CATs test pupils' abilities in spatial, numerical and verbal reasoning, and are a guide to pupils' cognitive abilities. These abilities are thought to be relevant to pupils' ability to achieve in a wide range of school subjects.

■ **Table 6.1.3** Marking and motivation: helping pupils understand feedback

---

Marking to encourage learning includes:

* helping pupils to take responsibility for their learning
* discussing the purpose of learning with pupils
* encouraging pupils to judge their work by how much they have learned and how much progress they have made
* helping pupils to understand the criteria by which their learning is assessed and to assess their own work
* giving feedback that enables pupils to know the next steps and how to succeed in taking them
* encouraging pupils to value effort and a wide range of attainments
* encouraging collaboration among pupils and a positive view of each others' achievements.

In addition to NCTs and CATs, schools use other large datasets recording pupil performance over time, including MIDYIS (Middle Years Information System) and YELLIS (Year Eleven Information System) (CEM, 2008). Other schools use the Fischer Family Trust dataset (FFT, 2008). In addition there is the RAISE (Reporting and Analysis for Improvement through School Self-Evaluation) dataset published by the government in England (Department for Children, Schools and Families (DCSF), 2008j).

These datasets allow teachers and schools to compare their own cohort of pupils in relation to the national cohort. The FFT dataset makes it possible to disaggregate the relative performance of different groups of pupils, such as boys, pupils on free school meals, pupils for whom English is an additional language and so on. In addition comparisons can be made between pupils in one school and all other schools of a similar type or composition. The dataset enables teachers to discern whether particular pupils are coasting or generally underachieving; or explore the performance of individual teachers, departments, schools and LAs. Points scores for individual pupils on, for example, their Key Stage 2 NCTs results compiled according to a statistical formula based on the whole dataset can be used to predict performance in secondary school and in the General Certificate of Secondary Education (GCSE). The prediction is saying 'all things being equal/in average conditions, a pupil who got this score at this stage will probably go on to get a B (or whatever) at GCSE'.

The data sets a parameter or expectation on future pupil attainment. This data can be used by the teacher, in negotiation with the pupil to discuss progress and try to boost performance beyond the predicted grade. The teacher (school, department or LA) has the challenge of trying to get the pupil to surpass the predicted grade to show that they have 'added value' beyond what might be expected for that pupil. Schools can use the data to see how they compare to similar institutions, how well they are performing with different types of pupils and where their strengths and weaknesses lie in terms of pupil attainment.

Teachers or departments who have been particularly successful in terms of the 'value' they have added to their pupil's performance can be invited to talk to other teachers and departments and share good practice. Conversely where teacher or departmental 'residuals' are lower than those for pupils in other subjects or cohorts, a degree of anxiety and concern is created. Assessment data is used not just for formative purposes but also for accountability and comparison.

Some teachers welcome the use of such assessment data as it takes into account pupils' prior accomplishments in ways that raw examination results and league tables do not. Others feel that the imprecision of such testing and its susceptibility to manipulation make it inappropriate for such 'high stakes' testing.

Because schools use different combinations of these data resources a pragmatic way forward would be for you to first familiarise yourself with the assessment data in your school experience school and, at a later stage, try to develop an understanding of other assessment data which you may encounter. A lucid and more developed explanation of the assessment instruments described above is in Husbands (2007). There is also a booklet for teachers in England on managing data in secondary schools (Department for Education and Skills (DfES), 2002b).

An example of pupil assessment data given in Figure 6.1.9 provides information on the prior attainment of three Year 8 pupils, in terms of their performance in the Cognitive Ability Tests taken at the start of Year 7, their performance in mathematics and reading tests, their spelling and reading 'ages' and their performance in Key Stage 2 NCTs for English, maths and science (see Figure 6.1.9).

This information was provided for all class teachers at an English secondary school, so the teachers would have this data 'at their fingertips' as they were teaching their classes. Such information may also be helpful to the teacher in arranging pupil groupings for collaborative work. The information gained can help teachers to plan their teaching to match pupils' individual needs.

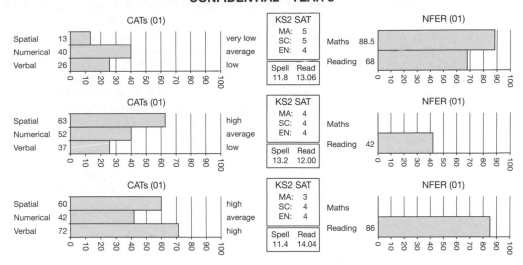

■ **Figure 6.1.9** Chart providing subject teacher with a profile of the prior attainment of three of the pupils in a teaching group

The aspiration is that the massive increase in the amount of assessment data available to the teacher on each pupil helps deliver a more 'personalised' learning experience for pupils.

MARKING AND YOUR PROFESSIONAL DEVELOPMENT

We discuss here two features of assessment: first, how to keep your marking 'load' under control, and second, to identify your professional development needs now and in the longer term.

Keeping assessment manageable

You need to think about how to manage your time, so that assessment does not detract from the rest of your teaching (see Unit 1.3). Commentators who operate at some distance from the classroom sometimes bemoan the fact that not all weaknesses and mistakes in pupils' work are systematically and comprehensively corrected, including every error of spelling, punctuation and grammar. You may agree with this view. Measure how long it takes to do a comprehensive, all-embracing diagnostic dissection of a full set of thirty exercise books for a piece of extended writing and then review your position.

Teachers have to manage their time effectively; it is not practical to mark every piece of pupils' work in the same way without compromising the time available for lesson planning, etc. There are no magic answers to make assessment burdens disappear but there are ways to improve your time management. Table 6.1.4 explains these.

The list in Table 6.1.4 is not intended to suggest that you leave yourself without marking but to make room, on occasions, for you to conduct a rigorous diagnosis of your pupils' written work and provide your pupils with constructive advice. These occasions allow you to give detailed and helpful comments which encourage your pupils and improve their learning. All marking should let your pupils feel that you care about their progress in your subject.

Learn to mark flexibly. Some pieces of work require detailed attention and diagnosis; sometimes a light touch monitoring is more appropriate.

Plan marking time. When you are planning lessons, think about setting fewer written tasks which require you to mark them in a time-intensive way. Student teachers sometimes set written tasks to take the pressure off themselves in the early stages of first placement or for class management purposes. Student teachers tend to set more written tasks than experienced teachers and this produces a heavy marking load. Since lesson planning takes a long time, extra pressure builds. In your planning, try to prepare some lessons which have as one of the objectives not requiring marking to be done in their aftermath for example by pupils marking their own work, or that of their peers.

Marking codes. Develop a shorthand code for signalling marking corrections, such as symbols for omissions, development, non sequitur, irrelevant, spelling errors or clumsy phrasing.

Common errors. Make brief notes on common errors as you are marking pupils' work so that you can report on them orally to the group as a whole, rather than writing the same comment in many books.

Pupil marking. On occasions it is appropriate and helpful for pupils to mark each others' work, or their own work. You can use this strategy to help pupils understand the criteria for marking.

Oral feedback. It takes more than five times as long to write something down compared to saying it so sometimes prepare detailed oral feedback to pupils.

Pupil response to tasks. Structure tasks so that sometimes pupils present their work as a poster, a group ICT task or display work. This response may not require you to mark their books although you need to keep records of their performance.

Using ICT. Consider ways in which ICT might help to reduce the administration involved in assessment, recording and reporting. The facility to 'copy', 'cut and paste' and build up an archive of models and exemplars of good and bad responses can save considerable amounts of time.

Acquiring a 'grounding' in assessment

Assessment can seem a very complicated area. It takes time to become familiar with the battery of acronyms and the range of assessment procedures, but assessment is an integral part of your day-to-day classroom teaching. By the end of your course you should make sure that you have acquired 'a grounding' in assessment issues.

You should consider your immediate needs (short-term goals) and your developing needs (long-term goals). The short-term agenda is for you to become effective in your responses to pupils' work, adapting to the demands of the assessment policies of the department in which you are working. The longer-term agenda is for you to develop a sound understanding of the nature and purposes of assessment, how to use the different techniques appropriately and to interpret assessment data, and above all, how to use assessment to improve your teaching, and pupils' learning. These skills are one of the hallmarks of accomplished teachers. A checklist or aide-memoire of questions to ask yourself about your grasp of assessment issues can be accessed at http://www.uea.ac.uk/~m242/historypgce/assess/welcome.htm.

SUMMARY AND KEY POINTS

Although teachers and schools do have to be accountable for their work, the most valuable and important function of assessment is to move pupils forward in their learning.

Assessment needs to be built into your planning for teaching and learning, not tacked on as an afterthought. You should be familiar with the idea of assessment as part of the cycle of planning, teaching, and assessing leading into revised planning and teaching. A key facet of your ability as a teacher is your ability to devise and execute activities which help to move pupils forward in their learning, and which enable you to gain insight into the extent to which the pupils have learned what you were trying to teach. Ascribing levels, marks or grades to pupils' work is not unproblematic but it is not the most difficult part of assessment. The most challenging and important part of assessment is saying and doing things which help pupils to make progress in their learning.

As well as thinking about how to provide constructive comments to pupils when responding to their work, you need to use a wide range of assessment strategies to develop pupils' learning. By the end of your course you must be familiar with the range of assessment methods and terminology used in schools. You should be aware of the tensions that arise when one form of assessment is used for a variety of purposes, e.g. CATs, and of the tension between validity and reliability of an assessment instrument when trying to maintain both at a high level.

Some of the issues raised in this unit can be developed further through the Further Reading at the end of this unit. You may find helpful the publication *Formative Assessment: Implications for Classroom Practice* (QCA, 2008j). We have indicated the change in assessment practice in England through the abolition of Key Stage 3 national testing. The implications of this development have yet to be realised.

Check which requirements of your course have been addressed through this unit.

FURTHER READING

Assessment Reform Group (ARG) (2002) *Testing, motivation and learning*, Cambridge: University of Cambridge Faculty of Education.
> A summary of a systematic review of research on the impact of summative assessment and testing on pupils' motivation for learning and its implications for assessment policy and practice.

Black, P. and Wiliam, D. (1998) *Inside the Black Box*, London: King's College.
> A succinct introduction (20 pages) to the importance of formative assessment, and the ways in which formative assessment might lead to gains in learning.

Black, P., Harrison, C., Lee, C., Marshall, B. and Wiliam, D. (2003) *Assessment for Learning: Putting it into Practice*, Maidenhead: Open University Press.
> Addresses the practical implications of putting assessment for learning into practice.

Clarke, S. (2005) *Formative Assessment in Action: Weaving the Elements Together*, London: Hodder and Stoughton.
> A helpful chapter on developing effective questioning techniques and a good source for implementing assessment for learning generally.

DfES (Department for Education and Skills) (2002b) *Releasing Potential, Raising Attainment: Managing Data in Secondary Schools*, London: DfES.
> A thorough explanation of the ways in which secondary schools use assessment data to raise pupil attainment.

Miller, D. and Lavin, F. (2007) '"But now I feel I want to give it a try": formative assessment, self-esteem and a sense of competence', *The Curriculum Journal*, 18(1): 3–25.
> Helpful for giving pupils' views on assessment for learning in practice.

Additional resources for this unit are available on the companion website:
www.routledge.com/textbooks/9780415478724

UNIT 6.2

EXTERNAL ASSESSMENT AND EXAMINATIONS

Bernadette Youens

INTRODUCTION

Principles of assessment were introduced in Unit 6.1, which highlighted the importance that this aspect of education has assumed since the Education Reform Act of 1988 (ERA, 1988). This unit looks at the particular role, function and nature of external assessment and examinations. We suggest you read Unit 6.1 before studying this unit. The public examination system at 16 (General Certificate of Secondary Education (GCSE)) reflects closely the content and principles of the National Curriculum (NC) for England and you may need to refer to Unit 7.3 as you read this unit.

This unit aims to provide you with an overview of the framework for external assessment and examinations in secondary schools. Although you are familiar with the public examinations that you took in school, in recent years there have been significant developments in assessment methods and in the range of external examinations taken by pupils in secondary schools. In England this has been particularly true in the 14–19 sector of education, which has seen the growth of vocationally related courses which are taught in secondary schools alongside the GCSE and the General Certificate of Education (GCE) at Advanced (A) level. New vocationally focused examination courses called Diplomas were introduced in September 2008.

As well as knowing how pupils are assessed throughout their secondary education, it is important to be aware of the many purposes of external assessment and examinations. These purposes can be usefully divided into those associated with candidates and those that have more to do with educational establishments and public accountability (see Unit 6.1). Two important, recurrent themes that arise when discussing external assessment and examinations are validity and reliability. These two concepts, together with the agencies, regulations and processes involved in ensuring consistency in these two areas, are discussed in Unit 6.1. Teaching externally examined classes is a challenge for any teacher and demands particular teaching skills and strategies in addition to the routine elements of good lesson planning and teaching. This aspect of teaching is discussed in the final part of this section.

We use the term Standard Assessment Task (SAT) throughout but the name is changing to National Curriculum Task (NCT).

OBJECTIVES

At the end of this unit you should be able to:

■ identify the range of external assessments in secondary schools and access the national framework for qualifications

■ understand the relationship between the National Curriculum in England and external examinations

■ explain the main purposes of assessment

■ recall the processes involved in the setting of external examinations and the institutions involved

■ start addressing the issues relating to teaching examination classes.

Check the requirements for your course to see which relate to this unit.

YOUR OWN EXPERIENCE

A good starting point for this unit is your own experience of external assessment and examinations. Think back to your time at school. What did you think was the purpose of sitting examinations and did preparing for examinations impact on your motivation as a learner? Did the teaching strategies of examination classes differ from non-examination classes? Recalling these impressions may provide a good starting point from which to develop your understanding of the issues pertaining to external assessment and examinations.

TYPES OF ASSESSMENT

In Unit 6.1, we discussed formative and summative assessment and we remind you of those terms. Formative assessment can be defined as assessment for learning, and summative assessment as assessment of learning (Stobart and Gipps, 1997). External assessment and examinations are generally considered to be forms of summative assessment. There are two important methods used extensively in summative assessment that you need to be familiar with, namely, norm-referenced assessment and criterion-referenced assessment.

Norm-referenced assessment has been used extensively throughout the British education system. In norm-referenced assessment, the value of, or grade related to, any mark awarded depends on how it compares with the marks of other candidates sitting the same assessment. The basis for this form of assessment is the assumption that the marks are normally distributed, i.e. if you plot the marks awarded against the number of candidates, a bell-shaped curve is produced, providing the sample is big enough. This curve is then used to assign grade boundaries based on predetermined conditions, e.g. that 80 per cent of all those sitting the examination pass and 20 per cent fail. In this way an element of failure is built into the examination. The system of reporting by grades is essentially norm-referenced. For example, if you are awarded the highest grade (A*) in a subject at GCSE this grade does not give any specific information about what you can do in that subject, simply that you were placed within the top group of candidates.

Criterion-referenced assessment, on the other hand, is concerned with what a candidate can do without reference to the performance of others, and so provides an alternative method to address the limitations of norm-referencing. A simple example of criterion-referenced assessment is that of a swimming test. If a person is entered for a 100-metre swimming award, and swims 100 metres, then she is awarded that certificate irrespective of how many other people also reach this standard. Academic courses, such as GCSE science, use criterion-referencing when assessing practical skills as part of the coursework element of the course but the grades are awarded and the results are reported by comparing pupils with other pupils, i.e. norm-referenced. The overall assessment in most examinations is a mixture of criterion-referenced marking together with norm-referenced grading and reporting of the level of achievement. In addition, the setting of GCSE examination questions uses as a guide the Attainment Targets described for that age group in the subject specification of the NC in England and so the setting of questions is also norm referenced.

Vocational courses use criterion-referenced assessment for marking and grading of achievement, to which we return later.

THE FRAMEWORK OF EXTERNAL ASSESSMENT IN SECONDARY SCHOOLS

We consider in turn Key Stage 3, GCSE and post-16 education in England. Different arrangements are in place in Wales, Scotland and Northern Ireland. Until autumn 2008 pupils all in England at Key Stage 3 were statutorily assessed by external examination, the Standard Assessment Tasks (SATs).

Key Stage 3 assessment

On transfer to secondary school pupils bring with them information obtained from the end of Key Stage 2 assessment, arising from both internal teacher assessment and external assessment. This information is usually reported as NC levels (see Unit 7.3). Pupils are teacher-assessed again at the end of Key Stage 3 to determine their attainment and to measure progression, again reported as NC levels. National Curriculum Tests (formerly SATs) in English, mathematics and science continue to be available to schools to take up as they choose. Within these new arrangements more frequent reporting by schools to parents is anticipated (National Assessment Agency, 2008).

In England, until 2008, external assessment at the end of Key Stage 3 took the form of written tests, called SATs, for the core NC subjects of English, mathematics and science. The SATs were first envisaged as activity based tasks embedded in everyday teaching contexts; this recommendation was rejected after trials showed that this process was complex, time consuming and intended to sample the pupil population. They were expensive also on teacher time. Written tests were adopted as simpler to write and administer, could be marked and graded externally and could be used to test the whole age cohort. The SATs were then set by external agencies and marked by paid, external markers.

The SATs were written with reference to the subject Programmes of Study of the NC (see Unit 7.3); the questions in the tests were designed so that the demand on the pupil links closely to the level descriptions. Papers were set for each of the core subjects and tiered assessment was used to allow pupils to be entered for the paper most suited to their achievement. However, the written SATs were limited in what they tested and so had reduced validity in relation to the overall aims of NC subject specifications.

Schools are statutorily required also to provide teacher-assessed levels for all attainment targets and an overall assessment level for each of the core and foundation subjects. In this way, the progress of each pupil can be identified and reported. In 2007, the proportion of all pupils achieving NC Level 5 or above in written SATs at Key Stage 3 in English, with 2004 figures in brackets, was 74 (71) per cent, in mathematics 70 (73) per cent and in science 73 (66) per cent (Department for Children, Schools and Families (DCSF), 2008m).

GCSE

The GCSE examination was introduced in 1986 to be taken at the end of Key Stage 4 to replace the GCE Ordinary (O) level and Certificate of Secondary Education (CSE) to provide certification for about 90 per cent of candidates.

The GCSE examination offers eight pass grades, A* to G; the A* grade recognises exceptional performance. The GCSE examinations are designed to test recall, understanding and skills. The GCSE has a tiered assessment pattern, as do the NC SATs, to enable pupils to be entered for the paper most appropriate for their current achievement. Teachers assess pupils' progress and potential and advise pupils on which tier of the examination to enter. Tiered papers carry grade limits, thus narrowing the opportunities of pupils. For example, a higher paper may enable pupils to be awarded grades A*–D, while a foundation paper may allow only the award of grades C–G. Furthermore, if pupils fail to achieve the marks required for the lowest grade in their tier then they receive an unclassified grade.

Coursework was also introduced with the GCSE and in the early days of GCSE some courses offered were entirely coursework-based. However, coursework is now restricted by legislation by the Qualifications and Curriculum Authority (QCA) through the subject specifications and the amount allowed depends on the subject. In the more practically based subjects (e.g. music, physical education) a larger proportion of teacher-based assessment contributes to the final mark than in other subjects. In music the upper limit in practice is about 60 per cent, whereas in physical education a maximum of 25 per cent of the total mark is commonly awarded for coursework. No subject can be assessed only by teacher-assessed coursework.

Although a good motivator for students, coursework has also attracted significant criticism, focusing on plagiarism issues, principally the practice of some pupils buying ready-made answers from the Internet, and also the amount of parental contribution to the coursework. In response to these criticisms significant changes to coursework have been introduced. From September 2007 all coursework in mathematics GCSE was abolished, and from September 2009 coursework and tasks will be replaced by 'controlled assessment' tasks. In practice this is likely to mean that pupils can undertake research and investigations in preparation for a task outside the classroom, but all written work must be completed in a classroom under supervised conditions. 'Controlled assessments will increase public confidence in the GCSE and allow the integration of new sources of data and information, including the Internet, under supervision' (Boston, 2007). A further change from 2010 to GCSE teaching is the incorporation of functional skills, in English, mathematics, and information and communications technology (ICT).

The GCSE examination combines norm-referenced methods with criterion-referenced methods. Coursework is an integral feature of the GCSE examination and it is important that you have a clear understanding of this part of the examination; Task 6.2.1 (p. 356) asks you to address this issue.

POST-16 ASSESSMENT

Of all of the changes that have followed the ERA 1988, the area that has continued to undergo reorganisation is the provision of post-16 courses and their assessment.

GCE A level courses

GCE A level courses were first introduced in 1951 and since then have been regarded as the academic 'gold standard' by successive governments. Substantial changes were made to the framework of the 16–19 qualifications following a review and implemented in September 2000 (Dearing, 1996).

Task 6.2.1 **GCSE COURSEWORK**

Working with another student teacher in your own subject specialism, obtain a copy of a recent GCSE specification. Read the general introduction and familiarise yourself with the aims, assessment objectives and assessment patterns of the specification. Now turn to the coursework section and find out:

- how much of the overall mark is allocated to coursework
- the form the coursework takes
- how the coursework is assessed
- what information about coursework criteria is given to candidates.

Use this information to discuss your findings with an experienced teacher in your school experience school. Ask if you can have access to some samples of coursework from pupils at your school and use them to find out:

- how the coursework was set
- the range of tasks set
- how these tasks were made accessible and relevant to pupils.

Now try assessing the coursework samples yourself using the criteria provided by the awarding body and any internal marking scheme provided by the school. Discuss your marking with a member of staff and find out how the department internally moderates the coursework. Find out, too, from the coursework specifications the processes the awarding body uses to moderate the marking.

Check the requirements for your course to see which relate to public examinations. A record of your work could be placed in your professional development portfolio (PDP).

Each GCE A level examination subject is composed of six discrete units of approximately the same size. The first three units make up an Advanced Subsidiary (AS) course, representing the first half of an advanced level course of study. The other three units, which make up the second part of the GCE A level, are known as A2. One of the aims of the AS proposal was to provide a more appropriate and manageable 'bridge' between GCSE and GCE A level. The AS course may be taken as a qualification in itself or it may be used as a foundation to study the A2 section of the course. GCE A levels, like GCSE examinations, may be assessed in stages, as in a modular course, or terminally. The introduction of modular assessment has proved to be very popular with candidates. The main advantages of modularity are:

- motivating pupils to maintain a high, constant commitment throughout their course
- providing valuable diagnostic information from early results
- providing the opportunity to have achievement recognised (Dearing, 1996).

All GCE A2 courses must include an element of synoptic assessment designed to test a candidate's ability to make connections between different aspects of the course. There is no synoptic assessment at AS level. The synoptic element must normally contribute 20 per cent to the full A level and take the form of external assessment at the end of the course. GCE A level pass grades range from A to E with A the highest grade. As with GCSE examinations there is also teacher assessment of coursework but there is no general upper limit set by the QCA. In practice, the amount of teacher-assessed coursework is set by the QCA subject specification. For example, in GCE A level English there is up to 30 per cent teacher-assessed coursework. Advanced Extension Awards (AEAs), introduced in 2002 to replace the Special papers at GCE A level, were designed to challenge the most able advanced level students at a standard comparable with the most demanding tests found in other countries. A further stated aim of the AEAs is to assist universities to differentiate between the most able candidates, particularly in subjects with a high proportion of A grades at advanced level.

However, after eight years, significant changes are once more being made to GCE A levels. These changes include a reduction from six units to four units for most subjects, with the aim of reducing the burden of assessment. The particular changes to the A2 specifications include the setting of a broad range of question types to ensure that a wide range of skills is assessed, a requirement for extended writing to give candidates the opportunity to demonstrate the depth and breadth of their knowledge and understanding, and synoptic assessment. A further development is the Extended Project qualification, first taught in 2008. This is a separate qualification that pupils may add to their study programme and is designed to be distinctive from GCE A level coursework and sits within a common framework between the stand-alone qualification for GCE A levels and its equivalent for Level 3 Diplomas (see 'Vocational or applied courses' below).

In recent years there has been an increase in the number candidates achieving Grade A in the GCE assessment giving rise to concerns about the standard of GCE A level. In response to this concern the QCA introduced new requirements for this examination (QCA, 2007d). An A* grade is introduced to the grading at A level for candidates who achieve 90 per cent or more on the Uniform Mark Scale across A2 units. The first A* grades will be awarded in 2010.

Vocational or applied courses

The main vocational qualifications traditionally encountered in secondary schools in England have been the General National Vocational Qualifications (GNVQs) that were introduced into schools in 1992 and developed from National Vocational Qualifications (NVQs).

The NVQs are work-related, competence-based qualifications and the courses were designed for people in work or undertaking work-based training. NVQ courses provide job-specific training, the assessment of which takes place in the work environment and is criterion-referenced. Central to the NVQ model is the idea of competence to perform a particular job, where competence is defined as the mastery of identified performance skills.

GNVQs were introduced to provide pupils with an introduction to occupational sectors through school- or college-based courses. Indeed one of the principal aims of the GNVQ was to provide a middle road between the general academic route and occupational courses, such as the NVQs described above.

One of the main stumbling blocks to the uptake of vocational qualifications by schools was the difference in assessment practice and terminology between academic and vocational courses identified in the report on 16–19 qualifications (Dearing, 1996). The report identified as well the need for a coherent qualifications framework encompassing all national qualifications and one that provides a status for vocational qualifications equivalent to the academic subjects at GCE A level. Following that report, vocational GCSE and vocational GCE A levels were introduced, accompanied by a timetable for the phased withdrawal of GNVQ courses. The vocational GCSE and GCE qualifications have been renamed and are known as GCSE and GCE in applied subjects with pupils at both levels maintaining their study of core curriculum subjects. The purpose of these developments together with the other reforms outlined is to encourage Key Stage 4 pupils and post-16 students to broaden their programme of study to include vocational courses. The National Qualifications framework in Table 6.2.1 shows how the three qualification strands discussed are intended to overlap.

The framework for national qualifications was revised in September 2004 and now has a total of nine separate levels of qualification (entry level to level 8). The Framework is available online (QCA, 2006a); see also Brooks and Lucas (2004).

GCSE qualifications in vocational subjects were first introduced in September 2002 and are offered in eight subjects, e.g. applied art and design, leisure, applied science and tourism. Each course consists of three common, compulsory and normally equally weighted units in each subject. The qualification is equivalent to two GCSEs and is graded, like GCSEs, from A*–U, covering both level 1 and level 2 of the national qualifications framework (Table 6.2.1).

Applied A levels are available at three different levels:

■ Advanced Subsidiary Applied Certificate of Education (3 units)
■ Advanced Applied Certificate of Education (6 units)
■ Advanced Applied Certificate of Education (Double Award) (12 units).

The units assigned to each qualification are intended to be comparable to the GCE A level units described earlier in this section. Applied GCE A levels are reported using the grade range A–E so that direct comparisons with GCE A level can be made (see Table 6.2.1).

Applied courses are assessed through an internally assessed portfolio of evidence and externally set tests, projects or case-study work. As with other examinations, the internal assessment is externally moderated. In general, the portfolio contributes two-thirds of the final mark and the external assessment the remaining third. Each unit is graded and these grades are aggregated to produce a mean grade for the whole qualification. Further discussion of vocational (applied) courses can be found in Brooks and Lucas (2004).

The future of Applied A levels is likely to be limited because of yet further development by the QCA of 14–19 qualifications through the Diploma route. By 2013, fourteen new Diploma qualifications are planned, offered at three levels (Table 6.2.1), designed as an alternative to traditional GCSEs and A levels. Diplomas are to be multi-component qualifications themed around an industry sector (QCA, 2006b).

Diplomas are expected to combine applied content with rigorous theoretical learning, set in the context of a particular industry, known as a 'line of learning'. They are designed to provide an education as well as training and to guarantee learning in practical skills, functional skills in English, mathematics and ICT, as well as concentrating on personal, learning and thinking skills.

Diplomas are based on units and credits, are flexible and allow pupils to complete a GCSE or a GCE A level as part of their Diploma programme. Post-16 pupils can still choose from occupational qualifications, such as apprenticeships and NVQs.

■ **Table 6.2.1** Framework of national qualifications (entry level to level 3 only)

Level of qualification	General qualifications
3	GCE A level grades A–E
2	GCSE grades A*–C
1	GCSE grades D–G
Entry	Certificate of (educational) achievement

In 2008 five Diplomas were introduced, covering information technology; society, health and development; engineering; creative and media; construction and the built environment. Details are available on the Diploma website (DCSF, 2008k).

THE PURPOSES OF EXTERNAL ASSESSMENT

There is a long-standing history in the United Kingdom of externally examining pupils at particular stages in their education, which is quite different from the practice in some other countries (QCA, 2008l). For example, Australia has not had public examinations for many years. In recent years in England this practice has extended to the external assessment of pupils at the end of Key Stages 2 and 3. If the time and resources spent on this form of assessment are to be justified then it is important that the purposes of external assessment are fully understood. The purposes of external, summative assessment can be thought of in terms of certificating candidates and the public accountability of teachers and schools. Another purpose is to categorise candidates in order to select people for higher education, employment; or to recognise achievement. The outcome of national examinations provides pupils with a grade so that they and other people can compare them with other candidates.

One recognised function of external assessment is that of certification. If you hold a certificate then it is evidence that competence in particular skills or a level of knowledge has been achieved. For example, if you hold a driving licence this is evidence that in a driving test you successfully performed a hill start, completed a three-point turn, reversed around a corner and so on. The significance to pupils, of both the categorising and certification purposes, is evidenced by the fact that an impending examination provides an incentive for pupils to concentrate on their studies and to acquire the relevant knowledge and skills required by the examination for which they are entered. Thus a consequence of external examinations is to provide motivation for both pupils and teachers. Motivation is discussed in greater detail in Unit 3.2.

Public accountability

Assessment is high on the political agenda at present because it is inextricably linked with the notion of raising standards and school improvement and to make schools publicly accountable. Since 1982 schools have been statutorily required to publish examination results, with the aim of providing parents with more information. The statutory requirements were extended in 1992 to include a detailed breakdown not only of examination performance but additional statistics such as number of pupils on the roll with special educational needs (SEN), but without a Statement (see Unit 4.6). This information is published each year as Achievement and Attainment Tables, often referred to simply as the 'league tables' (see Unit 6.1). Although a variety of statistical information is published each year, currently schools in England are placed in rank order based on just one variable, the percentage of pupils gaining five or more GCSE passes at grades A*–C. The high profile given to the Achievement and Attainment Tables by the different stakeholders in the education process has led schools to implement strategies to increase the percentage of their pupils achieving five or more A*–C grades. For example, schools often target pupils predicted to achieve grade Ds at GCSE for additional mentoring and academic support as part of the school's strategy to increase the number of pupils achieving grade Cs in their GCSE exams. There is also evidence that schools are selecting particular qualifications that make it easier for students to reach the key target of 5 GCSEs at A*–C. In an effort to stop this, the government has recently strengthened the league tables to include English and mathematics qualifications as part of the 'five grade A*–C grades' when reporting results. Since September 2007 the tables have reported also the number of pupils gaining the equivalent of a Double Award in science at GCSE. The introduction of external assessment of pupils at the end of Key Stages 2 and 3 has also had a significant impact on the teaching of pupils in this age range, providing further evidence of the effect that so-called 'high stakes' external assessment has on classroom practice. This impact is reported as a concentration on the curriculum to be assessed through SATs or GCSE at the expense of a broader curriculum (Office for Standards in Education (Ofsted), 2006a; Josefkowicz, 2006). The abolition in 2008 of statutory national testing at the end of Key Stage 3 may be a response to that criticism.

The term 'high stakes' is used to describe assessment that has significant consequences for either the candidate or the school (see Unit 6.1). In addition to school achievement and attainment, since September 1998 all schools in England and Wales have had to set and publish targets for all pupils aged 11–16 in order to demonstrate year-on-year improvements. To assist schools with this process, from 2007 all schools in England have been able to access the programme RAISEonline, (replacing the Ofsted Performance and Assessment (PANDA) reports and DCSF's Pupil Achievement Tracker (PAT)). This programme provides interactive analysis of school and pupil performance data across the four Key Stages of the National Curriculum. This package contains benchmark data so that schools can compare whole-school performance with that of schools with similar intakes and profiles: also Contextual Value Added (CVA) data, which aims to take into account a wide range of variables which may affect a school's performance, e.g. number of pupils with SEN or number of pupils entitled to free school meals.

The central government in England also sets national targets to encourage progress in raising attainment and to narrow the achievement gap between groups of pupils. To help schools to achieve the national targets set by them, the government has introduced a raft of initiatives aimed at raising pupil attainment at Key Stage 3. For example, over a number of years concern was expressed about the perceived overall lack of progress made by pupils at Key Stage 3 (Ofsted, 2000b; QCA, 2000) In response to these observations, the Key Stage 3 Strategy was introduced (DCSF, 2008l; DfES, 2001d, 2003c; see also Unit 7.4). A further, separate set of tests in English and mathematics, the Year 7 progress tests, were available for pupils who did not achieve level 4 at the end of Key Stage 2, and who are assessed to be working at level 3 or 4 during Year 7. These tests are set by QCA and marked externally. At the time of writing this unit the status and availability of these tests are unclear.

A further example of an initiative in the drive to raise national standards in England was that, from September 2005, schools could purchase additional tests from the QCA, in order to assess pupils' progress in the school years not covered by statutory tests, i.e. Years 7 and 8, These tests were marked by teachers. By collecting information on pupils' progress in English and mathematics in this way as well as SATs, schools were supported in monitoring pupils' progress in the years between the statutory tests. The tests were designed to contribute to teaching and learning strategies and enable schools to meet their targets by the end of Key Stage 3 (QCA, 2004c). Again, at the time of writing the status and availability of these tests are unclear.

VALIDITY AND RELIABILITY

The concepts of validity and reliability are central to understanding an assessment process and are also discussed in Unit 6.1. For all external assessments and examinations, frameworks of regulations have been developed to ensure that the examination process and the results produced are both valid and reliable. To understand this framework you need to be aware of the institutions and processes involved in this regulation.

The QCA is the government agency that approves all course specifications, as well as monitoring examinations through a programme of scrutinies, comparability exercises and probes, and is accountable to the DCFS. At the end of 2003 the government announced the introduction of a new body, the National Assessment Agency (NAA). This agency is a subsidiary agency of the QCA and its remit is to modernise the examination system and to work alongside the Unitary Awarding Bodies (usually called awarding bodies). There are three awarding bodies in England authorised by the government to offer GCSE, GCE A and AS and applied (vocational) courses. These are:

■ Assessment and Qualifications Alliance (AQA): http://www.aqa.org.uk
■ Edexcel: http://www.edexcel.com/
■ Oxford Cambridge and RSA Examinations (OCR): http://www.ocr.org.uk/index.html.

Following the 'Guaranteeing Standards' consultation (Department for Education and Employment (DfEE), 1997), the formation of a single awarding body was considered but a group of three was thought useful to retain a measure of competition. The key recommendations of the standards' consultation report were:

■ for each externally examined course there is a specification for the core material
■ the publication of a detailed code of practice designed to ensure that grading standards are consistent across subjects and across the three awarding bodies, in the same subject, and from year to year

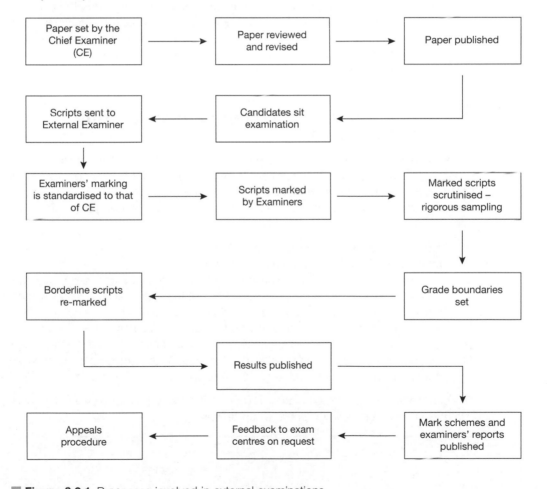

■ **Figure 6.2.1** Processes involved in external examinations

■ that this code of practice should also set out the roles and responsibilities of those involved in the examining process and the key procedures for setting papers, standardising marking and grading.

The processes employed by the awarding bodies to address the recommendations above and to ensure that the examinations are valid and reliable are outlined in Figure 6.2.1.

TEACHING EXTERNALLY ASSESSED COURSES

All teachers have to think beyond the particular lesson they are teaching to the end of the unit of work, to ensure that pupils can respond successfully to any assessment scheduled to take place. When pupils are assessed externally the same considerations apply, i.e. how to maximise pupils' achievement. However, you do need to be fully aware of the nature of the external assessment for which you are preparing your pupils. It is important not just to teach to the examination but to hold on to the principles of good classroom practice.

In preparing your pupils for external assessment you need to be familiar not only with the subject content, but also with the particular demands of the assessment process, such as the types of questions set and the language used in setting questions. Questions are set which often employ words with a specific meaning, e.g. they ask candidates to describe, or explain, or use short notes or summarise. Candidates need to know what these words mean in examination conditions. Task 6.2.2 is designed to help you become familiar with types of questions currently set in examination papers in your own subject and the corresponding reports of examiners.

Task 6.2.2 **USING EXAMINATION PAPERS IN YOUR TEACHING**

Collect a number of GCSE examination papers for your subject together with the mark schemes and specification. Where possible obtain the relevant examiners' report which are sent to schools entering candidates to that awarding body. Read through the specification for the examination arrangements. Then address the questions in the paper in the following way:

■ answer the questions on the paper yourself
■ mark your answers using the mark scheme
■ evaluate your answers and marking and identify the key knowledge and concepts needed to gain maximum marks. Look back at the examination questions, identify the key words and phrases most often used in the questions
■ use the examiners' report to review and refine your findings
■ identify any ideas that might be useful to consider in your day-to-day teaching
■ repeat the exercise for another year of the same paper, or repeat the exercise using papers from a different level, e.g. GCE A level .

The completed task should be placed in your PDP. Check the requirements for your course related to preparing pupils for public examinations.

Once you are familiar with the structure and language used in past papers you can then integrate this information into your teaching throughout the course. Another important aspect to consider is the development of study skills both in your lessons and throughout the school (see Task 6.2.3). For further advice on developing study skills see Balderstone and King (2004).

Task 6.2.3 **STUDY SKILLS**

Discuss with your tutor or other experienced teacher in your school experience school the whole-school and departmental approaches available to support the development of pupils' study skills. These skills include, for example, managing coursework, planning and supporting revision and time management. Use the information you gain to integrate the teaching of study skills in your own teaching.

Check the requirements for your course which relate to developing study skills.

Finally we ask to review the influence of public assessment on the way pupils are grouped and taught (see Task 6.2.4, p. 364).

SUMMARY AND KEY POINTS

In this unit we have linked the framework for external assessment and examinations with the nature and purposes of summative assessment. In England, both the National Curriculum assessment and the external examinations utilise aspects of both norm-referenced and criterion-referenced methods, and this is an important feature of assessment of which you need to be aware and understand. Norm-referencing and criterion-referencing are factors used in discussions seeking to explain the steady increase in the proportion of candidates achieving A*–C grades. The changes recently introduced to the post-16 sector of education aim to encourage students to broaden their studies to include applied courses alongside traditional academic courses. As this book went to press significant changes to the qualifications offered to 14- to 19-year-old pupils were in place for implementation in September 2008. First, all GCE specifications have been revised to take account of new criteria which includes a reduction in the burden of assessment and also to help universities differentiate between candidates. A potentially much more significant change is the introduction of the first five Diploma courses offered alongside the GCSE and GCE. It will be interesting to and important for you to monitor the impact of these reforms of the 14–19 curriculum in the coming years.

There are likely to be further innovations in approaches to assessment and, as long as there remains a political focus on raising standards in our schools, external assessment and examinations will maintain their present high profile and powerful influence in educational practice.

Check which requirements for your course have been addressed through this unit.

Task 6.2.4 **LEAGUE TABLES AND PUPIL GROUPING**

Achievement and Attainment Tables ('league tables') have been said to influence not only what is taught in a school, but also how pupils are grouped and subjects they are offered for study, thus influencing their final qualifications. Investigate if pupil grouping in the school is affected by the 'league tables'; that is, is teaching and learning being driven by public accountability. You may find sharing this work with another student teacher helpful.

Check first your understanding of setting, banding, streaming and mixed-ability grouping (see Unit 4.1).

In your school experience school:

1   For each year cohort 7–11, identify how tutor groups are formed for pastoral purposes and the criteria used to do this.
2   Identify in broad terms which subjects teach tutor groups and those that reorganise the tutor groups for teaching purposes.
3   For your teaching subject and either English, mathematics or science in Year 7 find out how pupils new to the school are assessed and placed into teaching groups. Compare the criteria used for each subject. How flexible is the placement, can pupils move between groups? If so, how is this movement managed in practice?
4   For your teaching subject and either English, mathematics or science in Year 10 identify how pupils are grouped for teaching, and which courses the different groups are offered, and the criteria used to do this.

With the information gained from tasks 1–5 above arrange to interview a number of pupils in Year 10 to elicit their views about subject and course choice and grouping. Choose your subject area and, if time, one other subject. How do pupils make their subject/course choices at the end of Year 9? Are pupils:

1   provided with information to help make choices?
2   guided towards particular courses by teachers at the school?
3   aware of the long-term consequences of the courses they select?
4   encouraged to choose courses which help them to achieve five 'good' GCSE passes?

Within the same subject areas find out what pupils think about the groups they are taught in. Are pupils aware of:

1   the reasons they were allocated to a group in that subject?
2   any advantages to the grouping system?
3   any limitations to their particular group? For example, limit on grade at GCSE?

Finally interview a head of department for one or more subject areas chosen to find out their views about pupil choice of subject in Year 9 and how pupils are grouped for teaching purposes in their subject. You could use the questions you have put to the pupils.

Write a summary of your findings for discussion with your tutors, indicating whether the importance of league tables has had any effect on the way pupils are grouped. Evidence should be kept anonymous.

Background reading
Black, P. (1998); DfES (2006a); Wiliam, D. and Bartholomew, H. (2004).

FURTHER READING

Brooks, V. (2002) *Assessment in Secondary Schools: The New Teacher's Guide to Monitoring, Assessment, Recording, Reporting and Accountability*, Buckingham: Open University Press.
Written specifically for new teachers, this is a comprehensive introduction to all aspects of assessment in the secondary school.

DCFS (Department for Children, Schools and Families): http://www.dcsf.gov.uk/.
An informative website that is particularly useful for looking at school achievement and attainment tables, autumn package information and for up-to-date information about government initiatives.

Fautley, M. and Savage, J. (2008) *Assessment for Learning and Teaching in Secondary Schools*, Exeter: Learning Matters.
Targeted specifically at student teachers, this text links explicitly to the new QTS Standards, and the tasks provide opportunities for reflection and for practising the range of skills involved in assessing pupils.

QCA (Qualifications and Curriculum Authority): http://www.qca.org.uk/.
A useful website for detailed information on the range of qualifications available in secondary schools, such as Diplomas, as well as current information on NC tests.

Stobart, G. (2008) *Testing Times: The Uses and Abuses of Assessment*, Abingdon: Routledge.
A critical review of aspects of current assessment practice; addresses five issues that currently have high-profile status: intelligence, testing, learning skills, accountability and formative assessment.

Additional resources for this unit are available on the companion website:
www.routledge.com/textbooks/9780415478724

7 THE SCHOOL, CURRICULUM AND SOCIETY

This chapter takes you away from the immediacy of teaching to consider the aims of education, how those aims might be identified and, more importantly perhaps, how the curriculum reflects those aims.

In the day-to-day urgency of teaching the given curriculum it is easy to push the 'why' into the background and simply get on with the 'how'. As a prospective teacher you need to be able to explain to pupils and parents what the personal benefit to them is from learning your subject.

In Unit 7.1, a comparative and analytical approach is taken to examine assumptions about education. Unit 7.2 examines the school curriculum in terms of aims and purposes. In the remaining units, we examine the structure of the curriculums in England, Northern Ireland, Scotland and Wales. The units for Northern Ireland and Wales are included on the website.

Different criteria for 'qualified teacher status' are laid down for England, Northern Ireland, Scotland and Wales and comparing these different expectations helps you develop your philosophy of education.

If education is 'what is left after most of what you have learned in school is forgotten', then what is education for and who decides? Young people, between the ages of 5 and 16, spend a substantial part of the formative period of their lives in school and a significant slice of the national budget is channelled into education so you can expect to be held accountable for what you teach.

> **Readings for Learning to Teach in the Secondary School: A Companion to M Level Study**
> brings together essential readings to support you in your critical engagement with key issues raised in this textbook.

AIMS OF EDUCATION

Graham Haydon

INTRODUCTION

Education is a value-laden activity; this unit is designed to help you reflect on the values you encounter in your work and the values you yourself bring to it. People's ideas about the aims of education may, in part, be simply 'read off' from the educational traditions of their own society, which already incorporate certain shared values; and they may be formed through an individual's own reflection on their personal values. Not surprisingly, then, in a complex society there is room for differences in views about educational aims. Individuals with, for example, different educational and life experiences, different religious beliefs and cultural traditions, and different political tendencies may all differ in their conceptions of the aims of education. In Britain views about the aims of education in the past have remained often more implicit than explicit, but in recent years there has been some conscious attention to aims at government level. Thus what may at first sight seem rather an abstract question – what should the aims of education be? – is in fact an unavoidable part of the context in which you are working as a teacher.

OBJECTIVES

At the end of this unit you should be able to:

- list a variety of actual and possible aims for education
- reflect on and formulate your own aims on being a teacher
- discuss aims of education with other teachers and with parents

Check the requirements for your course to see which relate to this unit.

THE SOCIAL AND POLITICAL CONTEXT OF AIMS

One difference between education systems is that the aims which teachers are expected to pursue may be decided at different political or administrative levels. Many countries today have a national education system, at least partly state-funded and state-controlled. In some cases, as once in the Soviet Union, a clearly defined ideology sets aims which the whole education system is meant to promote. Even in a more decentralised system in which many decisions are taken at a local level, there may be across the whole society a more or less widely shared sense of what the aims of education should be. Thus in the USA there seems to have been in the early decades of the twentieth century a widely shared sense that one aim of the national education system was to make a single nation out of diverse communities.

In Britain, both historically and today, the picture is mixed. Through much of the twentieth century schools had a good deal of autonomy, from a legal point of view, in setting and pursuing their aims, though in many cases the aims of a particular school were not made explicit. There was also room for some variation at local authority (LA)[1] level. For instance, in the 1970s and 1980s there were cases in which particular local education authorities (LEAs) pursued more radical equal opportunities policies than were supported centrally. A case in point (though it is not relevant to go into the details here) is the political controversy over 'Clause 28', which began in the 1980s and lasted into the present century. Some LEAs were perceived, rightly or wrongly, as aiming to promote homosexuality through programmes and curriculum materials used in schools, and legislation was brought in to rule this out (section 28 of the Local Government Act 1988). There was a good deal of confusion in the debate over this (particularly since the influence of LAs over individual schools had been very much reduced), until eventually the relevant legislation was repealed as part of the Local Government Act 2003. Whatever your own view of the rights and wrongs of the issue, it illustrates that the question of whether schools should or should not be pursuing certain aims is potentially controversial.

The more general question of how far there is to be scope for diversity between different localities and different schools is still being played out in Britain in the twenty-first century. A system which in some respects sets up competition between different schools, each aiming to attract and hold onto pupils, encourages each school to make its own aims clear to parents and prospective pupils, and perhaps to present itself as being in some way distinctive from other schools at the level of aims. As obvious examples, we might expect that schools labelled (and funded) as technology colleges, or schools labelled as sports colleges, would give explicit attention in their public statements of aims to the focus of their activities. At the time of writing, the tendency in England, in line with government policy, is towards the majority of secondary schools having a curricular specialisation (while still teaching the National Curriculum (NC)). There are also likely to be increasing numbers of faith schools, and it could be expected that if a school has been set up by a particular religious community and is dedicated to offering education within a particular faith, this would be reflected in its aims. If you have school experience within a specialist or faith school, or if you yourself attended one, keep this in mind in the tasks that follow.

While in some respects the last two decades have seen increasing room for variation in aims between schools, there has also been increasing attention to aims at the level of national politics. Though there has not been widespread public debate about the aims of education, politicians often express their views on the matter. These views are not always put explicitly in the terminology of aims, but when politicians say that schools should enable Britain to compete economically with other nations, or that schools should inculcate moral standards or should promote active citizenship, they are in effect recommending certain aims for schools. At this broad level it is generally assumed that the same aims are shared by all schools.

Thus there is a potential tension between (a) the possibility of a diversity of aims in different schools, perhaps because they are serving rather different communities, and (b) the promotion of common aims across the school system as a whole. At the end of the twentieth century the focus on common aims became more prominent in England, for reasons connected with the revision of the NC, as we shall see in a later section.

As a student teacher, then, you are working in a context in which many expectations about aims are already in place. Even if aims were not mentioned in legislation, or in the prospectus of your school, you would still have to recognise that other teachers, your pupils, their parents, and the wider society all have views about what you should be doing, and a legitimate interest in what you are doing. What aims you pursue as a teacher are clearly not just up to you.

Is there, in that case, much point in your doing your own thinking about the aims of education? It is a premise of this unit that there is a lot of point in this, in fact that any good teacher has some view about the aims of education. (This does not mean that your view is different from anyone else's, but it does mean that it is a view that you have thought through and endorsed for yourself.)

Here are two reasons why your own thinking about aims is relevant (you may well think of further reasons). First, within the constraints, your own thinking about aims influences the way you approach your task as a teacher of a particular subject (this aspect of aims of education is discussed further in Unit 7.2). Second, as a citizen, you have the same right as any other citizen to form and express your own view about the aims of education in general; but at the same time other people might reasonably expect that as a member of the teaching profession you are in a better position than the average citizen to make your views clear and be prepared to argue for them.

THINKING ABOUT AIMS

Tasks 7.1.1 and 7.1.2 are intended to give you some insight into the nature and variety of aims in education, as well as some experience in thinking about aims and their implications and discussing this with others.

Task 7.1.1 **SCHOOL AIMS: A COMPARISON**

When you have carried out this task by yourself, try to compare your findings with those of other student teachers. Identify the aims of two schools with which you are familiar, for example: the school in which you received your own secondary education (or the majority of it, if you changed schools) and your current school experience school.

For the first school, your data will be wholly or largely from your own memory. Answer the following questions as far as you can:

■ Did your school have an explicit statement of its aims?
■ Were you as a pupil aware of the school's aims?
■ In what ways did the particular aims of your school impinge on your experience as a pupil?

For your school experience school, ask:

■ Does your school have an explicit statement of aims – if so, what does it say?
■ Are the pupils you are teaching aware of the school's aims?
■ Does the existence of these aims appear to make any difference to the pupils' experience in the school?

If you are a parent, you could also identify the aims of your child's school, using the school's documentation and, perhaps, discussion with staff.

Answering these questions in the case of your present school experience school gives scope for some small-scale empirical research. Depending on your subject, you may be able to incorporate some research into your teaching, e.g. in a discussion about school aims or through pupils themselves conducting a survey into how far their fellow pupils are aware of the school aims. You should discuss first with your school tutor any inquiry you plan.

Compare your findings for the two schools. Do you find that aims have a higher profile in one school or the other? Is there any evidence that the existence of an explicit policy on aims enhances the education the school is providing? Check the standards for your course which identify with the aims of schools.

Task 7.1.2 THE GOVERNING BODY: AIMS FOR A NEW SCHOOL

This is a group task involving role play. It is suitable for a group of several student teachers in the same school, or for a tutorial session with student teachers from several schools.

With other student teachers, role-play a governors' meeting which is intended to put together a statement of aims for a new school (imagining that it is a new school allows you to start with a relatively clean sheet). Within the allotted time (say, one hour) you must try to produce a statement of aims to be included in the prospectus, to help show prospective pupils and their parents what is distinctive about the new school and its educational priorities.

Before you start the role play, you should agree on any special characteristics of the area in which the school is located. It may be best to make it a school which has to serve a wide range of interests, i.e. a comprehensive school with a socially and ethnically varied intake.

Depending on the number in your group, you can assign individuals to some of the following roles as governors. (You may think there is some stereotyping in the brief descriptions of these roles. If you have experience of role play, you should be able to distance yourself from the stereotypes.)

- A Conservative-voting company director
- A Labour-voting trade union leader
- A Church of England vicar
- A spokesperson for the main local ethnic community
- A parent of a bright child, with high academic ambitions for their child
- A teacher-governor
- The headteacher

One of you should be elected to chair the meeting and another to take notes on the points made and record anything which is agreed.

After the role play, if you have not arrived at an agreed statement, talk about what it was that prevented agreement. In what ways does the disagreement within your group reflect the actual diversity of interests and cultures within our society?

Keep your notes from Task 7.1.1 and 7.1.2 in your professional development portfolio (PDP) for use when you are considering what schools to apply to work in and to preparation for an interview for a teaching post (see Unit 8.1).

WHY BOTHER WITH AIMS?

> Education as such has no aims. Only persons, parents and teachers, etc., have aims, not an abstract idea like education
>
> (Dewey, 1916: ch. 8, 'Aims of education')

Your experience in doing the tasks may have backed up Dewey's point. You may have considered how much difference aims can make. You may also have seen that different people can have different aims for education. A statement of aims on paper does not, of course, make any difference by itself (there is an example of this below in the context of the English NC). But what people do and how they do it is certainly influenced by what they are themselves aiming at. Aims, at their different levels, can affect:

■ How a whole school system is organised. (For example, the movement towards comprehensive education which began in the 1960s was driven at least partly by explicit aims of breaking down class barriers and distributing opportunities for education more widely.)

■ How an individual school is run. (For example, various aspects of a school's ethos and organisation may be motivated by the aim that pupils should respect and tolerate each other's differences.)

■ How curriculum content is selected and taught (there is more on this aspect of aims in Unit 7.2).

MAKING SENSE OF THE VARIETY OF AIMS

Because aims can be so diverse, it is useful to be able to categorise aims in some way. There is no single right way of dividing different aims into categories; in fact it is more helpful to be able to work with different categorising schemes.

Some approaches assume that education is aiming to develop personal qualities and capacities of one sort or another, and therefore divide aims up into the categories of concepts, attitudes, skills and knowledge. A distinction between academic, personal and vocational aims is related to this, but does not coincide exactly with it. Part of the importance in practice of the academic/personal/ vocational division is that it can be recognised to some extent in ways in which different types of school historically have conceived their task.

But even if you are confident that education should be developing certain qualities or capacities in individuals, there is the wider question – still one about aims – of why this should be done. Is it just for the benefit of each individual, or for the general good of society? In other words, what should the aim be of the educational system as a whole? Should it be to do the best that can be done for each individual? Or should it be to promote and maintain a certain kind of society – perhaps a democratic society, or a just society, or an economically successful society?

This question introduces another distinction which has its uses but which, like any categorisation, can be misleading if not used carefully. If you had to make a sharp choice, say between developing in people the capacities which enable each individual to lead a fulfilling life, and giving them the skills and attitudes which fit them to be cogs in an impersonal system, then there would be a real divide between aiming at the good of the individual and aiming at the good of society. But it is not necessarily like that. If, for instance, your view of a good society is that it is the kind of society in which all individuals have the capacity to lead fulfilling lives and there are no obstacles in the way of people exercising those capacities, then there need be no contradiction between aiming at the good of the individual and aiming at the good of society.

In fact, even without being idealistic, many aims do cut across the individual/society division. Giving people skills which enable them to get productive jobs, for example, is in many instances of benefit to the individuals concerned and to others in society. Other cases may be more difficult. Certain types of academic knowledge might benefit individuals who have that knowledge, if only because they happen to find it interesting, without having any spin-off for others. Certain types of moral socialisation might benefit others while on balance having a negative effect for the self-fulfilment of the individuals concerned. Historical examples of this might include working-class boys being brought up to assume they would follow in their father's kind of employment, or girls being brought up to be always deferential to the male members of their families. Few people now would explicitly endorse such aims as these (see the section below 'Equal aims for everyone?') but prejudices can still survive and make a difference. This is one reason both for making aims explicit and for paying attention to the 'hidden curriculum' (see Unit 7.2).

JUSTIFYING AIMS

In your discussions and role plays people have been trying to defend their own conceptions of what the aims of education should be. What sorts of argument have they been using?

One approach which used to be favoured by philosophers was to say that certain aims are incorporated into the *concept* of education. So, if someone aimed at inculcating in pupils particular religious or moral beliefs, this could be rejected on the ground that inculcating unquestioned beliefs is simply not part of our concept of education. In fact (it might be said) it is part of our concept of *indoctrination*, whereas the concept of *education* implies the promotion of rationality and critical thinking.

You may agree with this. Its limitation as an argument, though, is that it does not allow you to meet on their own ground people who might argue that what they want teachers and schools to do *is* to inculcate certain unquestioned beliefs. They may not mind if you don't *call* this education; it is what they want you to aim at.

In the end, argument about educational aims is not about concepts, or how people use words. It is also not a matter that can be settled empirically by surveying what people actually think the aims of education should be. The history of education shows how much ideas about the aims of education can change from one generation to another (as the next section illustrates); if certain ideas about educational aims are widely accepted now, that does not show they are beyond criticism (see also Bottery, 1990: chs 1 and 2). Discussion about educational aims is fundamentally discussion about values, an ethical discussion about the responsibilities of adult members of society towards the young members of society and towards the next generation. In some way probably almost everyone would agree that education should be seeking to improve the quality of life of individuals or of the society in general – otherwise why bother about it? – but there is room for dispute over what is important in a good quality of life. Do we do more for someone's quality of life by enabling them to earn a good income, or by developing, say, their scientific curiosity or their appreciation of art (even if these do not help them to work productively)?

Argument about the aims of education, then, may come down in the end to questions of what matters most in life. But it also has to be about the distinctive contribution that teachers and schools can make to promoting what is important in life. If we agree, for instance, that health is important in everyone's life, this does not mean that the aim of teachers in relation to health is the same as the aim of nurses and doctors, but it may be that there are particular kinds of contribution that teachers can make to improving people's chances of having a healthy life.

The individual/social dimension affects the issue of justification. If some aspect of education is seen as being of value only for (some) individuals, there may be questions about why society – people in general – should support it. If some aspect is seen as being of value only for the majority of people, but not everyone, there may be questions about whether it should be imposed on individuals if they do not freely choose it. Such questions, though, may often stem from a simplistic contrast between society and individual; for a more sophisticated view, see Dewey (1916).

Some writers today would argue that aims for education should be derived explicitly from a conception of the kind of society in which young people will be living. If this is to be a liberal, democratic and multicultural society, then education should be preparing people to live in that kind of society, and other more particular aims will follow from this.

The last two sections of this unit discussed the variety of aims for education and ways in which aims can be justified. To explore further these two aspects of aims of education we suggest you address Task 7.1.3.

> ## Task 7.1.3 WHY TEACH INFORMATION AND COMMUNICATIONS TECHNOLOGY (ICT)
>
> ICT is a relatively new subject in the curriculum but is much more influential than just a subject. ICT is conceived of as contributing to other subjects, to introducing new ways of learning and influencing the way teachers teach.
>
> Discuss the place of ICT in the school curriculum in terms of possible aims, such as the personal, academic and vocational. How might those aims be justified to the pupil, the parents or the wider society? You may wish first to re-read the previous two sections of this unit.

EQUAL AIMS FOR EVERYONE?

Through much of the history of education, it would have been an unquestioned assumption that the aims of education should be different for different people. Plato built his conception of an ideal state (*The Republic*) on the argument that the people in power would need a much more thorough education than anyone else. A similar position was apparent in Victorian Britain, where the expansion of education was driven in part by the aim that the mass of the population should be sufficiently well educated to form a productive workforce but not so well educated that they might rebel against the (differently educated) ruling classes. In the mid-twentieth century, within a system selecting by ability, there were different aims behind the education offered in different types of schools: secondary modern, technical and grammar. Also, through much of the twentieth century, differences in aims were apparent between boys' and girls' schools, and between religious foundations and schools which were effectively secular.

Today the unquestioned assumption is often the reverse: that the basic aims of education are the same for everyone, even if different methods have to be used with different people in pursuing the same aims, and even if some people go further in the process than others. This assumption underlies many important developments in the promotion of equal opportunities. One of the basic reasons for being concerned with equal opportunities is that, if what you are aiming at is worthwhile, no one should be excluded from it because of factors, like race or gender, which ought to be irrelevant to achieving these worthwhile aims. But this basic assumption is still not without its problems.

In the area of special educational needs (SEN) in England and Wales, for instance, the Warnock Committee, which was set up in the late 1970s to look into the education of pupils with physical and mental disabilities, argued that the fundamental aims of education are the same for everyone (Department of Education and Science (DES), 1978). This was part of the thinking which led in the 1980s to the integration of an increasing proportion of pupils with SEN into mainstream schools, rather than their segregation in special schools. (See Unit 4.6 for the current position.) Schools require, for example, to be specially equipped to respond to some needs and to appoint additional specialist staff. This requirement has led to funding problems and to inadequate provision due to a shortage of suitably trained teaching staff.

As regards gender, few people would now suggest that the aim of education for girls should be to produce wives and mothers while the aim of education for boys should be to produce breadwinners. When people today argue for single-sex schooling or for dividing teaching groups according to gender, it is usually not because they think there are separate aims for the education of boys and of girls, but because they recognise that giving boys and girls an equal opportunity to achieve those aims requires attention to practical conditions, and this may make a difference to the means though not to the ends. So, for instance, the teaching of boys and girls may be more effective if they are taught without the distractions or pressures present in mixed-sex groups (whether this actually is so is a matter for research). Even so, some might argue that a degree of differentiation in aims is needed; perhaps, for instance, there should be an attempt to develop assertiveness in girls and sensitivity in boys. See Unit 4.4 for further discussion of gender issues.

Turning to different cultural, religious or ethnic groups, it is not surprising if governments expect the same aims to be pursued for all groups; anything else would seem grossly discriminatory. But at the same time the members of particular groups may have special aims they would like to see pursued for their own children. To some religious believers it may be more important that their children are brought up within the faith of their community than that they are brought up as citizens of a secular society; such differences lie behind the demands that some religious groups make for separate schools.

These examples illustrate again the point made earlier, that while at one level statements of the aims of education can appear rather platitudinous and bland, there is the potential for controversy when aims are considered in more detail and an attempt is made to see how the pursuit of certain aims can be implemented in practice.

AIMS OF THE NATIONAL CURRICULUM

You may be working within the constraints of a NC, such as that for England (http://curriculum. qca.org.uk). How much scope does this leave you and your colleagues in deciding on your aims?

A first stage in answering this is that you should be aware of what the documentation of your NC actually says about aims. In the case of the NC for England there have been several changes since it was first introduced in 1988. Does this mean that the aims the government sets for schools in England have changed substantially over a twenty-year period? Probably not. What has changed is the amount of explicit attention given to aims; the wording of specific aims; and, arguably, the extent to which the details of the curriculum are actually grounded in the stated aims. In the 1988 documentation the explicit aims were limited to the following: 'to promote the spiritual, moral, cultural, mental and physical development of pupils at the school and of society; to prepare pupils for the opportunities, responsibilities and experiences of adult life'.

As a statement of aims, this was not very controversial (with the possible exception of the idea of spiritual development). Its problem was that it was so broad and general that it gave very little guidance. And in fact there was no indication within the rest of the original documentation of the NC for England and Wales that its content had been influenced at all by the statement of aims. That statement seemed to be an example of an error which it is easy for government agencies, and also schools, to slip into: setting out a statement of aims which looks good, but which makes no apparent difference to what actually happens.

In the 1999 revision of the NC there were again two overall aims: to provide opportunities for all to learn and achieve; and to promote pupils' spiritual, moral, social and cultural and emotional well-being. But there were also two more significant changes: that it was explicitly said that the aims of the curriculum were rooted in certain values that were held to be widely shared in the society; and that a listing of more specific aims was given under each of the two overall aims. In the 2008 version of the NC the idea that aims rest in underlying values is still there, but there are now three overall aims:

■ Education influences and reflects the values of society, and the kind of society we want to be. It is important, therefore, to recognise a broad set of common purposes, values and aims that underpin the school curriculum and the work of schools.

■ Clear aims that focus on the qualities and skills learners need to succeed in school and beyond should be the starting point for the curriculum. These aims should inform all aspects of curriculum planning and teaching and learning at whole-school and subject levels.

■ The curriculum should enable all young people to become:
- ■ successful learners who enjoy learning, make progress and achieve
- ■ confident individuals who are able to live safe, healthy and fulfilling lives
- ■ responsible citizens who make a positive contribution to society.

(http://curriculum.qca.org.uk/key-stages-3-and-4/aims/index.aspx)

Each of the three main aims is broken down into more detail; see http://curriculum.qca.org.uk/uploads/Aims_of_the_curriculum_tcm8-1812.pdf?return=/key-stages-3-and-4/aims/index.aspx.

As a statement of aims, it could be said that this has 'something for everyone'. There is probably not much there that anyone could dissent from. At the same time there are ideas which are open to interpretation, and there is scope for balancing one aim against another in all sorts of ways. As a teacher in a school in England (and also elsewhere, even if the same legislation does not apply to you), you could consider this statement of aims as a resource you can draw on. Unit 7.2 suggests ways in which you may do this in relation to your own curriculum subject.

Is this a statement of aims that actually makes a difference in schools? At least we can say that there is enough detail there to make it possible to ask whether a school is actually working in a way that helps to promote these aims. To take, for instance, just one subordinate aim from each of the three overall aims, we can ask whether enough attention is being paid in a school to enabling pupils to be 'creative, resourceful and able to identify and solve problems', to helping them to 'have a sense of self-worth and personal identity', and to ensuring they become citizens who 'challenge injustice, are committed to human rights and strive to live peaceably with others'. These are clearly ambitious aims, and they are just three among twenty-nine in the NC statement.

One interesting point about this statement is that it is explicitly presented as setting out 'The aims of the curriculum'. Units 7.2 and 7.3 give you a basis to begin thinking about how far the curriculum you are working with is likely to promote such aims as these. At the same time it is clear that there are other aspects of the life of a school, besides the formal curriculum, that have to be involved if such aims are to be achieved. In the end it is for you and other teachers to determine how much difference this or any other statement of aims makes to what actually goes on in your school. Task 7.1.4 invites you to compare the values and aims for schooling set out by government.

Task 7.1.4 **AIMS AND EDUCATION**

Working from the full statement of values and aims for schooling in one or more country (see website information at the end of the unit), consider:

■ Do these aims seem to you to fit together into a coherent idea of what education is about?

■ Is there is anything in these statements of aims which you would not have expected to see there? Is there anything which is likely to prove controversial? Is there anything you think should be mentioned which is not mentioned?

Discuss your responses with other student teachers, or with your tutor. Keep these notes together with those from Task 7.1.1 and 7.1.2.

SUMMARY AND KEY POINTS

In working as a teacher you necessarily have some aims, and these are more likely to be coherent and defensible if you have thought them through. At the same time, you are operating within the context of aims set by others. Aims can exist at different levels, local or national. In Britain, in recent years, the dominant tendency has been towards a common conception of aims for everyone, and most recently in England a broad set of aims has been incorporated into the documentation for the NC. But this still leaves room for you to form your own view as to the most important priorities for education, and to discuss with others how these aims can best be realised.

It is always possible to raise questions about the justification of educational aims. Ultimately our aims for education rest on our values – our conceptions of what makes for a good life both for individuals and for our society as a whole. Because we do not share all of our values with each other, there always is room for debate about the aims of education.

Check which requirements for your course have been addressed through this unit.

FURTHER READING

Bottery, M. (1990) *The Morality of the School*, **London: Cassell.**
A wide-ranging discussion relevant to any teacher thinking about aims. Chapter 1 gives you an exercise to help you classify your own beliefs about education into one of four categories: the cultural transmission model, the child-centred model, the social reconstruction model, and the 'GNP (gross national product) code'.

Brighouse, H. (2006) *On Education*, **Abingdon: Routledge.**
Part One, 'Educational Aims' argues for four aims: educating for self-government; educating for economic participation; educating for human flourishing; and educating for citizenship.

Dewey, J. (1916) *Democracy and Education*, **New York: Free Press.**
A classic book (often reprinted) which is still well worth reading. Though Dewey is often thought of simply as an advocate of child-centred education, his educational theory is part of a well-worked-out theory of the relation between individual and society and of the nature of knowledge and thought. See especially Chapters 1 to 4, 8 and 9.

White, J. (2007) *What Schools Are For and Why,* **Philosophy of Education Society of Great Britain (IMPACT series no. 14).**
Contains a sympathetic critique of the statement of aims in the NC for England, and proposes an improvement on it.

RELEVANT WEBSITES

English National Curriculum for Key Stage 3 and 4: http://curriculum.qca.org.uk/key-stages-3-and-4/index.aspx

Northern Ireland Curriculum: http://www.nicurriculum.org.uk/

Scottish Curriculum: http://www.ltscotland.org.uk/5to14/guidelines/

Welsh National Curriculum: http://old.accac.org.uk/eng/content.php?cID=6

Additional resources for this unit are available on the companion website:
www.routledge.com/textbooks/9780415478724

NOTE

1   The term LA replaced LEA in 2004 when social care and education departments within councils were merged.

UNIT 7.2

THE SCHOOL CURRICULUM

Graham Haydon

INTRODUCTION

This unit discusses the curriculum as one of the most important 'tools' through which educational aims can be realised. But we need first to be clearer about what the term 'the curriculum' refers to.

The *curriculum* is an important part of the context within which you work as a teacher. The planned or formal curriculum is the intended content of an educational programme set out in advance. We refer later to the informal and hidden curriculum. Like other aspects of the context of your work (the school buildings, say, or the administrative organisation of the school), the curriculum forms a 'frame' to what you are doing even when you are not explicitly thinking about it. But often you find that you do refer to the curriculum, in your everyday conversations with colleagues, and less frequently perhaps in meetings with parents or in talking to pupils in a pastoral role. It might seem that the curriculum is so clearly part of the context of your work that it must be obvious what the curriculum is. In which case, why does a book of this nature need units on the curriculum?

The purpose of this unit is to show you that, once you think about it, it is not so obvious what the curriculum is, and that it is not something you should, as a teacher, take for granted. Rather than relying on implicit assumptions about the curriculum, you should be able and willing, as part of your professional role, to think about the curriculum, about its role in education and about ways in which it is controversial and might be open to challenge. In doing this, you will, of course, need to keep in mind the relevant legislation and government documentation for the country within which you are working. The 1988 National Curriculum (NC) was for England and Wales, but the 1999 legislation made changes specifically for England. Scotland and Northern Ireland have their own curricula, and since devolution the curriculum in Wales has also increasingly been diverging from that in England: for the position in Wales in 2008, see websites at the end of this unit.

OBJECTIVES

At the end of this unit you should be able to:

- distinguish a number of different conceptions of the curriculum
- discuss ways in which the curriculum may or may not help to realise educational aims
- see why the content of the curriculum, even if often taken for granted, is potentially controversial
- discuss the place of your particular teaching subject within the broader curriculum.

Check the requirements for your course to see which relate to this unit.

THE CURRICULUM IN GENERAL AND WITHIN PARTICULAR SUBJECTS

It helps to avoid confusion in the rest of this unit (and hopefully in your thinking more generally) if we distinguish between the curriculum of a school (or even of schools in general) and the curriculum within a particular subject. Sometimes this distinction is marked by speaking of the 'syllabus', rather than curriculum, of a particular subject. The term 'syllabus' usually refers to a specific programme in a specific subject set out in detail in advance, possibly designed by a particular teacher, but often laid down by an examination board or other body external to the school. But it is common now to speak of, say, 'the science curriculum' or 'the arts curriculum'; in the details of the NC for England (http://curriculum.qca.org.uk) the main use of the term 'syllabus' is a rather specialised one in Religious Education, where there is a system of locally agreed syllabuses.

For most of this unit the focus is on the broad curriculum. Questions are raised about the role of particular subjects within the curriculum in general, more than about what goes on *within* the teaching of particular subjects. But we shall have to say something about the latter point as well, because the role of a subject within the curriculum partly depends on what is done within that subject (see the subject-specific and practical books which are a part of this Learning to Teach series, p. ii). (So far, the term 'the whole curriculum' has been avoided because that too may carry some ambiguity.) In Task 7.2.1 we ask you to compare the two curricula.

Task 7.2.1 SCHOOL CURRICULA: A COMPARISON

(This task is deliberately parallel to Task 7.1.1 on aims in Unit 7.1.)

When you have carried out this task by yourself, try to compare your findings with those of other student teachers.

Select two schools with which you are familiar, for example:

■ the school in which you received your own secondary education (or the majority of it, if you changed schools)
■ your current school experience school.

From memory, write down briefly what was in the curriculum of the school you attended as a pupil. Then (without referring to documentation at this stage) write down what is in the curriculum of your school experience school. Compare the two accounts.

THE FORMAL CURRICULUM

There *could* be considerable variety in what you and other student teachers have written for Task 7.2.1, because the term 'curriculum' can be used in various ways. But it is likely that what you have written down, for both schools in the comparison, is a list of subjects. What this illustrates is that when people refer to 'the curriculum' without qualification, most often they think of what we can usefully label 'the formal curriculum'. This is the intended content of an educational programme, set out in advance.

At a minimal level of detail, the formal curriculum can be stated as a list of names of subjects. Thus the statement that we expect pupils to learn 'reading, writing and arithmetic' would itself be a curriculum statement (about the thinnest possible). So would be the extended and modified version adopted in the school attended by Lewis Carroll's Mock Turtle:

Reeling and Writhing, of course, to begin with . . . and then the different branches of Arithmetic

– Ambition, Distraction, Uglification and Derision . . . Mystery, ancient and modern, with Seaography: then . . . Drawling, Stretching, and Fainting in Coils (but the Mock Turtle didn't do 'Laughing and Grief' though they were on the curriculum; they must have been options).
(Gardner, 1965: 129–130, with explanations)

While the formal curriculum can be listed simply as a set of subjects, it is always possible to set out in more detail the content which is supposed to be taught and learned. Even when the curriculum is stated simply as a list of subjects, those who write it and those who read it have some implicit understanding of what goes into each subject. It is important to keep this in mind when comparing the curriculum offered in schools at different times. It is likely that your lists contain a number of items, probably the majority, that are common to both lists. In fact we can go further. The NC in England at the beginning of the twenty-first century can be compared with the secondary school regulations for England of 1904 (Aldrich, 1988: 48). Even with the mid-nineteenth-century curriculum which Lewis Carroll was parodying, we find many items in common (the most noticeable absence from Carroll's list, which appears in all the subsequent ones, is science).

This finding illustrates something about the extent to which the formal curriculum in schools in England has *not* gone through revolutionary changes. At the same time, it would be a mistake to conclude from the similarity of the lists that the curriculum has hardly changed at all. Even if we could set aside all changes in teaching method and concentrate solely on the content of the subjects, what is taught under the heading of history or science in the early twenty-first century is obviously going to be very different in many ways from what was taught under the same headings in the early twentieth century.

Another point to note under 'formal curriculum' is that the curriculum may contain parts which are optional. Even before the introduction of an NC in England and Wales in 1988 made certain subjects compulsory, it was normal for most of the curriculum in a secondary school to consist of subjects which all pupils were expected to take. But there may also be options within the curriculum, particularly in the later years of secondary school.

Related to the idea of a compulsory curriculum are the notions of a 'common curriculum', that is, one taken by everyone in practice, whether or not it is actually compulsory; and a 'core curriculum', the part of a whole curriculum which everyone takes, around which there is scope for variations.

THE INFORMAL CURRICULUM AND THE HIDDEN CURRICULUM

The notion of the *formal curriculum* refers to the content which is, quite deliberately, taught by teachers in a school, usually in periods structured by a timetable and labelled according to subject. So the fact that something is on the curriculum means that it is taught (or at least that the intention of the curriculum planners or of the school management is that it shall be taught). But since some pupils may fail to learn what teachers are intending to teach, the fact that something is a non-optional part of the formal curriculum does not guarantee that pupils learn it.

On the other hand, pupils may learn things in school which are not taught as part of the formal curriculum. There are two ways in which this can happen. One is that the school intends that pupils should learn things which cannot be directly taught in lessons. Many of the possible aims of a school, which you were thinking about in Unit 7.1, involve matters of this kind. If a school wants, for instance, to promote co-operation and consideration for others, then (if these are to be more than pious aspirations) it needs to do something to try to bring about co-operation and to encourage pupils to behave in considerate ways. (See also Unit 4.5 for further discussion of promoting common values.) Teachers might agree to build co-operative work into their lessons, whatever the subject; teachers and pupils might draw up a code of behaviour; there may be some system of rewards and sanctions; the school management may pay attention to the way that pupils move around the school during break times; and so on. All such arrangements can be counted as part of the *informal curriculum* of the school. The *curriculum* can be defined taking into account both the formal and informal curriculum in some way such as this: 'The school curriculum comprises all learning and other experiences that each school plans for its pupils' (Department for Education and Employment/ Qualifications and Curriculum Authority (DfEE/QCA), 1999a: 10).

But pupils may also learn things at school that the school does not intend them to learn. For several decades sociologists have pointed out that many pupils at school were learning, for instance, to accept passively what they were told or to see themselves as failures; while some were learning to identify with and follow the mores of a rebellious subculture; and some were learning racist and sexist attitudes, and so on. Such learning was not normally part of what the school was intending its pupils to learn, and the school may not have been aware of many of the things that its pupils were learning; from the school's point of view, these outcomes were side-effects of the pupils' time in school. So the term 'hidden curriculum' was invented to cover such learning.

The side-effects just mentioned are undesirable ones, and it has often been undesirable effects that people have in mind when they used the term 'hidden curriculum'. But side-effects could also be desirable ones; for instance, a side-effect of pupils of different ethnic backgrounds learning and playing together might be the development of understanding and respect. The point about the idea of the hidden curriculum is not that its content is necessarily bad, but that the school is not aware of it. Today, teachers are far more likely to be aware of the likely side-effects of all aspects of the school's activity. In that way, what might once have been part of a hidden curriculum comes to be hidden no longer. This does not mean that schools today have no hidden curriculum; it means that a school has to try consciously to uncover and become aware of side-effects of what it does in its teaching and its organisation.

If these side-effects are unwelcome – if, say, they work against the school achieving its intended aims – then the school may make deliberate attempts to counteract them. Often a school does this by paying attention to aspects of its teaching and organisation outside of the formal curriculum. So, where the learning of racist or sexist attitudes might once have been part of the *hidden curriculum* in some schools, it is more likely today that the *informal* curriculum includes anti-racist and anti-sexist policies. And it may also be that such policies alter what is done within the *formal* curriculum, e.g. within a Personal, Social and Health Education (PSHE) course or citizenship education (an area often referred to as the *pastoral curriculum*).

Mention of the informal curriculum shows that the curriculum as a whole is not, for any teacher, a rigid framework within which there is no room for flexibility or planning. Even when the formal curriculum is determined largely in advance, as in the NC for England, there is still scope open to the school to design the details of the curriculum and the way that links between curriculum subjects are (or are not) made; and there is space outside the NC since that is not supposed to occupy the whole timetable.

You should, then, see it as part of your professional role as a teacher that you can take an overview of the curriculum, have a sense of 'where it comes from' and be able to engage in discussion on whether it could be improved and, if so, in what ways.

CURRICULUM AS A SELECTION FROM CULTURE

A number of writers have referred to the curriculum as a selection from the culture of a society. 'Culture' here refers to 'everything that is created by human beings themselves: tools and technology, language and literature, music and art, science and mathematics – in effect, the whole way of life of a society' (Lawton, 1989: 17). Any society passes on its culture to the next generation, and in modern societies schooling is one of the ways in which this is done. But obviously no school curriculum can accommodate the whole of human culture; so a selection has to be made.

A natural question to ask next is how do we make that selection? Different curriculum theories give different answers.

A first move is to recognise that, since some aspects of culture are passed on or picked up independently of schools, it may make sense for schools in general to concentrate on matters which will not be learned if they are not included in the school curriculum; and secondary schools in particular have to try to build on, but not to duplicate, what pupils have learned by the end of primary school. Even these points give rise to many questions. For example, many young people of secondary age pick up much of what they know about computers, sport or popular music independently of school; does this mean there is no point in including study of these areas in the curriculum?

After putting on one side things which pupils learn independently of school (if we can identify such things), there are principles by which we might try to make a selection from culture. In this unit there is space to mention just three: to select what is *best*, or what is *distinctive* of a particular culture, or what is in some way *fundamental*.

The idea of selecting, and enabling people to appreciate, what is *best* goes back at least to Matthew Arnold. Arnold was not only a Victorian poet and a commentator on the culture of his day, but also a school inspector. Historically, this principle has been linked with the idea of whole areas of culture – 'high culture', centred on arts and literature, being of greater value than the rest of culture, and also perhaps being accessible only to a minority of society. The principle does not have to be interpreted in that way (see Gingell and Brandon, 2000, for an updated interpretation). Whatever area of culture we are dealing with, including for instance football and rock music, we may well want people to be able to appreciate what is good rather than what is mediocre. It does not follow, though, that the school curriculum should always be focused on what is best in any area. If we suppose, for instance, that the greatest science is that of Einstein or Stephen Hawking, it does not mean we place this science at the centre of the school curriculum. In many areas, if people are ever to be able to appreciate the best, they need to start by understanding something more basic.

Another principle of selection which is sometimes favoured is to pick from the whole of human culture what is *distinctive* of a particular culture – the way of life of a particular nation, or ethnic group, or religion. This may apply more to the detailed content within areas of the curriculum than to the selection of the broad areas. We do not just learn language, we learn a particular language; and while it is possible to study historical method, any content of history is that of particular people in a particular part of the world. One question for curriculum planning, then, is how far to select from what we see as 'our' culture, and how to interpret what is 'our' culture. That question, in England, has to be resolved in a context of a multicultural society, within a world in which there is increasing interaction between different cultures.

Rather than looking to what is best, or what is distinctive, we may try to look to what is *fundamental*. This idea may apply both across the curriculum and within subject areas of the curriculum. Within the sciences and mathematics, for instance, the idea of what is culturally distinctive may have little application (which is not to say that these subjects as actually taught are culture free), and the idea of teaching the best may be inappropriate. We need to think about what is fundamental. In the educational context, this does not mean what is fundamental in the whole structure of human knowledge; it means what people need to learn if they are to have a foundation on which further knowledge or skills can be built.

Thinking about the curriculum in general, we can also try to ask what is fundamental in the whole human culture in which people are living. But this question depends in turn on some particular understanding of what is important in human life. Is it the development of the capacities for rational thought and judgement? Then we might argue, as the philosopher of education Paul Hirst (1974) once did, that there are certain basic forms of human understanding – science, mathematics, interpersonal understanding and so on – which are not interchangeable and each of which is necessary in its own way to the development of rational understanding.

Or is human life more fundamentally about providing the material necessities of life? Then we might stress what can be economically useful, and our curriculum might be primarily a vocational one. Or is the essential aspect of human life, so far as education is concerned, the fact that people live together in groups and have to organise their affairs together? Then preparing people to be citizens might turn out to be most fundamental.

So far, none of these approaches looks as if it takes us very far, by itself, in selecting which aspects of culture should make up a school curriculum. Of course, there is much more to the arguments than can be considered in detail here. But even this much discussion suggests that the attempt to select from human culture does not take us far without an explicit consideration of the aims of education. We do not, after all, have to transmit culture just as it stands (and in any case it is constantly changing). We may have views about what kind of society we wish to see, or about what kind of persons we hope will emerge from education.

RELATING CURRICULUM TO AIMS

Unit 7.1 raised the question about the aims of education in general terms, suggested some approaches to answering it, and looked briefly at the current stated aims of the NC in England. The important point to emphasise is that the curriculum of schools is a major part of the way in which we attempt to realise educational aims. So, rationally, the planning of a curriculum should depend on how the overall aims of education are conceived. Historically, as was pointed out in the last unit, this has not always happened. Even when, as now, there is an NC in England incorporating an extended statement of aims, there is room to question how far the required curriculum is actually likely to realise the stated aims.

Among the aims mentioned in Unit 7.1, and in the statement of aims for the NC in England (http://curriculum.qca.org.uk/uploads/Aims_of_the_curriculum_tcm8-1812.pdf?return=/key-stages-3-and-4/aims/index.aspx) there are many aims which it would be difficult to relate to the content of any specific curriculum subject. In addition to those mentioned at the end of Unit 7.1, think for instance of the aims that pupils should 'relate well to others and form good relationships'; be 'willing to try new things and make the most of opportunities'; or 'appreciate the benefits of diversity'. It is clear that the *informal curriculum* has a large role to play here; any school which gave its attention *only* to the *formal curriculum* could not guarantee that these aims would be realised at all. In Task 7.2.2 we ask you to consider how the details of a curriculum might be developed from a set of general aims.

Task 7.2.2 **LINKING CURRICULUM CONTENT TO AIMS**

Look again at the statement of aims of the National Curriculum as used for Task 7.1.4. If you were responsible for planning a curriculum which could realise these aims, what would you think should go into that curriculum?

Discuss with other student teachers or your tutor.

Notice that this question about the National Curriculum is asking you to consider the formal curriculum with which schools in England now operate (apart from the small amount of time available to schools for pursuing activities outside the National Curriculum). But we can also think about ways in which a school's informal curriculum may further the same aims.

How far, then, *can* the formal curriculum promote the kinds of aims set out? It is clear that the traditional curriculum subjects have a role in promoting pupils' learning and achievement (achievement within those subjects, that is, although there are other kinds of achievement as well), and in promoting their intellectual development. What is not so clear is the role of the traditional subjects in promoting the broader aims of an emotional, moral and social kind (which were always recognised in NC documents since 1988 and have now gained greater prominence than before).

The first version of the NC in England and Wales, brought in by the Education Reform Act 1988, attempted to address such aims by incorporating a number of cross-curricular themes: health education, citizenship education, careers education and guidance, environmental education, and education for economic and industrial understanding. These themes did not have the statutory force of the core and foundation subjects, and it was left largely to individual schools (with limited published guidance) to decide how to teach them. In fact, in many schools the cross-curricular themes were not systematically taken up at all. Of the original cross-curricular themes, only citizenship has gained the status of a statutory subject, and the other areas are no longer there as distinct themes. In relation to the original cross-curricular themes, citizenship education carries a heavy burden, as does PSHE, even though that is not statutory. The 2008 NC (see Unit 7.3) also puts considerable emphasis on personal development (http://curriculum.qca.org.uk/key-stages-3-and-4/personal development/index.aspx). On the relevance of these areas see also Unit 4.5.

Though it does not have the original themes, the 2008 NC does have a set of cross-curriculum dimensions (http://curriculum.qca.org.uk/key-stages-3-and-4/cross-curriculum-dimensions/index. aspx). Whatever your own curriculum subject, you are likely to find that it has the potential to contribute to at least some of these dimensions. It is still the case, and likely to remain so for the foreseeable future, that the bulk of curriculum expectations revolve around a limited list of subjects, many of which, as we have seen above, have been in the curriculum of schools in Britain (and many other countries) for a long time. How far can the inclusion of these subjects be justified, not just because they have traditionally been in the curriculum, but because they can actually be shown to contribute to the stated aims of the curriculum? Each subject in the documentation of the NC for England contains a statement of importance; but it is not always the case that these statements refer explicitly to the stated aims of the curriculum overall. It still appears to be the case in the 2008 curriculum as in the 1988 version that the list of subjects was already in place before the aims were formulated.

Nevertheless, questions can still be raised about the contribution of individual subjects to the overall aims of the curriculum. How such questions are answered not only bears on the justification of particular subjects being in the compulsory curriculum at all, but can also make a difference to the aims of a teacher of a particular subject. In science and mathematics, what is the balance between equipping pupils with skills which they can put to practical use (thus furthering training and employment opportunities) and trying to show pupils something of the sheer fascination which science and mathematics can hold quite apart from their applications? In history, what is the balance between trying to promote a sense of a common British inheritance and exploring the history which has led to Britain being the multicultural society it now is? Similar questions can be raised about other subjects. The book *Rethinking the School Curriculum: Values, Aims and Purposes* (White, 2004) devotes a chapter to each subject of the National Curriculum, with the exception of PSHE, citizenship and information and communications technology (ICT), and also to religious education. The discussions and further references in that book will help you in thinking about the role of your own subject within the whole curriculum (see also Task 7.2.3).

Task 7.2.3 JUSTIFYING YOUR SUBJECT IN THE SCHOOL CURRICULUM

This can be a two-part task, with an individual stage followed by a group stage.

The task is to contribute to a school prospectus (it might be for the same imaginary school which you used in Task 7.1.2.). Suppose that the school has adopted the statement of aims from a national curriculum.

Your individual task is to write a paragraph of not more than one hundred words setting out for prospective parents the ways in which your teaching subject fits into the whole curriculum, and thus contributes to realising the overall aims of the curriculum. (Remember that some parents – and pupils – may wonder what the point of studying certain subjects is at all.)

The group task for you and your fellow student teachers, representing different subjects, is to make sure that the individual subject statements fit together into a coherent description of a whole curriculum, complementing and not competing with each other.

Reflect on this task and check your development against the standards for your course.

SUMMARY AND KEY POINTS

The curriculum is an important part of the context in which you work as a teacher. It includes both the *formal curriculum*, which sets out in detail the subjects to be taught; the *informal curriculum*, which covers the variety of ways in which a school can attempt to achieve the kinds of aims which cannot be captured in the content of timetabled subjects and the *hidden curriculum* which is the way the school relates to pupils and parents, sometimes referred to as the ethos of a school.

The curriculum is perhaps the most important means through which educational aims can be pursued. Although much of the curriculum exists already as a framework within which you work, you may well have the opportunity to contribute to discussion about the curriculum, and you should be able to take and argue a view both on the whole curriculum and on the place of your own subject within it.

Any curriculum is a selection from the culture of a society, but any way of selecting elements from culture – the attempt to select what is best, or what is distinctive or what is fundamental – may not by itself be adequate without a view of the overall aims of education.

While the current National Curriculum for England includes a statement of aims, there is still room for discussion over the contribution which individual subjects make to the achievement of those overall aims.

In this unit no more has been done than to introduce a few of the questions that can be raised and approaches that can be taken. Within educational research and theory, 'curriculum studies' has come to be a subject area in its own right on which there is a large literature; some of this literature is in the further reading below.

Check which requirements for your course have been addressed in this unit.

FURTHER READING

Aldrich, R. and White, J. (1998) *The National Curriculum beyond 2000: The QCA and the Aims of Education*, London: Institute of Education.

With contributions from an historian and a philosopher, this argues for basing the curriculum on an explicit consideration of aims, and for deriving these aims from democratic values.

Lawton, D. (1996) *Beyond the National Curriculum: Teacher Professionalism and Empowerment*, Sevenoaks: Hodder and Stoughton.

From one of the major British contributors to curriculum studies, this, as the title implies, considers not just the National Curriculum, but how the curriculum impinges on teachers and how teachers can be involved in curriculum planning.

White, J. (ed.) (2004) *Rethinking the School Curriculum: Values, Aims and Purposes*, London: Routledge Falmer.

The first two chapters and the conclusion are about the relationship between aims and curriculum, going in more depth into many of the issues raised in Units 7.1 and 7.2. The intervening chapters are about individual subjects, covering the NC for England (except PSHE, citizenship and ICT) and RE.

RELEVANT WEBSITES

See Unit 7.1.

Additional resources for this unit are available on the companion website:
www.routledge.com/textbooks/9780415478724

THE NATIONAL CURRICULUM FOR ENGLAND

Graham Butt

INTRODUCTION

The National Curriculum, as implemented in state maintained primary and secondary schools in England and Wales from September 1989, represented a bold step towards standardising the school curriculum. Introduced as part of the 1988 Education Reform Act (ERA) the National Curriculum has dominated state education in England and Wales in a way that contrasts sharply with earlier practice, for governments had previously made few interventions into the curriculum taught in schools. Indeed, up until the late 1980s, schools themselves largely determined what and how they taught – the only statutory requirement being that religious education was provided for children, as a consequence of the 1944 Education Act. The 1988 Act was therefore a very significant piece of legislation, the effects of which are seen in schools to this day. It paved the way for the publication of schools' examination results, the creation of national 'league tables' based on schools' performance, the greater availability of information on pupils' attainment, and the restructuring of the subject-based curriculum taught in all state schools. The National Curriculum defines the educational entitlement for pupils of compulsory school age – as a student teacher you will have to teach lessons that meet the requirements of the National Curriculum and support pupils in line with the national standards it sets. It describes each subject's content, the expected levels of pupil performance at different stages of their schooling, and the means of assessing and reporting on such performance. The National Curriculum applies to all pupils in state-maintained schools but is not statutory for the independent sector.

More flexibility in the National Curriculum is being gradually introduced although the purpose of the National Curriculum remains – having recently been restated in terms of offering a 'curriculum of the future' (Qualifications and Curriculum Authority (QCA), 2007a). This unit provides you with an overview of the implementation, modification and impact of the National Curriculum on state education in England since its inception in 1988. It also poses questions about the scope, content and structure of this curriculum (see Beck, 2000).

OBJECTIVES

At the end of this unit you should be able to:

■ understand the key terms associated with the National Curriculum
■ understand the nature, scope and content of the National Curriculum
■ understand why changes have been made to the National Curriculum since its inception
■ understand the National Curriculum as it is experienced by pupils in state schools
■ appreciate the place of your own subject within the National Curriculum framework.

Check the requirements for your course to see which relate to this unit.

BACKGROUND

The original rationale for the National Curriculum was that it should: provide teachers with clear objectives for their teaching; pupils with identifiable targets for learning; parents with accurate, accessible information about what their children can be expected to know, understand and be able to do and what they actually achieve (National Curriculum Council (NCC), 1989). Working groups were established in the late 1980s and early 1990s to create the National Curriculum. The curriculum structure is in four key stages with, at the time, formal assessments reported at the end of each stage. Expectations of levels of performance for pupils in each key stage are outlined in Table 7.3.1. Religious education, based on a locally agreed syllabus, has to be included in the school curriculum, but is not formally part of the National Curriculum.

Associated with each subject was a framework of *Attainment Targets* (ATs), *Programmes of Study* (PoS) and *Statements of Attainment* (SoAs) (see Table 7.3.2) appropriate for each of the key stages.

SCOPE OF THE NATIONAL CURRICULUM

The ERA 1988, and the subsequent Education Act 1997, required all state schools to provide a curriculum for pupils which:

■ was broad and balanced
■ promoted spiritual, moral, cultural, mental and physical development
■ prepared pupils for the opportunities, responsibilities and experiences of adult life
■ included religious education and, for secondary pupils, sex education.

In essence, the National Curriculum is therefore a framework of subjects and other requirements representing a minimum curriculum entitlement for pupils. Schools are allowed to develop their own, fuller curriculum which reflects their particular ethos, needs and local requirements but will normally have to deliver the statutory curriculum.

It was soon apparent in the early 1990s that the first version of the National Curriculum was overloaded with content and creating considerable problems for teachers, particularly with respect to assessment. These issues, alongside limited industrial action by teachers in response to the

■ **Table 7.3.1** Key Stages, pupil ages, year groups and range of levels of performance determined by the National Curriculum (England)

Key Stage	Ages	Year groups	Current range of levels within which the majority of pupils are expected to work	Expected attainment for the majority of pupils at the end of the Key Stage
Foundation	3–5	Reception	A separate curriculum has been developed.	
1	5–7	1–2	1–3	At age 7 = 2
2	7–11	3–6	2–5	At age 11 = 4
3	11–14	7–9	3–7*	At age 14 = 5/6
4	14–16	10–11	National qualifications are the main means of assessing attainment	National qualifications are the main means of assessing attainment

* a level 8 description and description for 'exceptional performance' also exists.

■ **Table 7.3.2** Some terms associated with the National Curriculum (England)

Attainment Target (AT)

The knowledge, skills and understanding that pupils of different abilities and maturities are expected to have by the end of each key stage.

Programmes of Study (PoS)

The matters, skills and processes that should be taught to pupils of different abilities and maturities during the key stage.

Statement of Attainment (SoA)

Statements used to define subject content more precisely than the Attainment Target within which they were contained. SoAs were removed after the Dearing Report in 1994.

Level Description (LD)

Level descriptions describe the types and range of performance that pupils working at that level should characteristically demonstrate. They indicate progression in the knowledge, skills and understanding as set out in the programmes of study.

introduction of Key Stage 3 tests in 1993, led to the Dearing review (1996) which reduced the original requirements.

Dearing flagged the increasing importance of providing vocational options for pupils. The subsequent introduction of Diplomas (2008) has reasserted the importance of vocational education. The new Diploma framework has implications both for the National Curriculum and for the revision of existing academic and vocational examinations in the future.

With the creation of the National Assembly for Wales in 1998 education policy and practice for England and Wales began to move in somewhat different directions, including a separate document for the National Curriculum and its assessment in Wales.

The National Strategies

From the late 1990s the national literacy (NLS) and national numeracy strategies (NNS) have taken government-led curriculum initiatives further than ever before, changing the whole emphasis of the National Curriculum (see Unit 7.4). In some respects, the Strategies not only prescribe the teaching approaches to be adopted by teachers – something which the National Curriculum had never explicitly attempted before – but also specify the amount of time to be spent on the teaching of literacy and numeracy. Although not statutory, most maintained schools adopt these strategies in an attempt to meet the ambitious achievement targets laid down for them.

The sharp focus on literacy and numeracy means that the time most primary schools spend on teaching other subjects is now substantially reduced, with an obvious 'knock-on' effect when pupils come to study them again at Key Stage 3. As a result, many pupils may undertake very little work in the non-core subjects before transferring to Key Stage 3, creating obvious pressures on teachers to achieve expected attainment levels for their pupils at the end of that Key Stage.

Further reviews of the National Curriculum such as Curriculum 2000 have made it more flexible and less prescriptive.

SCHEMES OF WORK

The government department, the Department for Children, Schools and Families (DCSF) and the government agency, the QCA, publish schemes of work related to the National Curriculum requirements in all subjects at Key Stages 1, 2 and 3. These are optional and exemplify how the programmes of study for each subject might be taught in the medium term and how they could be modified for pupils with particular needs. Although it was the intention that schools would draw upon these schemes to meet their own needs, rather than adopting them wholesale, there is evidence that some teachers apply them as though they are a statutory part of the National Curriculum.

Task 7.3.1 REVIEWING THE NATIONAL CURRICULUM (ENGLAND)

Review and become familiar with the National Curriculum subject documents and supporting information available online at http://www.nc.uk.net and http://www.qca.org.uk/curriculum. These help to outline the values, aims and purposes of the National Curriculum.

■ To what extent do you consider the National Curriculum to be successful in educating young people in the values, beliefs and cultural aspects of the society in which it is based? What aims do you think underpin the National Curriculum?

■ Is it a curriculum which specifically prepares pupils for adopting useful roles in society based upon achieving gainful employment or are the aims focused on education more broadly?

■ Does the National Curriculum achieve a sensible balance between centralised control and local interpretation?

■ Is the curriculum primarily driven by what is 'good for the state' or by what is 'good for the individual'?

■ Ideally, what should a National Curriculum aim to do?

Answering these questions will help you develop your personal philosophy of education.

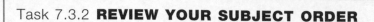

Task 7.3.2 **REVIEW YOUR SUBJECT ORDER**

Look closely at your subject order online at http://www.nc.uk.net and http://www.qca.org.uk/curriculum. Does it adequately reflect what you want pupils to learn about your subject at Key Stage 3? Discuss this with two other colleagues and try to achieve a consensus about what the aims of education within your subject should be at Key Stage 3. Talk to your tutor or mentor about what you perceive to be the strengths and weaknesses of your existing subject order at Key Stage 3.

General teaching requirements apply across all the National Curriculum subjects and cover equalities, diversity and inclusion; use of language; ICT in the curriculum; and health and safety. Table 7.3.3 provides further detail.

■ **Table 7.3.3** National Curriculum guidance on equalities, diversity and inclusion; use of language; ICT in the curriculum; and health and safety (England)

Equalities, diversity and inclusion

The provision of effective learning opportunities for all pupils – including those who are gifted and talented, those with learning difficulties and disabilities, learners who are learning English as an additional language, children who are in care, learners with social, emotional and behavioural difficulties, who are from different ethnic and cultural backgrounds – is a foundation of the National Curriculum. Guidance is given about the ways in which teachers can modify the Programmes of Study to make learning relevant and appropriately challenging at each Key Stage, whilst also recognising the different needs of boys and girls.

Use of language

Pupils in all subjects should be taught to express themselves correctly and appropriately and be able to read accurately and with understanding. Since standard English is the predominant language in which knowledge and skills are taught and learned, pupils are expected to be taught to recognise and use this.

Use of ICT

Supporting information on the National Curriculum explains that pupils should be given opportunities to apply and develop their ICT capabilities through the use of ICT tools to support their learning in all subjects.

Health and safety

The health and safety guidance applies mainly to science, design and technology, ICT, art and design and PE. It highlights hazards, risks and risk control by and for pupils.

THE CURRENT NATIONAL CURRICULUM

From September 2008 a further revised version of the National Curriculum was introduced in phases to state secondary schools in England. Designed primarily to enable schools to 'help all their learners meet the challenges of life in our fast changing world' (QCA, 2007b) it seeks to promote greater flexibility for schools to create their own curricula. Subject content again is less prescribed, whilst the curriculum is designed to focus more clearly on the key concepts and processes that underlie each subject. The major intention is that the National Curriculum framework helps schools to respond more to their local contexts and the needs, capabilities and aspirations of their learners, whilst also making the links between subjects easier. It is also hoped that teachers will gain more time to personalise their teaching, to offer 'catch-up' classes in the basics, and to create opportunities for pupils to deepen and extend their learning in areas where they have particular aptitudes.

A new set of aims, which incorporate the *Every Child Matters* agenda, were the starting point for these changes. These were created to help young people become:

- successful learners who enjoy learning, make progress and achieve
- confident individuals who are able to live safe, healthy and fulfilling lives
- responsible citizens who make a positive contribution to society (QCA, 2007b)

Amongst a raft of new 'opportunities' the revised National Curriculum is expected to further develop pupils' personal, learning and thinking skills, whilst also including dimensions such as sustainability, cultural diversity and enterprise. The scope of these revisions is ambitious – particularly given claims that they aid transition between Key Stages 3 and 4, whilst allowing pupils to access new opportunities in the 14–19 curriculum such as specialised diploma courses. In some schools this has resulted in the condensing of the Key Stage 3 curriculum into two years, with the 'free year' being used to enable some pupils to take certain General Certificate of Secondary Education (GCSEs) examination early, for some to spend an extra year on their GCSEs, and for others to experience a wider curriculum containing 'enrichment activities'.

In terms of assessment the National Curriculum is now expected to give teachers more opportunities to focus on personalised 'assessment for learning' strategies, as a vehicle for offering greater support and/or challenge for pupils according to their particular needs and abilities. The level descriptions that were introduced in 1994 were modified to complement changes to the subject orders, whilst a new set of descriptions have been introduced to accompany the Citizenship orders.

STRUCTURE OF THE SUBJECT ORDERS

Each of the *subject orders* in the 2008 National Curriculum follows a common format. This ensures that subjects are structured around set headings, including:

- an importance statement, which describes why the subject matters and how it contributes to the National Curriculum
- key concepts, that underpin the subject
- key processes, as well as essential skills
- range and content, outlining the breadth of the subject matter from which the teacher can develop knowledge, concepts and skills
- curriculum opportunities to enhance and enrich learning, including linking to the wider curriculum (based on QCA, 2007a).

As a result of these revisions, from September 2008, the Key Stage 3 National Curriculum consists of the following subjects:

- Art and Design
- Citizenship
- Design and Technology
- Personal, social health and education*
- English
- Geography
- History
- ICT
- Mathematics
- Modern foreign languages
- Music
- Physical education
- Religious education*
- Science
(*non-statutory programme of study)

The Key Stage 4 curriculum is made up of:

- Citizenship
- Personal, social health and education*
- English
- ICT
- Mathematics
- Physical education
- Religious education*
- Science
(*non-statutory programme of study)

Cross-curricular dimensions, designed to provide unifying areas of learning to help young people make sense of the world and give education relevance and authenticity, were also included in the following areas: identity and cultural diversity; healthy lifestyles, community participation; enterprise; global dimension and sustainable development; technology and the media; and creativity and critical thinking.

Two non-statutory programmes of study were introduced covering personal well-being and economic and financial capability. These effectively draw together previous themes of personal, social and health education; sex education, careers education, and enterprise and work-related learning.

SUMMARY AND KEY POINTS

The original remit for the National Curriculum was ambitious. This has now changed, although the broad aims remain. The curriculum currently provides an entitlement for pupil learning; the development of literacy and numeracy skills; creativity; and the flexibility for teachers to plan ways to teach that will inspire pupils to make a long-term commitment to learning. Since 2000 it has made explicit the rights of all pupils to be included in the learning process and to be given chances to succeed – whatever their individual needs or barriers to learning. Principles of equality of opportunity, shared values, recognition of diversity and the need for sustainable development are also mentioned, while education for citizenship was a notable addition to the National Curriculum in 2002.

There is also an increasing emphasis on a 14–19 orientation of the curriculum. The government is keen to develop 'world class' academic routes and better vocational options, with stronger partnerships between schools, colleges and employers. Here the increase in curriculum initiatives will inevitably loosen the grip of the National Curriculum even further.

Teachers are now broadly positive about the National Curriculum. The reasons for this have been outlined in the unit – the relaxation of the hold that the National Curriculum once had on the whole school curriculum and the flexibility for greater teacher-led curriculum development is widely supported. However, a question arises as to how far the National Curriculum can be allowed to become reformulated and customised by schools before it ceases to be a National Curriculum in any real sense. The provision of a common learning experience for all pupils in state schools was always ambitious, but now appears to be further away than at any time since 1988.

Check which requirements for your course have been addressed through this unit.

FURTHER READING

Beck, J. (2000) 'The school curriculum and the National Curriculum', in J. Beck and M. Earl (eds) *Key Issues in Secondary Education*, London: Cassell.

A useful overview of what we mean by the term 'curriculum', as well as the nature of its scope, content and structure. Beck raises issues about the control of the curriculum in schools and focuses on the various 'ideologies' that influenced how the National Curriculum originally developed, and continues to develop.

RELEVANT WEBSITES

National Curriculum: http://www.nc.uk.net

The National Curriculum online. Contains information on elements of the National Curriculum which will continue to be taught in English schools until 2011. This site provides direct links to the QCA/DCSF schemes of work and other online resources relevant to each part of the programmes of study for each subject.

National Curriculum examples: http://www.ncaction.org.uk

NC in Action uses pupils' work to exemplify the National Curriculum at Key Stages 1 and 3. New materials are added to the site as they become available. Work exemplifying creativity and the use of ICT are also included.

Qualifications and Curriculum Authority: http://www.qca.org.uk

Website for the Qualifications and Curriculum Authority. Provides guidance on the curriculum, assessment and the range of qualifications offered in schools and colleges.

Secondary curriculum for England: http://www.qca.org.uk/curriculum

Details the secondary curriculum for England implemented in schools from 2008 to 2011. Programmes of study, support and guidance and case studies are all available on this website.

Additional resources for this unit are available on the companion website:
www.routledge.com/textbooks/9780415478724

THE SECONDARY NATIONAL STRATEGY IN ENGLAND

Rob Batho

INTRODUCTION

The Secondary National Strategy for school improvement (http://www.standards.dcsf.gov.uk/secondary/framework/) is a vital part of the national drive to transform secondary education to enable children and young people to attend and enjoy school, achieve personal and social development and raise educational standards in line with the *Every Child Matters* agenda. It began life in 2001 as the Key Stage 3 Strategy where the challenges were particularly demanding as these are vitally important years for laying the foundations for life-long learning and influencing the crucial decisions pupils will make at age 14. It became the Secondary National Strategy in 2006 to reflect the work it had been doing and continues to do in Key Stage 4 and beyond. Its main aim is to raise standards by strengthening teaching and learning across the curriculum for all 11- to 19-year-olds.

The English and mathematics strands of the Strategy were introduced to all schools in England in 2001 with science, foundation subjects and information and communications technology (ICT) following in 2002 and behaviour and attendance in 2003. In addition, the Strategy also addresses a number of elements such as literacy, numeracy and ICT across the curriculum and assessment for learning. The content and coverage of the secondary school curriculum is the responsibility of the National Curriculum (NC), not the Strategy. The Strategy's role is to advise and guide school senior managers, subject leaders and teachers about effective strategies, techniques and skills to improve teaching and learning in classrooms. It is designed to support schools and teachers in addressing the learning needs of 11- to 19-year-old pupils. The support for schools and teachers is twofold; first, through local authority (LA) consultants who work directly with schools and teachers to improve teaching and learning in the classroom, and, second, through providing high-quality training and guidance materials, most of which you should be able to find in your school if you are in England or alternatively on the Strategy website (see further reading).

OBJECTIVES

At the end of this unit you should be able to:

- understand the Strategy's main principles of teaching and learning
- understand and use learning objectives and learning outcomes in your planning and teaching to aid pupil progress
- know the main features of a good lesson
- know where and to whom to go to find guidance and materials on teaching and learning.

Check the requirements for your course to see which relate to this unit.

MAIN PRINCIPLES OF THE SECONDARY NATIONAL STRATEGY FOR ENGLAND

The Strategy's main principles of teaching and learning are as follows:

- *Focus on learning objectives*. Teaching needs to focus on and be led by learning objectives so that teachers and pupils have a clear direction and purpose for their teaching and learning.
- *Start with what pupils already know*. It is usually important to begin any new teaching by finding out and building on what pupils already know so that they can make a bridge into new learning, and then to move forward at a pace that ensures their understanding and application
- *Use modelling and demonstration*. To make concepts and conventions explicit. Pupils often find difficulty in understanding the concepts within a subject or the writing conventions of a particular type of text used in a subject. Teachers can help by explicitly demonstrating or modelling (which includes thinking out loud to give pupils access to their thought processes) the convention or the explanation of the concept.
- *Structure the learning*. To structure the learning explicitly teachers can, through planning and careful explanation, help pupils see and understand the links between the different parts of a lesson and how they contribute to their learning.
- *Make the learning active and interactive*. Pupils are more likely to be motivated and gain genuine understanding if the teacher plans for them to be interactive with their peers and with the teacher in their discussions and tasks, and if the activities demand that pupils actively seek meaning and knowledge rather than simply receive it.
- *Make learning enjoyable*. Secondary age pupils can easily become bored and reluctant to learn. The teacher needs to consider such matters as the suitability of the content, stimulating and enjoyable activities, instigating and maintaining an appropriate pace for learning and setting an appropriate degree of challenge for all pupils.
- *Scaffold and support learning*. When pupils are confronted by a new task and new learning, they may often be unsure of how to proceed if the leap is too big from their existing understanding. The teacher can help by providing suitable scaffolds and support (e.g. writing frames*, guided group work) to enable those pupils to take small, secure steps towards their learning goal.

■ *Use assessment for learning.* Teachers can provide opportunities for pupils to reflect on their learning and to identify and take responsibility for future learning. Effective learners are usually able to reflect on their own learning, what they have learned and how they have learned, and are able to use that facility to identify what they need to do next to improve their learning. Many pupils do not have that facility but teachers can teach them how to be able to reflect on their learning and to plan the next steps in their learning, so that they become more independent learners.

The Strategy advises teachers to adopt these principles to inform their planning and teaching of lessons in order that their pupils are motivated, suitably challenged and make progress in their learning.

* Note: A writing frame is commonly an outline of a text type that the teacher expects the pupils to write. Writing frames help pupils to structure a text and often contain the opening words for key paragraphs and indications as to how to connect each paragraph. Writing frames, as Beard (2000: 113) explains, 'are intended to be used as a kind of prompt, eventually to be dispensed with, as children begin to adopt the features of the genre for themselves'.

WHAT MIGHT A GOOD SECONDARY SCHOOL LESSON LOOK LIKE?

We want you to consider what a good secondary school lesson might look like where these principles have been applied (see Task 7.4.1).

Task 7.4.1 APPLYING NATIONAL STRATEGY PRINCIPLES TO LESSONS

Look back at the main principles of the Strategy and consider what a lesson might look like where they have been applied.

■ What would you expect the teacher to be doing at the start, middle and end of the lesson?
■ What would you expect the pupils to be doing at those stages?
■ Observe a lesson and if possible the scheme of work and lessons plans for it and note down the key features and where they match the main principles.
■ Now compare your thoughts and findings with Table 7.4.1 and consider if there are any elements that you could usefully introduce into your own teaching.

Although Table 7.4.1 indicates four phases, the Strategy does not advocate a certain number of phases or a particular structure. Instead, it advises that lesson design should be flexible but carefully and clearly structured to match the learning needs of pupils.

PLANNING AND SETTING LEARNING OBJECTIVES AND LEARNING OUTCOMES

The Strategy is keen to emphasise the importance of teachers setting clear *learning objectives* and outcomes for their pupils so that they, the pupils, know and understand what it is they are expected

■ **Table 7.4.1** Features of a good lesson that applies Strategy principles

Phase	Features
Start	**Starter activity** The first 5 minutes or so are spent on a brisk activity to catch pupils' imagination. This could be linked directly with previous work or the work to come. This is the place to establish early teaching points or to position 'little and often' objectives which require revisiting and practising skills or consolidating knowledge. It also allows the teacher to quickly establish any gaps in pupils' knowledge or understanding.
Introduction	**Learning objectives and outcomes** The teacher introduces the lesson by sharing with the pupils what they are going to learn and what outcomes are expected. The focus is on the learning: 'Today we're going to look into the factors that affect how quickly enzymes work', or 'learn how to substitute an integer into a simple formulae', or 'consider how the writer builds suspense' or 'look at the benefits and drawbacks of using ICT to automate processes'.
	Active teaching New learning is introduced by teacher input. For example, the teacher may model or demonstrate on the board how to compose a particular type of writing, and the pupils are drawn in to contribute. The teacher is not afraid to be an expert.
	Pupil participation Pupils are expected to participate; to respond to questions, to think, make suggestions and explain. For example, they might have a moment to talk to a partner and come up with an idea. They might have to work on a problem and hold up answers on individual whiteboards.
Development	**Pupils applying what they've learned** Pupils may apply what they've learned in group work or paired work, or an individual exercise. The teacher does not wait for pupils who 'get stuck' to put their hands up; rather they are more likely to sit with one group for several minutes and guide them through the work, helping them to apply new skills.
	Teaching assistant usefully deployed Any teaching assistant is well prepared, has helped the teacher to plan the lesson and is familiar with their special role. The assistant may be sitting with a group of pupils to help them keep up with the work, or making notes for the teacher on how pupils are managing a task.
End	**Plenary session** The lesson closes with a plenary session in which the teacher draws out the key points. Learning is reviewed and there is an opportunity to reflect on the learning process itself. Pupils do most of the work. They are encouraged to explain what they've learned and how it can be used in the future, perhaps in other lessons. (There may have been a series of shorter plenaries to review learning at the end of specific episodes throughout the lesson).
	Homework Regular homework helps individuals to reinforce and consolidate what they have learned in the lesson or to prepare for the next one. Homework tasks can be differentiated as appropriate to the needs of individuals. They should be planned for and used to enhance or extend work in the lesson, not as a fill-in or a finishing-off exercise.

to learn (the learning objectives) and how that might be achieved, monitored and assessed (the learning outcomes). The learning objective and the outcome are not synonymous, but the outcome is usually derived from the objective. For example, a learning objective for writing for a Year 7 class might be:

■ To plan writing and develop ideas to suit a specific audience and purpose by adapting familiar forms and conventions.

And the learning outcome for that could be:

■ To produce a plan for a letter to the local council to outline the main points for and against the proposed new housing estate.

Learning outcomes are commonly written outcomes but they do not need to be. They could for instance be oral; a talk to the whole class or to a small group, or a spoken answer to the teacher in a plenary session. Alternatively, the outcome could be visual such as a drawing, plan or diagram. Task 7.4.2 provides you with practice in setting learning objectives and outcomes.

Task 7.4.2 SETTING LEARNING OBJECTIVES AND LEARNING OUTCOMES

Locate a scheme of work and any associated lesson plans for a class that you are teaching or are likely to teach during your school experience (they could be the same ones that you used for Task 7.4.1).

Identify one particular learning objective that will apply to a particular lesson.

Ask yourself, 'Considering the learning objective, what is it that I want the pupils to have done (the outcome) that will illustrate that learning?' Now, either write a learning outcome, or amend the wording of the existing outcome so that it reflects what you want the pupils to learn.

To assist teachers in their planning and in their setting of learning objectives the Strategy has published frameworks for teaching in each of the core subjects, in ICT, modern foreign languages and design and technology. Those for the core subjects and ICT have been recently revised to reflect the changes to the new National Curriculum programmes of study (PoS). All of these frameworks of objectives are available on the Strategy website (see further reading at the end of the unit).We suggest you find the framework most relevant to the subject you teach and locate and consider some of those learning objectives for your own classroom teaching.

The Strategy advises teachers to use the learning objectives from the frameworks to assist them in planning for progression in pupils' learning in lessons and schemes of work. If pupils are to progress in their learning, it is more effective if teachers plan and pupils understand what they need to learn next and how that might be achieved, and the revised frameworks were designed with that in mind. Whilst the original framework objectives were for Key Stage 3 only, in the revised frameworks the objectives span Years 7 to 11 and so allow for and encourage long-, medium- and short-term progressive planning.

If in your planning, your scheme of work and lessons clearly state the learning objectives and the learning outcomes, then you and your pupils are sharply focused not only on what needs to be learnt and taught but also on what needs to be assessed, but your plans also need to address progression (see Unit 2.2) to ensure pupils' learning is built over a series of lessons.

SUPPORT FOR YOUR TEACHING FROM THE SECONDARY NATIONAL STRATEGY

As has been mentioned, the Secondary National Strategy provides a wealth of practical support and guidance for teachers to improve teaching and learning in the classroom, and you may be wondering how you can access that support and guidance. There are a number of ways. Most of the materials are available for downloading on the Secondary National Strategy website (http://www. nationalstrategies.standards.dcsf.gov.uk/secondary/). In addition, you should find that all of your school experience schools have hard copies of many of the Strategy training and guidance materials.

Your subject leader in your school experience school will have some understanding of the Strategy, its resources and training, but the other person in the school who will know where those hard-copy materials are and who has a responsibility for organising Strategy training in the school is the strategy manager. The strategy manager should be an excellent source of information and advice for you about the Strategy. They can tell you about the training that has taken place and is about to take place in your subject area as well as providing you with, or showing you where to find, data and information concerning the achievement and attainment of pupils you might be teaching. The data gives detailed information about the past attainment and achievement of the pupils you will be teaching to help you plan more effectively. It is important therefore that you make the acquaintance of the Strategy manager in your school experience school. Task 7.4.3 provides suggestions for the questions you might ask.

Task 7.4.3 MEETING THE SCHOOL STRATEGY MANAGER

1   Find the name of and arrange to meet the school's strategy manager during your school experience.
2   At the meeting ask the strategy manager the following:

   ■ What Strategy training has your subject department taken part in over the past two years, what are they likely to be involved in during your time at the school and would it be possible and appropriate for you to be included?
   ■ What Strategy whole-school initiatives and training (such as assessment for learning or literacy across the curriculum) has the school been involved in, what are they likely to be involved in during your time at the school and again would it be possible and appropriate for you to be included?
   ■ Where are any hard-copy Strategy materials held?
   ■ Where can you find pupil achievement data concerning the pupils you are scheduled to teach that would be useful in helping you to know the pupils and in supporting your planning?

Your notes might come in useful and your next school. We suggest you file them in your PDP.

THE IMPACT OF THE STRATEGY

There is research evidence that indicates that the Strategy has had an impact on teaching and learning in secondary schools. For example, the Office for Standards in Education (Ofsted) in its evaluation of the third year of the Strategy points to the following successes (Her Majesty's Inspectorate (HMI), 2004):

■ the Strategy is helping to improve teaching, in particular through the use of specific objectives for learning and better planning, which add greater challenge and purpose to lessons

■ the improvements in teaching are leading to better attitudes to work, especially, but not only among boys

■ there is an increasingly positive effect, albeit uneven, on pupils' attainment.

In its evaluation of the fourth year (HMI, 2005):

■ pupils' engagement has improved, because they find the teaching more enjoyable, they like the better structured lessons, the broader range of teaching approaches and the brisk pace of learning.

In its evaluation of the fifth year (HMI, 2006):

■ The Secondary National Strategy ... continues to have a positive influence on pupils' attainment.

■ The quality of teaching and learning are also improving as teachers continue to apply Strategy techniques. The best lessons include a wider range of teaching strategies, with more emphasis on pupils thinking for themselves. They respond well to the variety and the brisk pace.

A National Foundation for Education Research (NFER, 2004) study of the perceptions of LA officers and teachers of the implementation and effectiveness of the Key Stage 3 National Strategy reported the following:

■ the Key Stage 3 National Strategy was well received by LAs and schools

■ it was an important component of the School Development Plan, a tool to improve teaching and learning in the classroom and, thereby, ultimately raise attainment

■ LAs and schools felt well supported during the implementation of the Strategy and described the Strategy materials as good quality, useful resources

■ LAs and schools perceived that the Strategy was making a significant contribution to the improvement of teaching and learning and had contributed to the improvement of lesson structures including pace of lessons, the variety of teaching methods used and the focus on the learner.

The following tasks are designed to help you consolidate the ideas in this unit:

Task 7.4.4 **LEARNING OBJECTIVES AND OUTCOMES**

Arrange to interview a small group of pupils you have been teaching recently. Ask them what they have learned during the last three or four of your lessons. Also ask them if they can remember what the learning outcomes were for those lessons. Make a note of what they say.

Now compare what the pupils said with the learning objectives and outcomes from your lesson plans, plus any work of theirs which would give evidence of their learning. Are there any significant differences between what the pupils said and their work and your expectations as expressed in the learning objectives and outcomes? If there is, how do you account for this and what could you do in future to narrow that difference?

Task 7.4.5 **FEATURES OF AN EFFECTIVE LESSON**

Look again at Table 7.4.1. Arrange to observe a teacher or another trainee teaching (it doesn't need to be your subject). Discuss and agree which two phases from the table will be the focus of your observation. In the lesson, note down where the features of the teaching in each of the two chosen phases are either similar to or different from those in the table, and the impact that they have on pupils' learning.

Soon after the lesson, discuss with the teacher the similarities and differences that you noticed as well as the impact they had on pupils' learning, in order to identify the strengths of the teaching on pupils' learning.

SUMMARY AND KEY POINTS

In this unit, we have introduced you to the Secondary National Strategy's main principles for teaching and learning and to the key features of a lesson which engages and motivates pupils and develops their learning. We have also emphasised the importance of careful planning to set pupils clear learning objectives and learning outcomes so that they are able to understand what they are expected to learn and produce. To support your own professional development, we have also indicated the key personnel (such as the school's strategy manager) in any school who will be able to advise and guide you on matters to do with teaching and learning. And remember, there are also plenty of Strategy training and guidance materials which should be readily accessible to you in your training institution and school experience, most of which are freely available on the Strategy website (http://www.nationalstrategies.standards.dcsf.gov.uk/secondary/).

Check which requirements for your course have been addressed through this unit.

FURTHER READING

Department for Children, Schools and Families (DCSF) (2008n) *Framework for Teaching English: Years 7 to 11*, London: DCSF.

DCSF (2008o) *Framework for Teaching Mathematics: Years 7 to 11*, London: DCSF.

DCSF (2008p) *Framework for Teaching Science: Years 7 to 11*, London: DCSF.

DCSF (2008q) *Framework for Teaching ICT: Years 7 to 11*, London: DCSF.

Department for Education and Skills (DfES) (2003b) *Framework for Teaching Modern Foreign Languages: Years 7, 8 and 9*, London: DfES.

DfES (2002e) *Training Materials for the Foundation Stage*, London: DfES Publications.
These training materials contain some excellent ideas and advice for teachers of all subjects on areas such as assessment for learning, planning, questioning, explaining, modelling, starters and plenaries.

DfES (2004k) *Framework for Teaching Design and Technology: Years 7, 8 and 9,* London: DfES.

RELEVANT WEBSITE

The Strategy frameworks can all be found on the Strategy website: http://www.standards.dcsf.gov.uk/secondary/framework/
They not only provide yearly progressive learning objectives for each of the subjects, but they also include suggestions and advice on approaches to teaching and learning, planning and assessment.

Additional resources for this unit are available on the companion website:
www.routledge.com/textbooks/9780415478724

UNIT 7.5

SECONDARY SCHOOLS AND CURRICULUM IN SCOTLAND

Allen Thurston and Keith Topping

INTRODUCTION

Scotland's social and cultural identity is defined largely by its geography and history. Its political, educational, legal and religious systems make Scotland unique within the UK (Humes and Bryce, 2003). This unit aims to identify the distinguishing structures and characteristics of the Scottish education system. It explores the nature and functioning of the various stakeholders in Scottish education. It illustrates the systems of governance of education and examines the structure and nature of the curriculum and assessment in Scotland. It also explores the standards for initial teacher education (ITE) and discusses the preparation necessary for a newly qualified teacher (NQT) entering the profession in Scotland.

OBJECTIVES

At the end of this unit you should be able to:

- describe the distinctive features of the structures of governance in Scottish schools
- explain what principles have shaped curriculum design in Scotland in the recent past
- describe the structure, balance and distinctive features of the Scottish secondary school curriculum
- have a clear understanding of the standards for ITE and a basic understanding of the standards for full registration of teachers in Scotland.

Check the requirements for your course to see which relate to this unit.

SCOTTISH EDUCATION

A separate legislative framework seems one of the most potent expressions of the distinctiveness of Scottish education (Humes and Bryce, 2003). This has allowed and indeed encouraged a number of differences to develop between the Scottish education system and that of England. Of particular note is the introduction and subsequent adaptation of Curriculum Guidelines for pupils in Scotland within the age ranges of 5–14 in the 1990s (Paterson, 2003), quite different from the 'National' Curriculum in England. No core subjects are identified, which gives more breadth and flexibility in curriculum and pedagogical methods. The guidelines are advisory recommendations and not prescriptive.

Provision for assessment is also different between England and Scotland. Teachers are free to conduct assessments at a time of their own choosing.

The Scottish Schools (Parental Involvement) Act 2006 established a new body for parental involvement, the 'Parent Councils'. Whilst not having as much influence and control over schools as the previous school boards, the Parent Councils are expected to play an active role in supporting parental involvement in the work and life of the school. They provide opportunities for parents to express their views on children's education and learning. The Parent Council, as a statutory body, has the right to information and advice on matters which affect children's education. The school and the education authority must consult with the Parent Council and take their views into account whenever decisions are being taken on education provided by the school (Learning and Teaching Scotland (LTS), 2007a).

In her opening address to the Scottish Government on 20 June 2007 Fiona Hyslop, the Cabinet Secretary for Education and Lifelong Learning, indicated that the Scottish Government had a commitment to improving the learning experience in schools. Social dimensions to learning including 'building self-confidence, social skills and an awareness of impact on others . . . [would] . . . create the foundations for good health and positive economic and civic engagement later in life' (Hyslop, 2007). In this statement the commitment to employing an additional 300 teachers in 2007 with the aim of reducing class sizes was made.

A cornerstone of the Scottish Government's vision for education is 'equality and inclusion linked to active citizenship'. 'Education for All' is a fundamental principle. This is at the core of the 'National Priorities' for Scotland (dealt with in greater detail later in this unit). The Scottish Government Education Directorate (SGED) sees three keys to the maintenance of quality in education (Scottish Executive Education Department (SEED), 2004a) namely:

■ all school teachers must be qualified and registered with the General Teaching Council for Scotland (GTCS)
■ examinations and qualifications are nationally accredited through the Scottish Qualifications Authority (SQA)
■ all schools and colleges are open to Her Majesty's Inspectorate in Education (HMIE) who are employed by, and report to, SGED.

There are four years of secondary school education (S1–S4). At the end of S4, pupils finish compulsory education, but may stay on for fifth and sixth years to gain further qualifications. These years are termed S5 and S6.

Pupils study for awards at two broad levels in Scotland, Standard Grade and Higher Grade, conferred by the SQA. The Scottish Certificate of Education (known as Standard Grade) is normally gained after S4. It is available at three levels – Foundation, General and Credit. Higher Grades (known as Highers and Advanced Highers) follow on from Standard Grade, generally taken in S5 and S6. Both Higher and Advanced Higher awards use a combination of external examination and internal assessment. Additional national qualifications are available to pupils in S5 and S6, including courses at Access 1, 2 and 3, and Intermediate 1 and 2 levels. More details of these are found later in this unit.

Overview of the curriculum in Scotland

In S1/S2 the curriculum generally follows SGED's 5–14 Programme (LTS, 2004a). Pupil choice is a fundamental principle at this stage. The transition from S1 to S2 is often characterised by increasing specialism, e.g. a move from general science to courses more aligned with chemistry, physics and biology (Gavin, 2003). The S3/S4 curriculum is strongly influenced by study for Standard Grade awards. A typical pupil studies for seven or eight subjects at Standard Grade. In S5/S6 (the 16–18 age range), schools offer a varied range of Higher Grade awards, sixth year study awards and Scottish Vocational Education Council (SCOTVEC) awards.

SGED's 5–14 Programme, monitored by HMIE, lays down guidelines for the progress and performance of pupils at different stages through primary and early secondary school. Each stage allows for different levels of development. Each school is free to create its own curriculum with advice from the guidelines, local authority education departments and specialist agencies. The detail of this is dealt with later in this unit. The stated aims of the 5–14 curriculum (SEED, 2000) are that it should help each pupil to acquire and develop:

■ knowledge, skills and understanding in literacy and communication, numeracy and mathematical thinking
■ knowledge, understanding and appreciation of themselves and other people and of the world around them
■ the capacity to make creative and practical use of a variety of media to express feelings and ideas
■ knowledge and understanding of religion and its role in shaping society and the development of personal and social values
■ the capacity to take responsibility for their health and safe living
■ capability in information and communications technology (ICT) and an awareness of the uses of ICT in the world at large
■ the capacity to treat others and the world around them with care and respect
■ the capacity for independent thought through enquiry, problem solving, information handling and reasoning
■ positive attitudes to learning and personal fulfilment through the achievement of personal objectives.

The structure and nature of the 5–14 curriculum in Scotland appears to offer teachers much greater freedom and professional responsibility than the National Curriculum in England. Even within curriculum areas the guidance is not prescriptive, and there is latitude for teachers to exercise professional judgement in deciding what to teach and when. Programmes of study for individual pupils vary greatly, but pupils up to the age of 16 study English, mathematics, science, one modern foreign language and a social subject. In the fifth and sixth years, study expands to include a wide range of modular courses in vocational subjects. Scotland's school-based qualifications are awarded by SQA, a quango answerable to SGED. The SQA develop assessments at Standard Grade and Higher Grade.

THE STRUCTURE AND GOVERNANCE OF EDUCATION IN SCOTLAND

Local authorities

There are 32 local authorities in Scotland. The most recent reorganisation of local government resulted in a larger number of smaller authorities. Each has a degree of autonomy over educational issues. To a large extent the provision of education in Scotland is still governed by the Education Scotland Act 1980 which emphasised:

■ the right of pupils to be educated as far as is reasonable in accordance with the wishes of their parents
■ the duty of parents to provide efficient education by ensuring their children regularly attend school or receive efficient education by other means
■ the duty of education authorities to secure adequate and efficient provision of school education.

Scottish Government Education Directorate (SGED)

SGED is the central government agency administering education in Scotland. It promotes quality in education through three principal channels. First, it develops guidance on curriculum in Scotland in the form of the 5–14 curriculum guidelines. Second, it has an agency that has a cross-sector responsibility for standards and quality assurance throughout the Scottish education system (HMIE). Third, it supports a nationally accredited qualifications and assessment framework managed by the SQA. The roles and responsibilities of SGED in relation to the other stakeholders in Scottish education are represented in Figure 7.5.1. Further information on the structure of governance of Scottish Education is available from: http://www.scotland.gov.uk/Resource/Doc/923/0052945.pdf.

Her Majesty's Inspectorate of Education (HMIE)

HMIE is responsible for undertaking inspections of the standards and quality of services in pre-school centres, schools and post-compulsory education. HMIE places great emphasis on self-evaluation by schools and the focused professional development of teaching professionals, seeking to offer support in addition to critique.

Learning and Teaching Scotland (LTS)

LTS offers advice to Scottish Ministers on issues affecting the learning experiences of children and young people aged 0–18 years. It is supported by five Advisory Council Reference Groups which develop advice on specific areas of education. It is funded by SGED. LTS also aims to provide support, resources and staff development opportunities for teachers to enhance achievement and attainment and promote life-long learning. It is LTS which produces the National Curriculum Guideline documentation for SGED and the additional supporting material for teachers.

NATIONAL PRIORITIES FOR EDUCATION IN SCOTLAND

Following extensive consultation throughout 2000 (the 'Great Debate'), National Priorities for School Education in Scotland were defined and incorporated into the Education (National Priorities) Order 2000. The National Priorities were defined under the following headings:

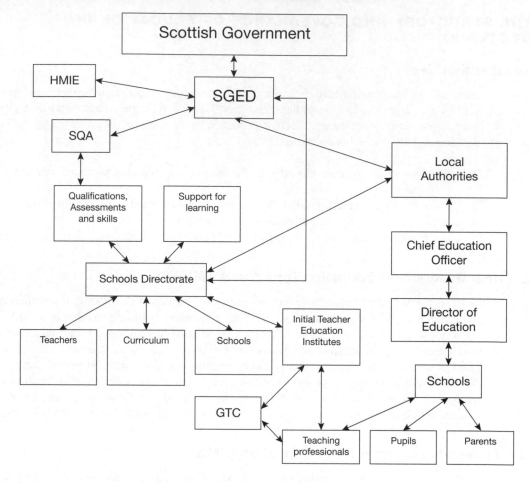

Figure 7.5.1 Interrelationships between stakeholders in Scottish education

- ■ Achievement and Attainment
- ■ Framework for Learning
- ■ Inclusion and Equality
- ■ Values and Citizenship
- ■ Learning for Life.

The National Priorities (SEED, 2003) are:

1   to raise the standards of educational attainment for all in schools, especially in the core skills of literacy and numeracy, and to achieve better levels in national measures of achievement including examination results

2   to support and develop the skills of teachers, the self-discipline of pupils and to enhance school environments so that they are conducive to teaching and learning

3   to promote equality and help every pupil benefit from education, with particular regard paid to pupils with disabilities and special educational needs, and to Gaelic and other lesser used languages

4   to work with parents to teach pupils respect for self and one another and their interdependence with other members of their neighbourhood and society and to teach them the duties and responsibilities of citizenship in a democratic society

5   to equip pupils with the foundation skills, attitudes and expectations necessary to prosper in a changing society and to encourage creativity and ambition.

Task 7.5.1 asks you to consider how to integrate the National Priorities into your professional practice.

Task 7.5.1 INTEGRATING THE NATIONAL PRIORITIES INTO PROFESSIONAL PRACTICE (SCOTLAND)

Examine the National Priorities above. Choose one to think about in more detail. How could you integrate the aims of this National Priority into your professional practice? What critical factors would need to be taken into account? Use the National Priorities website to help you: http://www.standards.dcsf.gov.uk/secondary/framework/.

THE SCHOOL CURRICULUM

In Scotland the school curriculum aims to develop certain dispositions in pupils towards learning. These dispositions (SEED, 2000) are stated as:

■ a commitment to learning as a life-long process
■ a respect and care for self and others, having a positive self-image as well as respect for diversity of people, cultures and beliefs
■ a sense of social responsibility and a willingness to take a positive role in society
■ a sense of belonging and a feeling that one is a stakeholder in society.

The 5–14 curriculum

The 5–14 curriculum identifies core skills and capabilities that should be developed by the formal curriculum in school. These include:

■ personal and interpersonal skills including working with others
■ language and communication skills
■ numeracy skills
■ ICT skills
■ problem-solving skills
■ learning and thinking skills.

In addition to the learning dispositions and the development of the core skills the 5–14 curriculum aims to foster the acquisition of knowledge and the development of understanding in the main curriculum areas.

Structure and balance of the 5–14 curriculum

The 5–14 national guidelines are non-statutory guidelines for Scottish local authorities and schools. They cover the structure, content and assessment of the curriculum in primary schools and in the first two years of secondary education. The stated aim is to promote the teaching of a broad, coherent and balanced curriculum that offers all pupils continuity and progression as they move through school (LTS, 2004a).

The 5–14 curriculum aims to offer a breadth of experience and a balance of opportunity for learning for every pupil through clear pathways though the areas of learning which provide the framework for personal growth and progression in learning for pupils. The curriculum aims to build on pupils' experience and learning and tries to be responsive to their needs. The 5–14 guidelines (SEED, 2000) state that the curriculum is intended to:

■ relate to events and facets of pupils' everyday lives
■ help pupils develop intellectually, aesthetically, socially, emotionally, spiritually, imaginatively and physically
■ prepare pupils to face the challenges of life in a rapidly changing society. It should help guide them through the transition from childhood to adulthood.

To these ends the 5–14 curriculum in Scotland has been developed so that:

■ Breadth ensures coverage of a sufficiently comprehensive range of areas of learning.
■ Balance ensures that appropriate time is allocated to each area of curricular activity and that provision is made for a variety of learning experiences.
■ Coherence emphasises links across the curriculum so that pupils make connections between one area of knowledge and skills and another.
■ Continuity ensures that learning builds on pupils' previous experience and attainment and prepares them for further learning.
■ Progression provides pupils with a series of challenging but attainable goals.

Breadth of the 5–14 curriculum

The 5–14 curriculum is divided into a number of areas:

■ English language
■ mathematics
■ environmental studies
■ expressive arts
■ religious and moral education
■ health education
■ personal and social development
■ modern languages
■ ICT.

For each area there are attainment outcomes, each with a number of strands or aspects of learning that pupils experience. Most strands have attainment targets at five or six broad levels: A–E or A–F.

Task 7.5.2 asks you to audit your knowledge of the content of the curriculum as part of your preparation to teach.

Task 7.5.2 **AUDITING CURRICULUM CONTENT KNOWLEDGE IN PREPARATION TO TEACH**

Are you ready to teach? Look at the 5–14 curriculum guidelines for the area you teach. Examine the attainment targets at levels E and F. Do you have the subject knowledge necessary to teach all the subject matter. Conduct a brief audit and identify any areas that are a development need for you.

Balance of the 5–14 curriculum

Time allocation for each subject in schools has an element of flexibility built into it. Within secondary schools in S1 and S2 the suggested time allocations are represented in Table 7.5.1.

Schools can use the remaining 20 per cent of time flexibly to meet development needs of pupils and the development priorities of the school.

Coherence in the 5–14 curriculum

The 5–14 guidelines suggest that links between the curricular areas should be developed in order to give the curriculum coherence. Cross-cutting themes suggested by the 5–14 guidelines as appropriate areas where this could be attempted include:

■ personal and social development
■ education for work
■ education for citizenship
■ the culture of Scotland
■ ICT.

Each cross-curricular aspect requires careful planning and teaching. This allows its distinctive contribution to be made both within and across the curriculum areas. It is important that these are taught in ways that allow pupils to make connections to other parts of the curriculum. This strengthens coherence in pupils' learning and avoids unhelpful fragmentation of the curriculum. It also provides an important basis for the personal and social development of all pupils.

■ **Table 7.5.1** Time allocations for curriculum subjects (Scotland)

Curriculum areas	Minimum time allocation (%)
English language	20
Mathematics	10
Environmental studies	30
Expressive arts	15
Religious and moral education	5
Total core	80

Continuity within the 5–14 curriculum

The principle of continuity within 5–14 is that the curriculum should build on pupils' experience and attainment and prepare them for further learning. Continuity also deals with important issues relating to transition from primary to secondary school, issues within the secondary school curriculum, and transition from secondary school to life after school. Scotland is similar to the rest of the UK in that transition from primary to secondary has often been associated with a dip in attainment (Galton *et al.*, 1999). Transition programmes have been established between secondary and feeder primary schools, and have been reported to have a beneficial effect in counteracting attainment dips (Waldon, 2000). However, evidence is growing that since the introduction of 5–14 curriculum more pupils are coping better with the primary–secondary transition, and report positively on preparation programmes (Graham and Hill, 2003).

SGED suggest that bridging topics, jointly taught by both primary and secondary staff, can strengthen continuity during transition. In addition, the guidelines suggest that continuity is enhanced if curriculum planning and programmes in feeder primary schools and receiving secondary schools are based on the 5–14 guidelines. It is common to have regular cluster group meetings between staff of feeder primary and associated secondary schools. It is desirable that there is continuity in expectations and approaches to learning and teaching within and between schools. Liaison meetings between schools and pre-transition visits by primary pupils to high school help in preparing them for the transition. The introduction of the 5–14 curriculum, the consistent approaches to assessment and shared understanding of levels of pupils' level of attainment means that information is more easily passed on as pupils move class either within or between schools.

Within the secondary school, continuity in pupils' learning is based (SEED, 2000) on:

■ an active and ongoing partnership between home and school, where the distinctive contributions of home and school to pupils' learning are recognised and valued
■ a positive school/classroom ethos being established where praise, encouragement, challenge and the celebration of achievement increase motivation and create a learning environment in which pupils become confident and enthusiastic learners ready to make the most of school
■ teachers' planning recognising prior learning and whole-school programmes being matched tightly to national guidelines and syllabuses of standard grades and national units/courses
■ consolidation of learning and an understanding that teachers make time for pupils to review and consolidate their prior learning at the beginning and end of lessons and topics
■ sustaining appropriate intervention strategies to support learners by giving, as appropriate, more direction, opportunities to talk about problems and solutions, and time for reflection, further experience, or more practice
■ evaluation of lessons, topics and teaching strategies with assessment information, feedback from pupils (through self-assessment activities or learning diaries) and discussions with colleagues involved in collaborative teaching, providing important sources of evidence for evaluating the effectiveness of teaching and learning.

Progression within the 5–14 curriculum

There is a link between high attainment in schools and a pupil's ability to see progression in the curriculum (Harland *et al.*, 2002). The curriculum is a route plan for pupils' learning. The 5–14 guidelines feature strands, pathways and attainment targets. Within the strands progressively more demanding attainment targets have been developed (A–F). These provide specific statements of what pupils should know and be able to do at each level for each of the attainment outcomes. In each subject area each of the six levels of progression of the attainment targets are based on the following descriptions of levels:

Level A: should be attainable in the course of P1–P3 by almost all pupils.
Level B: should be attainable by some pupils in P3 or even earlier, but certainly by most in P4.
Level C: should be attainable in the course of P4–P6 by most pupils.
Level D: should be attainable by some pupils in P5–P6 or even earlier, but certainly by most in P7.
Level E: should be attainable by some pupils in P7–S1, but certainly by most in S2.
Level F: should be attainable in part by some pupils, and be completed by a few pupils, in the course of P7–S2.

Curriculum for Excellence

In 2007 the guidance on the new 'Curriculum for Excellence' was published by SGED for adoption during the following year. At its heart is a programme of work that is reviewing the 5–14 curriculum guidelines. It is the hope of SGED that the Curriculum for Excellence will aim to offer better educational outcomes for all young people and should provide more choices and more chances for those young people who need them, helping Scotland to meet the needs of pupils in school. Curriculum for Excellence will provide:

■ a coherent and inclusive curriculum from 3 to 18 wherever learning is taking place, whether in schools, colleges or other settings
■ a focus on outcomes
■ a broad general education
■ time to take qualifications in ways best suited to the young person
■ more opportunities to develop skills for learning, skills for life and skills for work for all young people at every stage
■ a focus on literacy, numeracy and health and well-being at every stage
■ appropriate pace and challenge for every child
■ connections between all aspects of learning and support for learning.

A stated aim is to enable all young people to become effective contributors with an enterprising attitude, resilience and self-reliance; confident individuals with self-respect, a sense of physical, mental and emotional well-being, secure values and beliefs and ambition; successful learners with enthusiasm and motivation for learning, determination to reach high standards of achievement and openness to new thinking and ideas; and finally responsible citizens with respect for others and a commitment to participate responsibly in political, economic, social and cultural life. It is based on the seven principles as outlined by SGED below (Learning and Teaching Scotland, 2007b, 2008):

1   Challenge and enjoyment: Pupils should find their learning challenging, engaging and motivating. The curriculum aims to encourage high aspirations and ambitions for all. At all stages, learners of all aptitudes and abilities should experience an appropriate level of challenge, to enable each individual to achieve his or her potential. The curriculum is designed to promote active engagement of learners in their learning and develop creativity.
2   Breadth: All pupils are required to have opportunities for a broad, suitably balanced range of experiences.
3   Progression: The curriculum will provide the opportunity for children and young people to experience continuous progression in their learning from 3 to 18 within a single curriculum

framework. Each stage should build upon earlier knowledge and achievements. Pupils should be able to progress at a rate which meets their needs and aptitudes.

4   Depth: The curriculum structure is designed to allow learners to develop and apply increasing intellectual rigour, draw different strands of learning together, and explore and achieve more advanced levels of understanding.

5   Personalisation and choice: The curriculum will respond to individual needs and support particular aptitudes and talents. It should give each child increasing opportunities for exercising responsible personal choice as they move through their school career.

6   Coherence: Taken as a whole, pupils' learning activities should combine to form a coherent experience. There should be clear links between the different aspects of pupils' learning, including opportunities for extended activities which draw different strands of learning together.

7   Relevance: Pupils should understand the purposes of their activities. They should see the value of what they are learning and its relevance to their lives, present and future.

How learning is organised

The current curriculum areas and subjects have been grouped into eight new curriculum areas that have been refreshed and re-focused in accordance with the purposes of the curriculum, the principles underlying the curriculum, and with an added emphasis on cross-curricular activities. Through cross-curricular activities, young people can develop their organisational skills, creativity, teamwork and the ability to apply their learning in new and challenging contexts The new curriculum areas are expressive arts, health and well-being, languages, mathematics, religious and moral education, sciences, social studies and technologies.

LTS indicate that 'Curriculum for Excellence' will challenge teachers to think differently about the curriculum and plan and act in new ways (LTS, 2007b). LTS plans that implementation of Curriculum for Excellence will go beyond the provision of guidance on curriculum content. It will also have implications for:

■   the teaching profession and other staff
■   the organisation of the curriculum in our schools and centres
■   the qualifications system
■   the recognition of wider achievement
■   the improvement framework.

ASSESSMENT

National tests

SQA have developed an online bank of national assessment materials which are based on Assessment of Achievement Programme (AAP) and Scottish Survey of Achievement tests and tasks. These national assessments are used by teachers to confirm their judgements about pupils' levels of attainment in the key areas of reading, writing and mathematics, for each of 5–14 levels, A–F. The professional judgement of the teacher is very important. The national assessments should support valid and reliable judgements of attainment against published attainment targets, when results are considered alongside other available evidence. The basic premise is that pupils should be tested when a teacher deems them ready to pass the test.

Formative assessment

One of the major assessment initiatives in Scotland has been the Assessment for Learning (AFL) programme, emphasising formative assessment to shape teaching in process (rather than summative assessment to assess the effect of teaching after it has happened). The programme has tried to develop a unified process of assessment and feedback so that pupils, parents, teachers and other professionals can plan effectively for pupils' developmental needs. The specific aims of the programme (LTS, 2004b) are to:

■ develop one unified system of recording and reporting, the Personal Learning Plan (PLP), which brings together the current PLP, progress file, transition records and Individualised Educational Programmes (IEPs)

■ bring together current arrangements for assessment, including the Assessment of Achievement Programme, Scottish Survey of Achievement, national tests and the annual 5–14 Survey of Attainment

■ provide extensive staff development and support through its project-based approach.

The programme is designed to promote better and timelier feedback for pupils, leading to improved achievement. It is intended that simplified systems of assessment are adopted, making it possible to provide more effective support for teachers (and therefore a reduction in workload). Teachers should also be able to provide clearer information for parents.

The AFL programme as adopted has three main components. These components emphasise:

■ assessment *for* learning
■ assessment *as* learning
■ assessment *of* learning.

The programme has provided guidance in each of these key areas. AFL explores the collection of evidence by assessment to inform onward planning. It also explores how to form effective partnerships with parents and pupils. In addition it explores the roles of pupils as learners, teachers as learners and the management and role of assessment (5–14 national tests, the Assessment of Achievement Programme (AAP) and alternatives). The AAP has monitored the attainment of a representative sample of pupils in P3, P5, P7 and S2 in specified areas of the Scottish 5–14 curriculum through the use of tests since 2002. The curricular areas monitored have included social subjects (2002), science (2003), mathematics (2004) and English language (2005). Pupils' performance in the core skills of communication, numeracy, ICT, problem solving and working with others has been assessed in the surveys since 2002. There is evidence that the use of formative assessment led to pupils taking more responsibility for their learning, contributing to improved motivation, confidence and classroom achievement, especially for lower attainers. Employing such strategies has also developed teachers' conceptual understanding of formative assessment, moving some from a teacher-centred pedagogy to one which placed pupils and their learning needs at the heart of teaching (Kirton, Hallam, Peffers, Robertson and Stobart, 2007). Further information about the AFL programme is available from http://www.ltscotland.org.uk/assess/index.asp.

QUALIFICATIONS

The SQA offers a number of national qualifications. Within the secondary school sector these are generally Standard Grades, National Units and National Courses. Standard Grade is generally taken over the third and fourth year of secondary school. An exam usually takes place at the end of fourth year. There are three levels of study from credit (the highest), to general and foundation (the lowest). Students usually take examinations at two of the three levels, e.g. credit and general or general and foundation. This enables pupils to try and get the highest level of qualification that they can achieve.

National Courses take place at seven levels: Access 1, Access 2, Access 3, Intermediate 1, Intermediate 2, Higher and Advanced Higher.

Access 1 courses are designed for pupils who require considerable support with their learning, while Access 2 courses are designed for those with moderate support needs. Access 3 courses are comparable to Standard Grade Foundation level, but there is not a requirement to sit an exam. Intermediate 1 and 2 courses are for pupils who have completed Standard Grades or Access 3 courses. Intermediate 2 is used sometimes as a stepping stone to Higher Grade by pupils (or even by pupils who are starting a new subject in school). Highers are generally taken by pupils who have passed subjects at Standard Grade Credit level or have successfully completed a course at Intermediate level.

Highers are normally needed to allow entry to a university or college course that leads to a degree or Higher National Diploma. Advanced Highers are generally for those pupils who have passed Highers. Normally taken in sixth year, they are additional qualifications for entry into higher education or the workplace. National Units are qualifications in their own right. They are usually of about 40 hours' duration and are assessed by the student's own teacher. Assessment for National Courses from Intermediate 1 to Advanced Higher includes an external examination. Performance in the external assessment determines the grade awarded for the course. Pass grades are awarded at grades A (top grade), B and C (lowest 'pass grade'). Grade D is awarded to pupils who just fail to obtain a grade C.

Task 7.5.3 asks you to review the requirements for national assessments and consider how you could prepare pupils for these.

Task 7.5.3 PREPARING PUPILS FOR NATIONAL ASSESSMENTS (SCOTLAND)

Obtain a copy of the Standard Grade and Higher Grade (and 5–14 National Text if you teach mathematics or English language) examinations for last year. Look at the nature and structure of the examination papers. How could you ensure pupils are prepared for undertaking examinations of this nature?

A teaching profession for the twenty-first century

In January 2001 the findings of the McCrone Report into teaching in Scotland, *A Teaching Profession for the 21st Century*, were published (SEED, 2001a). The report detailed the recommendations of an enquiry into the pay and conditions of teachers in Scotland. The report recommended the following career structure for teachers in Scotland:

> Progression to Chartered Teacher is via qualification at masters level. Progression to principal teacher, depute headteacher and headteacher is by application for specific jobs. Promotion to depute headteacher and headteacher is normally preceded by study towards the 'Standard for Headship'.

The report recommended a working week of 35 hours for teachers, of which not more than 22.5 hours should be class contact time. Importantly, probationer teachers were guaranteed one year's

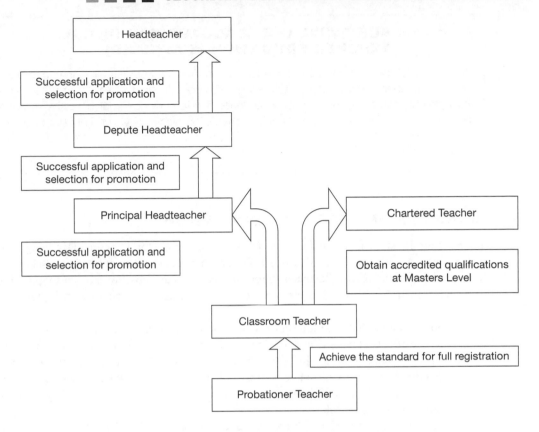

■ **Figure 7.5.2** Summary of the career structure for teachers in Scotland

on-the-job training leading to full registration with the GTCS. During this probationary period their maximum class contact time should be no greater than 0.7 of the working week (one and a half days of non-contact time). Further details of the McCrone Report can be found at: http://www. scotland.gov.uk/Resource/Doc/158413/0042924.pdf.

General Teaching Council for Scotland

In order to teach in a Scottish state sector school teachers are expected to maintain a current valid registration with the GTCS. The GTCS has three levels for registration:

■ Standard for Initial Teacher Education – which providers of initial teaching qualifications must build into their programme design. Students must then demonstrate competence against the benchmarks of education practice in order to register as a probationer teacher in Scotland.

■ Standard for Full Registration – which individual probationer teachers must attain to achieve full registration with the GTCS.

■ Standard for Chartered Teacher – to be attained by individual teachers. Aspects of this are covered earlier in this unit.

Task 7.5.4 sets you the task of providing evidence that you are meeting the standards required of teachers in Scotland.

Task 7.5.4 **ACHIEVING THE STANDARD FOR INITIAL TEACHER EDUCATION (SCOTLAND)**

Look at the Standard for Initial Teacher Education. As you progress through your programme of initial teacher education, your tutors are likely to ask you to keep brief notes regarding your development in each of these Standards. Try to evidence your notes by developing a personal portfolio drawing in personal thoughts, examples of plans and pupils' products from school experience and examples of summative and formative feedback from tutors, teachers and mentors.

CHALLENGES FACING SCOTTISH EDUCATION

The education system in Scotland faces a number of challenges for the future. Declining populations in rural areas of Scotland mean that it is becoming financially difficult to sustain schools that can have a roll in single figures. Balanced against this are the huge costs of transporting rural pupils to centralised facilities. Delivering effective CPD to teachers in rural locations presents similar problems.

Another major challenge for Scottish education is the development of effective strategies to counteract the effects of poverty and low income. Scotland has one of the highest rates of childhood poverty in the Western world, in both urban and rural communities. One-third of Scotland's households live in poverty. One in five Scottish children is entitled to a free school meal. In Scotland many children under 16 become homeless every day and 360,000 children live in property affected by dampness (Mackenzie, 2003).

Mainstreaming of pupils with special education needs as the default expectation was made explicit in the Education Act 2000. The legislative framework that deals with SEN in Scotland focuses on rights and responsibilities (pupils' rights to be educated and educators' responsibilities to educate them) (Scottish Office Education and Industry Department, 1999). Pupils with SEN have by Co-ordinated Support Plans (CSPs). The CSPs emphasise multi-agency co-operation. They document the capabilities of pupils, the educational objectives and additional support required. In addition the persons responsible for providing the additional support are identified and a timetable for implementation established. More can be read regarding the CSP at: http://www.opsi.gov.uk/legislation/scotland/ssi2005/20050518.htm.

Like the rest of the UK, Scotland been tackling behavioural problems in schools. A task group on discipline in Scottish schools reported to SEED (2001b), and as a result £10 million per year over four years was allocated to establish restorative justice schemes (SEED, 2004b). Effective educational provision for children of asylum seekers and refugees is a related issue. A survey of 2,000 Scots revealed that a high proportion self-reported themselves as having racist views and over one-quarter openly admitted to being racist (Humes and Bryce, 2003). In 2002 the Scottish Executive launched a major initiative to combat racism in Scotland – the 'One Scotland, Many Cultures' campaign. Problems of bullying associated with racism and minority language use have been recognised by SEED (2002). Support materials have gone to all schools and an anti-bullying network has been established (see http://www.antibullying.net).

Given its historical tradition of excellence in education, Scotland pays attention to international comparisons of educational performance. In the results published by the Programme for International Student Assessment (PISA) Scottish secondary pupils performed reasonably well in mathematics and science (PISA, 2000). In mathematics Scottish pupils ranked equal fifth amongst the Organisation for Economic Co-operation and Development (OECD) countries and in science they ranked ninth amongst the OECD countries (SEED, 2004c). However, Scotland did not perform as well as England in either mathematics or science. The Progress in International Reading Literacy Study (PIRLS) with pupils at the end of primary education showed that while Scotland was well above the international average in literacy, it had lower scores than England (Department for Education and Skills (DfES), 2003c). This suggests the Scottish education system has little room for complacency and considerable further room for improvement. High hopes are held that 'Curriculum for Excellence' will have a positive impact in Scottish schools and help teachers improve standards and pupils realise potential.

SUMMARY AND KEY POINTS

This unit has sought to identify key features that make the Scottish education system distinctive from that in the rest of the UK. It has also tried to raise awareness of the structure and balance of the curriculum in Scotland and highlight the preparation that may be required to teach within that framework. The intention has been to stimulate thinking about some of the issues that are likely to impact upon the professional practice of teachers in Scotland.

The challenge for NQTs in Scotland is to teach their subject well, but not to lose sight of the overview of how they fit into the larger picture of education in Scotland. In addition to this, in order to enhance standards in Scottish education it is imperative that research is used to inform practice. In order to facilitate articulation with research to allow it to inform and shape professional practice in this way, making early connections with interested research groups such as Scottish Education Research Association (SERA) or Scottish Council for Research in Education (SCRE) is desirable. This will allow research evidenced based practice to develop rather than more intuitive methods of working in the classroom. This should lead to more effective learning and teaching.

Check which requirements for your course have been addressed through this unit.

FURTHER READING

Bryce, T. G. K. and Humes, W. M. (2003) *Scottish Education*, 2nd edn, Post-Devolution, Edinburgh: Edinburgh University Press.

Paterson, L. (2003) *Scottish Education in the Twentieth Century*, Edinburgh: Edinburgh University Press. This book gives a comprehensive overview of Scottish education.

RELEVANT WEBSITES

Scottish Executive Education Department (SEED) (2000) *The Structure and Balance of the Curriculum: 5–14 National Guidelines*: http://www.ltscotland.org.uk/5to14/htmlguidelines/saboc/intropage1. htm (last accessed 5 November 2007)

This website gives an overview of the rationale behind the 5–14 national curriculum guidelines. It also explores the principles underpinning the design of 5–14.

Scottish Government (2007) *Curriculum for Excellence*: http://www.curriculumforexcellencescotland.gov.uk/ (last accessed 6 November 2007)

This website provides an introduction to the aims, purposes and principles of Curriculum for Excellence. It contains news about the initiative and links that explore the curriculum in greater depth and provide resources to help education professionals embed the new curriculum into their professional practice.

Quality Assurance Agency website: http://www.qaa.ac.uk/academicinfrastructure/benchmark/iteScotland/introduction.asp

The Standards for Initial Teacher Education in Scotland are available on this website.

Additional resources for this unit are available on the companion website:
www.routledge.com/textbooks/9780415478724

YOUR PROFESSIONAL DEVELOPMENT

In this chapter we consider life beyond your student teaching experience. The chapter is designed to prepare you for applying for your first post and to be aware of the opportunities available to continue your professional development as a teacher after you have completed your initial teacher education course. It contains three units.

Getting a job at the end of your initial teacher education is important, time-consuming and worrying for student teachers. Unit 8.1 is designed to help you at every stage of the process of getting your first post. It takes you through the stages of deciding where you want to teach, looking for suitable vacancies, sending for further details of posts that interest you, making an application, attending an interview and accepting a post.

The success of any school depends on its staff. However, although you have successfully met the requirements to qualify as a teacher at the end of your initial teacher education course, you still have a lot to learn about teaching to increase your effectiveness. Unit 8.2 considers the transition from student teacher to newly qualified teacher, immediate induction into the school and the job, ongoing induction throughout the first year and continuing professional development which helps you continue to learn and develop professionally throughout your career.

Unit 8.3 is designed to give you an insight into the system in which many of you will be working as teachers. We look briefly at the structure of the state education system in England and then at teachers' accountability: professional, moral and contractual. This leads into a slightly fuller consideration of the legal and contractual requirements and statutory duties that govern the work of teachers.

> *Readings for Learning to Teach in the Secondary School: A Companion to M Level Study*
> brings together essential readings to support you in your critical engagement with key issues raised in this textbook.

UNIT
8.1

GETTING YOUR FIRST POST

Alexis Taylor, Julia Lawrence and Susan Capel

INTRODUCTION

Obtaining your first teaching post involves a number of stages, each of which is equally important. You need to be clear about why you want to enter the teaching profession, and then decide where you want to teach, look for suitable vacancies, select a post which interests you and send for further details, prepare your curriculum vitae (CV), write a generic and then specific letter of application, contact potential referees to make sure that they are prepared to act for you and to confirm their contact details, make an application, prepare for an interview (and undertake a mock interview, if possible), attend the interview and accept the post. This unit is designed to help you with that process.

OBJECTIVES

At the end of this unit you should be able to:

■ articulate why you want to enter the teaching profession
■ follow the procedure for, and process of, applying for your first teaching post
■ make a written application
■ prepare for an interview for a teaching post.

Check the requirements for your course to see which relate to this unit.

CONSIDER WHY YOU WANT TO ENTER THE TEACHING PROFESSION

Obtaining your first teaching post may be one of the most important decisions of your life, so it needs to be taken carefully. Teaching is an attractive choice of career for many reasons. However, now you have undertaken some school experience, you are also aware that in reality teaching is a demanding career. Even though student teachers are already committed to a programme of initial teacher education (ITE) it is not automatic that all wish to enter the teaching profession. Think carefully why you want to enter the teaching profession; altruistic and intrinsic reasons appear to be more important and motivating to those committed to joining the teaching profession than extrinsic reasons (Reid and Cauldwell, 1997). This may influence your choice of post.

There are many factors that you have to consider when thinking about applying for your first post. The remainder of the unit explores these. It focuses on procedures and processes in England. If you are applying for teaching jobs in Scotland, Northern Ireland or Wales, you should obtain further information from your higher education institution (HEI), and/or the relevant government office (see useful websites at the end of the unit).

DECIDING WHERE YOU WANT TO TEACH

For most student teachers, deciding where to apply is your first major decision. If you are committed to living in one place because, for example, you have family commitments, you need to consider the distance it is possible to travel to a job in order to determine the radius in which you can look for a post. The travel time to and from school is important, as you probably will not want a long journey in your first year of teaching when you are likely to be tired at the end of each day or when you have had school commitments in the evening.

If you opt for a popular area it could be difficult to obtain a post. The reasons for popularity may be that there are few schools, turnover from schools is low or there are a number of applications from student teachers at a local HEI. It is therefore worth considering if your preferred areas are popular and, if so, whether there are other areas to which you could go or whether you could be totally flexible as to where you teach. Unless you are committed to a restricted geographical area due to personal reasons, it is worth doing some research about other areas of the country rather than basing your decision on assumptions about certain areas.

Alternatively, you may want to teach abroad, either in a paid or voluntary capacity. You need to decide in which area and/or country you wish to teach. The regulations for each country vary. For example, some teaching jobs abroad require you to have teaching experience before you can apply. Also, there may well be an assessment of whether your teaching qualification is suitable for the chosen county. You can get further advice from agencies such as Voluntary Service Overseas (VSO), Teaching Projects Abroad and also the British Council. These bodies normally support suitable candidates in obtaining registration as teachers. If you wish to obtain registration in a specified country and do not wish to work through a voluntary agency, you can independently contact the Embassy or High Commission of your chosen country (you probably need to do this anyway to obtain a work permit and other necessary documentation). If you do decide to teach abroad you need to consider issues regarding your induction period.

Now complete Tasks 8.1.1 and 8.1.2.

Task 8.1.1 WHERE DO YOU WANT TO TEACH?

Think about where you would like to teach and how flexible you are able to be in where you can look for a post. List all areas in which you would consider working and find out something about them. If you know anyone from the area, talk to them about it. Visit the area if at all possible to get a general 'feel of the place' and further information.

Task 8.1.2 **WHAT TYPE OF SCHOOL DO YOU WANT TO TEACH IN?**

Think back to the types of schools you experienced/are experiencing during your ITE. List the aspects you find positive about working in them and list the opportunities you were/are not able to experience in these different types of schools.

List criteria for the types of schools you would be happy to teach in (e.g. primary, middle, secondary; maintained or voluntary; specialist school; sixth form college). Identify why you want to teach in that type of school. However, it is advisable not to close your mind to other options. During your school experiences you see or hear about a range of types of school and you may surprise yourself by enjoying teaching in a type of school that you have not previously considered.

Find out what types of school there are in those areas in which you would like to teach. You can start by looking at the *Education Authorities Directory and Annual* and the *Education Year Book* (see Further Reading at the end of this unit). If you visit the area(s) in which you would like to teach, try to arrange to visit some schools.

LOOKING FOR SUITABLE VACANCIES

The majority of advertisements for teaching posts are for specific posts in specific schools. Advertisements generally start around January or February. However, the majority of advertisements are around April or May because teachers who are leaving at the end of the academic year are required to hand in their notice by the end of May. Independent schools may advertise earlier, from December onwards.

Teaching posts are advertised in a number of different places: in the national press, sometimes the local press (especially for part-time posts). The major source of information about teaching posts is the *Times Educational Supplement* (http://www.tes.co.uk published every Friday). However, jobs are also advertised in other national newspapers, such as the *Guardian* (http://www.guardian.co.uk Tuesday edition). There are also advertisements in religious and ethnic minority newspapers such as the *Asian Times*, *Catholic Herald* and the *Jewish Chronicle*.

Online there are agencies with which you can seek for jobs, for example, the jobs go public website (http://www.jobsgopublic.com/). For student teachers registered on an ITE course, some online recruitment agencies match up preferences for posts (e.g. location, type of school and type of contract) with vacancies at schools, and forward information about suitable posts. Local Authorities (LAs) sometimes advertise posts on their websites.

Information about specific jobs may be sent to your HEI.

SELECTING A POST WHICH INTERESTS YOU AND SENDING FOR FURTHER DETAILS

If an advertisement interests you then write for further details. Write briefly and to the point, for example:

Dear Sir/Madam (or name if given in advertisement)

I am interested in the vacancy for a (subject) teacher (quote reference number if one is given) at ABC School, advertised in (publication, e.g. the *Times Educational Supplement*) of (date) and would be grateful to receive further details of this post.

Yours faithfully (*if you use Sir/Madam* or Yours sincerely *if you use a name*)

MAKING AN APPLICATION

As you read details of all posts to which you are interested in applying, highlight key words and phrases, and the requirements specified, which indicate whether the post is suitable for you as a first post. Details of teaching posts normally focus in the main on knowledge, skills and experience. They should not ask about age, date of birth, martial status, ethnicity, disability or religious affiliation (although these aspects may well be requested for the specific purpose of monitoring equality matters). Consider whether you have the knowledge, skills, experience and qualities the school is looking for and whether the school meets some or all of your requirements. If you decide to apply, remember that first impressions are very important and applications are the first stage in the selection process. You need to present yourself effectively on paper. Plan the content of your application before you complete an application form or CV or write a covering letter for a specific post. You use the same basic information for all applications; however, you cannot have a standard application form, CV or letter of application which you use for every application. Each application needs to be slightly different as you want to match your experience and qualifications to the requirements of the post, highlighting different points and varying the amount of detail you provide according to specific requirements of the post and the school. You should find it useful to look back at the key words, phrases and requirements you underlined in the details for the post. These help you to customise and personalise the application. A customised application shows that you have taken the time to find out about a specific post in a specific school, and should help your application to stand out from the others. An application which fails to explain why you are interested in the specific post in the specific school is unlikely to be considered further. Thus, completing an application form for each post takes time. Two hours is probably the minimum time to complete an application properly without rushing it if you have prepared beforehand and have all the information available. It takes longer if you have not prepared in advance. It is a good idea to keep all your information on a specific file on your computer. This helps you to customise your applications more easily.

Referees

When applying for teaching posts you are normally asked to supply the names and addresses of at least two referees. Before you complete an application, contact potential referees to make sure that they are prepared to act for you, to confirm their contact details and if there are any dates when they are away and unable to respond should a request arrive. Your first referee is normally someone associated with your teacher education course. Check if there is one particular person within your institution who you should name as the first referee and, if not, decide who you would like this to be and then ask that person. This reference covers all areas of your professional and academic work on the course. It is often helpful for the person compiling your reference to have additional information about you which might be included in a reference, e.g. other activities in which you are involved. Therefore, check whether it would be helpful for your referee to have a copy of your CV. Your second referee should be someone who knows you well and is able to comment on your character, qualities, achievements and commitment to teaching as a career. Your mentor (or other member of staff) at the school where you were placed for your final block school experience may well be an appropriate person, otherwise an employer from any work you have undertaken. It is not normal practice to include open testimonials with your application as schools or LAs value confidential references more highly. Some LAs have a policy of open references, i.e. the reference is shown to the applicant in certain circumstances. The referee knows this at the time of writing the reference.

Methods of application

Schools normally require job applicants to submit a letter of application and completed application form or a CV.

Address
(at top right hand of page)
Date

Name of headteacher
Address

Dear Sir/Madam or Name of headteacher

Paragraph 1
In reply to your advertisement in (name of publication) of (date) I would like to apply for the post of (subject(s)) teacher (quote reference number if one is given) at (name of school). *Or* I have been informed by my institution that, in September, you will have a vacancy on your staff for a teacher of (subject(s)) and I would like to apply for this post.

Paragraphs 2/3
This section should begin by explaining why you are applying for this particular post. It should then carefully match your qualifications, experience, particular skills and personal qualities to those required by the school, indicating what you could contribute as a teacher of the subject(s) specified.

Paragraphs 3/4
These might begin: The enclosed curriculum vitae provides details of the content of my teacher education course. I would also like to draw your attention to . . . (*here outline any special features of your course and your particular interest in these, anything significant about your teaching and any other work experience, anything else you have to offer above that required specifically by the post, including being able to speak a language other than English, extra-curricular activities, a second subject; skills in ICT; pastoral work; etc. that you wish the school to be aware of and any other information about interests and activities related to the post or to you as a teacher, including additional qualifications, awards and positions of responsibility you have held*).

If you are unavailable for interview on any days, this is the point to mention it. You might indicate this by including a statement such as 'It may be helpful to know that my examinations (or other event) occur on the following date(s) (quote actual dates). Unfortunately this means that I am not able to attend for an interview on those dates. I hope this does not cause inconvenience as, should you wish to interview me, I could come at any other time.'

Yours faithfully (or Yours sincerely if you use the headteacher's name)
(signature)

■ **Figure 8.1.1** Format for a letter of application

An editable version of Figure 8.1.1 is available on the companion website:
www.routledge.com/textbooks/9780415478724

CURRICULUM VITAE

Name: Date of birth:
Home address: Term time address: (if different)
Telephone number: Telephone number:
e-mail address: e-mail address:

(*indicate dates when you are at your term time address and when your home address should be used*)

Academic qualification(s): (your first and higher, with subject, institution and class)

Professional qualification (for teaching): *If you are yet to qualify write* I am currently on a PGCE/PGCert/BEd (or other) course and expect to qualify in July 200?.

Previous relevant experience:

(*provide only very brief details here to highlight the most important points to help the reader; expand on these later in the CV*)

EDUCATION

(*list institutions from Secondary School on, in reverse chronological order*)

Institution   Dates attended

(*you should include some detail about your degree and/or teacher education course, particularly emphasising those aspects of your course which match the requirements of the post*)

QUALIFICATIONS

(*list qualifications from O levels/GCSEs on, in reverse chronological order*)

Qualification gained   Date awarded (with subject(s),
                                              grades or classification)
                                              (or date to be awarded)

TEACHING EXPERIENCE

(*list any prior teaching experience and the school experiences on your course, in reverse chronological order*)

School and subject(s) taught Year(s) and length and focus of experience

e.g. ABC school. Final block school experience of 7 weeks' duration, comprising: hockey (Year 7); gymnastics (Year 8); swimming (Year 9); GCSE PE (Year 10).

OTHER WORK EXPERIENCE

(*list permanent full- or part-time jobs and holiday jobs separately, each in reverse chronological order*)

Job   Dates (start and finish)

(include anything special about each job, particularly where it relates to children and/or teaching)

INTERESTS AND ACTIVITIES

(*e.g. membership of clubs or societies, details of offices held, achievements, e.g. sport, music, hobbies; group these together if appropriate, with the most relevant first and if giving dates, in reverse chronological order*)

ADDITIONAL QUALIFICATIONS

(e.g. additional languages, music grades, coaching or first aid awards)

OTHER INFORMATION

(*include anything else that you think is important here in relation to the post for which you are applying. For example, do state if you are a qualified driver as the ability to drive the school mini-bus may be useful to the post*)

REFERENCES

First referee   Position   Address   Telephone number   e-mail address
Second referee   Position   Address   Telephone number   e-mail address

■ **Figure 8.1.2** Sample curriculum vitae

An editable version of Figure 8.1.2 is available on the companion website:
www.routledge.com/textbooks/9780415478724

LETTER OF APPLICATION

A letter of application should state clearly your reasons for applying for the post, matching your qualifications, experience, particular skills and personal qualities to the post as described in the information sent to you from the school. The letter is normally between one and two sides of A4 in length, on plain white notepaper. A suggested format for a letter of application is given in Figure 8.1.1.

AN APPLICATION FORM

The information required on an application form closely matches that identified for a CV (see below). Usually an electronic copy of the application form is available. If not, we recommend that you make a photocopy of the blank form and complete this in pencil as a practice before completing the original form. (You may also do this to help you focus on what you are going to write and check that the information fits into the space provided.) Read through an application form before you write anything on it. Follow exactly any instructions given. Check that there is no missing information, dates or other details or questions which have not been answered. Do not leave any sections of the form blank. If there are sections which you cannot complete write N/A (not applicable).

One page of the form is often blank and in the space provided you are required to explain why you are applying for the post and to elaborate on the skills and experience that equip you for it. This section should be written in continuous prose as if it is a section of a letter, following the suggested format and containing the type of information given for a letter of application (see Figure 8.1.1). It is usually acceptable to use additional sheets of paper and attach them to the form. This section requires information that would otherwise be included in a letter of application; therefore a letter of application with such an application form is normally very brief, indicating that you have included your application for the post of (subject) teacher as advertised in (publication). A longer letter of application would be needed if the application form does not include such a section.

CURRICULUM VITAE (CV)

A CV should always be accompanied by a longer letter of application. A CV summarises your educational background, qualifications, teaching and other work experience, interests and activities and any other relevant qualifications and information. A sample format for a CV is provided in Figure 8.1.2.

Now complete Task 8.1.3.

Task 8.1.3 **YOUR CURRICULUM VITAE**

Draft a specimen letter of application and CV and obtain and complete an application form. Ask your tutor to check these for you. Use these as the basis for all your job applications.

NOTES ABOUT APPLICATIONS

Do remember that you are not the only person applying for a post, therefore it is important that you do your best to make your application stand out. Consider how you can do this in what and how you write. For example, try to use phrases in your letter that show your good qualities. For example, 'I successfully led an initiative to introduce a new system for checking pupils' homework' is better than 'I checked pupils' homework'. However, do not exaggerate claims. Applications should be laid out well and presented clearly and without using jargon (you should also type any non-electronic application). Check your application to ensure that there are no basic errors such as typing errors, mistakes in spelling, grammar or punctuation; and that the information is accurate and consistent. If there is time, it is worthwhile asking a colleague, friend or tutor to read through your final application.

Indicate clearly any dates that you are unable to attend for interview, e.g. because you have an examination. Examinations must normally take precedence over interviews. However, holidays do not take precedence and most schools do not wait until you return from holiday to interview you; therefore do not book holidays at times when you are likely to be called for interview. Remember that if you put down additional skills or experiences, for example, that you can sing, you may be invited to use those skills in school, e.g. in the school choir. Therefore, do not make exaggerated claims about your skills or additional experiences.

Always send the original application, but keep a copy of every letter of application, application form or CV so that you can refresh your memory before an interview.

ATTENDING AN INTERVIEW

If you are offered an interview, acknowledge the letter at once, in writing if there is time, indicating that you are pleased to attend for interview on that date. If you are offered two interviews on the same day, you probably have to choose which one you attend, unless they are at different times and close enough together to enable you to attend both. Write and decline the interview you decide not to attend. If there is a problem with an interview date, for example, it coincides with an examination, let the school know immediately. Expenses (including basic overnight accommodation where necessary) are usually paid for attending an interview (although not if you are offered and decline a post). However, you might want to check in advance as rules vary between schools and LAs. You should receive a travel and expenses claim form with the letter notifying you of the interview (or at the interview itself).

Preparing for the interview

Prepare for an interview in advance. Read through the advertisement, job description and any other information about the school and post again. Also try to find out if there is anyone at your institution or placement school who knows the school. You may want to gather some further information about the school, for example, the latest Ofsted inspection report and performance data (http://www.ofsted.gov.uk). If possible, visit the school beforehand. This helps you to check your journey in preparation for the interview day, but also allows you to find out more about the school and about the local area. Most schools welcome this – as long as you ask in advance. Do not just turn up at the school and expect to look around. Decide what to look for when shown round the school. If possible, talk to a newly qualified teacher (NQT) in the school. You might find it helpful to reflect on why you applied for this particular post, so that you can put across the relevant information convincingly at the interview. Read through your application again so that you can communicate effectively the information and evidence you consider to be relevant to the post. It also helps you avoid any contradictions between what you say and what you wrote in your application, as each member of the interview panel has a copy of your application and so can compare answers. If you are not reading the *Times Educational Supplement* on a regular basis, we recommend that you do so before your interview so that you can talk about and answer questions on the latest educational issues and debates. It is useful to have a portfolio of your teaching-related activities, for example, good lesson plans, examples of pupils' assessed work, worksheets, evaluations, photographs of displays or trips, review of resource(s), information and communications technology (ICT) skills. This is derived from the professional development portfolio you have been keeping throughout your ITE course (see also Unit 8.2).

Plan what you are going to wear to the interview as your appearance is important. Knowing something about the school is useful, for example, if the staff dress formally you should dress formally. If you are unsure, it is advisable to be conservative in your dress.

Attending the interview

It is difficult to generalise about interviews because these vary considerably, although normally a formal letter of invitation explains exactly what the day will comprise. It may, for example, comprise a tour of the school, teaching a lesson/part of a lesson, an informal talk or interview with the head of department or a senior teacher, lunch and a panel interview. In many schools, all people invited to interview arrive at the school at the same time, are shown round the school in a group, then sit and wait in an allocated space while everyone is interviewed in turn, for a decision to be made and for the successful candidate to be told. In other interviews, candidates are invited at different times so that they do not meet. In some interview situations you may also be asked to take part in a group discussion or do an activity, such as an 'in-tray' exercise or drafting a letter to parents. Some also include a second interview stage where a small number of candidates are selected from the initial long list interviewed previously. Another recent initiative is to gather the views of pupils on the interviewee, so at some schools you may be asked to talk with a group of pupils at some stage during the day.

Because of these variations, the format and length of interview days varies. An example of an interview day is shown in Figure 8.1.3.

Teaching a lesson as part of the interview

It is now regular practice for candidates to be asked to teach a lesson or section of a lesson as part of the interviewing process. If required, make sure that you know the age and size of the class, what you are expected to teach and the pupils' prior knowledge, the length of the lesson, what resources and equipment are available, i.e. all the information you require before teaching any class. You should be told this in the letter of invitation to the interview. If not, telephone and ask. Plan this lesson carefully, giving attention to learning outcomes, purpose of content and activities, teaching strategies including questioning technique and resources. It is useful to have copies of the lesson plan available to give to those observing you. This is an opportunity to show the quality of your preparation and planning. Lessons taught as part of the interviewing process also provide you with the opportunity to demonstrate the level of your subject content knowledge, so, again, prepare well, particularly if you have been asked to teach a topic with which you are not totally familiar. It is probably best to try to base your interview lesson on something that has been successful on a previous occasion with similar classes. The lesson is also an opportunity to show your enthusiasm for teaching and pupils' learning. Try to appear confident and relaxed, although those observing you understand that you probably feel a little nervous!

You may be asked to discuss your teaching as part of the formal interview but if not it is a good idea to talk with relevant staff about the lesson; for example, how you feel it went; what the pupils learnt; how you know that they learnt this; and what you might change. Do not be anxious about mentioning if some things have not gone to plan. For example, your timing might have gone astray, or your instructions were not as clear as you had anticipated. Use the opportunity to show

9.15 a.m.	Arrive at school	At this stage you are normally welcomed, along with other candidates, by the headteacher and are given the schedule for the day, if this has not already been sent to you.
9.30 a.m.	Meeting with head of faculty and/or head of department	At this stage information about the school and department is explained to you, for example, structure and organisation; curriculum; roles and responsibilities; assessment policies; examination results; procedures about school routines and expectations. You may well receive further documentation. If there is more information you would like to have, please take the opportunity to ask questions.
10.00 a.m.	Coffee	This might be with members of the department or other staff. Again, use this as an opportunity to learn more about the school and the post.
10.15–10.45 a.m.	Tour of the school	You may well be escorted by a member of staff or by pupils. This provides an opportunity to take note of what may be your working environment, for example, the layout of the school and department; display work; facilities; the learning and working atmosphere in the lessons.
10.45–12.00 p.m.	Teaching a lesson/part of a lesson	There is normally a rota for this with other candidates. You should be prepared for this part of the interview in advance. It is not normal practice for this to be sprung on you without warning! For further information see the section on teaching a lesson as part of an interview.
12.00 p.m.	Lunch	This may well be in the school canteen with other members of the department. Again, use this period of the day as an opportunity to ask questions.
12.45 pm	Group interview	Here you take part in a discussion with other candidates about a selected topic (usually given in advance). You are observed on aspects such as your knowledge of the topic, consideration of others, listening, articulating your views, the extent of your contribution, your level of confidence etc
1.30 p.m.	Formal interviews	Candidates are interviewed individually, normally in alphabetical order. If you have a legitimate commitment (e.g. a train to catch) and need to leave before your allocated time, it is best to say so, and the school usually accommodates this request.

■ **Figure 8.1.3** Possible interview schedule

that you have realised this and analyse why it happened and how you might change this in the future. This demonstrates that you are reflective and thoughtful and serious about your own practice.

The interview process

As the format for interview days varies so does the panel interview. In some interviews you are faced by a panel comprising anything between two to three and six to seven people, in others you have a series of interviews with different people. In either case these people normally include some of the following: the headteacher, a (parent) governor, another senior member of the school staff, head of department. The length of time for a panel interview can vary from about half an hour to one-and-a-half hours.

An interview is a two-way affair. At the same time as being interviewed you are, in effect, interviewing the school and deciding if this is a school in which you could work and therefore if this is a post for you. Take the opportunity to learn as much as you can about the school, the post and the working environment. This requires you to be alert to what is being said and to be prepared to ask as well as to answer questions. If not included as part of the interview day, be firm in requesting an opportunity to look round the school prior to interview, including sitting in on a lesson if possible. If at any stage during the day you feel this is not the right post/school for you, you can withdraw and it is fair to do so.

The initial impact you make is very important as interviewers tend to form an overall impression early. The interview starts as soon as you walk into the school and you are assessed throughout the day. Your performance, including your verbal and non-verbal communication, in each activity is therefore important and could make the difference between being offered the job or not. Particular attention is paid to the impression you create in the formal interview. For example, do not sit down until you are invited to do so and then sit comfortably on your chair looking alert; do not sit on the edge of your chair looking anxious or slouch in your chair looking too relaxed. Look and sound relaxed and confident (even if you are not). Try to be yourself. Try to smile and direct your answers to the person who has asked the specific question as well as the panel during discussion. Do try not to speak too quickly and try to keep your answers to the question posed. If you are unsure about how much information to give when answering questions it is probably better to keep an answer brief and then ask the panel if they would like further information. If you are not sure what the question means then do not hesitate to clarify this. Avoid repetition but do not worry if you repeat information included in your application, as long as you do not contradict what you wrote. Interviewers have various degrees of specialist knowledge and understanding. Avoid jargon in explanations but assume interviewers have some knowledge and understanding of your subject area. Aim to provide a balanced picture of yourself, being on the whole positive and emphasising your strengths, but being aware of areas for development.

Interviewers are trying to form an impression of you as a future teacher and as a person and have a number of things they are looking for. These include:

■ Your knowledge and understanding of your subject and your ability to teach it. Interviewers assess your ability to discuss, analyse, appraise and make critical comment about ideas, issues and developments in your subject and subject curriculum, your personal philosophy about and commitment to the teaching of your subject(s).

■ Your professional development as a teacher. This is based partly on your school experiences. Interviewers assess your ability to analyse observations of pupils' behaviour and development, your own development and your involvement in the whole life of the school on school experiences, and your ability to discuss, analyse, appraise and make critical comment about educational issues.

■ Your ability to cope with the post. Interviewers assess how you would approach your teaching, for example, your understanding of the different roles you are required to undertake as a teacher, how you have coped or would cope, in a number of different situations, for example, disciplining a pupil or class experiencing difficulties, dealing with a concerned parent or with teaching another subject.

■ Your ability to fit into the school and the staffroom and to make contact with and relate to colleagues and pupils. Interviewers assess your interpersonal skills, verbal and non-verbal communication skills (your written communication skills have been assessed from your application).

■ Your commitment to the specific post. Interviewers assess the interest and enthusiasm you show for the post to try to find out if this is a post you really want or whether you see this post as a short-term stop-gap before you can find a post in an area or school where you really want to teach.

After introductions and preliminaries, most interviewers ask why you have applied for this particular post. They then focus on the information in your application, including your personal experiences, your education, qualifications, teaching skills gained from school experiences and other teaching and/or work experience, your interests and activities and other qualifications. You are normally also asked what you feel you can contribute to the school and general questions about professional or personal interests, ideas, issues or attitudes. Therefore, think about areas you want to emphasise or any additional evidence of your suitability for the post that you did not have room to include in your application. Draw on both your teacher education course and school experience and on other experiences, e.g. other work with children such as work in a youth club or voluntary work. This demonstrates your commitment to working with children.

You also need to show that you realise you still have things to learn and that you are committed to continuing your development as a teacher. You should be able to talk about areas for development. For example, you need to consolidate your learning in your first year, perhaps gain further experience of other areas, such as teaching Years 12 and 13 or taking on a tutor group. There may also be specific areas of subject content knowledge you need to enhance. Depending on when your interview takes place, if you are learning to teach in England, it might be possible for you to refer to your Career Entry and Development Profile which indicates your strengths towards the end of your ITE and also areas for development in your induction year. It is helpful to have a career plan, but not to appear so ambitious that you give the school the impression that you will leave at the first opportunity.

Possible interview questions

Questions asked at interview vary considerably; therefore it is not possible for you to prepare precisely for an interview. However, it is helpful if you identify possible questions in your preparation and prepare some possible outline responses to such questions. It is useful to give a general response to the question to show you are aware of some of the principles and issues and also to refer to examples of your own practice. For example,

Interviewer: How did you set about planning differentiated learning for a class you have taught recently?

Candidate: This is an important way of enabling all pupils to have equal access to the curriculum so that they learn as much as possible and fulfil their individual potential. There are a number of strategies that can be used; for example, differentiation by outcome, by task or by rate of work. During my last school experience I was teaching a Year 7 class about religious festivals. I did not want to give out several different worksheets as this might have embarrassed some pupils, so I made one worksheet which had some core tasks for all pupils and also some option tasks, which involved different levels of work and different types of activities. I also developed differentiation through my use of questioning . . .

Some further questions which might be asked

YOUR COMMITMENT TO TEACHING

- Why did you choose teaching as a career?
- Why did you choose to teach the secondary (middle/upper) age range?
- Why did you choose this particular course?
- Tell me what you have learnt about teaching from your school experiences.
- What have been the most difficult aspects of your school experiences and why?
- Tell me about any other experiences of working with pupils which you think are relevant
- What have you learned from these?

YOUR KNOWLEDGE AND UNDERSTANDING OF YOUR SUBJECT AND SUBJECT APPLICATION

- What experience do you have of teaching your subject(s)?
- Which aspects of your subject have you taught on school experience and to what years?
- How would you introduce topic X to a Year 9 class?
- Recall a lesson that went well and/or one that went badly. Describe why this lesson went well or badly.
- How would you improve the lesson that went badly?
- How would you deal with, say, three pupils misbehaving during a lesson in which there are safety implications?
- Do you think that your degree subject prepares you to teach GCE A level?
- How has your development in your subject during your course contributed to your work in the classroom?
- How do the theory and practice on your course relate?
- How can you tell if pupils are learning in your subject?
- Can you describe one incident where a pupil was not learning and what you did about it?
- What experience have you had of setting targets for pupils?
- How have you incorporated ICT into your teaching? Please can you give an example.
- What other subjects could you teach and to what level? What background/teaching experience do you have in these subjects?

YOUR VIEWS ABOUT EDUCATION, PHILOSOPHY OF EDUCATION AND EDUCATIONAL IDEAS

- What do you think education is about (individual development or to acquire skills to get a job)?

YOUR PROFESSIONAL DEVELOPMENT ■ ■ ■ ■

- Do you view yourself as a teacher of children or of X subject?
- Why should all pupils study X subject?
- What do you think the aims of secondary (middle/upper) education should be?
- How did you set about planning differentiated learning for a class you have taught recently?
- How do you think your subject can contribute to the education of all pupils?
- What are your views about the way that the National Curriculum should develop?
- What are your views on assessment?
- On what should pupils' achievements be based?

YOUR ABILITY TO COPE AS A TEACHER

- What do you think are the qualities of a good teacher?
- What do you consider to be your strengths and areas for development as a teacher at this stage in your career?
- How are you working to address areas for development?
- How would you maintain good discipline in the classroom?
- How would you motivate a group of Year 9 pupils who do not have much interest in your subject?

OTHER ROLES YOU MAY BE ASKED TO UNDERTAKE

- What experience do you have of being a form tutor?
- How do you feel about taking on the responsibilities of a form tutor?
- How do you feel about taking extra-curricular activities? Which ones?
- What experience do you have of working with parents?

YOUR FUTURE DEVELOPMENT AS A TEACHER

- What are your targets for development during your first year of teaching?
- How do you see your career developing?
- How do you think you will go about achieving your career goals?
- How do you aim to widen your experience as a teacher?

OTHER INTERESTS, ACTIVITIES ETC.

- What interests/hobbies do you have and how involved are you in these? Do you see yourself being involved with any of these at school?

OTHER QUESTIONS

At the end of the interview you may be asked: if you were to be offered this post would you be in a position to accept it? (At some interviews you may be asked this question earlier. You can say that you decline the offer to respond at that point but will respond after the interview.)

At the end of the interview you are normally asked if you have any questions. Asking one or two questions shows a genuine interest in the school and the post; therefore do ask questions (not too many), if you have any. You are likely to forget the questions you wanted to ask if you are nervous; therefore, do not be afraid to take a checklist of questions with you to an interview. It is quite acceptable to refer to this at this stage in the interview. You may also wish, at this stage, to clarify issues that have arisen during the interview day. You should enquire what arrangements there are for induction of NQTs in the school and what you might expect. You may also need to check about the type of contract (for example, fixed term or temporary contract) which is being offered and if this is what you were expecting and want, and whether, if you are starting work in September, you will be paid over the summer period. However, do not ask questions just to impress. If all your questions have been answered during the course of the day and you do not have any questions, just say politely that all the questions you wanted to ask have been answered during the day (or in the interview).

At some point you may want to ask about your starting salary. In a private school and in some situations in the state sector you may have to negotiate your salary. In the state sector you are on a national rate of pay, but any previous relevant experience may also be taken into account, and you may be able to negotiate your starting salary. If you feel you are in a strong position, you may want to negotiate your starting point on the scale during your interview and ask for confirmation of this before you accept a post. In other situations, for example, if you feel you are not in a very strong position to be offered the post, but really want the post, it may be appropriate to discuss the starting salary at a later date. How you describe experience in an application and at interview, therefore, is very important as it may be used to support any claim for increments above the starting salary.

We suggest you undertake a mock interview (see Task 8.1.4).

Task 8.1.4 **MOCK INTERVIEWS**

Arrange for a mock interview with your tutor or another student teacher. If possible, either have an observer or video the interview so that you and the interviewer can use this to analyse your verbal and non-verbal communication after the interview. If, on analysis, you or the interviewer feel that there is a great deal on which you can improve, arrange for another interview after you have worked at improving your weaknesses.

ACCEPTING A POST

Where all candidates are invited for interview at the same time, you may be offered a post on the same day as the interview. You are normally expected verbally to accept or reject the offer at that time. Schools rarely give you time to think about an offer. Therefore, it is important that you consider all the implications of accepting the post before you attend the interview. On rare occasions it may be that you feel you really need some time to think about the offer. You may want to ask if you can think about the offer overnight and telephone first thing in the morning. If your request is refused, you have to make a decision there and then or be prepared for the post to be offered to another candidate. Your decision depends on how much you want a particular job and how strong a position you think you are in.

If candidates are invited for interview at different times, you may have to wait for a few days before being offered a post. However, normally you do not know which format an interview is going to take until you arrive at the school, so you cannot rely on being able to do this.

If you are offered a post verbally at or following the interview, it is normal practice to be asked to confirm your verbal acceptance of a post in writing. It is unprofessional to continue to apply for other teaching posts after you have verbally accepted a post, even if you see one advertised that you prefer. However, the offer of a post should be followed by a written confirmation.

Do check your contract carefully and if you have not already done so you should check with the school your salary and details of your induction programme. Offers of a post are made on certain conditions. You may be asked to have a medical before you start a job, in which case, the school or the LA send you details.

There are two matters with which you must legally comply if you wish to accept the offer of a teaching post. The first is that all teachers wishing to teach in a maintained school are required to register with the General Teaching Council (GTC). You pay an annual fee for this which normally your employer deducts from your salary. There are different Councils for England, Wales, Scotland and Northern Ireland. The function of each is to keep a register of qualified teachers and govern the profession through competence and disciplinary procedures. The second is that as you have access to children and young people, you are required to disclose all previous criminal convictions as part of the check by the Criminal Records Bureau (CRB). (Although you completed a CRB as a student teacher you are required to complete a new one as you start a teaching post.)

The transition from student teacher to teacher is not easy. Being a teacher involves you coming to terms with your new role within a new institutional context and an awareness of the complexities of its organisation, structures and routines, as well as ethos and expectations. This transition may be easier if you make arrangements (with your future mentor, for example) to visit the school again to make preparations to start your post. This is not dissimilar to the preparations you make when you go on school experience. You are able to meet members of the department and find out about facilities in the school. You also need to collect information about classes you will be teaching and what you will be teaching, as well as your teaching timetable. You may find it useful to make a few visits to get to know the school culture. For example, you may become involved in some school activities such as sports day or perhaps observe some classes or teach some lessons before you start your post.

IF YOU ARE NOT OFFERED THE POST

It is disappointing when a post is offered to another candidate. However, try not to think of this in terms of failure on your part. There may be many legitimate reasons why the post was offered to the other candidate in preference to you. For example, the other candidate might have relevant teaching experience (which you did not have) in an aspect of the curriculum required for the post. If you are not successful, build this into your learning experience. Most interview panels routinely offer feedback to candidates. If not, you can ask if this is possible as it helps you identify strengths and areas to develop in preparation for your next interview.

Also, if you are still looking for a job when September arrives, do not be too disappointed and give up. Ask yourself whether you have been too set on obtaining your ideal job and perhaps re-assess your criteria for your search, for example, in terms of area, type of school and age-range. You may also want to consider other options, for example, a job-share, temporary post or supply work. These can give you valuable experience while you continue your search for a permanent full-time post, and can sometimes lead to permanent work in a school. You can find out details about supply work in your chosen area from the LA, schools and supply agencies. However, you should check carefully the implications of supply work in terms of contracts, pay and implications for your induction period. Do remember that for any of these options you need to be registered with the GTC and also have a successful CRB check completed. It is also worth contacting your HEI as they might be aware of some posts, or perhaps visiting the job-seekers area of the TeacherNet (http://www.teachnet.gov.uk) if you haven't already done so. You may also wish to consider jobs outside teaching but related to this profession. The government's extended schools agenda (http://www.tda.gov.uk/remodelling/extendedschools) may help you consider other opportunities.

Tasks 8.1.5 and 8.1.6 are designed to help you reflect on the change from student to newly qualified teacher.

Task 8.1.5 **MOVING FROM STUDENT TO NEWLY QUALIFIED TEACHER**

How do student teachers learn to become teachers? Consider this critically in relation to the literature.

Identify similarities and differences between being a student teacher and being an NQT. Discuss your perceptions with a small number of NQTs. Critically compare their biographical stories with your own considerations for entry into the teaching profession.

Task 8.1.6 **ENTERING AND STAYING IN THE PROFESSION**

Design and conduct a piece of empirical research with selected established teachers in your placement school about the factors which (a) influenced them to become teachers and (b) make them stay in the teaching profession. Report critically on your analysis, identifying issues related to the philosophical assumptions underlying the methods of data collection and analysis. Give a critical evaluation of their choice, design and effective use and well as a critical analysis of the factors identified and comparison with your own personal circumstances.

SUMMARY AND KEY POINTS

This unit is designed to help you realise that, just as with your teaching, you must prepare for obtaining your first post; you cannot leave it to chance or rely on your innate ability to perform well at interview. In this unit we have tried to lead you through the steps, skills and techniques you need to prepare actively for obtaining your first post.

Check which requirements for your course have been addressed through this unit.

FURTHER READING

Hammond, M. (2002) 'Why Teach ? A case study investigating the decision to train to teach ICT', *Journal of Education for Teaching* 28 (2), 135–148; Priyadharshini, E. and Robinson-Pant, A. (2003) 'The attractions of teaching: an investigation into why people change careers to teach', *Journal of Education for Teaching*, 29 (2), 95–112.

These two articles look at attractions of teaching, reasons for training as a teacher. These articles may be useful in embarking on Task 8.1.6.

The Education Year Book (annual), London: Pearsons.

This book lists all the LAs in England and Wales, along with names, addresses and telephone numbers of secondary schools and sixth form colleges in their areas.

Times Educational Supplement (annually, around the middle of January) *First Appointments Supplement*, TES.

YOUR PROFESSIONAL DEVELOPMENT ▪ ▪ ▪ ▪

This supplement is published yearly. It contains articles and features on processes and procedures to help you get your first post and what to expect when you start your first post. It also contains many advertisements from LAs about general applications.

RELEVANT WEBSITE

ATL: http://www.atl.org.uk/
NASUWT: http://www.nasuwt.org.uk/
NUT: http://www.teachers.org.uk/
voice: http://www.voicetheunion.org.uk/

These are some examples of teaching unions where you can find information about obtaining your first post.

Additional resources for this unit are available on the companion website:
www.routledge.com/textbooks/9780415478724

UNIT

8.2

DEVELOPING FURTHER AS A TEACHER

Julia Lawrence, Alexis Taylor and Susan Capel

INTRODUCTION

The success of any school depends on its staff. Staff need to keep abreast of changes in education, in order to develop their knowledge and teaching skills. Continuing professional development (CPD) helps you continue to learn and develop professionally throughout your career. The first part of your CPD is your induction. Although you have successfully met the requirements to qualify as a teacher at the end of your initial teacher education (ITE) course, you need to demonstrate successful achievement against standards for newly qualified teachers (NQTs) (the requirement for practising teachers in England are set out in the Core Standards (Training and Development Agency for Schools (TDA), 2007c). During this period you will realise that there is still a lot to learn about being an effective teacher. However, your professional development continues throughout your career. This sequence of development can be shown as follows:

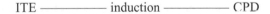

ITE ——————— induction ——————— CPD

You must take responsibility for your own professional development. During your induction period, support is provided by an induction tutor, whilst as your teaching career progresses you are supported by school managers. If you are teaching in England your Career Entry and Development Profile (CEDP) (http://www.tda.gov.uk/induction) is the beginning of active planning for your future career, identifying appropriate areas for development and activities and/or experiences appropriate to achieve these.

This unit considers life beyond your student teaching experience. It considers the transition from student teacher to NQT, induction into the school and the job and during the first year of teaching and CPD beyond the first year.

PROFESSIONAL DEVELOPMENT PORTFOLIO (PDP)

Throughout your ITE course you should develop a PDP (this may have another name, e.g. portfolio of achievement). Whilst many ITE courses require these to be completed as part of their assessment procedures, you might find it useful to compile a diary of reflective practice in which you record evidence of reflection and evaluations of yourself and those you teach (for further information refer to the introduction to this book).

Your PDP contains evidence of your developing professional knowledge and judgement to complement your subject knowledge. Specifically, it documents your performance in relation to the requirements to qualify as a teacher, your strengths and successes and areas for development, and gives examples of your work as part of your ITE course. Guidance on evidence you may wish to include in your PDP comes from a number of sources, including your tutors and, in England, the TDA.

Many ex-student teachers of ours have commented that they did not see the relevance of certain information, theories or activities while they were student teachers, but began to see their relevance as they developed as teachers. They then referred back to the notes they took or work they did as student teachers. Therefore the information contained in your portfolio is useful in its own right, as well as allowing you to gather materials that may prove useful to take to interview (if needed) and, for those of you learning to teach in England, to complete your self-evaluation tool. As you progress through your teaching career, the use of a PDP continues as you are required to demonstrate progress against the professional standards. Examples of what your PDP may contain are in Figure 8.2.1.

You may find that as you progress through your career you look to enhance your academic qualifications. Student teachers learning to teach in England by undertaking a Postgraduate Certificate in Education earn some M level credits alongside qualified teacher status which can be counted towards a masters degree. If you are not earning masters credits as part of your ITE course your PDP may potentially be used to provide evidence of prior learning (often called accreditation of prior experiential learning (APEL)) if you apply to study for a higher degree or further professional qualification. Therefore if you are in the habit of keeping your PDP up to date, you will have the evidence ready to hand for consideration. Task 8.2.1 is designed to encourage you to keep your PDP and diary of reflective practice throughout your ITE course.

Notes/reflections on HEI-based work

Notes/reflections on tasks undertaken as you read this book.

Notes/reflections on any aspect of your course as it happens

Written assignments

Notes/reflections on lessons observations

Notes/reflections on any micro-teaching, team/group teaching

Observation notes of your teaching

Schemes of work and lesson plans and any notes/reflections

Targets set for pupils and their progress towards meeting them

Individual Education Plans you have helped prepare and review

Pupils' work assessed by you

Any communications (including reports) with parents and carers (note: confidentiality issues)

Learning logs maintained

Reflections on professional development undertaken outside your course and evaluations of their impact on pupils' learning

Reflections on/evaluations of other aspects of teaching and learning and your development as a teacher in the broader school context.

Any other items you identify as appropriate/important to include.

■ **Figure 8.2.1** Items to include in your PDP

Task 8.2.1 **YOUR PDP**

If you are reading this unit near the beginning of your ITE course, start to develop your PDP, including items listed in Figure 8.2.1 and other evidence suggested by your tutors or which you identify as important. (If you are near the end of your course and have not been developing a PDP, collect together as much evidence as you can). Reflect on the evidence you have collected and write down what you perceive to be your strengths and your areas for development as a student teacher. Develop and implement an action plan to develop your strengths as well as areas identified for development.

TRANSITION FROM STUDENT TEACHER TO NEWLY QUALIFIED TEACHER

When you complete your ITE course successfully and get your first teaching post, you are likely to feel immense relief at having 'made it'. You may feel very confident and believe that you are going to be able to solve any problem you are faced with, e.g. motivating an unmotivated pupil or class or changing the teaching approaches in the department to encourage more active learning. You may also fear failing in your new job. Different people have different fears, e.g. fear of not being able to control pupils, of being thought to be lacking skill or ability, of not being accepted by other members of staff, of not liking the school or the people you work with.

As the new person in a school and department, you may not be sure of how to behave or of the rules or procedures to follow. You will have some successes and some failures and will soon realise that you cannot solve every problem or change the world. As a result, your confidence may decrease and you may not be fully effective until you are settled in the school and the job. A well-structured induction programme should help you make this transition. (Chapter 1 in Capel *et al.* (2004) gives further information about the transition from student teacher to NQT and on your immediate professional needs.)

Task 8.2.2 asks you to look at this transition whilst you are on your ITE course.

Task 8.2.2 **PREPARING FOR THE TRANSITION FROM STUDENT TO NQT**

Conduct a small-scale research project to look at experiences of NQTs in your school (see action research in Unit 5.4). On completion, record in your PDP how you could use these experiences during your own induction year.

INDUCTION

Induction can be divided into two main parts: immediate induction into the school and the job, which gives you vital information to help you through the early days; and ongoing induction throughout the first year, providing the link between ITE and CPD. We consider immediate induction first.

Immediate induction

Immediate induction applies only to the first few days and weeks of any job and focuses on general familiarisation and welfare aspects that all teachers (not only NQTs) in that school need. Immediate induction should, therefore, help you to understand, as quickly as possible, how you fit into the school and the department, building on information gained previously from the school literature, your interview and any further visits to the school after you were appointed. This covers a range of issues, such as:

■ management and administrative arrangements, e.g. staff responsibilities, terms and conditions of employment, sickness policy, issuing of keys
■ structure and departments, e.g. staff responsibilities, line managers, meeting other members of staff
■ rules, regulations and procedures within the school and your department, e.g. behaviour policy, equal opportunities, lesson planning
■ health and safety requirements, e.g. fire drill, medical policy.

Ongoing induction throughout the first year

Whilst in England an induction period is statutory (see http://www.tda.gov.uk/induction), you should find that wherever you first teach some form of induction occurs. Ongoing induction throughout the year should be linked to and develop from your ITE. At the end of your ITE course you should reflect on your experiences over your course, identify strengths and areas for further development, prioritising the latter. At the beginning of your induction year you should agree with your induction tutor (or head of department) the priorities for development and targets for these. At the end of your induction year you should reflect on and evaluate progress against the targets, reflect on your induction year more generally and identify further CPD requirements.

In our experience many NQTs report that school experience gave them an indication of the demands of teaching, but had not prepared them fully for the demands of a full-time post. They had felt that, as they would not be constantly observed, evaluated and assessed, tiredness and stress would reduce. However, they discovered that the first year of teaching was just as tiring and stressful as their school experience, if not more so.

In one way, being an NQT in your first teaching post is not much different from being a student teacher: you are still a beginner, albeit a beginner with more experience. In other ways, however, your first teaching post is a very different experience from school experience. You may feel differently about yourself as a 'real' teacher, which may influence the way you behave. Further, staff and pupils may treat you differently as a full member of staff.

Although your ITE course prepares you to enter the teaching profession you still have a lot to learn and inevitably feel unprepared for some aspects of the teacher's role. It is likely that as a student teacher you do not undertake all the activities that teachers undertake, for example, you are unlikely to be involved with developing schemes of work for a Year or a Key Stage or with administering examinations. In your first year of teaching you undertake a greater range of responsibilities than as a student teacher, for example, you have your own groups and classes and can establish your own procedures and rules for classroom management right from the beginning of the school year. You therefore undertake the full role of the teacher in your classroom.

During the first few weeks or first term in your new post, you will probably find that you concentrate mainly on becoming confident and competent in your teaching so as to establish yourself in the school. You are busy getting to know your classes, planning units of teaching from the school's schemes of work, preparing lesson plans, teaching, setting and marking homework, undertaking pastoral activities with your form and getting to know the rules, routines and procedures of the school.

Over the course of the first year you face situations that you did not experience as a student teacher. This includes undertaking activities for the first time, for example, discussing progress with pupils, setting questions for examinations, undertaking supervisory duties, or sustaining activities that you have not had to sustain over such a long period of time previously, e.g.:

■ planning and preparing material for a year to incorporate different material, teaching strategies and approaches to sustain pupil interest and motivation.

■ as you should be aware from your school experiences, you cannot plan one set of material and deliver it in exactly the same way to different groups of pupils. Adapting your planned unit of work and lessons to meet the needs of different groups of pupils requires careful planning and being able to think on your feet in order to meet the needs of particular pupils and classes.

■ setting targets to maintain progress in learning over the period of a year.

■ maintaining discipline over a whole year. This is very different to maintaining discipline over a short period of time on school experience. You cannot 'put up with' things that you may have been able to put up with for a relatively short period of time on school experience.

Although taking extra-curricular activities may be expected of you, we advise you not to take on too many (certainly not every lunch time and evening as many physical education teachers do). In your first year you need to concentrate initially on developing into an effective teacher.

However, as an NQT, you may not be expected to undertake the full range of roles and responsibilities of teachers, for example, you may not be expected to deal with some of the more serious pastoral problems or to undertake the full range of administrative demands. Thus, this:

> confirms and contradicts the assertion that probationers are invariably thrown in at the deep end of teaching. They might be thrown in, but it is a rather small pool in which they have to swim, since most of the administrative and managerial responsibilities do not come their way. Nonetheless, to continue the metaphor, it is possible to drown in a very small pool and . . . the classroom is notoriously hazardous. The major consolation is that much of the classroom-based work will have been encountered during the teaching practice term.
>
> (Marland, 1993: 191)

Thus, you may feel that teaching is more difficult than you first thought and realise that you still have a lot to learn. As a result, you may become frustrated and have doubts about whether you can teach and what you are achieving with the pupils. You may need help and understanding from other members of staff to overcome these doubts and continue to develop as a teacher. Although some support may be informal, your induction tutor (or head of department) should provide support. You can draw on your tutor's experience to help you to answer the numerous questions you have as new situations arise, and to overcome problems with aspects of your teaching. Your tutor can help you to learn as part of your normal job, by identifying and using opportunities available in your everyday work to develop further your skills, knowledge or understanding. You may discuss a problem and then go away and try to put some of the suggestions into practice. In an ideal world a tutor is proactive, making a conscious effort to look for opportunities for development. However, your tutor is busy and you spend much of your time in a classroom on your own with pupils, therefore there may be limited opportunities to work with your tutor. You may therefore use your tutor reactively by identifying areas where you feel you would benefit more from further development or where something has gone wrong. You can set up a situation where your tutor can help to address or correct that particular issue, for example, ask your tutor to observe a lesson and comment on a particular aspect of your teaching; or observe a lesson taken by your tutor or another member of staff, or team-teach particular topics on which you lack confidence.

Most other staff are helpful and understanding, especially if you establish good relationships with them. Relationships take time to develop and you need to be sensitive to the environment you are in. You will not get off to a good start if, for example, you sit in someone's usual chair in the staffroom, try to impress everyone with your up-to-date knowledge, ideas and theories, try to change something immediately because you think things you have seen in other schools could work better, ask for help before you have tried to solve a problem yourself or do not know when to ask for help, or do not operate procedures and policies and enforce school rules. If you do not operate procedures and policies or enforce school rules, you undermine the system and create tensions between pupils and teachers and between yourself and other members of staff.

However, as you settle into the job and work with your classes and learn the procedures, rules and routines, other staff may forget that you are new. As they become ever more busy with their own work as the term and year progress, they treat you as any other member of staff and do not offer help and advice. If you need support, approach staff and talk to them about your concerns and ask for help. You may form a support group with other NQTs. You can share your concerns and problems, support and learn from each other and remind each other that, despite the amount you still have to learn, you also have much to offer and are enthusiastic.

Before you begin your first teaching post complete Task 8.2.3.

Task 8.2.3 **PREPARING FOR INDUCTION**

At the end of your ITE course and before you begin your first teaching post, find out what is expected of you by the end of your induction period. Using your PDP identify your strengths and areas that you wish to focus on during your induction year (in England this is formalised in your self-evaluation tool). This should help you to gain an idea of what you need to prepare for your discussions with your induction tutor or head of department.

CPD (BEYOND THE FIRST YEAR)

At the end of your ITE course you identified areas for development during your induction year. However, successfully addressing these areas of development and completing a statutory induction period does not mean the end of your learning; indeed, it marks the start of a new period of your career, your CPD in the second and third years of teaching and beyond. CPD is part of your professional accountability as a teacher. Although not comprehensive, several aspects of your CPD to be considered are identified below. Before reading on, complete Task 8.2.4.

Task 8.2.4 **REFLECTING ON THE TRANSITION FROM STUDENT TO NQT**

During your first year of teaching, maintain your PDP. Within it identify your strengths and areas for development and how you might work to overcome them. At the end of the year reflect upon your transition from student to NQT, answering the following questions:

a   What was positive about the experience?
b   What would you change about the experience?
c   How will you use this experience: during the first few years of your CPD? To support newly qualified colleagues in the future?

Monitoring and evaluation

In order to continue to learn in the teaching situation, as well as get the most out of your CPD, continue the active, reflective approach to learning that you started during your ITE course. Monitor and evaluate your development as a teacher against specific objectives identified for development and continue to question what you are doing and identify alternative approaches. Record your progress in your PDP. Discuss this informally and/or through the appraisal process.

Your own individual development needs

As an NQT you spend the first couple of years in teaching establishing yourself. However, as you develop you will probably want to develop areas of expertise and take on posts of responsibility, either within the subject area or department or within the school. For example, you may wish to remain in the classroom and support other teachers to develop their teaching further (in England this is an Advanced Skills Teacher (AST); the requirements that demonstrate your excellence in the classroom can be accessed from http://www.teachernet.gov.uk/ast). Alternatively, you may wish to become involved in pastoral work or work placements, or even aspects of ITE. In the final assessment meeting of your induction period, you target areas for development for your second (and early) years of teaching.

It is important to recognise that, just as when you started your first post, when you take up a new responsibility or post, you go through a period of transition as you adjust to the new situation. You are likely to adjust more quickly if you have identified areas for development in a new post. This enables you to undertake appropriate CPD (and any formal qualifications or requirements) to develop an understanding of the role and the skills you need to undertake the role successfully. There are many CPD opportunities, for example, short or long courses, a higher degree or a further professional qualification and being involved in development and change activities in the school.

Performance management and appraisal

Performance management and appraisal are part of making explicit teacher accountability and are crucial to the school's performance management arrangements. The information recorded in your PDP during your induction year should inform your induction review meeting at the end of your induction period. Targets set at your induction review meeting form the basis of your annual review, a mandatory part of the performance management process. Appraisal normally consists of observation of your teaching and an appraisal interview. An appraisal interview should provide you with valuable dialogue. It may start with discussion of your observed teaching performance, then progress to your performance over the past year (particularly in relation to pupils' progress). In addition to your teaching, other topics may be discussed, for example, pastoral work, curriculum development work, management and administrative activities and membership of committees and working parties. In all of these areas you discuss your strengths and areas for development, CPD undertaken to address these or ways in which any identified needs might be met, e.g. by attending conferences, studying for a higher degree or other opportunities for CPD within the school.

Linking your CPD to the priorities of the school

Although not discussed in detail here, a valuable opportunity for CPD is provided through involvement in school development planning. Discuss with your tutor the school development plan and how you might be involved in this in relation to your development targets and your career aspirations.

Tasks 8.2.5 and 8.2.6 aim to help your further professional development.

Task 8.2.5 **FURTHER PROFESSIONAL DEVELOPMENT**

Despite the help of this book and other sources, there is a tendency as a student and NQT to focus on meeting the requirements of your course or induction year. However, it is important that you extend your horizons beyond this. Thus, at this stage we ask you to undertake a range of readings and use the information in your PDP to explore your own professional values and how they influence not only your own priorities for the aims of education and of your subject but also your day-to-day practice as a student and NQT. You may want to start this task by referring back to Unit 7.1. Write a reflective statement of your values, the reasons you hold these and how these values and practices are reflected in your own teaching, providing evidence (including practical examples) where appropriate. Identify, through the use of an appropriate action plan, how your knowledge and understanding of professional attributes can be enhanced. As part of your professional development you can use this as the basis for undertaking a small scale action research project to develop a deeper understanding of education etc.

Task 8.2.6 **CONDUCTING A SMALL-SCALE RESEARCH PROJECT**

Conduct a small research project to look at experiences of newly qualified staff in your school. Reflect on how you could use these experiences during your early development and how you could use these experiences to support newly qualified staff as you become more experienced.

SUMMARY AND KEY POINTS

This unit has considered the sequence of development as a teacher through:

ITE — — induction — — CPD

You are already on the professional development road and it is your professional responsibility throughout your career to seek opportunities for development. Your PDP helps you to build on your strengths in your first year of teaching and to target and address areas for development. It also helps you to take responsibility for your own professional development from the beginning of your career, by establishing the practice of reflecting, target setting and review. Your targets may include targets for wider aspects of your role as a teacher. For example, do you know what to do if a pupil has an epileptic fit or an asthma attack? Would you feel more in control if you knew how to deal with such situations in your teaching? If so, would you benefit from undertaking a first aid course?

The practice of reflecting, target setting and review establishes a good foundation for appraisal and your ongoing CPD. However, your first year of teaching is very demanding and you will want to ensure that you are fully established as a teacher first. Therefore, although you need to identify CPD opportunities to enable you to develop your knowledge and skills so you can successfully take responsibility for achieving your career goals, it is important not to take on too much at once and to pace yourself when planning for your CPD.

Check which requirements for your course have been addressed through this unit.

FURTHER READING

Capel, S., Heilbronn, R., Leask, M. and Turner, T. (2004) *Starting to Teach in the Secondary School: A Companion for the Newly Qualified Teacher*, 2nd edn, London: RoutledgeFalmer.
This book is designed to support NQTs in their first year of teaching to provide a firm foundation for ongoing development throughout their career.

Holmes, E. (2009) *The Newly Qualified Teacher's Handbook*, 2nd edn, London: RoutledgeFalmer.
This book provides guidance and support for newly qualified teachers and is linked to the current induction requirements.

RELEVANT WEBSITES

Websites containing information about induction and CPD in each of the countries of the United Kingdom:

England: *Supporting the Induction Process: TDA Guidance for Newly Qualified Teachers*, (2007c, TDA): http://www.tda.gov.uk
This website provides support material and guidance for induction in England. Much of the information is in a downloadable form.

Wales: *Induction and Early Professional Development for Newly Qualified Teachers in* Wales (revised September 2006) Circular Number: 21/2006 (2006, National Assembly for Wales): http://new.wales.gov.uk
This website provides an overview of induction procedure in Wales, identifying the roles and responsibilities of those involved.

Scotland: *The Standards for Full Registration* (2006, General Teaching Council for Scotland): http://www.gtcs.org.uk
This identifies the competences that must be met in Scotland to complete probation and become fully registered as a teacher.

Northern Ireland: *The Code of Values and Professional Practice* (2004, General Teaching Council of Northern Ireland): http://www.gtcni.org.uk
This website provides details of the competence required of teachers within Northern Ireland. A more recent publication *Teaching – the Reflective Profession* provides an overview of the competence and how they can be developed further.

We advise you to join your subject association. In the UK information about these can be found under http://www.subjectassociations.org.uk/.

Additional resources for this unit are available on the companion website:
www.routledge.com/textbooks/9780415478724

ACCOUNTABILITY, CONTRACTUAL AND STATUTORY DUTIES

Marilyn Leask and
Andrew Green

INTRODUCTION

Teachers' work is guided and regulated in different ways by national and local government, the school, school governors, parents and pupils – so you are accountable to a whole range of interested parties for the quality of your work.

To help you understand the context in which teachers and students work and are held accountable for their work, this unit outlines how the education system operates in the state system in England. See Chapter 7 for information about Scotland and Northern Ireland and Wales and fuller documents on the website for the book (www.routledge.com/textbooks/9780415478724).

OBJECTIVES

At the end of this unit you should be able to:

■ describe the structure of the state education system in which you are working
■ explain the legal and contractual requirements that govern the work of the teacher
■ explain the health and safety requirements for your subject.

Check the requirements for your course to see which relate to this unit.

WHERE DO TEACHERS FIT WITHIN THE EDUCATION SYSTEM?

The structure of the education system in England is set out in Figure 8.3.1. This shows the relationships between the education system and teachers. The Secretary of State, ministers and staff at the Department for Children, Schools and Families (DCSF) (previously Department for Education and Skills (DfES)) do not usually have teaching experience. They are provided with professional advice by advisory bodies such as the Qualifications and Curriculum Authority (QCA), the Training and Development Agency for Schools (TDA) and the British Education Communications Technology Agency (BECTA) which have some staff with a wide range of expertise in the profession. However, Local Authority (LA) education officers and professors and lecturers in higher education institutions normally start their careers as student teachers like yourself and so are likely to have been classroom teachers for some time before taking on these other roles.

Whilst the responsibilities within the school, listed in Figure 8.3.1, are shared out differently in different schools and the terminology used is in some cases different, the general structure is not too dissimilar. There are also numerous support staff whose contribution to school life is essential to the smooth running of the school, e.g. teaching assistants; school's premises' officer; nurse; administrative staff; technical staff; cleaners; lunch time supervisors; the bursar.

Staff from other professions are also linked with the school, e.g. the education welfare officer, school psychologists, and some pupils have social workers who are responsible for overseeing their progress. Parents and members of local communities will often have roles in schools – as governors or in providing support in a wide range ways.

Task 8.3.1 asks you to map out the system in which you work.

Task 8.3.1 **THE STRUCTURE OF THE EDUCATION SYSTEM**

Draw and label your own version of Figure 8.3.1 to fit the school and context in which you are. Check you are aware of any documents government agencies produce relevant to your work, by asking your tutor and checking key national websites. See for example, those referenced on the book website (www.routledge.com/textbooks/9780415478724) and in the units in Chapter 7.

ACCOUNTABILITY

Within the structure of the education system and individual schools, teachers are accountable for what they do and the Office for Standards in Education (Ofsted) plays a major part in monitoring the work in schools. As an individual teacher, however, you are also accountable – to parents, to colleagues, to pupils, to your employer. The Bristol Guide (2008) provides you with detailed information about the teacher and the law and membership of a teaching union provides you with access to a range of services both legal and professional.

Bush (in Goddard and Leask, 1992: 156–158) identifies three ways in which a teacher experiences accountability. Bush calls these: moral accountability, professional accountability and contractual accountability. *Moral accountability* is your conscience about how you should carry out your work. You are 'morally accountable' to pupils, parents and to society. Your *professional accountability* is your responsibility to your colleagues and to the teaching profession, to do your work to the highest standard of which you are capable. *Contractual accountability* is defined by legal requirements set down by your employer as well as in legislation passed by Parliament.

Whilst this may seem an oversimplification of a teacher's accountabilities, these three aspects provide a useful framework for developing your own understanding of your accountability. However, the way moral and professional accountability is personally perceived depends on the values of the individual teacher and on the standards you set for yourself (see Task 8.3.2). The next section sets out in some detail your legal duties, both contractual and statutory.

Task 8.3.2 **MORAL ACCOUNTABILITY AND PROFESSIONAL ACCOUNTABILITY**

Consider what being morally and professionally accountable means for you and for the way you approach your work. Discuss this with other student teachers or your tutor.

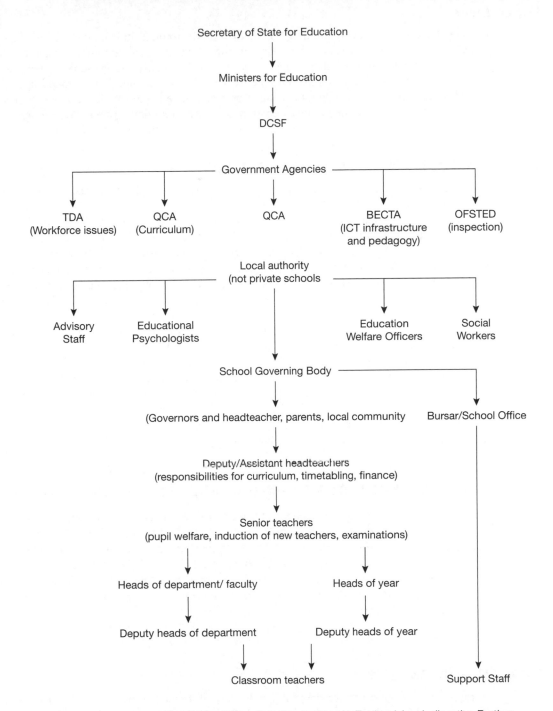

■ **Figure 8.3.1** Structure of aspects of the education system in England (excluding the Further Education (FE) sector)

Legal duties

Teachers have various legally binding contractual responsibilities and statutory duties. In addition you also have, as do all citizens, 'common law duties' which mean, among other things, that you have a duty of care towards other people. Teachers, again as citizens, are subject to criminal law. One aspect of criminal law you should note is that if you hit a pupil, or if a pupil hits you, this constitutes assault. It also is common sense to protect yourself against allegations by ensuring that you do not spend time alone in closed environments with individual pupils. You should exercise care in texting and emailing and otherwise communicating with pupils outside the school day.

CONTRACTUAL AND STATUTORY DUTIES

Contractual duties are negotiated between teachers and their employer. Statutory duties are those which the government has established through legislation.

In the case of teachers employed in state schools in England, the document that sets out teachers' contractual duties is *School Teachers' Pay and Conditions* which is produced by the DCSF and updated annually. In *School Teachers' Pay and Conditions* (http://www.teachernet.gov.uk/paysite/) guidelines are laid down for the exercise of your professional duties under the headings of teaching, other activities (which covers pastoral work), educational methods, assessment and reports, public examinations, appraisal, review, continuing professional development, discipline, health and safety, staff meetings, cover (for absent colleagues), management, administration and working time. Conditions of service, legal liabilities and responsibilities, child protection and other important aspects of a teachers' role are discussed in Cole (1999).

Additional conditions may apply in individual schools. There may also be 'implied terms' to your contract, i.e. terms which are not written down, for example, that you behave in a manner befitting your role – some schools operate a dress code. The *Headteacher's Guide to the Law* (updated annually, Croner) is recommended further reading for those with a particular interest in this area.

Specific advice on teaching contracts is available from the teaching unions. Task 8.3.3 asks you to review the contractual guidelines likely to apply to you.

Task 8.3.3 **YOUR CONTRACTUAL AND STATUTORY DUTIES**

Obtain a copy of the conditions likely to govern your employment as a teacher in the state maintained sector (independent schools' conditions are not necessarily the same). This may be available from the relevant government organisation or General Teaching Council. In England, *School Teachers' Pay and Conditions* contains this information. Discuss this with other student teachers or your tutor and ensure that you understand what is required of you when you become a qualified teacher. Keep these notes in your professional development portfolio (PDP) for reference when applying for jobs.

HEALTH AND SAFETY

All teachers are responsible for the health and safety of the pupils in their charge. Task 8.3.4 asks you to check the requirements in your school. Legally, as a student teacher you cannot take on that responsibility. Whenever you are teaching, the ultimate responsibility lies with the class teacher.

Nevertheless in planning your lessons you must take into account the health and safety of your pupils by appropriate planning, such as identifying activities that do not endanger pupils, e.g. climbing on chairs; or for science and related subjects following the COSHH (Control of

Substances Hazardous to Health) regulations. Sharing your lesson plans in advance with your class teacher is an essential feature of your responsibility to both the pupils and your teachers. If you have any doubts about the safety of the lesson, ask for advice. If advice is not available, then do not use that strategy.

Whilst you are teaching, an experienced teacher must always be available in the classroom or nearby. If the lesson has special safety considerations, e.g. in physical education or science, then if the class teacher or a suitably qualified teacher is not available, you must not proceed as if they were. Have an alternative lesson up your sleeve which does not require specific subject-specialist support but could be carried out with the support of another teacher. Sometimes you may have to cancel your planned lesson.

It follows from this situation that, legally, you cannot act as a supply teacher to fill in if the regular teacher is absent.

Task 8.3.4 **HEALTH AND SAFETY PROCEDURES**

What should you know and be able to do if you are to discharge your duties as a student teacher and as a teacher in your subject area? Find out who is responsible for health and safety in your school experience school. Find the school and departmental policies on health and safety which are relevant to your subject. Check the procedures you will be expected to apply – for example, in science, find out how you should check the safety of the chemicals or other equipment you may use, locate the eyewash bottle and gas, water and electricity isolating taps/switches; in physical education check that you know how to test the safety of any apparatus pupils might use. Find out the names of the first aiders in the school, where the first aid box is, what you are permitted to do if an incident occurs and what forms have to be filled in to record any accident. Discuss this with your tutor and other student teachers in your specialist area.

We suggest that you take a first aid course and find out how to deal with, for example, faints, nose bleeds, fits, asthma attacks, epilepsy, diabetic problems, burns, bleeding and common accidents. But you should not administer first aid yourself unless qualified and even then, only the minimum necessary. You should report any incident and make a written record. There will usually be a record book in school for this purpose. Your subject association should be able to provide you with subject specific safety information and St John Ambulance produce a first aid text for schools and provide first aid courses for school staff.

You also need to consider health and safety issues that affect you.

YOUR VOICE

A common problem for teachers new to the profession is voice strain. If you start developing hoarseness, a sore throat or problems in projecting your voice, ask your GP to refer you for specialist help. Drinking water between lessons will help.

See the end of this unit for details of the Voice Care Network from which you can obtain further free advice on protecting your voice.

VDUS

You may find yourself needing to spend a substantial amount of time working with a visual display unit (VDU), particularly when report writing time comes round. It is important that you adjust your chair so that your spine is straight and that you take frequent breaks to stretch your limbs.

HEALTH AND SAFETY POLICY

The school has a legal 'duty of care' towards you as well as pupils with regard to your personal health and safety. Your school should have a health and safety policy and member of staff and a governor with a particular remit for ensuring that your working environment is safe and suitable for the activities you undertake. You also have a legal responsibility to make sure that you do not put yourself or your pupils at risk of injury by practising or allowing unsafe behaviour, or failing to report damage which could cause an accident e.g. a hole in stair covering. Equally, you need to make sure that you look after your own safety. For example, it is not a good idea to agree to lift heavy boxes or furniture on your own and without making sure you are using the correct technique for doing so. The school site manager and staff should be willing and able to deal with heavy lifting and it is reasonable for you to expect help.

It is also advisable take steps to protect your personal safety should you have reason to leave school or a work-related meeting late at night. As a student teacher you are not likely to visit a pupil's home. Teachers should always make sure someone, be it a family member, a friend or a colleague, knows the location of the meeting and their expected arrival home.

SUMMARY AND KEY POINTS

As a student teacher, you need to be aware of the full range of a teacher's duties. Whenever you are working in a school, you are acting with the agreement and support of qualified teachers. When you take over their classes, you are accountable for the work in the classroom in the same ways they are. You are responsible for health and safety and need to be aware of the requirements in any school in which you teach.

Check which requirements for your course you have addressed through this unit.

FURTHER READING

Cole, M. (ed.) (1999) *Professional Attributes and Practice*, London: David Fulton.
This text contains chapters on a wide range of important aspects of a teacher's role.

The Bristol Guide (2008 edition) *Professional Responsibilities and Statutory Frameworks for Teachers and Others in Schools*, School of Education, University of Bristol. Available online at http://www.bristol.ac.uk/education/enterprise/docsum/guide.

RELEVANT WEBSITE

Department for Children, Schools and Families (DCSF) (updated annually) *School Teachers' Pay and Conditions*, London: HMSO: http://www.teachernet.gov.uk/
This document is well worth reading. It sets out the statutory conditions for the employment of teachers in England and provides information about salary scales and conditions of work.

Additional resources for this unit are available on the companion website:
www.routledge.com/textbooks/9780415478724

AND FINALLY

Marilyn Leask

INTRODUCTION

The introduction to this text set you the challenge of considering what you want to achieve by becoming a teacher and what it means to be a teacher. Each unit has been designed to help you understand the different elements of teaching. Here, in the last unit, we return to the bigger picture of the purpose of education, how your values shape your work as a teacher and the whole role of the teachers. This 'whole–part–whole approach' to introducing you to teaching through this text mirrors best practice in teaching – introducing the whole, undertaking work on the parts, returning to review the learning in the context of the whole.

This text has drawn on the evidence base for educational practice in providing information and background linked with tasks. Other tasks are intended to provide opportunities to reflect on and examine the practice of other teachers, of yourself and the organisation of schools. Some of the tasks are inquiry based and generate data or ideas for reflective thinking upon which an understanding of and an explanation for the complex world of teaching and learning in schools are built.

The relationship between explanation and practice is a dynamic one; explanations are needed to make sense of experience and inform practice. Some explanations will be your own, to be tried and tested against the theories of others, often more experienced teachers. At other times you may use directly the explanations of others. Explanations in turn generate working theories, responsive to practice and experience. Theory is important, it provides a framework in which to understand the complex world of the classroom and to direct further research into improving the quality of learning. It provides, too, a reference point against which to judge change and development, both of yourself and schools. It is the encompassing of these ideas, the interplay of theory and practice, which underpins the notion of the reflective practitioner.

Your view of the purpose of education and your view of your responsibilities to the young people in your care have a major impact on your teaching. Education is probably the most powerful influence on the development of our society. The education which young people receive through schooling goes beyond knowledge about a subject, it includes the values of community and individual rights and responsibilities which make our society distinctive. So you can expect that the form and content of education will always be under scrutiny and will be contested. As you develop your experience as an educator, you may wish to join such debates and shape the future of our society.

WHAT VALUES WILL YOU PASS ON?

What values will you pass on to the young people you teach? We ask you, as one last task, to consider the messages in the letter and poem in Table 9.1 and Table 9.2. The letter *Dear Teacher* was written by a US high school principal who was a World War 2 concentration camp survivor, having been separated from her sister and mother whom she never saw again. She gives the letter to all new staff at her school (Pring, 2004). The poem 'Children learn what they live' (Dorothy Law Nolta) is often displayed on staffroom walls.

Task 9.1 **WHAT WILL YOUR PUPILS LEARN FROM YOU?**

Consider the letter *Dear Teacher* and the poem *Children learn what they live*.

What kinds of members of society do you expect the young people you teach to become?

Review your values and beliefs and consider what you hope your pupils learn from you beyond your subject area.

■ **Table 9.1** A letter to new teachers

Dear Teacher

I am the survivor of a concentration camp. My eyes saw what no man should witness.

Gas chambers built by learned engineers,

Children poisoned by educated physicians,

Infants killed by trained nurses,

Women and babies shot and burned by high school and college graduates.

So I am suspicious of education.

My request is: Help your students become human.

Your efforts must never produce learned monsters, skilled psychopaths, educated Eichmans.

Reading writing and arithmetic are important only if they serve to make our children more human.

■ **Table 9.2** Children learn what they live

'Children learn what they live'

If a child lives with criticism,
he learns to condemn,
If a child lives with hostility,
he learns to fight,
If a child lives with ridicule,
he learns to be shy,
If a child lives with shame,
he learns to feel guilty,

If a child lives with tolerance,
he learns to be patient,
If a child lives with encouragement,
he learns confidence,
If a child lives with praise,
he learns to appreciate,
If a child lives with fairness,
he learns justice,

If a child lives with security,
he learns to have faith,
If a child lives with approval,
he learns to like himself,
If a child lives with acceptance and friendship,
he learns to find love in the world.

Source: Dorothy Law Nolta

SUMMARY AND KEY POINTS

As a teacher you will have an impact – beyond what you will ever know – on people's lives and thus on the community and society. We hope too that what your pupils learn from you will help them make positive contributions to their world. We hope that you will help pupils to build personal self-confidence and skills to cope with adult life and to become autonomous learners and caring members of society.

To achieve these goals, you should expect to carry on learning throughout your professional life. As a first step, joining your subject association will ensure you receive publications outlining good practice and attending your annual subject conference will introduce you to the network of educators taking thinking in your subject forward. You can find your subject association by asking colleagues and tutors or through the Council for Subject Associations (http://www.subjectassociation.org.uk).

Teachers are expected to continue to undertake reflective practice and learning beyond their initial teacher education and to achieve Masters' qualifications within a few years. The issues raised in this unit about the purpose of education and how children learn are areas you may wish to follow up through this further study.

FURTHER READING

The texts below provide an introduction to some of the debates and thinking which have shaped educational practice in the UK over more than 100 years.

Aldrich, R. (2006) *Lessons from History of Education: The Selected Works of Richard Aldrich*, Abingdon, Oxon: Routledge.

Bruner, J. S. (2006a) *In Search of Pedagogy Volume 1: The Selected Works of Jerome S. Bruner*, Abingdon, Oxon: Routledge.

Bruner, J. S. (2006b) *In Search of Pedagogy Volume 2: The Selected Works of Jerome S. Bruner*, Abingdon, Oxon: Routledge.

Dewey, J. (1933) *How We Think*, Boston, MA: Houghton Miffin.

Gardner, H. (2006) *The Development and Education of the Mind: The Selected Works of Howard Gardner*, Abingdon, Oxon: Routledge.

Additional resources for this unit are available on the companion website:
www.routledge.com/textbooks/9780415478724

Readings for Learning to Teach in the Secondary School: A Companion to M Level Study brings together essential readings to support you in your critical engagement with key issues raised in this textbook.

APPENDIX 1
Guidance for writing
Susan Capel and John Moss

INTRODUCTION

In order to gain qualified teacher status at the end of your initial teacher education (ITE) course you are required to meet both teaching and other course requirements, e.g. written assignments through which you demonstrate your ability to produce academic work worthy of a graduate or postgraduate qualification. These assignments are designed to encourage you to make connections between educational theory and practice. This is to ensure that when you complete your ITE course you have begun to develop an understanding of how to:

- make use of educational research and theory to inform and improve your practice
- develop your own theories about teaching and learning, reflecting on and evaluating established practice
- develop and assess innovative practice in your own classroom
- respond creatively and critically to local and national educational initiatives.

It is very likely that your education to date has provided you with many opportunities to find out what processes you need to work through in order to produce good written assignments. One literacy expert, Margaret Meek, has often argued that if we want to know what literacy practices will benefit our pupils, we should start by considering what works for us. Consequently, working on the assignments for your ITE course also provides you with an opportunity to reflect on how you can support your pupils in undertaking the writing tasks you set them. If something helps you, it may well help them. We now look at some of the issues research on writing suggests you need to pay attention to in your writing, i.e. understanding the genre; understanding the purpose and mode of writing; preparation and planning; academic referencing.

UNDERSTANDING THE GENRE

Genre is the term used to define a type of text which has a set of agreed conventions. These conventions apply at different levels, including the kind of vocabulary and voice it is appropriate to use, the structure and organisation of sentences and paragraphs, and larger required structural elements. In academic writing these elements may include:

■ an *introduction*, which should normally identify the focus of the assignment and outline the structure of what is to follow, i.e. the sequence in which material will be presented in the assignment

■ a *literature review*, which usually analyses the academic writing which is relevant to the topic, identifying common or contrasting views and current issues

■ a *statement of a question or hypothesis* to be investigated, which explains the scope of the investigation and the reasons for it

■ a *description and analysis of the methodology used*, which explains how an issue was investigated, the reasons for this and the advantages and disadvantages of the methods chosen

■ *data presentation and analysis*, which should enable a reader to see what is established fact in the work, where interpretation begins and how and why it has been made

■ *conclusions* should arise from the investigation and its results, and compare findings with established theory or positions; matters which are unresolved should be identified; recommendations for future practice may be included if appropriate

■ a *bibliography* which lists all references used in the assignment.

You bring preconceptions about academic writing from your earlier studies to assignments on your ITE course. However, the genre 'essay' contains many 'subgenres', just as the genre 'novel' includes science fiction, romance and detective stories among many other kinds of novel. You should establish which of the elements of academic writing listed above, and others, must be contained in the genre(s) you are expected to write in on your ITE course.

At the same time as focusing on the purpose of writing, in England an increasing number of student teachers on postgraduate ITE courses are being required, or offered the opportunity, to submit assignments for credits towards a master's qualification, which you can then complete in the early years of your career. Your tutors will provide you with the detailed information you need to be able to work and write assignments that meet master's level criteria.

However in general terms, to be successful, you have to meet some of the national general requirements for a master's qualification. You do not have to meet all of these in one or two assignments, so you should not be too daunted by them, but they are, nevertheless, challenging.

You are likely to be asked to show cutting edge knowledge of the topic you write about, which means reading and evaluating the most recent research and practice. You may also be expected to show understanding of at least one recognised educational research method and to apply it effectively in investigating your topic, and to evaluate its usefulness and effectiveness in helping you to understand the topic and the practice you saw. You are also expected to show a high level of conceptual understanding, and a degree of originality in your work and thinking. This is demanding, particularly in the relatively short assignments you are typically asked to write as part of an ITE course, so the fundamental principles of planning and writing outlined in this chapter will be even more important in helping you to ensure that you meet all the requirements.

Just as it helps your school pupils if you do this, it is good practice for your tutors to teach you explicitly what is appropriate. Some methods which may be used for this include modelling and providing examples of successful work. Good modelling involves a practical demonstration of how an assignment can be constructed, e.g. by developing a plan or a paragraph on an overhead transparency or interactive whiteboard. Formative feedback on plans or first drafts also helps writers to understand what is required of them, and consequently to improve. The assessment criteria for the assignment should be made explicit from the outset: these should always make reference to the kind of writing that is expected as well as the required content. It will also help you as a writer if you are given more than one opportunity to write a particular kind of assignment, so that you develop experience in using the genre.

UNDERSTANDING THE PURPOSE AND MODE OF WRITING

Writing may, of course, have many different purposes. In the pupil National Curriculum for English in England, for example, these range from description and narration to argument and persuasion. The primary purpose of many written assignments in ITE is to develop a capacity for *reflection*. This means that you are encouraged to use your writing to think about your experiences in school; your own teaching; your observations of, or discussions with, others; your analysis of school documents or activities or events that occur in school; and the content of lectures, seminars and academic literature, in order to reach a higher standard of awareness about your teaching and the ways in which pupils learn. (You should find the sections on observation and action research in Units 2.1 and 5.4 useful for gathering relevant information.)

Reflection requires you to stand back from a specific lesson, observation, activity or event and question it. You draw on your knowledge about, and understanding of, the work of others, research and theory, and consider it in relation to the practice on which you are reflecting. Reflection is as integral to your teaching as your ability to manage a class and pupil behaviour. It is a process which should continue into your first year of teaching and throughout your teaching career.

Evaluation of your teaching at the end of each lesson is an example of reflection. Evaluation can be carried out alone or with your tutor. When you evaluate a lesson, you question what went well in addition to what did not go well, what might have worked better and what you might do next time that will enable you to improve what you are doing. As evaluation and reflection involve recall of what took place, you need to focus your evaluation and reflection in advance. So much is going on in a lesson that unless you select one or two things on which to focus you are unlikely to collect the detailed data necessary for an in-depth evaluation. You also need to write a few notes and identify points of relevance as soon as you can after the lesson or observation so that you record your perceptions when they are fresh in your mind. There are checklists, guidelines and series of appropriate questions included in many of the texts concerned with helping teachers to develop into reflective practitioners (see further readings at the end of Unit 5.4).

Your reflections can then be used as the basis for explicitly linking the work of others to your practice in your assignments. You should aim to show how the work of others, research and theory, informs your practice and vice versa, rather than leaving the two separate, isolated from and not informing each other. Reflection allows you to give your own opinions, but you should underpin these with reference to research and theory. An assignment title which requires you to reflect does not imply that you should not include theory; nor does it mean that you should not include references. Rather, the expectation is that you should include examples from practice to support or challenge the research and theory. When drawing on examples from your practice in assignments, you should not use the name of the school or any individual. All such references should be made anonymous.

Common pitfalls in reflective writing include discussing the issues entirely at a theoretical level so that the work fails to engage with your practical experience in school or discussing only practical experience without making any reference to the theoretical context in which your own work is inevitably set. To avoid these and other pitfalls, detailed planning and preparation, and thorough drafting and academic referencing are necessary. These processes take time in themselves, and writing is also often improved if there are time gaps between the different stages of its production. These gaps help you, as the writer of an assignment, to approach it as a reader will do, and to judge your work to date more objectively. It follows that an important first step is to plan how you will pace your coverage of the planning, preparation and drafting stages indicated below, to allow you to meet the assignment deadline comfortably.

PREPARATION AND PLANNING

Identifying the title, focus and audience

Before you begin an assignment, you need to be very clear about what you are writing about, i.e. the title and focus of the assignment. You may be given a title or you may have to choose your own topic or focus. In either case, you should be very clear about the topic and focus before you start. It is often helpful to have a short title which clearly focuses on the issue you are going to address. The title can be written as a problem or a question and may be intentionally provocative. Phrasing the title in this way clarifies and reminds you of the focus when you are writing the assignment. Ask yourself: 'What am I seeking to find out by writing this assignment?'

You should also establish who the intended audience for the work is. Although many assignments in ITE are primarily reflective, you may be asked to produce work which can benefit the schools in which you undertake school experience or your peers on your ITE course. If there is an intended audience other than your tutor, as a monitor of your development as a reflective practitioner, this will influence how the assignment is written.

Collecting information

You should make a list of the sources of information that will help you to address the topic. You will be used to identifying and collecting together relevant notes from higher education institution (HEI) based sessions, books and articles from journals. However, you should also collect together information from school, such as lesson plans and evaluations of lessons you have taught or observed, and records of observations or interviews conducted in school, so that your written work can refer to, and link, published sources and your developing practice in school. It is helpful to keep this in your professional development portfolio (PDP).

Make sure that you are familiar with the library system at your HEI. You will probably have an induction briefing about this. If not, find out what books and journals are available to support your course. Ask what databases for education are available (e.g. BEIndex (British Educational Index)) and find out how to use them.

DRAFTING

Drafting is given high status in the National Curriculum for English. You should give it high status also. However, it is sometimes poorly interpreted as 'writing it out' and 'writing it out again with the spelling (and perhaps some grammar) corrected'. In fact, effective drafting is a complex process involving several stages.

Gathering ideas

Brainstorm ideas and collect key points from your reading notes (from books and journals), seminar or lecture notes, teaching file and material from school. At this point, include everything you think of that appears to be even slightly relevant. A spider diagram (see Unit 5.2) or arrows may be used, perhaps at a second stage, to begin to group points and ideas. It is a good idea also to identify examples and illustrations and note them next to key points and ideas: this will help you to work out which points you have most to say about and whether others should be eliminated or need more research before you start writing.

Selecting and sequencing ideas in a plan

Produce a plan which supports visually the organisation of ideas in a way which is appropriate to the content and structure of the assignment. For example, flow charts are useful if the core of the assignment is concerned with a process; columns can be used to list pairs of contrasting points or ideas and illustrations; a matrix provides a means of charting complex matters in which combinations of factors need to be taken into account. It can help to add numbers as you work out the order in which you will deal with the material. At this stage, it is possible to check whether you have appropriate amounts of information for each stage of a process to be discussed, each side of an argument, or each combination of factors. If not, you may decide you need to research or develop your thinking about part of the subject matter.

Drafting

Write parts of the assignment as fluidly as possible. Allow ideas to develop rather than stopping too often to check detail or refine wording. It is not necessary to write in sequence: in fact, it may help to work on the different main sections of the assignment concurrently. This can help you to work out what each section needs to include and, for example, to sort out what belongs appropriately in the introduction and conclusion. If you have a section reviewing literature or discussing examples of practice, writing about one text you have read or lesson you have taught can help you to see how much space you have for each example, and so help you to balance the whole assignment. Clearly, you need to sequence and re-sequence the writing as it develops. The aim of this stage is a continuous draft.

Redrafting

This stage should involve careful checking of the structure and balance of the assignment. A good exercise is to make notes from your own writing, paragraph by paragraph. This will draw attention to repetition, arguments which are left unsupported or incomplete, and the overall balance or imbalance in the work. Some questions to consider include: does the introduction cover what the reader needs to know about the context of the work? Are the arguments, illustrations and examples given appropriate space? Do the points in the conclusion really arise from the earlier discussion? Are there some unresolved issues which could only be addressed in another piece of work? Where you have used information selectively, is it clear that it has been selected from a larger set of information and for a particular reason? It may be appropriate to move or re-sequence whole paragraphs or larger sections of the text during this stage.

Editing and proofreading

Editing should always involve some consideration of what can be deleted from the work to improve it. Many assignments exceed word limits and you may be penalised for this. A GCSE English chief examiner once said how impressed he was by a candidate who had the courage to cross out a long paragraph from an answer in examination conditions: because the piece was better balanced and the arguments flowed better as a result, this action had made the answer worthy of an A* grade rather than an A.

Proofreading, or checking for technical accuracy, is often enhanced by reading aloud, which forces you to slow down, and draws attention to matters such as overlong sentences and word omissions. Reading each paragraph in turn from the end of the assignment to the beginning can also help you to focus on technical accuracy rather than the development of the content. Spelling and grammar checkers are useful tools, but remember, if this is appropriate, to use UK rather than US spelling which may be the preset option. Grammar checkers have a preference for active rather than passive constructions which may not always be appropriate in academic writing. It is also easy to click on 'change' rather than 'ignore' by mistake and vice versa when working quickly, so a final read through is still needed. We know this does not always happen from the number of times we find, for example, 'offside' in assignments rather than 'Ofsted'.

Some other technical points to consider are as follows. Academic writing is normally written mainly in the third person. However, if you are reflecting on your own practice you may want to use the first person – I – when referring to your own practice. This helps to distinguish your practice and thoughts from those of other writer(s) to whom you refer.

The first sentence of a paragraph usually identifies the main point of the whole paragraph; this is a useful signal for readers. Make sure that each sentence contains a main verb. You should use a new paragraph to mark the introduction of a new point or idea. It is important to use a variety of sentence structures. Words and expressions like: 'however', 'nevertheless', 'on the other hand', and 'in addition' provide useful openings which help a reader to understand how sentences are related to each other.

The word 'the' tends to be overused. A statement that includes 'the', for example, 'the fact that' or 'the answer is', is a categorical statement that portrays certainty or definitiveness which may not be appropriate. Words such as 'a', 'one of ', 'there is a suggestion that' or 'one possible answer is' may be better instead because they are less definite. Another overused word is 'very', for example, 'very large' when 'large' would do.

Choose your words carefully. Do not use slang, jargon, colloquialisms, abbreviations or words that need to be put into quotation marks, unless quoting somebody else. For example, the use of 'kids' is slang.

Be careful about punctuation, i.e. full stops, commas, colons, semi-colons, dashes (e.g. for an aside) and brackets (e.g. for an explanation). By using punctuation appropriately, you help the reader to make sense of what you are saying. Underuse of commas is probably the most common error. Leaving out a comma can alter the meaning of a sentence. Overuse of the word and to link more than one thought/idea in a sentence is also a common error. Different thoughts/ideas should be in separate sentences. Dashes or brackets (normally curved brackets), used at the beginning and end of a phrase or clause, may help a reader to better understand the relative importance of different parts of a sentence.

Many writers make errors when using or omitting apostrophes. Remember that apostrophes are used to indicate possession (e.g. where a word does not end in a letter s, apostrophes are used as in one pupil's or a group of pupils' work; one teacher's or teachers' in general book; where the plural does not end in an s, apostrophes are used as in the children's work; or where the word ends in a letter s, an apostrophe is used after the last s, as in Jones' book or with an additional s, as in Jones's book) or letter omission (e.g. can't instead of cannot; don't instead of do not; it's instead of it is; who's instead of who is) but not plural nouns. If there is no possession or no missing letter then an apostrophe should not be used.

ACADEMIC REFERENCING

In academic writing, ideas, descriptions and explanations should not be taken for granted, even if everything you have read about the issue seems to provide a consensus. Ideas, assertions, descriptions, explanations or arguments should be supported by evidence from the work of others – from research and theory, appropriately referenced (e.g. you might say that: 'the research undertaken by

Bloggs (2008) suggests that . . . This supports/contradicts the findings of Smith (2003)'). However, alongside references to the work of others, research and theory from texts or articles, you should draw on your own teaching or observations in school, giving examples of activities or events you have observed or participated in to provide evidence from practice, where appropriate.

By 'the work of others' and 'theory' we mean explanations of teaching and learning, and descriptions of research, as well as any theories which have been developed from such work. You should use the evidence of others and your own evidence to advance your own understanding and formulate your own theories. Your own theories evolve by bringing your critical faculties to bear on the work of others and what has been happening in your teaching.

When referring to other research or to texts, a reference should always be given, together with a page number if quotations are used. It is important to show, by appropriate citation, what is your work and what is the work of others. Direct quotations should be used sparingly; otherwise they disrupt the flow of the assignment. Do not put in a quotation for the sake of it: only use quotations with a clear purpose. You need to explain whether a quotation is being used as evidence which supports or disagrees with your point of view. If you are not clear why you are using a quotation, you may be better to paraphrase the point.

References cited in the text must be included in a *bibliography* at the end of the assignment. If you are not given information about how to present your references and bibliography, ask the library which system is used in the HEI/your department. The Harvard system is a common system used in educational writing. Details of this should be available in the library.

STRATEGIES FOR MAKING NEUTRAL REFERENCE TO GENDER

Your writing is governed by social customs and conventions. One such custom and convention in the UK is to make sure that there is no gender bias in your writing, that men and women are regarded as equal. This translates into not making assumptions that certain jobs (e.g. lawyer, doctor, lorry driver, primary teacher, nurse) are either male or female and, therefore, not using words such as *he*, *him*, *his* to refer to someone in one of these jobs.

Thus, in your writing you need to avoid such gender bias. Three methods of overcoming the difficulty have been identified: (i) *avoidance*, (ii) *inclusion* or (iii) *disclaimer*. Unless your HEI prescribes one method for you to use, you may use any of the three methods. Avoidance is probably the most popular and effective of these methods; inclusion, the least popular method.

Avoidance

'Avoidance' avoids using male or female words altogether when reference is made to a person who could equally be male or female. Some strategies for avoidance include:

Strategy	Example	
	Avoid this	by writing this
Change to plural	. . . when a teacher meets his class for the first time	. . .when teachers meet their classes for the first time
Change to an article	. . . so that every member of the group can give his opinion	. . . so that every member of the group can give an opinion
Recast the clause so that a different noun becomes the subject	if the student cannot understand his feedback . . .	If the feedback is difficult to understand . . .

continued . . .

Strategy	Example	
	Avoid this	by writing this
Use neutral words like *the other, an individual, the author*	Each student must read what his partner has written	In pairs, each student must read what the other has written
Omit the male/female word	. . . working with each individual to ascertain her needs	. . . working with each individual to ascertain needs

Inclusion

Some people prefer to get round the difficulty by using the phrase *he or she*, or *he/she* or *s/he*. If used occasionally, such phrases may be appropriate, but they affect the style and interrupt the flow of the assignment when used frequently. It may be best to avoid using this approach to avoid such difficulties.

Disclaimer

A disclaimer is a statement at the beginning of the assignment that you are using words like *he*, *him* and *his* throughout without wanting to convey gender bias. In other words you claim that the male words are to be read as neutral in their reference. Alternatively you might state that you are using *she*, *her* and *hers* throughout the assignment for some particular purpose, without wanting to convey gender bias. This might be appropriate, for example, when referring to primary teachers, as most primary teachers are female. Thus, the statement disclaims that any gender bias is intended. Without such a statement, you are likely to offend others. Even with a disclaimer, some readers may find this strategy offensive.

PLAGIARISM AND CHEATING

You should be familiar with what is meant by plagiarism and cheating, so that you avoid them. A simple definition of plagiarism is that it is cheating by passing off someone else's work as your own. This can include putting quotations into your work without acknowledging them, closely paraphrasing a source, copying another student or someone else's work, submitting work that is not entirely your own, or reusing material from one assignment for another. Higher Education Institutions (HEIs) regard plagiarism as a very serious offence – it can potentially result in the failure of a whole course, which could put your whole career in jeopardy – and ignorance of the rules is not seen as a defence.

Check the general rules of your HEI and also the particular rules for an assignment (for example, you may sometimes be allowed, or even required, to work together on collaborative assignments or presentations). However, in general you will be safe if you acknowledge all sources, whether directly quoted or paraphrased, including material other than words (diagrams and website material for example), make sure that you do not include so much material from other sources that the work could not be properly described as your own, and avoid using the same work in more than one assignment.

Finally, HEIs are increasingly requiring students to submit important assignments electronically, so that software can be used to check for plagiarism. This is in your interests, as previously markers may have spotted plagiarism in one offender's work but not another's.

Check the course handbook for your institution for further guidance.

CONCLUSION

To sum up, assignments set as part of your ITE course require good academic writing, but are also designed to enable you to show that you are reflecting on research and theory of teaching and are applying this to your own developing practice as a teacher, i.e. that you are developing as a reflective practitioner. You therefore need to combine academic writing and evidence of reflection in your written assignments.

Teachers need good writing skills. You write on the board, write notices, write reports, send letters to parents and undertake numerous other written communications. Whatever your subject, you have a responsibility to promote the development of key literacy skills in your lessons. Therefore, you need good spelling, grammar, punctuation, sentence construction and paragraph formation. If your writing skills are not as good as necessary to write effectively for a variety of audiences in school, you should ask for help during your time in ITE. HEIs will be able to direct you to study support units and to self-study materials available on the Internet.

Good luck with writing your course assignments and the written communications you make as part of your teaching.

FURTHER READING

Burchfield, R. (ed.) (1996) *The New Fowler's Modern English Usage*, Oxford: Clarendon Press.
Creme, P. and Lea, M. (2006) *Writing at University*, 2nd edn, Buckingham: Open University Press.
Greenbaum, S. and Whituit, J. (1988) *Longman Guide to English Usage*, Harlow: Longman.
Palmer, R. (1993) *Write in Good Style: A Guide to Good English*, London: E & FN Spon.
Palmer, R. (1996) *Brain Train: Studying for Success*, London: E & FN Spon.
Redman, P. (2001) *Good Essay Writing: A Social Sciences Guide*, Milton Keynes: Open University Press/Sage.
Smith, P. (2002) *Writing an Assignment*, Oxford: How to Books.
Warburton, N. (2006) *The Basics of Essay Writing*, London: Routledge.
Woods, P. (2006) *Successful Writing for Qualitative Researchers*, 2nd edn, London: Routledge.

Additional resources for this unit are available on the companion website:
www.routledge.com/textbooks/9780415478724

Readings for Learning to Teach in the Secondary School: A Companion to M Level Study brings together essential readings to support you in your critical engagement with key issues raised in this textbook.

ACKNOWLEDGEMENT

The section on strategies for making neutral reference to gender is adapted from a paper produced by the Department of Language Studies, Canterbury Christ Church University College.

APPENDIX 2
Glossary of terms

Terms shown in bold within a definition have their own entry in the glossary.

Academies. All-ability, secondary schools established by sponsors from business, faith or voluntary groups in partnership with central government and LAs. See **Other State Schools in England**

ACCAC Awdurdod Cymwysterau, Cwricwlwm ac Asesu Cymru (Qualifications, Curriculum and Assessment Authority for Wales).

AEB Associated Examining Board. Part of the **AQA**.

Annual Review The review of a statement of special educational needs which an LA must make within 12 months of making the statement or from a previous review.

AQA Assessment and Qualifications Alliance. An Awarding Body, including C and G, for GCSE, GCE A and AS levels and Diplomas. Online: http://www.aqa.org.uk.

Assessment for learning Assessment for which the first priority is promote students' learning. See **formative assessment**.

Assessment of learning. Assessment covers all those activities that are undertaken by teachers and others to measure the effectiveness of their teaching. See **formative assessment** and **summative assessment**.

Attainment Targets (ATs) of NC for England The knowledge, skills and understanding that pupils of different abilities and maturities are expected to have by the end of each Key Stage. Attainment targets consist of eight **level descriptions** of increasing difficulty, plus a description for exceptional performance above level 8. **Citizenship** has non-statutory ATs for KS 3 and 4 only. See also **Programmes of Study**.

Awarding Body There are three Awarding Bodies which set public examinations in England: the Assessment and Qualifications Alliance (**AQA**); **Edexcel** and the Oxford and Cambridge Regional (**OCR**).

BA/BSc (QTS) Bachelor of Arts/Bachelor of Science with **QTS**. A combined course with route to QTS.

Banding The structuring of a year group into divisions, each usually containing two or three classes, on grounds of general ability. Pupils are taught within the band for virtually all the curriculum. See also **mixed ability grouping, setting** and **streaming**.

Baseline Testing Any process which sets out to find out what the learner can do now in relation to the next stage of learning. For example the assessment of practical skills and familiarity of pupils with equipment and tools prior to a D and T course. Or the assessment of pupils in Year 1 and reception classes for speaking, listening, reading, writing, mathematics, social skills. See also **benchmarking.**

Basic curriculum for England from 2007 the National Curriculum at Key Stage 3 comprises thirteen statutory subjects including core subjects of English, mathematics and science plus **PSHE;** Religious Education (RE) has a non-statutory curriculum. At Key Stage 4 there are seven statutory subjects, including English, mathematics, science and RE as at KS3, plus PSHE.

BEd Bachelor of Education (a route to **QTS**).

Benchmarking. A term used to describe a standard against which comparisons can be made. Can be used by schools, e.g. to measure success of the school in public examinations relative to a national norm.

BTEC Business and Technician Education Council. Part of **Edexcel Foundation** which offers courses called **BTEC Nationals**.

C and G City and Guilds; part of **AQA**.

Career Entry and Development Profile (CEDP) All **ITT** providers in England are required to provide newly qualified teachers with a CEDP required by the **TDA** to help newly qualified teachers in their first teaching post schools and support **induction**.

Careers Education. Designed to help pupils to choose and prepare for opportunities, responsibilities and experiences in education, training and employment.

CDP Continuing professional development.

CEDP See **Career entry and development profile**.

CEHR Commission for Equality and Human Rights; sometimes Equality and Human Rights Commission.

Certificate of Achievement (COA) An examination designed to give a qualification to pupils who may not gain a GCSE grade, offered by the Awarding bodies. Also called **Entry Level**; see *Directgov* website: www.direct.gov.uk/en/EducationAndLearning/QualificationsExplained.

Church and faith schools Follow a religious education curriculum and religion-centred admissions criteria and staffing policies.

Citizenship A statutory subject of the English National Curriculum at Key Stages 3 and 4. See also **cross-curricular** elements.

Collaborative group A way of working in which groups of children are assigned to groups or engage spontaneously in working together to solve problems; sometimes called co-operative group work. See DCFS Standards website *Grouping pupils for success*.

Combined course A course to which several subjects contribute while retaining their distinct identity (e.g. history, geography and RE within humanities).

Community and Foundation special schools. For children with specific special educational needs, such as physical or learning difficulties.

Community of practice. Groups of people who share a concern for something or have knowledge and skills to share. For example, a subject association network or a network of teachers working on solving a particular problem.

Community schools run by the local authority (LA), which employs the staff, owns the school land and buildings and has primary responsibility for deciding on pupil admission criteria.

Comprehensive school A secondary school which admits pupils of age 11 to 16 or 19 from a given catchment area, regardless of their ability.

Continuity and progression Appropriate sequencing of learning which builds on previous learning to extend and develop pupils' capabilities.

Core skills Skills required by all students following 14-19 courses. See **Functional Skills** and **Personal, Learning and Thinking Skills.**

Core subjects. Foundation subjects which are taught at both KS3 and KS4 comprising English, ICT, mathematics and science in the National Curriculum for England.

Coursework Work carried out by pupils during a course of study marked by teachers and contributing to the final examination mark. Usually externally moderated.

CPD Continuing professional development.

CRE Commission for Racial Equality; now **CEHR.**

Criterion-referenced assessment A process in which performance is measured by relating candidates' responses to pre-determined criteria.

Cross-curricular elements Additional elements of a curriculum which includes careers, citizenship, key skills, personal and learning skills, personal social and health education and thinking skills which run across the whole curriculum.

CTC City Technology College, publicly funded independent schools for pupils aged 11 to 18 with a specific focus on science, mathematics and technology.

Curriculum A course of study followed by a pupil.

Curriculum guidelines Written school guidance for organising and teaching a particular subject or area of the curriculum. See also **Programmes of Study**.

D and T Design and technology in National Curriculum for England.

Dearing Report A review of the 'National Curriculum and its assessment' (1994).

Department Section of the curriculum/administrative structure of a (secondary) school, usually based on a subject.

DES Department of Education and Science in England (became **DfE** in 1992).

DfE Department for Education (previously **DES**); became **DfEE** in 1995.

DfEE Department for Education and Employment (previously **DfE**) (became **DfES** in 2001).

DfES Department for Education and Skills (previously **DfEE**) (became **DCSF** in 2007).

DCSF Department for Children, Schools and Families (previously DfES).

DCSF Circular Advice issued by the Department for Children, Schools and Families to **LAs**. Circulars do not have the status of law.

Differentiation The matching of work to the differing capabilities and learning needs of individuals or groups of pupils in order to extend their learning.

Diploma A 14–19 qualification involving academic qualifications, vocational qualifications and key skills.

Disapplication Arrangement for lifting part or all of the National Curriculum in England requirements for individuals or for any other grouping specified by the Secretary of State.

EAL English as an Additional Language.

EBD Emotional and behavioural difficulties and disorders. Used with reference to pupils with such difficulties or schools/units which cater for such pupils.

Edexcel Foundation An **Awarding Body**. Online http://www.edexcel.com.

Education welfare officer (EWO) An official of the LA concerned with pupils' attendance and with liaison between the school, the parents and the authority.

Entry Level: see Certificate of achievement.

EOC Equal Opportunities Commission; now **CEHR**.

ERA Education Reform Act (1988) for England and Wales.

ESL English as a second language.

Exclusion Headteachers of **state maintained schools** and **other state schools** in England are empowered to exclude pupils temporarily or permanently when faced with a serious breach of their disciplinary code. The exclusions are either a fixed term or permanent. Schools may send pupils to a **Pupil Referral Unit**.

Faculty Grouping of subjects for administrative and curricular purposes.

Formative assessment or assessment for learning, linked to teaching when the evidence from assessment is used to adapt teaching to meet pupils' learning needs.

Forms of entry (FE) The number of forms (of 30 pupils) which a school takes into its intake year. From this can be estimated the size of the intake year and the size of the school.

Foundation schools. The governing body employs the staff and sets the admissions criteria. The school land and buildings are owned by the governing body or a charitable foundation.

Foundation subjects. Subjects which **state-maintained schools** are required by law to teach. In England four **Foundation subjects** are designated **Core subjects**. Different subjects are compulsory at different Key Stages in England. See **Basic curriculum.**

Functional skills Functional skills in the NC for England are those core elements of English, mathematics and ICT that provide individuals with the skills and abilities they need to operate confidently, effectively and independently in life, their communities and work. See QCA 2008.

GCE General Certificate of Education

GCSE General Certificate of Secondary Education.

GNVQ General National Vocational Qualifications.

Grade-related criteria The identification of criteria, the achievement of which are related to different levels of performance by the candidate.

Grammar schools. Select all or almost all of their pupils based on academic ability.

Group work A way of organising pupils where the teacher assigns tasks to groups of pupils, to be undertaken collectively although the work is completed on an individual basis.

GTCE General Teaching Council for England. Northern Ireland, Scotland and Wales have their own teaching councils.

HEI Higher Education Institution.

HMCI Her Majesty's Chief Inspector of Schools in England.

HMI Her Majesty's Inspectors of Schools in England.

HOD Head of Department.

House system A structure for pastoral care/pupil welfare within a school in which pupils are grouped in vertical units, i.e. sections of the school which include pupils from all year groups.

HOY Head of Year.

IB International Baccalaureate. A post-16 qualification designed for university entrance.

ICT See **Information and Communications Technology**.

In-class support Support within a lesson provided by an additional teacher, often with expertise in teaching pupils with special educational or language needs.

Inclusion. Inclusion involves the processes of increasing the participation of pupils in, and reducing their exclusion from schools. Inclusion is concerned with the learning participation of all pupils vulnerable to exclusionary pressures, not only those with impairments or categorised as having special educational needs.

Independent school A private school that receives no state assistance but is financed by fees. Often registered as a charity. See also **public school**.

Information and Communications Technology (ICT) Computer hardware and software which extend beyond the usual word-processing, databases, graphics and spreadsheet applications to include hardware and software which allow computers to be networked across the world through the World Wide Web, to access information on the Internet and which supports other communication activities such as e-mail and video-conferencing.

Information Technology (IT) Methods of gaining, storing and retrieving information through microprocessors. Often encompassed within **Information and Communications Technology** **(ICT)**, a compulsory subject in the National Curriculum for England.

Integrated course A course, usually in a secondary school, to which several subjects contribute without retaining their distinct identity (e.g. integrated humanities, which explores themes which include aspects of geography, history and RE).

Integration Educating children with special educational needs together with children without special educational needs in mainstream schools: see **Inclusion**

Ipsative assessment A process in which performance is measured against previous performance by the same person. See also **criterion-referenced assessment**, **normative assessment**.

IT See **information technology**.

ITE Initial Teacher Education

ITT Initial Teacher Training.

Key Skills See **Functional skills** and **Personal Learning and Thinking skills**.

Key Stages (KS) The periods in each pupil's education to which the elements of the National Curriculum for England apply. There are four Key Stages, normally related to the age of the majority of the pupils in a teaching group. They are: Key Stage 1, beginning of compulsory education to age 7 (Years R (Reception), 1 and 2); Key Stage 2, ages 7–11 (Years 3–6); Key Stage 3, ages 11–14 (years 7– 9); Key Stage 4, 14 to end of compulsory education (Years 10 and 11). Post-16 is a further Key Stage.

LA see **Local Authority**

Language support teacher A teacher provided by the LA or school to enhance language work with particular groups of pupils.

LEA See **Local Education Authority**.

Learning objectives What pupils are expected to have learned as a result of an activity.

Learning outcomes Assessable learning objectives; the action or behaviour of pupils which provides evidence that they have met the learning objectives.

Learning support A means of providing extra help for pupils, usually those with learning difficulties, e.g. through a specialist teacher or specially designed materials. See also **learning support assistants**.

Learning support assistants Teachers who give additional support for a variety of purposes, e.g. general learning support for **SEN** pupils, **ESL**; most support is given in-class although sometimes pupils are withdrawn from class. See also **learning support**, **withdrawal**.

Lesson plan The detailed planning of work to be undertaken in a lesson. This follows a particular structure, appropriate to the demands of a particular lesson. An individual lesson plan is usually part of a series of lessons in a **unit of work**.

Level Description (National Curriculum for England) a statement describing the types and range of performance that pupils working at a particular level should characteristically demonstrate. Level descriptions provide the basis for making judgements about pupils' performance at the end of Key Stages 1, 2, 3. At Key Stage 4, national qualifications are the main means of assessing attainment in National Curriculum subjects.

Levels of attainment Eight levels of attainment, plus exceptional performance, are defined within the National Curriculum attainment targets in England. These stop at **Key Stage** 3; see **Level description**. In deciding a pupil's level of attainment teachers should judge which description best fits the pupil's performance (considering each description alongside descriptions of adjacent levels).

Local Authority (LA) has a statutory duty to provide education in their area.

Local Education Authority now **Local Authority.**

Maintained boarding schools. Offer free tuition, but charge fees for board and lodging.

Middle school A school which caters for pupils aged from 8–12 or 9–13 years of age. They are classified legally as either primary or secondary schools depending on whether the preponderance of pupils in the school is under or over 11 years of age.

Minority ethnic groups Pupils, many of whom have been born in the United Kingdom, from other ethnic heritages, e.g. those of Asian heritage from Bangladesh, China, Pakistan, India or East Africa, those of African or Caribbean heritage or from countries in the European Union.

Mixed ability grouping Teaching group containing pupils representative of the range of ability within the school. See also **banding**, **setting**, **streaming**.

Moderation An exercise involving teachers representing an **awarding body** external to the school whose purpose is to check that standards are comparable across schools and teachers. Usually carried out by sampling coursework or examination papers.

Moderator An examiner who monitors marking and examining to ensure that standards are consistent in a number of schools and colleges.

Module A definable section of work of fixed length with specific learning objectives and usually with some form of terminal assessment. Several such units may constitute a modular course.

National Curriculum (NC) in England. The **core** and other **foundation** subjects and their associated **attainment targets**, **programmes of study** and assessment arrangements of the curriculum.

National Curriculum Tests. See **Standard Assessment Tasks.**

National induction standards. Standards that all newly qualified teachers in England are required to demonstrate at the end of their induction period. See also **statutory induction**.

NFER National Foundation for Educational Research. Carries out research and produces educational diagnostic tests.

NAA National Assessment Agency The NAA supports the secure delivery of the public exam system and develops and delivers high quality national curriculum assessments. See also **QCA**.

Non-contact time Time provided by a school for a teacher to prepare work or carry out assigned responsibilities other than direct teaching.

Norm-referenced assessment A process in which performance is measured by comparing candidates' responses. Individual success is relative to the performance of all other candidates.

Normative assessment Assessment which is reported relative to a given population.

NQT Newly qualified teacher.

NSG Non-statutory guidance (National Curriculum in England). Additional subject guidance but which is not mandatory; to be found attached to National Curriculum Subject Orders.

NVQ National Vocational Qualifications.

OCR Oxford and Cambridge Regional **Awarding Body.** Online <http://www.ocr.org.uk>.

Ofsted Office for Standards in Education. Non-ministerial government department established under the Education (Schools) Act (1992) to take responsibility for the inspection of schools in England. Ofsted inspects pre-school provision, further education, teacher education institutions and **local authorities**. **Her Majesty's Inspectors (HMI)** form the professional arm of Ofsted. See also OHMCI.

OHMCI Office of Her Majesty's Chief Inspector (Wales). Non-ministerial government department established under the Education (Schools) Act (1992) to take responsibility for the inspection of schools in Wales. **Her Majesty's Inspectors (HMI)** form the professional arm of OHMCI. See also **Ofsted**.

Other State Schools in England These include **Academies, City Technology Colleges** (CTC), **Community and Foundation Special Schools,** Church and Faith Schools (see **Voluntary schools**), **Specialist Schools** and **Pupil Referral Units**. See 'Direct gov' website: www.direct.gov.uk. Search 'types of schools in England'.

PANDA See **Raiseonline**

Parent Under section 576 of the Education Act 1996 a parent includes any person who is not a parent of the child but has parental responsibility (see **Parental responsibility**), or who cares for him.

Parental responsibility: under section 2 of the Children Act 1989, parental responsibility falls upon:

■ all mothers and fathers who were married to each other at the time of the child's birth (including those who have since separated or divorced)

■ mothers who were not married to the father at the time of the child's birth

■ fathers who were not married to the mother at the time of the child's birth, but who have obtained parental responsibility either by agreement with the child's mother or through a court order. Under section 12 of the Children Act 1989 where a court makes a residence order in favour of any person who is not the parent or guardian of the child that person has parental responsibility for the child while the residence order remains in force. See *Code of Practice on the Identification and Assessment of Special Educational Needs* (DfEE, 2001a) for further details.

Partnership teaching An increasingly common means of meeting the language needs of bilingual pupils in which support and class teachers plan and implement together a specially devised programme of in-class teaching and learning. See also **Section 11 staff, In-class support** and **Learning support**.

Pastoral care Those aspects of a school's work and structures concerned to promote the general welfare of all pupils, particularly their academic, personal and social development, their attendance and behaviour.

PAT Pupil Achievement Tracker. Now part of **Raiseonline**.

Pedagogic knowledge the skills to transform subject knowledge into suitable learning activities for a particular group of pupils.

Personal Learning and Thinking Skills In the National Curriculum for England related to the development of pupils as independent enquirers, creative thinkers, reflective learners, team workers, self-managers, effective participants.

PGCE Postgraduate Certificate in Education. The main qualification for secondary school teachers in England and Wales recognised by the **DCFS** for **QTS**.

PG Cert Postgraduate Certificate in Learning and Teaching in Higher Education.

Policy An agreed school statement relating to a particular area of its life and work.

PoS See **Programmes of Study**.

Pre-vocational courses Courses specifically designed and taught to help pupils to prepare for the world of work.

Profile Samples of work of pupils, used to illustrate progress, with or without added comments by teachers' and/or pupils.

Programmes of study (PoS) for National Curriculum in England. The subject matter, skills and processes which must be taught to pupils during each **Key Stage** in order that they may meet the objectives set out in **attainment targets**. They provide the basis for planning **schemes of work**.

Project An investigation with a particular focus undertaken by individuals or small groups of pupils leading to a written, oral or graphic presentation of the outcome.

PSE Personal and social education.

PSHCE. Is **PSHE** with a specific additional **citizenship** component.

PSHE Personal, social, health and economic education, a non-statutory subject in the English National Curriculum. See also **PSHCE**.

PTA Parent-teacher association. Voluntary grouping of parents and school staff to support the school in a variety of ways.

PTR Pupil:Teacher Ratio. The ratio of pupils to teachers within a school or group of schools (e.g. 17.4:1).

Public school Independent secondary school not state funded. So-called because at their inception they were funded by public charity. See also **independent school**.

Pupil referral units. For children of compulsory school age who may otherwise not receive suitable education, focusing on getting them back into a mainstream school.

QCA The Qualifications and Curriculum Authority. Responsible for overview of the curriculum, assessment and qualifications across the whole of education and training, from pre-school to higher vocational levels. QCA advises the Secretary of State for Children, Schools and Families (DCSF) on such matters. Aspects of assessment are delegated by QCA to the **NAA**. The Welsh equivalent of QCA is **ACCAC**.

QTS Qualified teacher status. This is usually attained by completion of a Postgraduate Certificate in Education (**PGCE**) or a Bachelor of Education (**BEd**) degree or a Bachelor of Arts/Science degree with Qualified Teacher Status (**BA/BSc (QTS)**). There are other routes into teaching.

RAISEonline Reporting and Analysis for Improvement through School Self-Evaluation. In England it provides interactive analysis of school and pupil performance data. It replaces the Ofsted Performance and Assessment (PANDA) reports and DCSF's Pupil Achievement Tracker (PAT).

Record of achievement (ROA) Cumulative record of a pupil's academic, personal and social progress over a stage of education.

Reliability A measure of the consistency of the assessment or test item; i.e. the extent to which the test gives repeatable results.

RSA Royal Society of Arts.

SACRE The Standing Advisory Council on Religious Education in each LA to advise the LA on matters connected with religious education and collective worship, particularly methods of teaching, the choice of teaching materials and the provision of teacher training.

SATs See **Standard assessment tasks**.

Scheme of work A planned course of study over a period of time (e.g. a Key Stage or a Year). In England it contains knowledge, skills and processes derived from the **Programmes of Study** and **Attainment Targets**.

School Development Plan (SDP) A coherent plan, required to be made by a school, identifying improvements needed in curriculum, organisation, staffing and resources and setting out action needed to make those improvements.

SEN (Special Educational Needs) Children have special educational needs if they have a *learning difficulty*. Children have a *learning difficulty* if they:

a   have a significantly greater difficulty in learning than the majority of children of the same age; or

b   have a disability which prevents or hinders them from making use of educational facilities of a kind generally provided for children of the same age in schools within the area of the local authority

c   are under compulsory school age and fall within the definition at (a) or (b) above or would so do if special educational provision was not made for them. Very able or gifted pupils are not included in SEN.

SET: Self-Evaluation Tool: intended to replace CEDP (q.u.).

Setting The grouping of pupils according to their ability in a subject for lessons in that subject. See also **banding, mixed ability grouping**, **streaming**.

Short course A course in a National Curriculum foundation subject in Key Stage 4 which by itself does not lead to a GCSE or equivalent qualification. Two short courses in different subjects may be combined to form a GCSE or equivalent course.

Sixth Form College A post-16 institution for 16- to 19-year-olds. It offers GCSE, GCE A level and vocational courses.

SLD Specific learning difficulties.

SOA Statements of attainment (of National Curriculum subjects).

Special school. See **Community and Foundation special schools.**

Specialist schools. Teach the whole curriculum but with a focus on one subject area such as: arts; business and enterprise; engineering; humanities; language; mathematics and computing; music; science; sports; technology. See **Other state schools in England.**

Standard assessment tasks (SATs) Externally prescribed **National Curriculum for England** assessments which incorporate a variety of assessment methods depending on the subject and **Key Stage**. This term is not now widely used, having been replaced by **National Curriculum Tests.**

State maintained schools. In England these are **Community Schools, Foundation Schools, Trust schools, Voluntary-controlled** and **Voluntary-aided**. See also **Other State Schools in England.**

Statements of special educational needs Provided under the 1981 Education Act and subsequent Acts to ensure appropriate provision for pupils formally assessed as having **SEN**.

Statutory induction Induction is a statutory requirement for all newly qualified teachers; headteachers are responsible for ensuring statutory requirements are met. See **National induction standards**.

Statutory order A statutory instrument which is regarded as an extension of an Act, enabling provisions of the Act to be augmented or updated.

Streaming The organisation of pupils according to general ability into classes in which they are taught for all subjects and courses. See also **banding**, **mixed ability grouping**, **setting**.

Summative assessment Assessment linked to the end of a course of study; it sums up achievement in aggregate terms and is used to rank, grade or compare pupils, groups or schools. It uses a narrow range of methods which are efficient and reliable, normally formal, i.e. under examination conditions.

Supply teacher Teachers appointed by **LAs** to fill vacancies in maintained schools which arise as a result of staff absences. Supply teachers may be attached to a particular school for a period ranging from half a day to several weeks or more.

Support Teacher. See **In-class support** and **Learning support.**

TDA Training and Development Agency in 2005 superseded the **TTA** with an extended remit for overseeing standards and qualifications across the school workforce.

Teacher's record book A book in which teachers plan and record teaching and learning for their classes on a regular basis.

Team teaching The teaching of a number of classes simultaneously by teachers acting as a team. They usually divide the work between them, allowing those with particular expertise to lead different parts of the work, the others supporting the follow-up work with groups or individuals.

TES *Times Educational Supplement*. Published weekly and contains articles on educational issues, information about developments in education, job vacancies, etc.

Thinking skills Additional skills to be promoted across the **National Curriculum.** See also **cross-curricular elements** and **Personal learning and Thinking Skills.**

Traveller education The development of policy and provision which provides traveller children with unhindered access to and full integration in mainstream education.

Travellers A term used to cover those communities, some of which have minority ethnic status, and either are or have been traditionally associated with a nomadic lifestyle, and include gypsy travellers, fairground or show people, circus families, New Age travellers, and bargees.

Trust schools A **Foundation school** supported by a charitable foundation or trust, which appoints school governors. A trust school employs its own staff, manages its own land and assets, and sets its own admissions criteria.

TTA Teacher Training Agency. Established in 1994 and responsible for the quality of teacher education in England and the supply of teachers. In 2005, the TTA became the **TDA**.

Tutor group Grouping of secondary pupils for registration and pastoral care purposes.

Unit of work Medium-term planning of work for pupils over half a term or a number of weeks. The number of lessons in a unit of work may vary according to each school's organisation. A unit of work usually introduces a new aspect of learning. Units of work derive from **schemes of work** and are the basis for **lesson plans**.

Validity A measure of whether the assessment measures what it is meant to measure – often determined by consensus. Certain kinds of skills and abilities are extremely difficult to assess with validity via simple pencil and paper tests.

Voluntary-aided school Often religious schools. The governing body, often a religious organisation, employs the staff and sets admissions criteria. The school land and buildings are also owned by a charitable foundation.

Voluntary controlled school Mainly religious or 'faith' schools, but run by the LA. The land and buildings are often owned by a charitable foundation, but the LA employs the staff and has primary responsibility for admission arrangements.

Voluntary school. School which receives financial assistance from the **LA**, but which is owned by a voluntary body, usually religious. See **Voluntary school** and **Voluntary-aided school**.

Withdrawal Removal of pupils with particular needs from class teaching in primary schools and from specified subjects in secondary schools for extra help individually or in small groups. In-class support is increasingly provided in preference to withdrawal.

WJEC (Welsh Joint Education Committee) provides examinations, assessment, professional development, educational resources, support for adults who wish to learn Welsh and access to youth arts activities. It also provides examinations throughout England.

Work experience The opportunity for secondary pupils to have experience of a work environment for one or two weeks, usually within school time, during which a pupil carries out a particular job or range of jobs more or less as would regular employees, although with emphasis on the educational aspects of the experience. It may only take place after Easter in Year 10 (i.e. in the final year of statutory schooling).

World class tests Tests for high-achieving 9- and 13-year olds in 'Mathematics' and in 'Problem Solving in Mathematics, Science and Technology', sponsored by government. See http://www.nottingham.ac.uk/education/MARS/services/wct.htm.

Year system A structure for pastoral care/pupil welfare within a school in which pupils are grouped according to years, i.e. in groups spanning an age range of only one year.

Years 1–11 Year of schooling in England. Five-year-olds start at Year 1 (Y1) and progress through to Year 11 (Y11) at 16 years old. See **Key Stages** for details.

REFERENCES

Abercrombie, M. L. (1985) *The Anatomy of Judgement*, Harmondsworth: Pelican Books.

Addison, N. and Burgess, L. (eds) (2000) *Learning to Teach Art and Design in the Secondary School: A Companion to School Experience*, London: Routledge.

Addison, N. and Burgess, L. (eds) (2007) *Learning to Teach Art and Design in the Secondary School: A Companion to School Experience*, 2nd edn, Abingdon: Routledge.

Adey, K. and Biddulph, M. (2001) 'The influence of pupil perceptions on subject choice at 14+ in geography and history', *Educational Studies,* 27 (4), 439–450.

Adey, P. (1992) 'The CASE results: implications for science teaching', *International Journal of Science Education*, 14, 137–146.

Adey, P. (2005) 'Science teaching and the development of intelligence', in M. Monk and J. Osborne, *Good Practice in Science Education*, Buckingham: The Open University Press.

Adey, P. and Shayer, M. (1994) *Really Raising Standards*, London: Routledge.

Adey. P., Shayer, M. and Yates, C. (1989) *Thinking Science,* London: Macmillan.

Aldrich, R. (1988) 'The National Curriculum: an historical perspective' in D. Lawton and C. Chitty (eds) *The National Curriculum*, Bedford Way Papers No. 33, London: Institute of Education, University of London.

Aldrich, R. (2006) *Lessons from History of Education: The Selected Works of Richard Aldrich*, Abingdon, Oxon: Routledge.

Aldrich, R. and White, J. (1998) *The National Curriculum beyond 2000: The QCA and the Aims of Education*, London: Institute of Education, University of London.

All-Party Parliamentary Group on Scientific Research in Learning and Education. Online. Available online at: http://www.futuremind.ox.ac.uk/impact/parliamentary.html. Accessed 16.09.08.

Altrichter, H., Feldman, A., Posch, P. and Somekh, B. (2008) *Teachers Investigate Their Work: An Introduction To Action Research Across the Professions*, 2nd edn, London: Routledge.

Amade-Escot, C. (2000) 'The contribution of two research programs on teaching content: "Pedagogical content knowledge" and "didactics of physical education"', *Journal of Teaching in Physical Education*, 20, 78–101.

Ames, C. (1992a) 'Achievement goals and the classroom motivational climate', in D. H. Schunk and J. L. Meece (eds) *Student Perception in the Classroom*, Hillsdale, NJ: Erlbaum: 327–348.

Ames, C. (1992b) 'Classrooms: goals, structures and student motivation', *Journal of Educational Psychology*, 84, 261–271.

Amos, J-A. (1998) *Managing Your Time: What to Do and How to Do It in Order to Do More*, Oxford: How to Books.

Anderson, J. R., Reder, L. M. and Simon, H. A. (1996) 'Situated learning and education', *Educational Researcher*, 25, 5–11.

Anderson, L. W. and Krathwohl, D. (eds) (2001) *A Taxonomy for Learning, Teaching, and Assessing: A Revision of Bloom's Taxonomy of Educational Objectives*, New York: Longman.

Anderson, M. (1992) *Intelligence and Development; A Cognitive Theory*, London: Blackwell.

Ansari, D. (2006) 'Bridges over troubled waters: education and cognitive neuroscience', *Trends Cogn. Sci.*, 10 (4), 146–151.

ARG (Assessment Reform Group) (2002) *Testing, Motivation and Learning*, Cambridge: University of Cambridge Faculty of Education.

Arikewuyo, M. O. (2004) 'Stress management strategies of secondary school teachers in Nigeria', *Educational Research*, 46 (2), 196–207.

Arnold, R. (1993) *Time Management*, Leamington Spa: Scholastic Publications.

Association for Achievement and Improvement through Assessment (AAIA). Available online at: http://www.aaia.org.uk. Accessed 29.07.08

Atherton, J. (2005) *Learning and Teaching: Piaget's Developmental Theory*. Available online at: http://www.learningandteaching.info/learning/piaget.htm. Accessed 29.11.07.

Atkinson, J. W. (1964) *An Introduction to Motivation*, Princeton, NJ: Van Nostrand.

Atkinson, R. C. and Shiffrin, R. M. (1968) 'Human memory: a proposed system and its control processes', in K. W. Spence and J. T. Spence (eds) *The Psychology of Learning and Motivation: Advances in Research and Theory* (Vol. 2), New York: Academic Press: 742–775.

Ausubel, D. P. (1968) *Educational Psychology: A Cognitive View*, New York: Holt, Rinehart and Winston.

Ayers, H. and Prytys, C. (2002) *An A-Z Practical Guide to Emotional and Behavioural Difficulties*, London: David Fulton Publishers.

Balderstone, D. and King, S. (2004) 'Preparing pupils for public examinations: developing study skills', in S. Capel, R. S. Heilbronn, M. R. Leask and T. Turner (eds) *Starting to Teach in the Secondary School: A Companion for the NQT*, 2nd edn, London: RoutledgeFalmer.

Ball, S. (2003) *Class strategies and the education market: the middle classes and social advantage*, London: RoutledgeFalmer.

Ball, S. and Goodson, I. (1985) *Teachers' Lives and Careers*, Buckingham: Open University Press

Banks, F., Leach, J. and Moon, B. (1999) 'New understandings of teachers' pedagogic knowledge', in J. Leach and B. Moon (eds) *Learners and Pedagogy*, London: Paul Chapman Publishing: 89–110.

Barnes *et al.* (1987) *Learning Styles in TVEI: Evaluation Report No. 3*, Leeds: Manpower Service Commission.

Barnes, I. and Harris, S. (2006) *Special Series on Personalised Learning: An Overview of the Summary Report Findings*, Nottingham. National College for School Leadership NCSL (Spring).

Batchford, R., (1992) *Values: Assemblies for the 1990s*, Cheltenham: Stanley Thornes.

Battisch, V., Solomon, D. and Watson, M. (1998) 'Sense of community as a mediating factor in promoting children's social and ethical development'; Paper presented at the meeting of the American Educational Research Association, San Diego, CA, April 1998. Available online at: http://tigger.uic.edu/~lnucci/MoralEd/articles/battistich.html. Accessed 14.11.07.

Baumfield. V., Hall, E. and Wall, K. (2008) *Action Research in the Classroom*, London: Sage.

Bawden, A. (2007) 'Walking back to happiness', *Education Guardian*, 20 March: 8.

Beard, R. (2000) *Developing Writing 3-13*, London: Hodder and Stoughton.

Beck, J. (2000) 'The School Curriculum and the National Curriculum', in J. Beck and M. Earl (eds) *Key Issues in Secondary Education*, London: Cassell.

BECTA (British Educational Communications and Technology Agency) (2000) *A Preliminary Report for the DfEE on the Relationship between ICT and Primary School Standards*, Coventry: BECTA.

BECTA (2001) *Information Sheet: Parents, ICT and Education*, Coventry: British Educational Communications and Technology Agency. Available online at: http://partners.becta.org.uk/index.php?section=rh&&catcode=&rid=13684. Accessed 9.01.09.

BECTA (2002) *Educational Research into ICT and Pupil Motivation – A Selection of Abstracts*. Available online at: http://partners.becta.org.uk/upload-dir/downloads/page_documents/research/wtrs_bibs_motivation.pdf. Accessed 9.01.09.

BECTA (2003a) *Secondary Schools – ICT and Standards: An Analysis of National Data from Ofsted and QCA by BECTA*, Coventry, British Educational Communications and Technology Agency. Available online at: http://partners.becta.org.uk/upload-dir/downloads/page_documents/research/secschoolfull.pdf.Accessed 9.01.09.

BECTA (2003b) *What Research Says about Interactive Whiteboards, Coventry: British Educational Communications and Technology Agency*: Available online at: http://www.becta.org.uk/page_documents/research/wtrs_whiteboards.pdf. Accessed 1.01.08.

BECTA (2004a) *What Is ICT?* Available online at: http://schools.becta.org.uk/index.php?section =tl&catcode=as_cu_pr_sub_07&rid=1701. Accessed 1.01.08.

BECTA (2004b) *Getting the Most From Your Interactive Whiteboard: A Guide for Secondary Schools.* Available online at: http://foi.becta.org.uk/display.cfm?cfid=1476190&cftoken=29154&resID= 35754. Accessed 9.01.09.

BECTA (2005) *ICT Advice for Teachers: The Interactive Whiteboards Project*, Coventry: BECTA.

BECTA (2007a) *Learning and Teaching – Inclusion.* Available online at: http://schools.becta.org.uk/ index.php?section=tl&catcode=ss_tl_inc_02. Accessed 1.01.08.

BECTA (2007b) *Learning in the 21st century: The Case for Harnessing Technology*, Coventry: British Educational and Communications Technology Agency.

BECTA (2007c) *Harnessing Technology: Review 2007: Progress and Impact of Technology in Education*, Coventry: British Educational and Communications Technology Agency.

Bee, H. and Boyd, D. (2006) *The Developing Child*, 11th edn, London: Allyn and Bacon.

Benmansour, N. (1998) 'Job satisfaction, stress and coping strategies among Moroccan high school teachers', *Mediterranean Journal of Educational Studies*, 3, 13–33.

BERA (British Educational Research Association) (2004) *Revised Ethical Guidelines.* Available online at: http://www.bera.ac.uk/publications/pdfs/ETHICA1.PDF?PHPSESSID=4d00f1e23ff04c0d93c6 3a3e37969593. Accessed 16.10.08.

Beresford, J. (2003) 'Minding the gap', *Managing Schools Today*, 12 (6), 38–44.

Berkowitz, M., Gibbs, J and Broughton, J. (1980) 'The relation of moral judgment stage disparity to developmental effects of peer dialogues', *Merrill-Palmer Quarterly*, 26, 341–57.

Bernstein, B. (1977) *Class, Codes and Control, Volume 3: Towards a Theory of Educational Trans- missions*, 2nd edn, London: Routledge and Kegan Paul.

Berry, J. (2007). *Teachers' Legal Rights and Responsibilities: A Guide for Trainee Teachers and Those New to the Profession.* Hatfield: University of Hertfordshire Press.

Berryman, M., Glynn, T. and Wearmouth, J. (2006) HYPERLINK "http://www.routledge.com/shopping _cart/products/product_detail.asp?sku=&isbn=0415354021&parent_id=&pc=/shopping_cart/search /search.asp?"\t"_blank". London: Routledge.

BHF (British Heart Foundation) (2007) National Centre. *Physical Activity Patterns: Young People.* Available online at: http://www.bhfactive.org.uk physical activity patterns. Accessed 18.07.08.

Biggs, J. B. (1978) 'Individual and group differences in study processes', *British Journal of Educational Psychology*, 48, 266–279.

Biggs, J. B. (1987) *Student Approaches to Learning and Studying*, Hawthorne, Victoria: Australian Council for Educational Research.

Biggs, J. B. (1993) 'What do inventories of students' learning processes really measure? A theoretical review and clarification', *British Journal of Educational Psychology*, 63, 3–19.

Biggs, P. J. and Moore, P. (1993) *The Process of Learning*, London: Prentice Hall International.

Black, P. and Harrison, C. (2001) 'Feedback in questioning and marking: the science teacher's role in formative assessment', *School Science Review*, 82 (June), 43–49.

Black, P. (1998) *Testing*: *Friend or Foe? Theory and Practice of Assessment and Testing* (Master Classes in Education), London: RoutledgeFalmer

Black, P. and Wiliam, D. (1998) *Inside the Black Box: Raising Standards through Classroom Assessment*, London: King's College.

Black, P. and Wiliam, D. (2002) *Working Inside the Black Box: Assessment for Learning in the Classroom,* London: King's College.

Black, P., Harrison, C., Lee, C., Marshall, B. and Wiliam, D. (2003) *Assessment for Learning: Putting It into Practice*, Maidenhead: Open University Press.

Black, P., McCormick, R., James, M. and Pedder, D. (2006) 'Learning How to Learn and Assessment for Learning: a theoretical inquiry', HYPERLINK "http://www.ingentaconnect.com/content/routledg/ rred" \o "Research Papers in Education", 21 (2) 119–132.

Blackwell, L. S., Trzesniewski, K. S., and Dweck, C. S. (2007) 'Implicit theories of intelligence predict achievement across an adolescent transition: a longitudinal study and an intervention', *Child Development*, 78 (1) 246–263.

Blakemore, S. J. and Frith, U. (2005) *The Learning Brain: Lessons for Education,* Oxford: Blackwell Publishers.

Bleach, K. (2000) *The Newly Qualified Secondary Teacher's Handbook: Meeting the Standards in Secondary and Middle Schools*, London: David Fulton.

Bloom, A. (2007) 'Me level 4, you level 2 = end of friendship', *Times Educational Supplement*, 9 February: 13.

Bloom, B. S. (ed.) (1956) *Taxonomy of Educational Objectives*; *Handbook 1: Cognitive Domain*, New York: Longmans Green.

Boaler, J. (1997) *Experiencing School Mathematics: Teaching Style, Sex and Setting*, Buckingham: Open University Press.

Boston, K. (2007) (QCA). Quoted in BBC news broadcast June 17 'GCSE home coursework scrapped'. Available online at: http://news.bbc.co.uk/1/hi/education/6747973.stm. Accessed 30.06.08

Bottery, M. (1990) *The Morality of the School*, London: Cassell.

Bourdieu, P. (1974) 'The school as a conservative force: scholastic and cultural inequalities' in J. Egglestone (ed.) *Contemporary Research in the Sociology of Education*, London: Methuen and Co. Ltd.

Brandreth, G. (1981) *The Puzzle Mountain*, Harmondsworth: Penguin Books.

Bransford, J. D., Brown, A. and Cocking, R. C. (eds) (1999) *How People Learn: Brain, Mind, Experience, and School,* Washington, DC: National Academy Press.

Briggs, A. (1983) *A Social History of England*, London: Book Club Associates.

Brighouse, H. (2006) *On Education*, Abingdon: Routledge.

Bristol Guide (2008 edition) *Professional Responsibilities and Statutory Frameworks for Teachers and Others in Schools*, School of Education, University of Bristol. Available online at http://www. bristol.ac.uk/education/enterprise/docsum/guide. Accessed 09.01.09.

British Medical Association (BMA) (2003) *Adolescent Health*, London: BMA. Available online at: http://www.bma.org.uk (go to 'Health and Well-being', then search title). Accessed 18.07.08.

British Nutrition Foundation (BNF) (2003) *Establishing a Whole School Policy.* Available online at: http://www.nutrition.org.uk/upload/wholeschoolfoodpolicy.pdf. Accessed 18.07.08.

British Stammering Association *A Chance to Speak: Teachers Information Pack* (Video and leaflets). Available online at: http://www.stammering.org/teachers.html. Accessed 09.08.08.

Brody, L. (ed.) (2004) *Grouping and Acceleration Practices in Gifted Education*, Thousand Oaks, CA: Corwin Press and National Association for Gifted Children.

Bronfenbrenner, U. (1979) *The Ecology of Human Development*, Cambridge, MA: Harvard University Press.

Brooks, J. and Lucas, N. (2004) 'The school sixth form and the growth of vocational qualifications', in S. Capel, R. Heilbronn, M. Leask and T. Turner (eds) *Starting to Teach in the Secondary School: A Companion for the NQT*, 2nd edn, London: RoutledgeFalmer.

Brooks, V. (2002) *Assessment in Secondary Schools: The New Teacher's Guide to Monitoring, Assessment, Recording, Reporting and Accountability*, Buckingham: Open University Press.

Brown, A. L. (1994) 'The advancement of learning', *Educational Researcher*, 23, 4–12.

Brown, M. and Ralph, S. (2002) 'Teacher stress and school improvement', *Improving Schools*, 5 (2), 55–65.

Bruner, J. (1966) *Towards a Theory of Instruction*, New York: W.W. Norton.

Bruner, J. (1983) *Child's Talk: Learning to Use Language*, Oxford: Oxford University Press.

Bruner, J. (2006a) *In Search of Pedagogy Volume 1: The Selected Works of Jerome S. Bruner*, Abingdon, Oxon: Routledge.

Bruner, J. (2006b) *In Search of Pedagogy Volume 2: The Selected Works of Jerome S. Bruner*, Abingdon, Oxon: Routledge.

Bryce, T. G. K. and Humes, W. M. (2003) *Scottish Education*, 2nd edn, Post-Devolution, Edinburgh: Edinburgh University Press.

Bubb, S. and Earley, P. (2004) *Managing Teacher Workload: Work–Life Balance and Wellbeing*, London: Sage.

Bull, S. and Solity, J. (1987) *Classroom Management: Principles to Practice*, London: Croom Helm.

Bullock Report (1975) *A Language for Life*, London: HMSO.

Burchfield, R. (ed.) (1996) *The New Fowler's Modern English Usage*, Oxford: Clarendon Press.

Burgess, T. (2004) 'Language in the classroom and curriculum', in S. Capel, R. Heilbronn, M. Leask and T. Turner (eds) *Starting to Teach in the Secondary School: A Companion for the Newly Qualified Teacher*, 2nd edn, London: RoutledgeFalmer.

Burnett, P. (2002) 'Teacher praise and feedback and students' perception of the classroom environment', *Educational Psychology*, 22 (1), 5–16.

Burnham, S. and Brown, G. (2004) 'Assessment without level descriptions', *Teaching History*, 115, 5–15.

Burton, D. (2007) 'Psychopedagogy and personalised learning', *Journal of Education for Teaching*, 33 (1), 5–17.

Burton, D. and Bartlett S. J. (2006) 'Shaping pedagogy from popular psychological ideas', in D. Kassem, E. Mufti and J. Robinson (eds) *Education Studies Reader*, Maidenhead: Oxford University Press.

Burton, N., Brundrett, M. and Jones, M. (2008) *Doing your Education Research Project,* London: Sage.

Buzan, T. (2003) *Use Your Memory*, London: BBC Books.

Cains, R. A. and Brown, C. R. (1998) 'Newly qualified teachers: a comparative analysis of the perceptions held by Bed and PGCE-trained primary teachers of the level and frequency of stress experienced during the first year of teaching', *Educational Psychology*, 18 (1), 97–110.

Calderhead, J. and Shorrock, S. B. (1997) *Understanding Teacher Education: Case Studies in the Professional Development of Beginning Teachers*. London: Falmer Press.

Capel, S. (1994) 'Help – it's teaching practice again!' Paper presented at the 10th Commonwealth and International Scientific Congress, Victoria, BC, Canada.

Capel, S. (1996) 'Changing focus of concerns for physical education students on school experience', *Pedagogy in Practice*, 2 (2), 5–20.

Capel, S. (1997) 'Changes in students' anxieties after their first and second teaching practices', *Educational Research*, 39 (2), 211–228.

Capel, S. (1998) 'A longitudinal study of the stages of development or concern of secondary PE students', *European Journal of Physical Education*, 3 (2), 185–199.

Capel, S., Heilbronn, R., Leask, M. and Turner, T. (2004) *Starting to Teach in the Secondary School: A Companion for the Newly Qualified Teacher*, 2nd edn, London: RoutledgeFalmer.

Carr, D. (1993) 'Questions of competence', *British Journal of Educational Studies*, 41, 253–271.

Carr, W. and Kemmis, S. (1986) *Becoming Critical. Education, Knowledge and Action Research*, London: Falmer Press.

Carrington, B. and Tomlin, R. (2000) 'Towards a more inclusive profession: teacher recruitment and ethnicity', *European Journal of Teacher Education*, 23 (2), 139–157.

CEM (Curriculum, Evaluation and Management Centre) (2008) University of Durham. Available online at: http://www.cemcentre.org. Accessed 31.07.08.

Chalmers, G. (ed.) (2001) *Reflections on Motivation*, London: Centre for Information on Language Teaching and Research (CILT).

Cherniss, C. (2000) *Emotional Intelligence; What It Is and Why It Matters*. Available online at: http://businessballs.com/emotionalintelligenceexplanation.pdf. Accessed 24/09/07.

Child, D. (2004) *Psychology and the Teacher*, 7th edn, London: Continuum.

Child, D. (2007) *Psychology and the Teacher*, 8th edn, London: Continuum.

Clarke, S. (2001) *Unlocking Formative Assessment*, London: King's College.

Clarke, S. (2005) *Formative Assessment in Action*, London: Hodder Murray.

Claxton, G. (2002) *Building Learning Power: Helping Young People Become Better Learners,* Bristol: TLO.

Clayton, S., Burton, D. M., Campbell, A., O'Brien, M. and Qualter, A. (2008) 'How are the perceptions of Learning Networks amongst school professionals shaped at an early stage in their introduction? What does this mean for their implementation?', *International Review of Education* 54, 211–242.

Clough, J. (2001) Quoted in Duffy, M. 'How many strings to your bow?', *Times Educational Supplement*, 19 October. Available online at http://www.tes.co.uk.

Clough, P., Garner, P., Pardeck, T. and Yuen, F. (eds) (2004) *The Handbook of Emotional and Behavioural Difficulties*, London: Sage.

Cockburn, A. D. (1996) 'Primary teachers' knowledge and acquisition of stress relieving strategies', *British Journal of Educational Psychology*, 66, 399–410.

Cole, M. (ed.) (1999) *Professional Issues for Teachers and Student Teachers*, London: David Fulton.

Collins, J., Hammond, M. and Wellington, J. (1997) *Teaching and Learning with Multimedia*, London: Routledge.

Connolly (2006) 'The effects of social class and ethnicity on gender differences in GCSE attainment: a secondary analysis of the Youth Cohort Study of England and Wales 1997–2001', *British Educational Research Journal*, 32 (1), 3–21.

Conway, P. F. and Clark, C. M. (2003) 'The journey inward and outward: a re-examination of Fuller's concerns-based model of teacher development', *Teaching and Teacher Education*, 466–482.

Cooper, P. and Ideus, K. (1996) *Attention-Deficit/Hyperactivity Disorder – A Practical Guide for Teachers*, London: David Fulton Publications.

Covington, M. V. (2000) 'Goal theory, motivation and school achievement: an integrative review', *Annual Review of Psychology*, 51, 171–200.

Cox, M. J. (1999) 'Motivating pupils through the use of ICT', in M. Leask and N. Pachler (eds) *Learning to Teach Using ICT in the Secondary School*, London: Routledge.

Crace, J. (2007) 'Is this the end of SATs?' *Education Guardian*, 6 December: 1.

Crain, W (1985). *Theories of Development,* 2nd edn, Englewood Cliffs, NJ: Prentice-Hall. Available online at: http://faculty.plts.edu/gpence/html/kohlberg.htm. Accessed 28.11.07.

Creme, P. and Lea, M. (2006) *Writing at University*, 2nd edn, Buckingham: Open University Press.

Croner (updated annually) *The Headteacher's Guide to the Law*, New Malden: Croner Publications.

Crook, C. (1994) *Computers and the Collaborative Experience of Learning*, London: Routledge.

Crook, D., Power, S. and Whitty, G. (1999) *The Grammar School Question: A Review of Research on Comprehensive and Selective Education*, London: Institute of Education, University of London, 'Perspectives on Education'.

CSIE (Centre for Studies on Inclusion in Education) (2000*) Index for Inclusion: Developing Learning and Participation in Schools*, Bristol: CSIE with CEN (University of Manchester) and CER (Christ Church University College, Canterbury (T. Booth, ed.).

Cunningham, M., Kerr, K., McEune, R., Smith, P. and Harris, S. (2003) *Laptops for Teachers: An Evaluation of the First Year of the Initiative*, London: Department for Education and Skills/British Educational Communications and Technology Agency.

Davies, F. and Greene, T. (1984) *Reading for Learning in Science*, Edinburgh: Oliver and Boyd.

Davies, N. (2000) *The School Report: Why Britain's Schools Are Failing*, London: Vintage Books.

Davis, B. and Sumara, D. J. (1997) 'Cognition, complexity and teacher education', *Harvard Educational Review*, 67, 105–121.

Davison, J. and Dowson, J. (eds) (2003) *Learning to Teach English in the Secondary School: A Companion to School Experience*, 2nd edn., London: Routledge.

Day. C. (1999) *Developing Teachers: The Challenges of Lifelong Learning*. London: Falmer Press.

DCSF (Department for Children, Schools and Families) (2007a) (Jan) *GCSE and Equivalent Examination Results in England 2005/06 (Revised)*. Available online at: http://www.dfes.gov.uk/rsgateway/DB/SFR/s000702/index.shtml. Accessed 05.08.08.

DCSF (2007b) (Feb) *National Curriculum Assessments, GCSE and Equivalent Attainment and Post-16 Attainment by Pupil Characteristics, in England* 2005/06. Available online at: http://www.dfes.gov.uk/rsgateway/DB/SFR/s000708/index.shtml. Accessed 05.08.08.

DCSF (2008a) *First Statistical Release. Special Educational Needs in England, January, 2008*, London: DCSF.

DCSF (2008b) *Key Stage 3 National Strategy. Advice on Whole School Behaviour and Attendance Policy*, London: DCSF.

DCSF (2008c) *Bullying, Don't Suffer in Silence*. London: DCSF.

DCSF (2008d) *Primary National Strategy: Social and Emotional Aspects of Learning,* London: DCSF.

DCSF (2008e) *Career Development in EBD*. London: DCSF. Available online at: http://www.teachernet.gov.uk/wholeschool/behaviour/npsl_ba/. Accessed 26 06.08.

DCSF (2008f) *Guidance on Preventing Underachievement – A Focus on Exceptionally Able Pupils*, London: DCFS. Available online at: http://www.standards.dcsf.gov.uk/secondary/keystage3/all/respub/gt_prevent_udrachieve0006608. Accessed 01.03.08.

DCSF (2008g) *Pedagogy and Practice: Teaching and Learning in Secondary Schools*. Available online at: http://www.standards.dfes.gov.uk/secondary/keystage3/all/respub/sec_pptl0. Accessed 28.02.08.

DCSF (2008h) *Youth Cohort Study: Longitudinal Study of Young People in England: The Activities and Experiences of 16 Year Olds: England 2007*, London: DCSF. Available online at: http://www.dcsf.gov.uk/rsgateway/DB/SBU/b000795/index.shtml. Accessed 04.08.08.

DCSF (2008i) *Special Educational Needs: Inclusion Development Programme,* London: DCSF. Available online at: http://www.standards.dcsf.gov.uk/primary/features/inclusion/sen/idp. Accessed 09.08.08.

DCSF (2008j). *The Reporting and Analysis for Improvement through School Self-Evaluation (RAISE).* A dataset. Available online at: http://www.raiseonline.org.uk. Accessed 31.07.08.

DCSF (2008k) *The Diploma Website: Engineering.* Available online at: http://yp.direct.gov.uk/diplomas/diploma_details/what_can_i_learn/engineering/. Accessed 30.06.08.

DCSF (2008l) *The National Strategies, Secondary, Key Stage 3.* Available online at: http://www.standards.dfes.gov.uk/secondary/keystage3/. Accessed 01.07.08.

DCSF (2008m) *National Curriculum Assessments at Key Stage 3 in England.* Available online at: http://www.dfes.gov.uk/rsgateway/DB/SFR/s000776/index.shtml.Accessed 29 06.08.

DCSF (2008n) *Framework for Teaching English: Years 7 to 11*, London: DCSF.

DCSF (2008o) *Framework for Teaching Mathematics: Years 7 to 11*, London: DCSF.

DCSF (2008p) *Framework for Teaching Science: Years 7 to 11*, London: DCSF.

DCSF (2008q) *Framework for Teaching ICT: Years 7 to 11*, London: DCSF.

DCSF (updated annually) *School Teachers' Pay and Conditions*, London: HMSO. Available online at: http://www.teachernet.gov.uk/paysite/. Accessed 15.09.08

DCSF *The Research Informed Practice Site.* Available online at: http://www.standards.dfes.gov.uk/research/. Accessed 16.09.08.

de Bono, E. (1972) *Children Solve Problems*, London: Penguin Education.

Dearing, R. (1994) *The National Curriculum and Its Assessment. Final Report*, London: HMSO.

Dearing, R. (1996) *Review of Qualifications for 16–19 Year Olds (Full Report),* London: School Curriculum and Assessment Authority.

DES (Department of Education and Science) (1978) *Special Educational Needs*, London: HMSO.

DES (1985) *Education For All: The Final Report of the Committee of Inquiry into the Education of Children from Ethnic Minority Groups, Cmnd. 9469*, London: HMSO (The Swann Report).

DES (1989) *Discipline in Schools (The Elton Report)*, London: DES.

DES/WO (Department of Education and Science and the Welsh Office) (1988) *National Curriculum Task Group on Assessment and Testing* (the TGAT Report), London: DES/WO.

Desforges, C. (2002) *On Teaching and Learning*, Cranfield: NCSL (National College for School Leadership).

Dewey, J. (1916) *Democracy and Education*, New York: Free Press.

Dewey, J. (1933) *How We Think*, Boston, MA: Houghton Mifflin.

DfE (Department for Education) (1994a) *The Education of Children with Emotional and Behavioural Difficulties* (Circular 9/94). London: DfE.

DfE (1994b) *Code of Practice on the Identification and Assessment of Pupils with Special Educational Needs*, London: DfE.

DfEE (Department for Education and Employment) (1995) *Disability Discrimination Act (1995): A Consultation on the Employment Code of Practice Guidance on the Definition of Disability and Related Regulations*, London: DfEE.

DfEE (1997) *Guaranteeing Standards: A Consultation Paper on the Structure of Awarding Bodies,* London: DfEE.

DfEE (1998) *The National Literacy Strategy*, London: HMSO.

DfEE (1999a) *Social Inclusion: Pupil Support* (Circular 10/99), London: DfEE.

DfEE (1999b) *Youth Cohort Study: The Activities and Experiences of 16 Year Olds; England and Wales 1998*, London: Stationery Office. Issue 4/99, March Statistical Bulletin. Available online at: http://www.dfes.gov.uk/rsgateway/DB/SFR/s000230/contents.shtml. Accessed 06.08.08.

DfEE (2001a) *Framework for Teaching English: Years 7, 8 and 9,* London: DfEE.

DfEE (2001b) *Framework for Teaching Mathematics: Years 7, 8 and 9,* London: DfEE.

DfEE/QCA (Department for Education and Employment/Qualifications and Curriculum Authority) (1999a) *The National Curriculum for England. Handbook for Secondary Teachers: Key Stages 3 and 4,* London: The Stationery Office. Available online at: http://curriculum.qca.org.uk/key-stages-3-and-4/index.aspx. Accessed 10.08.08.

DfEE/QCA (1999b) *The National Curriculum: Handbook for Secondary Teachers in England*, London: DfEE/QCA.

DfEE/QCA (2001) *Guidelines for Teaching Pupils with Learning Difficulties.* Available online at: http://www.qca.org.uk/qca_1830.aspx. Accessed 16.10.08.

DfES (Department for Education and Skills) (2001a) *Special Educational Needs. Code of Practice*, London: DfES. Available online at: http://www.standards.dfes.gov.uk/eyfs/resources/downloads/sencodeofpractice.pdf. Accessed 09.08.08.

DfES (2001b) *Inclusive Schooling*, London: DfES.

DfES (2001c) *Education and Skills: Delivering Results – A Strategy to 2006*, Sudbury, Suffolk: DfES.

DfES (2001d) *The Key Stage 3 Strategy.* Available online at: http://www.standards.dfes.gov.uk/secondary/keystage3/. Accessed 02.07.08.

DfES (2002a) *Supporting Pupils Learning English as an Additional Language,* London: DfES.

DfES (2002b) *Releasing Potential: Raising Attainment. Managing Data in Secondary Schools*, London: DfES. Available online at: http://www.standards.dfes.gov.uk/ts/docs/RPRA.pdf. Accessed 31.07.08.

DfES (2002c) *Framework for Teaching Science: Years 7, 8 and 9*, London: DfES.

DfES (2002d) *Framework for Teaching ICT Capability: Years 7, 8 and 9*, London: DfES.

DfES (2002e) *Training Materials for the Foundation Stage*, London: DfES Publications.

DfES (2003a). *Every Child Matters.* Green Paper, Cm. 5860. London: The Stationery Office. Available online at: http://www.everychildmatters.gov.uk/_files/EBE7EEAC90382663E0D5BBF24C99A7 AC.pdf. Accessed 16.10.08.

DfES (2003b) *Framework for Teaching Modern Foreign Languages: Years 7, 8 and 9*, London: DfES.

DfES (2003c) *Key Stage 3 Strategy: Key Messages about Assessment and Learning*, London: DfES.

DfES (2003d) *Teaching and Learning in Secondary Schools: Pilot. Unit 4: Questioning*, London: Crown. (Ref: DfES 0344/2003).

DfES (2003e) *Primary National Strategy*, London: DfES

DfES (2003f) *Key Stage 3 Strategy: Key Messages about Assessment and Learning*, London: DfES.

DfES (2003g) *Aiming High: Raising the Achievement of Minority Ethnic Pupils.* Available online at: http://www.standards.dfes.gov.uk/ethnicminorities/links_and_publications/AHConsultationdocmar 03/. Accessed 22.09.08.

DfES (2004a) *Pedagogy and Practice: Teaching and Learning in the Secondary School, Unit 18 Improving the Climate for Learning*, London: DfES.

DfES (2004b) *Pedagogy and Practice: Teaching and Learning in the Secondary School, Unit 6 Modelling*, London: DfES.

DfES (2004c) *Pedagogy and Practice: Teaching and Learning in the Secondary School. Unit 7 Questioning*, London: DfES. Available online at: http://www. standards.dfes.gov.uk/keystage3/respub/sec_ppt10. Accessed 12.09.08.

DfES (2004d) *The 14-19 Gateway. Information on the 14-19 curriculum reforms*, London: DfES. Available online at: http://www.dfes.gov.uk/14-19/. Accessed 04.03.08.

DfES (2004e) *Pedagogy and Practice: Teaching and Learning in the Secondary School. Unit 10 Group work*, London: DfES. Available online at: http://www.standards.dfes.gov.uk/secondary/keystage3/downloads/sec_pptl043304u10groupwork.pdf.Accessed 06.03.08.

DfES (2004f) *Every Child Matters; Change for Children*, London DfES. Available online at: www.everychildmatters.gov.uk/aims/. Accessed 08.08.08.

DfES (2004g) *Removing Barriers to Achievement: The Government's SEN Strategy*, Nottingham: DfES Publications.

DfES (2004h) *Pedagogy and Practice: Teaching and Learning in the Secondary School, Unit 8 Explaining*, London: DfES.

DfES (2004i) *Pedagogy and Practice: Teaching and Learning in Secondary Schools: Unit 19 Learning Styles*, London: DfES.

DfES (2004j) *Pedagogy and Practice: Teaching and Learning in Secondary Schools Unit 5: Starters and Plenaries.* London: DfES. Available online at:.http://www.standards.dfes.gov.uk/secondary/keystage3/downloads/sec_pptl042804u5lstartplen_a.pdf. Accessed 01.01.08.

DfES (2004k) *Framework for Teaching Design and Technology: Years 7, 8 and 9,* London: DfES.

DfES (2004l) *Five Year Strategy for Children and Learners*, CM. 6272, London: DfES. Available online at http://publications/teachernet.gov.uk. Accessed 09.03.09.

DfES (2005a) *Implementing the Disability Discrimination Act in Schools and Early Years Settings.* Available online at: http://www.teachernet.gov.uk/wholeschool/sen/disabilityandthedda. Accessed 28.05.08.

DfES (2005b) *Learning Platforms: Making IT Personal – Secondary Version*, London: DfES.

DfES (2005c) *14–19 Education and Skills (White Paper)*, London: HMSO.

DfES (2005d) *National Curriculum Schemes of Work.* Available online at: http://www.standards.dfes.gov.uk/schemes3/. Accessed 01.09.08.

DfES (2006a) *National Strategies: Grouping Pupils for Success.* Available online at: http://www.standards.dfes.gov.uk/secondary/keystage3/all/respub/ns_grp_pup_succ. Accessed 01.03.08.

DfES (2006b) *Highlights: Personalised Learning*, London: DfES.

DfES (2006c) *Making Good Progress; How Can We Help Every Pupil to Make Good Progress at School?*, London: DfES.

DfES (2007). *Professional Standards for Teachers: Qualified Teacher Status.* London: Department for Education and Skills.

DfT (Department for Transport) (2006) *National Travel Survey 2005*, London: Transport Statistics.

Dickens, C. *Hard Times*, London: Penguin.

Dickenson, P (1999) *Whole Class Interactive Teaching.* Available online at: http://www.partnership.mmu.ac.uk/cme/Student_Writings/Masters/PaulDickinson.html. Accessed 17.10.08.

Dillon, J. and Maguire, M. (eds) (2001) *Becoming a Teacher: Issues in Secondary Teaching,* 3rd edn, Buckingham: Open University Press.

DirectGov (2008) *Aimhigher: Helping You into Higher Education.* Available online at: http://www.direct.gov.uk/en/EducationAndLearning/UniversityAndHigherEducation/DG_073697. Accessed 04.08.08.

Dix, P. (2007) *Taking Care of Behaviour: Practical Skills for Teachers*, Harlow: Pearson Education.

DoH (Department of Health) (2007) Department of Health Food in Schools. Pilot project 'Dining room environment'. Available online at: http://www.dh.gov.uk/en/policyandguidance/healthandsocial caretopics/foodinschoolsprogramme. Accessed 18.07.08.

DoH and DfES (Department of Health and Department for Education and Skills) (2004) *Food in Schools.* Available online at: www.dh.gov.uk/en/Publichealth/Healthimprovement. Accessed 09.03.09.

Doise, W. (1990) 'The development of individual competencies through social interaction', in H. Foot, M. Morgan and R. Shute (eds) *Children Helping Children*, Chichester: Wiley.

Donaldson, M. (1978) *Children's Minds*, Glasgow: Collins/Fontana (also London: Croom Helm).

Donaldson, M. (1992) *Human Minds: An Exploration*, London: Allen Lane.

Driver, R. (1983) 'The fallacy of induction in science teaching', in R. Levinson (ed.) (1994) *Teaching Science*, London: Routledge.

Driver, R. and Bell, J. (1986) 'Students' thinking and learning of science: a constructivist view', *School Science Review*, 67, 240: 443–456.

Dryden, G. and Vos, J. (2001) *The Learning Revolution: To Change the Way the World Learns,* Network Educational Press in association with Learning Web.

Dunham, J. and Varma, V. (eds) (1998) *Stress in Teachers: Past, Present and Future*, London: Whurr.

Dunne, J. (2003) 'Arguing for teaching as a practice; a reply to Alasdair MacIntyre', *Journal of Philosophy of Education*, 37 (2), 351–371.

Dussault, M., Deaudelin, C., Royer, N. and Loiselle, J. (1997). 'Professional isolation and stress in teachers'. Paper presented at the Annual Meeting of the American Education Research Association, Chicago, March.

Dweck, C. (1986) 'Motivational processes affecting learning', *American Psychologist*, 41, 1040–1048.

Dweck, C. (2007) 'Psychology professor discusses "growth" versus "fixed" mindsets', *Stanford Report,* February 7. Podcast. Available online at: http://news-service.stanford.edu/news/2007/february7/videos/179_flash.html. Accessed 12.09.08.

Dweck, C. (2008) 'Transforming students' motivation to learn', *Brainology* (Winter), Journal of the Independent Schools Association.

Dweck, C. S. and Leggett, E. (1988) 'A social-cognitive approach to motivation and personality', *Psychological Review*, 95, 256–273.

EBS (Educational Broadcasting Services) (2004) *Looking at Learning: Tactics of Questioning* (CD Rom), London: Teacher Training Agency.

Elliott, J. (1991) *Action Research for Educational Change*, Milton Keynes: Open University Press.

Elliott, J. (1998) *The Curriculum Experiment: Meeting the Challenge of Social Change*, Buckingham: Open University Press.

Entwistle, N. J. (1981) *Styles of Learning and Teaching*, Chichester: Wiley.

Entwistle, N. J. (1990) *Handbook of Educational Ideas and Practices*, London: Routledge.

Entwistle, N. J. (1993) *Styles of Learning and Teaching*, 3rd edn, London: David Fulton.

EOC (Equal Opportunities Commission) (2001) *Women and Men in Britain: Sex Stereotyping – From School to Work,* London: EOC. Available online at: http://83.137.212.42/sitearchive/eoc/PDF/wm_sex_stereotyping.pdf?page=15824. Accessed 06.08.08.

EOC (2006). *Facts about Women and Men in Great Britain.* Available online at: http://www.equalityhumanrights.com/Documents/Gender/Research/facts_about_GB_2006.pdf. Accessed 09.08.09.

Erikson, E. H. (1980) *Identity and the Life Cycle*, New York: Norton.

Etzioni, A. (ed.) (1969) *The Semi-Professions and Their Organisation*, New York: Free Press.

Evans, C. (2004) 'Exploring the relationship between cognitive style and teaching style', *Educational Psychology*, 24 (4), 509–530.

Evans, J., Harden, A., Thomas, J. and Benefield, P. (2003) *Support for Pupils with Emotional and Behavioural Difficulties (EBD) in Mainstream Primary Classrooms: A Systematic Review of the Effectiveness of Interventions*, London: EPPI.

Fautley, M. and Savage, J. (2008) *Assessment for Learning and Teaching in Secondary Schools*, Exeter: Learning Matters.

Fernandez-Balboa, J. M., Barrett, K., Solomon, M. and Silverman, S. (1996) 'Perspectives on content knowledge in physical education', *Journal of Physical Education, Recreation and Dance*, 67, 54–57.

Ferrigan, C. (2005) *Passing the ICT Skills Test*, Exeter: Learning Matters.

FFT (Fischer Family Trust) (2008). Available online at: http://www.fischertrust.org/. Accessed 31.07.08.

Fielding, M. (1996) 'Why and how learning styles matter: valuing difference in teachers and learners', in S. Hart (ed.) *Differentiation and the Secondary Curriculum: Debates and Dilemmas*, London: Routledge.

Fitzgerald, R., Finch, S. and Nove, A. (2000) *Black Caribbean Young Men's Experiences of Education and Employment, Report No. RR186*, London: DfEE.

Flavell, J. H. (1982) 'Structures, stages, and sequences in cognitive development', in W.A. Collins (ed.) *The Concept of Development: The Minnesota Symposia on Child Psychology*, 15: 1–28.

Fletcher-Campbell, F. (2004) 'Moderate learning difficulties', in A. Lewis and B. Norwich (eds) *Special Teaching for Special Children? Pedagogies for Inclusion*, Maidenhead: Open University Press.

Fontana, D. (1993a) *Managing Time*, Leicester: British Psychological Society Books.

Fontana, D. (1993b) *Psychology for Teachers*, London: Macmillan.

Foresight (2007) *'Tackling Obesities: Future Choices* – Project report, Government Office for Science. Online. Available online at: http://www.foresight.gov.uk. Accessed 24.11.07.

Freeman, P. (2002) *Teaching and Learning Styles for KS3*, Milton Keynes: Chalkface Project.

Freud, S. (1901) 'The psychopathology of everyday life', republished 1953, in J. Strachey (ed.) *The Standard Edition of the Complete Psychological Works of Sigmund Freud Vol. 6*, London: Hogarth.

Frost, J. and Turner, T. (2004) *Learning to Teach Science in the Secondary School: A Companion to School Experience*, 2nd edn, London: RoutledgeFalmer.

FSA (Food Standards Agency) (2000) *National Diet and Nutrition Survey: Young People Aged 4–18 Years,* London: The Stationery Office.

FSA (2004) *School Lunch Box Survey 2004*. Available online at: http://www.food.gov.uk. Accessed 18.07.08.

FSA (2007) *Vending Healthy Drinks*. Available online at: http://www.food.gov.uk. Accessed 18.07.08.

Fuller, F. F. and Brown, O. H. (1975) 'Becoming a teacher', in K. Ryan (ed.) *Teacher Education: The Seventy-Fourth Yearbook of the National Society of the Study of Education*, Chicago: University of Chicago Press.

Furlong, J and Maynard, T. (1995). *Mentoring Student Teachers: The Growth of Professional Knowledge*, London: Routledge.

Gagné, R.M. (1977) *The Conditions of Learning*, New York: Holt International.

Gallagher, S. A. and Stephen, W. J. (1996) 'Content acquisition in problem-based learning: depth versus breadth in American studies', *Journal for the Education of the Gifted*, 19: 257–275.

REFERENCES ▨ ▨ ▪ ▪

Galton, M., Gray, J. and Ruddock, J. (1999) *The Impact of School Transitions and Transfers on Pupil Progress and Attainment*, London, Department for Education and Employment.

Gardner, H. (1983) *Frames of Mind: The Theory of Multiple Intelligences*, New York: Basic Books.

Gardner, H. (1986) Quoted in *The New York Times Educational Supplement*, 19 November.

Gardner, H. (1991) *The Unschooled Mind*, London: Harper Collins.

Gardner, H. (1993a) *Frames of Mind: The Theory of Multiple Intelligences*, 2nd edn, London: Fontana.

Gardner, H. (1993b) *Multiple Intelligences: The Theory in Practice*, New York: Basic Books.

Gardner, H. (1998) 'Assessment in context' in P. Murphy (ed.) *Learners, Learning and Assessment*, Buckingham: Open University Press: 90–117.

Gardner, H. (1999) *Intelligence Reframed: Multiple Intelligences for the 21st Century*, New York: Basic Books.

Gardner, H. (2006) *The Development and Education of the Mind: The Selected Works of Howard Gardner*, Abingdon, Oxon: Routledge.

Gardner, H., Kornhaber, M. and Wake, W. (1996*) Intelligence: Multiple Perspectives*, Fort Worth, TX: Harcourt Brace.

Gardner, M. (1965) *The Annotated Alice*, Harmondsworth: Penguin.

Gavin, T. (2003) 'The Structure of the Secondary Curriculum', in T. G. K. Bryce and W. M. Humes (eds) *Scottish Education*, 2nd edn, Post-Devolution, Edinburgh: Edinburgh University Press.

Gessell, A. (1925) *The Mental Growth of the Preschool Child*, New York: Macmillan.

Giallo, R. and Little, E. (2003) 'Classroom behaviour problems: the relationship between preparedness, classroom experiences, and self-efficacy in graduate and student teachers', *Australian Journal of Educational and Developmental Psychology*, 3 (2), 21–34.

Gilbert, I. (2002) *Essential Motivation in the Classroom*, London: RoutledgeFalmer.

Gilham, B. (ed.) (1986) *The Language of School Subjects*, London: Heinemann.

Gillard, D. (2005) Rescuing Teacher Professionalism. *Forum* Vol. 47, 175-180. Available online at: http://www.wwwords.co.uk/rss/abstract.asp?j=forum&aid=2566. Accessed 17.09.08.

Gillborn, D. and Gipps, C. (1996) *Recent Research on the Achievements of Ethnic Minority Pupils: Ofsted Reviews of Research*, London: HMSO.

Gillborn, D. and Mirza, H. S. (2000) *Educational Inequality: Mapping Race, Class and Gender; A Synthesis of Research Evidence*, London: Ofsted. Available online at: http://www.ofsted. gov.uk. Accessed 01.11.07.

Gingell, J. and Brandon, E. P. (2000) *In Defence of High Culture*, Oxford: Blackwell.

Gipps, C. (1997) 'Principles of assessment', unpublished lecture, Institute of Education, University of London, 7 Feb.

Gipps, C. and Stobart, G. (1997) *Assessment: A Teacher's Guide to the Issues*, 3rd edn, London: Hodder and Stoughton.

Goddard, D. and Leask, M. (1992) *The Search for Quality: Planning for Improvement and Managing Change*, London: Paul Chapman Publishing.

Gold, K. (2003) 'Poverty *is* an excuse', *Times Educational Supplement* (07.03.03).

Gold, Y. and Roth, R. A. (1993) *Teachers Managing Stress and Preventing Burnout*, London: Falmer Press.

Goldberger, M. (2002) 'Mobility ability', in M. Mosston and S. Ashworth (eds) *Teaching Physical Education*, 5th edn, San Francisco: Benjamin Cummings.

Goleman, D. (1995) *Emotional Intelligence*, New York: Bantam.

Goleman, D. (1996) *Emotional Intelligence: Why It Can Matter More Than IQ*, London: Bloomsbury.

Good, T. and Brophy, J. (2000) *Looking in Classrooms*, 8th edn, New York: Addison-Wesley Longman.

Gordon, P. and Lawton, D. (2003) *Dictionary of British Education*, London: Woburn.

Goswami, U. (2006) 'Neuroscience and education: from research to practice,' *Nature Reviews Neuroscience*, 7 (5), 406–411.

Gould, S. J. (1981) *The Mismeasure of Man*, New York: Norton and Co.

Goulding, M. 'Pupils learning mathematics', in S. Johnston-Wilder, P. Johnston-Wilder, D. Pimm and J. Westwell (2005) (eds) *Learning to Teach Mathematics in the Secondary School: A Companion to School Experience*, 2nd edn, Abingdon: Routledge.

Graber, K. (1995) 'The influence of teacher education programs on the beliefs of student teachers: general pedagogical knowledge, pedagogical content knowledge and teacher education course work', *Journal of Teaching in Physical Education*, 14, 157–178.

Graham, C. and Hill, M. (2003) *Spotlight 89, Negotiating the Transition to Secondary School*, Glasgow: Scottish Council for Research in Education.

Greenbaum, S. and Whituit, J. (1988) *Longman Guide to English Usage*, Harlow: Longman.

Greenfield, S. A. (2008) 'Let your body do the thinking', *Times Educational Supplement*, 15 February.

Greenhalgh, P. (1994) *Emotional Growth and Learning*, London: Routledge.

Griffin, L., Dodds, P. and Rovegno, I. (1996) 'Pedagogical content knowledge for teachers: integrate everything you know to help students learn', *Journal of Physical Education, Recreation and Dance*, 67, 58–61.

Gross, R. (2001) *Psychology: The Science of Mind and Behaviour*, 4th edn, London: Hodder and Stoughton.

Grossman, P. L. (1990) *The Making of a Teacher: Teacher Knowledge and Teacher Education*, New York: Teachers College Press.

Grossman, P. L., Wilson, S. M. and Shulman, L. S. (1989) 'Teachers of substance: subject matter knowledge for teaching', in M. C. Reynolds (ed.) *Knowledge Base for the Beginning Teacher*, Pergamon Press: Oxford.

GTCE (General Teaching Council for England) (2002) *Code of Professional Values and Practice for Teachers*, London: GTC (England). Available online at: http://www.gtce.org.uk/gtcinfo/code.asp. Accessed 02.07.04.

GTCE (2003) *Teacher Survey*, London: General Teaching Council.

GTCE (2006) *The Statement of Professional Values and Practice for Teachers*. Available online at: http://www.gtce.org.uk/shared/contentlibs/92511/92572/statement_of_values.pdf. Acccssed 12.07.08.

Hall, D. (1996) *Assessing the Needs of Bilingual Pupils*, London: David Fulton Publishers.

Hallam, S. (2002) *Ability Grouping Schools: A Literature Review*, London: Institute of Education, University of London. In the series: 'Perspectives on Education'.

Hammond, M. (2002) 'Why teach? A case study investigating the decision to train to teach ICT', *Journal of Education for Teaching*, 28, 2, 135–148

Hansen, D. (1995) 'Teaching and the moral life of classrooms', *Journal for a Just and Caring Education,* 2. Available online at: http://tigger.uic.edu/~lnucci/MoralEd/articles/hansen.html. Accessed 17.09.08.

Hargreaves, A. (1984) *Improving Secondary Schools: Report of the Committee on the Curriculum and Organisation of Secondary Schools,* London: ILEA.

Harland, J., Moor, H., Kinder K. and Ashworth, M. (2002) *Is the Curriculum Working? The Key Stage 3 Phase of the Northern Ireland Curriculum Cohort Study*, Slough: National Foundation for Educational Research.

Harlen, W. (1995) 'To the rescue of formative assessment', *Primary Science Review*, 37, April: 14–15.

Harlen, W. (1997) 'Making sense of the research on ability grouping', *Newsletter 60* (Spring), Scottish Council for Research in Education (SCRE).

Harrison, C., Comber, C., Fisher, T., Haw, K., Lewin, C., Lunzer, E., McFarlane, A., Mavers, D., Scrimshaw, P., Somekh, B. and Watling, R. (2003) *The Impact of Information and Communication Technologies on Pupil Learning and Attainment: Full Report*, London: DfES/BECTA.

Hart, K. (1981) *Children's Understanding of Mathematics*, London: Murray.

Harter, S. (1985) 'Competence as a dimension of self-evaluation: toward a comprehensive model of self-worth', in R. L. Leay (ed.) *The Development of the Self,* Orlando, FL: Academic Press.

Hatcher, R. (1988) 'Class differentiation in education: rational choices?', *British Journal of Sociology of Education*, 19 (1), 5–24.

Hattie, J. (2003) 'Building teacher quality'. Paper given at the Australian Council for Educational Research Annual Conference.

Hay McBer (2000) *Research into Teacher Effectiveness*, London: DfEE. Available online at: http://www.teachernet.gov.uk/docbank/index.cfm?id=1487. Accessed 09.01.09.

Haydn, T. (2006) *Managing Pupil Behaviour*, London: Routledge.

Haydn, T., Arthur, J. and Hunt, M. (2001) *Learning to Teach History in the Secondary School: A Companion to School Experience*, 2nd edn, London: Routledge/Falmer.

Haydon, G. (2006a) *Values in Education,* London: Continuum.

Haydon, G. (2006b) *Education, Philosophy and the Ethical Environment*, London: Faber and Faber.

REFERENCES ▪ ▪ ▪ ■

Haydon, G. and Hayward, J. (2004) 'Values and citizenship education' in S. Capel *et al.* (eds). *Starting To Teach In The Secondary School: A Companion to the Newly Qualified Teacher*, 2nd edn, London: RoutledgeFalmer.

Hayes, D. (ed.) (2004) *The RoutledgeFalmer Guide to Key Debates in Education*, London: RoutledgeFalmer.

HCR (House of Commons Report) Health Committee (2004) *Obesity – Third Report of Session 2003–04*, London: The Stationery Office.

Head, J., Hill, F. and Maguire, M. (1996) 'Stress and the post graduate secondary school trainee teacher: a British case study', *Journal of Education for Teaching*, 22 (1), 71–84.

Heilbronn, R., Jones, C., Bubb, S. and Totterdell, M. (2002) 'School-based induction tutors: a challenging role', *School Leadership and Management*, 22(4), 371–389.

Hirst, P. (1974) *Knowledge and the Curriculum*, London: Routledge.

HMI (Her Majesty's Inspectorate) (1977) *Curriculum 11-16*, London: Department for Education.

HMI (Her Majesty's Inspectorate of Schools) (2004) *The Key Stage 3 Strategy: Evaluation of the Third Year*, London: Ofsted.

HMI (2005) *The Key Stage 3 Strategy: Evaluation of the Fourth Year*, London: Ofsted.

HMI (2006) *The Secondary National Strategy: Evaluation of the Fifth Year*, London: Ofsted.

Holmes, E. (2009). *The Newly Qualified Teacher's Handbook*, 2nd edn, London: RoutledgeFalmer.

Holt, J. (1964) *How Children Fail*, London: Pitman.

Holt, J. (1984) *How Children Learn,* 2nd edn., London: Penguin.

Hook, P. and Vass, A. (2000) *Confident Classroom Leadership*, London: David Fulton Publishers

Hopkins, D. (2002) *A Teacher's Guide to Classroom Research*, 3rd edn, Buckingham: Open University Press.

Howard-Jones, P. (2008) *Fostering Creative Thinking: Co-constructed Insights from Neuroscience and Education*, The Higher Education Academy – ESCalate. Available online at: http://escalate.ac.uk/4389. Accessed 12.09.08.

Howe, M. (1998) *Principles of Human Abilities and Learning*, Hove: Psychology Press.

Hoyle, E. and John, P. (1995) *Professional Knowledge and Professional Practice*, London: Cassell.

Hughes, H. and Vass, A. (2001) *Strategies for Closing the Learning Gap*, Stafford: Network Educational Press.

Huitt, W. and Hummel, J. (2003). 'Piaget's theory of cognitive development', *Educational Psychology Interactive*. Valdosta, GA: Valdosta State University. Available online at: http://chiron.valdosta.edu/whuitt/col/cogsys/piaget.html. Accessed 28.22.07.

Husbands, C. (2007) 'Using assessment data to support pupil achievement', in V. Brooks, I. Abbott and L. Bills (eds) *Preparing to Teach in Secondary Schools*, Maidenhead: Open University Press.

Hyslop, F. (2007). *Debate on making Scotland smarter*. Available online at: http://www.scotland.gov.uk/News/This-Week/Speeches/classsizes. Accessed 02.01.09.

ICAN (I Can 'Helps children communicate'). Available online at: http://www.ican.org.uk. Accessed 09.08.08. (I Can' is a charitable organisation for speech and language difficulties).

IFOM (Institute for the Future of the Mind), Oxford University. Available online at: http://www.futuremind.ox.ac.uk/impact/. Accessed 16.08.2008.

Intercom Trust (2008) *Joint Action Against Homophobic Bullying Project* (JAAHB). Available online at: http://www.intercomtrust.org.uk/goodschools/index.htm. Accessed 05.08.08.

Ireson, J. and Hallam, S. (2001) *Ability Grouping in Education,* London: Paul Chapman Publishing.

Ireson, J., Hallam, S. and Hurley, C. (2005) 'What are the effects of ability grouping on GCSE attainment?' *British Educational Research Journal*, 31, 443–458.

Ireson, J., Hallam, S., Hack, S., Clark, H. and Plewis, I. (2002) 'Ability grouping in English secondary schools: effects on attainment in English, mathematics and science', *Educational Research and Evaluation*, 8 (3), 299–318.

Jackson, C. and Warin, J. (2000) 'The importance of gender as an aspect of identity at key transition points in compulsory education', *British Educational Research Journal*, 26 (3), 375–391.

James, A. and Pollard, A. (eds) (2004) *Personalised Learning*, London: TLRP/ESRC.

Jensen, E. (1998) *Super Teaching*, 3rd edn, San Diego, CA: The Brain Store Inc.

Joyce, B. and Weil, M. (1986) *Models of Teaching*, 3rd edn, New Jersey: Prentice Hall.

Jozefkowicz, E. (2006) 'Too many teachers "teaching to the test"', HYPERLINK "http://www.educationguardian.co.uk/" (Thursday July 20).

Kagan, S. (1994) *Cooperative Learning*, San Clemente, CA: Kagan.

Kemmis, S. and McTaggert, R. (1988) *The Action Research Planner*, 3rd edn, Geelong: Deakin University Press.

Kerry, T. (2002) *Learning Objectives, Task-setting and Differentiation*, London: Nelson-Thornes.

Kerry, T. (2004) *Explaining and Questioning*, Cheltenham: Nelson Thornes.

Keys, W., Harris, S. and Fernandes, C. (1996) *Third International Mathematics and Science Study – First National Report – Part 1. Achievement in Mathematics and Science at Age 13 in England*, London: National Foundation for Educational Research.

Kirton, A., Hallam, S., Peffers, J., Robertson, P. and Stobart, G. (2007) 'Revolution, evolution or a Trojan horse? Piloting assessment for learning in some Scottish primary schools', *British Educational Research Journal*, 33 (4), 605–627.

Kluger, A. and DeNisi, A. (1996) 'The historical effects of feedback interventions on performance: a historical review', *Psychological Bulletin*, 119 (2), 254–284.

Kohlberg, L. (1976) 'Moral stages and moralization: the cognitive-developmental approach', in T. Lickona (ed.) *Moral Development and Behaviour: Theory, Research, and Social Issues*, New York: Holt, Rinehart and Winston.

Kohlberg, L. (1985). 'The just community: approach to moral education in theory and practice', in M. Berkowitz and F. Oser (eds) *Moral Education: Theory and Application*, Hillsdale, NJ: Lawrence Erlbaum Associates Inc.

Kolb, D. A. (1976) *The Learning Style Inventory: Technical Manual*, Boston: McBer and Co.

Kolb, D. A. (1985) *The Learning Style Inventory: Technical Manual*, revised edn, Boston: McBer and Co.

Koshy, V. (2005*) Action Research for Improving Practice – A Practical Guide*, London: Paul Chapman.

Kozulin, A. (1998) *Psychological Tools: A Sociocultural Approach to Education,* Cambridge, MA: Harvard University Press.

Kramarski, B. and Mevarech, Z. R. (1997) 'Cognitive-metacognitive training within a problem-solving based Logo environment', *British Journal of Educational Psychology*, 67, 425–445.

Kutnick, P., Hodgkinson, S., Sebba, J., Humphreys, S., Galton, M., Steward S., Blatchford, P. and Baines, E. (2006) 'Pupil grouping strategies at Key Stage 2 and 3: case studies of 24 schools in England', *DFES Research report RR796*, London: DfES.

Kutnick, P., Sebba, J., Blatchford, P., Galton, M. and Thorp, J. (2005) 'The effect of pupil grouping: a literature review', *DfES Research Report RR 688*, London: DfES (copyright the University of Brighton).

Kyriacou, C. (1998) *Essential Teaching Skills*, 2nd edn, Cheltenham: Stanley Thornes.

Kyriacou, C. (2000) *Stress-Busting for Teachers*, Cheltenham: Stanley Thornes.

Kyriacou, C. (2001) 'Teacher stress: directions for future research', *Educational Review*, 53 (1), 27–35.

Kyriacou, C. (2007) *Essential Teaching Skills*, 3rd edn, Cheltenham: Nelson Thornes.

Kyriacou, C. and Stephens, P. (1999) 'Student teachers' concerns during teaching practice', *Evaluation and Research in Education*, 13 (1), 18–31.

Kyriacou, C., Kunc, R., Stephens, P. and Hultgren, A. (2003) 'Student teachers' expectations of teaching as a career in England and Norway', *Educational Review*, 55 (3), 255–263.

Lam, S-f., Yim, P-s., Law, J. S. F. and Cheung, R. W. Y. (2004) 'The effects of competition on achievement motivation in Chinese classrooms', *British Journal of Educational Psychology*, 74, 281–296.

Lambert, D. and Balderstone, D. (2000) *Learning to Teach Geography in the Secondary School: A Companion to School Experience*, London: Routledge.

Langford, P. (1995) *Approaches to the Development of Moral Reasoning*, Hove: Erlbaum.

Lave, J. and Wenger, E. (1991) *Situated Learning: Legitimate Peripheral Participation*, Cambridge, UK: Cambridge University Press.

Lawton, D. (1989) *Education, Culture and the National Curriculum*, Sevenoaks: Hodder and Stoughton.

Lawton, D. (1996) *Beyond the National Curriculum: Teacher Professionalism and Empowerment*, Sevenoaks: Hodder and Stoughton.

Leadbetter, C. (2004) *Personalisation Through Participation: A New Script for Public Services,* London: DEMOS/DfES Innovations Unit.

Learning and Teaching Scotland (2004a) *About 5–14*. Available online at: http://www.ltscotland.org.uk/5to14/about5to14/index.asp. Accessed 02.01.09.

Learning and Teaching Scotland (2004b) *Assessment Is for Learning*. Available online at: http://www.ltscotland.org.uk/assess/. Accessed 02.01.09.

Learning and Teaching Scotland (2007a) *Parents as Partners in Learning*. Available online at: http://www.ltscotland.org.uk/parentsaspartnersinlearning/parentcouncils/index.asp. Accessed 02.01.09.

Learning and Teaching Scotland (2007b) *Introduction to Curriculum for Excellence*. Available online at: http://www.ltscotland.org.uk/curriculumforexcellence/. Accessed 02.01.09.

Leask, M. (ed.) (2001) *Issues in Teaching with ICT*, London: Routledge.

Leask, M. and Moorehouse, C. (2005) 'The student teacher's role and responsibilities', in S. Capel, M. Leask and T. Turner (eds) *Learning to Teach in the Secondary School*, 4th edn, Abingdon: Routledge: 18–31.

Leask, M. and Pachler, N. (2005) *Learning to Teach using ICT in the Secondary School*, 2nd edn, London: Routledge. See, for example, the chapter on the use of ICT for pupils with SEN.

Leask, M. and Williams, L. (2005) 'Whole school approaches: integrating ICT across the curriculum', in M. Leask and N. Pachler (eds) *Learning to Teach using ICT in the Secondary School*, 2nd edn, London: Routledge.

Leask, M. and Younie, S. (2001) 'Communal constructivist theory: information and communications technology pedagogy and internationalisation of the curriculum', *Journal of Information Technology for Teacher Education*, 10 (1 and 2), 117–134.

Leat, D. (1998) *Thinking through Geography*, Cambridge: Chris Kington Publishing.

Levinson, R. (2005a) 'Planning for progression in science', in J. Frost and T. Turner (eds) *Learning to Teach Science in the Secondary School: A Companion to School Experience*, 2nd edn, Abingdon: RoutledgeFalmer.

Levinson, R. (2005b) 'Science for citizenship', in J. Frost and T. Turner (eds) *Learning to Teach Science in the Secondary School: A Companion to School Experience*, 2nd edn, Abingdon: RoutledgeFalmer.

Levitin, D. J. (2006) *This Is Your Brain on Music: The Science of a Human Obsession*, New York: Dutton Books, Penguin USA.

Lewis, A. and Norwich, B. (2004) *Special Teaching for Special Children? Pedagogies for Inclusion*, Maidenhead: Open University Press

Lewis, M. (ed.) (2007) *The Bristol Guide: Professional Responsibilities and Statutory Frameworks for Teachers and Others in Schools*, Bristol: University of Bristol Press.

Lickona, T. (1983) *Raising Good Children*, New York: Bantam Books.

Liverpool Sportslinx Project (2003) *Report on the Health and Fitness of Liverpool Primary and Secondary School Children 01–03*, Liverpool: Liverpool City Council, Education, Libraries and Sports Services.

Lu, Mei-Lu (1998) *English-Only Movement: Its Consequences for the Education of Language Minority Children*, Bloomington, IN: University of Indiana. An ERIC Clearinghouse digest No 139 (EDO-CS-98-12 Nov 1998). Available online at: http://www.indiana.edu/~reading/ieo/digests/d139.html. Accessed 05.09.08.

Macdonald, B. (2000) 'How education became nobody's business', in H. Altricher and J. Elliott (eds) *Images of Educational Change*, Buckingham: Open University Press.

Mackenzie, J. (2003) 'Disaffection with schooling', in T. G. K. Bryce and W. M. Humes (eds) *Scottish Education*, 2nd edn, Post-Devolution, Edinburgh: Edinburgh University Press.

Mager, R. (1990) *Preparing Instructional Objectives – A Critical Tool in the Development of Effective Instruction*, London: Kogan Page.

Mager, R. (2005) *Preparing Instructional Objectives*, 3rd edn, Atlanta, GA: Center for Effective Performance.

Maitland, I. (1995) *Managing your Time*, London: Institute of Personnel and Development.

Maker, C. J. and Nielson, A. B. (1995) *Teaching/Learning Models in Education of the Gifted*, 2nd edn, Austin, TX: Pro-Ed.

Manouchehri, A. (2004) 'Implementing mathematics reform in urban schools: a study of the effect of teachers' motivational style', *Urban Education*, 39 (5), 472–508.

Marland, M. (1993) *The Craft of the Classroom*, London: Croom Helm.

Marsden, E. (2009) 'Observing in the classroom', in S. Younie, S., Capel and M. Leask (eds) *Supporting Teaching and Learning in the Secondary School*, London: Routledge.

Marshall, B. (2005) 'Testing, testing, testing', in T. Wragg (ed.) *The Future of Education: Letters to the Prime Minister*, The New Vision Group.

Maslow, A. H. (1970) *Motivation and Personality*, 2nd edn, New York: Harper and Row.

Maybin, J., Mercer, N. and Stierer, B. (1992) 'Scaffolding learning in the classroom', in K. Norman (ed.) *Thinking Voices: The Work of the National Oracy Project*, London: Hodder and Stoughton.

Mayer, J. D. and Salovey, P. (1997). 'What is emotional intelligence?', in P. Salovey and D. Sluyter (eds) *Emotional Development and Emotional Intelligence: Implications for Educators*, New York: Basic Books.

McCarthy, B. (1987) *The 4MAT System*, Barrington, IL.: Excel.

McClelland, D. C. (1961) *The Achieving Society*, Princeton, NJ: Van Norstrand.

McDiarmid, G. W., Ball, D. W. and Anderson, C. W. (1989) 'Why staying one chapter ahead doesn't really work: subject specific pedagogy', in M. C. Reynolds (ed.) *Knowledge Base for the Beginning Teacher*, Oxford: Pergamon Press: 193–205.

McGregor, D. (1960) *The Human Side of Enterprise*, New York: McGraw-Hill.

McLaughlin, T. (2004) 'Philosophy, values and schooling: principles and predicaments of teacher example', in W. Aiken and J. Haldanne (eds) *Philosophy and Its Public Role: Essays in Ethics, Politics, Society and Culture,* Exeter, UK and Charlottesville, VA: Imprint Academic.

McLean, A. (2007) 'Motivating every learner', *Raising Achievement Update*, July, London: Optimus.

McNiff, J. and Whitehead, J. (2002) *Action Research: Principles and Practice*, London: Routledge.

Mehrabin, A. (1972) *Non-verbal Communication*, New York: Aldine Atherton.

Mercer, N. (2000) *Words and Minds: How We Use Language to Think Together*, London: Routledge.

Miles, M. and Huberman, M. (1994) *Qualitative Data Analysis: A Sourcebook of New Methods*, Thousand Oaks, CA: Sage.

Millar, R., Leach, J., Osborne, J. and Ratcliffe, M. (2006) *Improving Subject Teaching*, London: RoutledgeFalmer.

Miller, D. and Lavin, F. (2007) 'But now I feel I want to give it a try': formative assessment, self-esteem and a sense of competence', *Curriculum Journal*, 18 (1).

Miller, O. (1996). *Supporting Children with Visual Impairment in Mainstream Schools*, London: Franklin Watts.

Mills, S. (1995) *Stress Management for the Individual Teacher*, Lancaster: Framework Press.

Mitchell, R. (1994) 'The communicative approach to language teaching: an introduction', in A. Swarbrick (ed.) *Teaching Modern Languages*, London: Routledge/Open University.

Montgomery, D. (1996) 'Differentiation of the curriculum in primary education', *Flying High*, 3, 14–28 (Worcester: The National Association for Able Children in Education).

Moore, A. (2004) *The Good Teacher: Dominant Discourses in Teaching and Teacher Education*, London: RoutledgeFalmer.

Morton, L. L., Vesco, R., Williams, N. H. and Awender, M. A. (1997) 'Student teacher anxieties related to class management, pedagogy, evaluation, and staff relations', *British Journal of Educational Psychology*, 67, 68–89.

Mosston, M. (1972) *Teaching: From Command to Discovery*, Belmont, CA: Wadsworth Publishing Co.

Mosston, M. and Ashworth, S. (2002) *Teaching Physical Education*, 5th edn, San Francisco: Benjamin Cummings.

Muijs, D. and Reynolds, D. (2005) *Effective Teaching: Evidence and Practice*, 2nd edn, London: Paul Chapman (Sage).

Murdock, T. B. (1999) 'The social context of risk: social and motivational predictors of alienation in middle school', *Journal of Educational Psychology*, 91, 1–14.

Murphy, P. (ed.) (1999) *Learners, Learning and Assessment,* Buckingham: Open University Press.

Murray-Harvey, R., Slee, P., Lawson, M., Silins, H., Banfield, G. and Russell, A. (2000) 'Under stress: the concerns and coping strategies of teacher education students', *European Journal of Teacher Education*, 23 (1), 19–35.

Myhill, D. (2006) 'Talk, talk, talk: teaching and learning in whole class discourse', HYPERLINK "http://www.ingentaconnect.com/content/routledg/rred" \o "Research Papers in Education", 21 (1), 19–41.

Namrouti, A. and Alshannag, Q. (2004) 'Effect of using a metacognitive teaching strategy on seventh grade students' achievement in science', *Dirasat*, 31 (1), 1–13.

NASEN (National Association for Special Educational Needs, UK) (2000) *Specialist Teaching for Special Educational Needs and Inclusion: Policy Paper 4 in the SEN Fourth Policy Options series*, London: NASEN. Available online at: http://www.nasen.org.uk. Accessed 17.09.08.

National Assembly for Wales (2006) *Induction for NQTs in Wales: Circular No: 24/05*. Available online at: http://www.nasuwt.org.uk/Shared_ASP_Files/UploadedFiles/07BAA1D6-B33A-44AD-A8C9-43929219B366_Induction_Wal2006.pdf. Accessed 16.10.08.

National Heart Forum (2007) *Lightening the Load: Tackling Overweight and Obesity*, London: Department of Health Publications.

National Research Council, USA (2000) *How People Learn: Brain, Mind, Experience, and School*, Washington, DC: National Academy Press.

Naylor, S. and Keogh, B. (1999) 'Constructivism in the classroom: theory into practice', *Journal of Science Teacher Education*, 10 (2), 93–106.

NCC (National Curriculum Council) (1989) *An Introduction to the National Curriculum*, York: National Curriculum Council.

Nelson, I. (1995) *Time Management for Teachers*, London: Routledge.

Newton, D. P. and Newton, L. P. (2001) 'Subject content knowledge and teacher talk in the primary science classroom', *European Journal of Teacher Education*, 24 (3), 369–379.

NFER (National Foundation for Educational Research) (2004) *The Key Stage3 National Strategy: LEA and School Perceptions*, LGA Research Report 11/04: ISBN 1 903880 75 0.

Nicholls, J. G. (1984) 'Achievement motivation: conceptions of ability, subjective experience, task choice and performance', *Psychological Review*, 91, 328–346.

Nicholls, J. G. (1989) *The Competitive Ethos and Democratic Education*, Cambridge, MA: Harvard University Press.

Noble, T. (2004) 'Integrating the revised Bloom's Taxonomy with multiple intelligences: a planning tool for curriculum differentiation', *Teachers College Record*, 106 (1), 193–211.

Nolte, Dorothy Law, (1972) *Children Learn what They Live* (poem). Available online at: http://www.empowermentresources.com/info2/childrenlearn-long_version.html. Accessed 16.10.08.

Norman, K. (ed.) (1992) *Thinking Voices: The Work of the National Oracy Project*, London: Hodder and Stoughton.

Noyes, A. (2003) 'Moving schools and social relocation', *International Studies in Sociology of Education*, 13 (3), 261–280.

Nucci, L. (1987) 'Synthesis of research on moral development'; *Educational Leadership* February 1987. Available online at: http://tigger.uic.edu/~lnucci/MoralEd/articles/nuccisynthesis.html. Accessed 28.11.07.

Nucci, L. (2007) *Moral Development and Moral Education: An Overview/Kohlberg's Theory*. Available online at: http://tigger.uic.edu/~lnucci/MoralEd/overview.html#kohlberg. Accessed 29.11.2007.

NUT (National Union of Teachers) (annual) *Your First Teaching Post*, London: NUT.

O'Brien, M., Burton, D. M., Campbell, A., Qualter, A., Varga-Atkins, T. (2006). 'Learning Networks for schools: keeping up with the times or a leap into the unknown', *Curriculum Journal*, 17 (4), 397–411.

OECD (Organisation for Economic Cooperation and Development) (2001) *Knowledge and Skills for Life: First Results from the OECD Programme for International Student Assessment (PISA) 2000*, Paris: Organisation for Economic Co-operation and Development. Available online at: http://www.pisa.oecd.org/document/4/0,3343,en_32252351_32236159_33668932_1_1_1_1,00.html. Accessed 01.01.08.

OECD (2007) *PISA 2006 Science Competencies for Tomorrow's World*, Paris: Organisation for Economic Co-operation and Development. Available online at: http://www.pisa.oecd.org/pages/0,2987,en_32252351_32235731_1_1_1_1_1,00.html. Accessed 01.01.08.

OFCOM (2006) *Television Advertising of Food and Drink Products to Children*, London: LSE. Available online at: http://ofcom.org.uk/research/. Accessed 18.07.08.

OFOM (Institute for the future of the mind) (undated) Transcript of Scientific Research in Learning and Education All-Party Parliamentary Group seminar, '*Brain-science in the classroom*'. Available online at: http://www.futuremind.ox.ac.uk/impact/parliamentary.html. Accessed 12.09.08.

Ofsted (Office for Standards in Education) (1993) *The New Teacher in School: A Survey by Her Majesty's Inspectorate in England and Wales*, 1992, London: HMSO.

Ofsted (1996) *The Annual Report of Her Majesty's Chief Inspector of Schools*, London, HMSO.

Ofsted (2000a) *Improving Schools: The Framework*, London: Ofsted.

Ofsted (2000b) *Progress in Key Stage 3 Science*, London: Ofsted.

Ofsted (2002) *Good Teaching; Effective Departments*, London: Ofsted.

Ofsted (2003a) *Boys' Achievement in Secondary Schools,* London: Ofsted.

Ofsted (2003b) *Good Assessment Practice in History*, HMI 1475, London: Ofsted.

Ofsted (2004a) *ICT in Schools: The Impact of Government Initiatives Five Years On*, London: Office for Standards in Education.

Ofsted (2004b) *Ofsted Subject Reports 2002/03: Information and Communication Technology in Secondary Schools*, London: Office for Standards in Education.

Ofsted (2004c, March) *Promoting and Evaluating Pupils' Spiritual, Moral, Social and Cultural Development.* Available online at: http://www.ofsted.gov.uk/assets/3598.pdf. Accessed 17.02 08.

Ofsted (2006a) *Evaluating Mathematics Provision for 14-19 Year Olds*, London: Ofsted.

Ofsted (2006b) *Inclusion: Does It Matter Where Pupils Are Taught?* An Ofsted report on the provision and outcomes in different settings for pupils with learning difficulties and disabilities. Available online at: http://sen.ttrb.ac.uk/viewarticle2.aspx?contentId=12629. Accessed 09.08.08.

Ofsted (2007a) *Food in Schools: Reference Number 070016,* London: Ofsted.

Ofsted (2007b) *The Annual Report of Her Majesty's Chief Inspector of Education, Children's Services and Skills 2006/07*, London: The Stationery Office. Available online at: http://www.ofsted.gov.uk/. Accessed 20.11.07.

Ogborn, J., Kress, G., Martins, I. and McGillicuddy, K. (1996) *Explaining Science in the Classroom*, Buckingham: Open University Press.

Olweus, D. (1993) *Bullying at School*, Oxford: Blackwell.

Owen-Jackson, G. (ed.) (2008) *Learning to Teach Design and Technology in the Secondary School: A Companion to School Experience*, 2nd edn, Abingdon: Routledge.

Pachler, N. and Field, K. (2002) *Learning to Teach Modern Foreign Languages*, Abingdon: RoutledgeFalmer

Paechter, C. (2000) *Changing School Subjects: Power, Gender and Curriculum*, Buckingham: Open University Press.

Pateman, T. (1994) 'Crisis, what identity crisis?', *First Appointments Supplement, Times Educational Supplement*, 14 January, 28–29.

Paterson, L. (2003) *Scottish Education in the Twentieth Century*, Edinburgh: Edinburgh University Press.

Peacey, N. (2005) 'Special Educational Needs and ICT', in M. Leask and N. Pachler (eds) *Learning to Teach using ICT in the Secondary School*, 2nd edn, London: Routledge.

Pearce, S. (2005) *You Wouldn't Understand: White Teachers in the Multiethnic Classroom*, Stoke on Trent: Trentham Books

Perrenoud, P. (1991) 'Towards a pragmatic approach to formative evaluation', in P. Weston (ed.) *Assessment of Pupils' Achievement, Motivation and School Success*, Amsterdam: Swets and Zeitlinger: 79–101.

Phan, H. and Deo, B. (2007) 'The revised learning process questionnaire: a validation of a Western model of pupils' study approaches to the South Pacific context using confirmatory factor analysis', *British Journal of Educational Psychology*, 78 (1), 719–739.

Piaget, J. (1932) *The Moral Judgment of the Child*, London: Routledge and Kegan Paul

Piaget, J. (1954) *The Construction of Reality in the Child*, New York: Basic Books.

Pickering. S and Howard-Jones. P. (2007) 'Educator's views on the role of neuroscience in education; findings from a study of UK and international perspectives', *Mind, Brain, and Education*, 1 (3), 109–113.

Pinney, A. (2004) *Reducing Reliance on Statements: An Investigation Into Local Authority Practice and Outcomes*, DfES Research Report 508, Norwich: HMSO. Available online at: http://www.dcsf.gov.uk/research/data/uploadfiles/RB508.pdf. Accessed 10.08.08.

PISA (Programme for International Student Assessment) (2000) *Programme for International Student 2000-Scotland Analysis*. Available online at: http://www.scotland.gov.uk/stats/bulletins/00343.pdf. Accessed 02.01.09.

Plato (1955) *The Republic*, tr. H.D.P. Lee, Harmondsworth: Penguin Books.

Pollard, A. (2002) *Reflective Teaching,* London: Continuum.

Postlethwaite, K. (1993) *Differentiated Science Teaching: Responding to Individual Differences and Special Educational Needs*, Milton Keynes: Open University Press.

Powell, E (2005) 'Conceptualising and facilitating active learning: teachers' video-stimulated reflective dialogues', *Reflective Practice*, 6(3), 407–418

Power, S., Edwards, T., Whitty, G. and Wigfall, V. (2003) *Education and Middle Class,* Buckingham: Open University Press.

Price Waterhouse Coopers (2001) *A Study into Teacher Workload*, London: Price Waterhouse Coopers.

Pring, R. (2004) *A Comprehensive Curriculum for Comprehensive Schools: The Fourth Caroline Benn Memorial Lecture*, London: Socialist Educational Association.

Priyadharshini, E. and Robinson-Pant, A. (2003) 'The attractions of teaching: an investigation into why people change careers to teach', *Journal of Education for Teaching*, 29 (2), 95–112.

Prosser, M. and Trigwell, K. (1999) *Understanding Teaching and Learning: The Experience in Higher Education*, The Society for Research into Higher Education, Buckingham: Open University Press.

QCA (Qualifications and Curriculum Authority) (2000) *Research into the Dip in Performance in English of Children Entering Key Stage 3 (Secondary School for Most). Report Commissioned by the QCA and Carried out by the University of Cambridge Local Examinations Syndicate Evaluation Team,* London: QCA.

QCA (2001a) *Planning, Teaching and Assessing Pupils with Learning Difficulties*, Sudbury: QCA Publications. Available online at: http://www.qca.org.uk/qca_11583.aspx. Accessed 28.05.08.

QCA (2001b) *Supporting School Improvement: Emotional and Behavioural Development*, London: QCA.

QCA (2004c) *Key Stage 3 Assessment and Reporting Arrangements*, London: QCA.

QCA (2004a) *Progression in Geography*. Available online at: http://www.qca.org.uk/geography/innovating/geography_matters/continuityandprogression/www.qca.org.uk/geography/innovating/geography_matters/continuityand progression/ (search 'progression'). Accessed 04.12.08.

QCA (2004b) *National Curriculum in Action*. Available online at: http://curriculum.qca.org.uk/index.aspx (search 'NC in Action'). Accessed 18.07.08.

QCA (2005) *Pupil Perceptions of History at Key Stage 3*, London:: QCA. Available online at: http://www.qca.org.uk/qca_6391.aspx. Accessed 18.07.08.

QCA (2006a) *National Qualifications Framework*. Available online at: http://www.qca.org.uk/libraryAssets/media/qca-06-2298-nqf-web.pdf. Accessed 01.02.08.

QCA (2006b) *Diplomas – Spearheading a New Era of Reform*. Available online at: http://www.qca.org.uk/qca_10325.aspx. Accessed 30.06.08.

QCA (2007a) *National Curriculum for England*. Available online at: http://www.nc.uk.net/index.html. Accessed 18.07.08.

QCA (2007b) *Statement of Values: National Curriculum for England* (updated May 2008). Available online at: http://curriculum.qca.org.uk/uploads/Statement-of-values_tcm8-12166.pdf. Accessed 22.07.08.

QCA (2007c) *National Curriculum for England: Inclusion*. Available online at: http://curriculum.qca.org.uk/ (search 'Inclusion statement'). Accessed 08.08.08.

QCA (2007d) *Revised GCE AS/A levels 2008*. Available online at: http://www.qca.org.uk/qca_13027.aspx. Accessed 03.07.08.

QCA (2007e) *National Curriciculum for England Statement of Values*. Available online at: http://curriculum.qca.org.uk/uploads/Statement-of-values_tcm8-12166.pdf. Accessed 12 07.08.

QCA (2007f) *National Curriculum for England A Set of Values* (updated May 2008). Available online at: http://curriculum.qca.org.uk/key-stages-3-and-4/aims/index.aspx. Accessed 12.07.08.

QCA (2008a) National Curriculum online: *General Teaching Requirements. Inclusion: Providing Effective Learning Opportunities for All Pupils* Available online at: http://www.nc.uk.net/nc_resources/html/inclusion.shtml. Accessed 29.02.08.

QCA (2008b) *Guidance on Teaching the Gifted and Talented*. Available online at: http://www.qca.org.uk/qca_2346.aspx. Accessed 29.02.08.

QCA (2008c) *National Curriculum in Action: Creativity*. Available online at: http://www.ncaction.org.uk/creativity. Accessed 20.11.07.

QCA (2008d) *National Curriculum for England: General Teaching Requirements: Inclusion*. Available online at: http://www.nc.uk.net/nc_resources/html/inclusion.shtml. Accessed 23.11.07.

QCA *National Curriculum Key Stages 3 and 4*. Available online at: http://curriculum.qca.org.uk/key-stages-3-and-4/index.aspx. Accessed 04.09.08.

QCA (2008f) *National Curriculum for England*. Available online at: http://curriculum.qca.org.uk/. Accessed 04.09.08.

QCA (2008g) *National Curriculum for England: Thinking Skills*. Available online at: http://curriculum.qca.org.uk/key-stages-3-and-4/skills/plts/index.aspx?return=/key-stages-3-and-4/skills/index.aspx. Accessed 04.09.08.

QCA (2008h) *What Are Ground Rules for Discussion*. Available online at: http://www.standards.dfes.gov.uk/schemes2/citizenship/cit01/01q2. Accessed 19.02.08.

QCA (2008i) *National Curriculum – Personal Development*. Available online at: http://curriculum.qca.org.uk/key-stages-3-and-4/personaldevelopment/index.aspx?return=/search/index.aspx%3FfldSiteSearch%3Dnational+curriculum+personal+development%26btnGoSearch.x%3D13%26btnGoSearch.y%3D13 Accessed 09.01.09.

QCA (2008j) *Formative Assessment: Implications for Classroom Practice*. Available online at: http://www.qca.org.uk/qca_4356.aspx. Accessed 31.07.08.

QCA (2008k) *The Aims, Values and Purposes of the National Curriculum (in England)*. Available online at: http://curriculum.qca.org.uk (search 'Key skills'). Accessed 29.07.08.

QCA (2008l) *International Review of Curriculum and Assessment Frameworks Internet Archive; Table 9 National Assessment and Public Examinations*. Available online at: http://www.inca.org.uk/pdf/table_9.pdf. Accessed 1.03.08.

RCP (Royal College of Physicians) (2004) *Storing Up Problems: The Medical Case for a Slimmer Nation. Report of a Working Party*, London: Royal College of Physicians.

Redpath, R. and Harker, M. (1999) 'Becoming Solution-Focused in Practice', *Educational Psychology in Practice*, 15 (2), 116–121.

Reid, I. and Caudwell, J. (1997) 'Why did PGCE students choose teaching as a career?', *Research in Education*, 58, 46–58.

Riding, R. (2002) *School Learning and Cognitive Styles*, London: David Fulton.

Riding, R. J. and Cheema, I. (1991) 'Cognitive styles – an overview and integration', *Educational Psychology*, 11, 193–215.

Riding, R. J. and Rayner, S. (1998) *Learning Styles and Strategies*, London: David Fulton.

Ridley, M. (2003) *Nature via Nurture: Genes, Experience and What Makes Us Human*, London: Fourth Estate. Book reviewed as 'Natural Conclusion' by Stephen Rose in Guardian *Review* (19.04.03), 13.

Ripley, K. Daines, B and Barrett, J (1997*) Dyspraxia: A Guide for Teachers and Parents*, London: David Fulton.

Robertson, J. (1996) *Effective Classroom Control: Understanding Teacher–Student Relationships*, 3rd edn, London: Hodder and Stoughton.

Robinson, K. (Chair) (1999) *All Our Futures: Creativity, Culture and Education. Report of the National Advisory Committee on Creative and Cultural Education*, London: DfEE.

Roffey, S. (2004) *The New Teacher's Survival Guide to Behaviour*, London: Sage.

Rogers, B. (2002) *Classroom Behaviour,* London: Paul Chapman Publishing.

Rogoff, B. (1990) *Apprenticeship in Thinking: Cognitive Development in Social Context*, Oxford: Oxford University Press.

Rose, R. (2004) 'Towards a better understanding of the needs of pupils who have difficulties accessing learning', in S. Capel, R. Heilbronn, M. Leask and T. Turner (eds) *Starting to Teach in the Secondary School: A Companion for the Newly Qualified Teacher*, Abingdon: RoutledgeFalmer.

Rosenthal, R. and Jacobson, L. (1968) *Pygmalion in the Classroom*, New York: Holt, Rinehart and Winston.

Rovegno, I. (1992a) 'Learning a new curricular approach: Mechanism of knowledge acquisition in preservice teachers', *Teaching and Teacher Education*, 8: 253–264.

Rovegno, I. (1992b) 'Learning to teach in a field based methods course. The development of pedagogical content knowledge', *Teaching and Teacher Education*, 8: 69–82.

Rudduck, J. (2004) *Developing a Gender Policy for Secondary Schools*, Buckingham: Open University Press.

Rudduck, J., Brown, N. and Hendy L. (2006) *Personalised Learning and Pupil Voice*, The East Sussex Project: DFES.

Russell, J. (2007) 'The folly of our text fixation is plain to all. Except ministers', *Guardian*, 7 February: 36.

Ryan, A. (2007) 'So what is the bias in the system?, *Times Higher Educational Supplement*, 26 October.

Sadler, P. (1994) *Simple Minds,* BBC 2 television broadcast, 19 September.

Sage, R. (2000) *Class Talk,* Stafford: Network Educational Press.

Sala, S. D. (1999) *Mind Myths: Exploring Popular Assumptions about the Mind and Brain*, Chichester: Wiley.

Salovey, P. and Mayer, J. D. (1990) 'Emotional intelligence', *Imagination, Cognition and Personality*', 9, 185–211.

Sammons, P., Day, C., Kington, A., Gu, Q., Stobart, G. and Smees, R. (2007) 'Exploring variations in teachers' work, lives and their effects on pupils: key findings and implications from a longitudinal mixed-method study', *British Educational Research Journal*, 33 (5), 681–701.

SCAA (School Curriculum and Assessment Authority) (1996) *Teaching English as an Additional Language: A Framework for Policy*, London: SCAA.

Schön, D (1983) *The Reflective Practitioner*, New York: Basic Books.

School Teachers' Review Body (2002) *A Review of Approaches to Reducing Teacher Workload*, London: School Teachers' Review Body.

Schwab, J. J. (1964) 'The structure of the disciplines: meanings and significance', in G. Ford, and L. Purgo (eds) *The Structure of Knowledge and the Curriculum*, Chicago: Rand McNally.

Scottish Consultative Council on the Curriculum (1996) *Teaching for Effective Learning*, Dundee. Available online at: http://www.ltscotland.org.uk/learningaboutlearning/assessment/research/assessmenttfel.asp. Accessed 16.10.08.

Scottish Executive (2004) *Ambitious Excellent Schools: Our Agenda for Action*, Edinburgh: Scottish Executive.

Scottish Office Education and Industry Department (1999) *A Manual for Good Practice in Special Educational Needs*, Edinburgh: Scottish Office Education and Industry Department.

Sector Skills Development Agency (2004) *A Flying Start for New Information Technology Qualification (ITQ)*. Available online at: http://www.ssda.org.uk/Default.aspx?page=250. Accessed 01.01.08.

SEED (Scottish Executive Education Department) (2000) *The Structure and Balance of the Curriculum: 5–14 National Guidelines*. Available online at: http://www.ltscotland.org.uk/5to14/htmlguidelines/saboc/intropage1.htm. Accessed 02.01.09.

SEED (2001a) *A Teaching Profession for the 21st Century*. Agreement reached following recommendations made in the McCrone Report. Available online at: http://www.scotland.gov.uk/Publications/2002/09/15501/11386. Accessed 02.01.09.

SEED (2001b) *Better Behaviour, Better Learning*, Edinburgh: Scottish Executive Education Department.

SEED (2002) *Welcoming Asylum Seekers to Schools*. Available online at: http://www.scotland.gov.uk/News/Releases/2002/01/910. Accessed 02.01.09.

SEED (2003) *National Priorities in School Education*. Available online at: http://www.scotland.gov.uk/Publications/2002/02/10678/File-1. Accessed 02.01.09.

SEED (2004a) *About the Scottish Executive Education Department*. Available online at: http://www.scotland.gov.uk/Topics/Education. Accessed 02.01.09.

SEED (2004b) *The Standard for Headship*. Available online at: http://www.scotland.gov.uk/Publications/2003/11/18435/28427. Accessed 02.01.09.

SEED (2004c) *Action to Tackle Discipline in Schools*. Available online at: www.scotland.gov.uk/News/Releases/2003/09/4212. Accessed 02.01.09.

SEED (2004d) *Programme for International Student 2000-Scotland Analysis*. Available online at: http://www.scotland.gov.uk/stats/bulletins/00343-00.asp. Accessed 02.01.09.

Selman, R. L. (1980) *The Growth of Interpersonal Understanding*, New York: Academic Press.

Shaffer, R.H. (1997) *Making Decisions about Children*, 2nd edn, Oxford: Blackwell Publishers.

Sharples, J.M. *et al.* (2008) 'The science of learning: a pilot-programme of professional development for teachers' *in preparation*.

Shayer, M. (2008) 'Intelligence for education: as described by Piaget and measured by psychometrics', *British Journal of Educational Psychology*, 78 (1), 1–29.

Shayer, M. and Adey, P. (eds) (2002) *Learning Intelligence: Cognitive Acceleration Across the Curriculum from 5 to 15 Years*, Buckingham and Philadelphia: Open University Press.

Short, G. (1986) 'Teacher expectation and West Indian underachievement', *Educational Research*, 27, 2, 95–101.

Shulman, L. (1986) 'Those who understand: knowledge growth in teaching', *Educational Researcher*, 15, 4–14.

Shulman, L. (1987) 'Knowledge and teaching: foundation of a new reform', *Harvard Review*, 57, 1–22.

Skinner, B. F. (1953) *Science and Human Behavior*, New York: Macmillan.

Smith, A. and Call, N. (2002) *The ALPS (Accelerated Learning in Primary School) Approach*, London: Accelerated Learning in Training and Education (ALITE).

Smith, A., Douglas, W., Campbell, J. and Topping, K. (2003) 'Cross-age peer tutoring in mathematics with 7- and 11-year-olds: influence on mathematical vocabulary, strategic dialogue and self-concept', *Educational Research*, 45 (3), 287–308.

Smith, R., and Standish, P. (eds). (1997) *Teaching Right and Wrong: Moral Education in the Balance*, Stoke on Trent: Trentham.

Snow, C.P. (1960) *The Two Cultures and the Scientific Revolution*, Cambridge: Cambridge University Press. (The Rede Lecture, 1959).

Somekh, B. and Davis, N. (eds) (1997) *Using Information Technology Effectively in Teaching and Learning*, London: Routledge.

Sousa, D. (1997) *How the Brain Learns*, Reston, VA: National Association of Secondary School Principals.

Sousa, D. A. (2006) *How the Brain Learns*, 3rd edn, Thousand Oaks, CA: Corwin Press.

Spear, M., Gould, K. and Lee, B. (2000) *Who Would Be a Teacher? A Review of Factors Motivating and Demotivating Prospective and Practising Teachers*, Slough: National Foundation for Educational Research (NFER).

Stenhouse, L. (1975) *An Introduction to Curriculum Research and Development*, London: Heinemann.

Stobart, G. (2008) *Testing Times: The Uses and Abuses of Assessment*, Abingdon: Routledge.

Stobart, G. and Gipps, C. (1997) *Assessment: A Teacher's Guide to the Issues*, London: Hodder and Stoughton.

Stokking, K., Leenders, F., De Jong, J. and Van Tartwijk, J. (2003) 'From student to teacher: reducing practice shock and early dropout in the teaching profession', *European Journal of Teacher Education*, 26 (3), 329–350.

Stones, E. (1992) *Quality Teaching: A Sample of Cases*, London: Routledge.

Strauss, R.S. (2000) 'Childhood Obesity and Self Esteem', *Pediatrics*, 105 (1), e15. Available online at: http://www.pediatrics.org/cgi/content/full/105/1/e15. Accessed 18.07.08.

Strauss, R. S. and Pollack, H. A. (2003) 'Social marginalisation of overweight children', *Archives of Pediatrics and Adolescent Medicine*, PubMed 12912779.

Sutton, C. (1981) *Communicating in the Classroom*, London: Hodder and Stoughton.

Tanner, J. M. (1990) *Foetus into Man,* Cambridge, MA: Harvard University Press.

TDA (Training and Development Agency for Schools) (2007a) *Professional Standards for Teachers; Qualified Teacher Status*. Available online at: http://www.tda.gov.uk/teachers/professional standards.aspx. Accessed 06.08.08.

TDA (2007b) *Guidelines for Professional Standards*. Available online at: http://www.tda.gov.uk/ teachers/professionalstandards/standards/attributes/relationships/core.aspx. Accessed 17.02.08.

TDA (2007c) *Supporting the Induction Process: TDA Guidance for Newly Qualified Teacher*. Available online at: http://www.tda.gov.uk/upload/resources/pdf/s/supporting_the_induction_process.pdf. Accessed 17.09.08.

TDA (2008a) *Numeracy Skills Test*. Available online at: http://www.tda.gov.uk/skillstests/numeracy.aspx. Accessed 04.09.08.

TDA (2008b) *Teacher Training Resource Bank*. Available online at: http://www.tda.gov.uk/partners/ quality/practiceresearch/ttrb.aspx. Accessed 09.08.08.

TeacherNet (2004) *Embedding ICT@ Secondary – Interactive Whiteboards*. Available online at: http://www.teachernet.gov.uk/wholeschool/ictis/infrastructure/iwb/. Accessed 01.01.08.

TeacherNet *Personalised Learning*. Available online at: http://www.teachernet.gov.uk/management/ newrelationship/personalisedlearning/. Accessed 17.09.08.

Telecsan, B. L., Slaton, D. B. and Stevens, K. B. (1999) 'Peer tutoring: teaching students with learning disabilities to deliver time delay instruction', *Journal of Behavioural Education*, 9 (2), 133–15.

Terry, P. M. (1997, April). 'Teacher burnout: Is it real? Can we prevent it?' Paper presented at the annual meeting of the North Central Association of Colleges and Schools, Chicago, IL. (ERIC Document Reproduction Service No. ED 408 258)

The Education Authorities Directory and Annual (annual) Redhill, Surrey: School Government Publishing Company Ltd.

The Education Year Book (annual) London: Longman.

Tilston, D.W. (2000) *10 Best Teaching Practices*, Thousand Oaks, CA: Corwin Press.

Times Educational Supplement (TES) (annual) *First Appointments Supplement*, TES.

Times Educational Supplement (TES) (2008) 'Gender/Race/Class: TES Series challenging education stereotypes', London: TES. Available online at: http://www.multiverse.ac.uk/ViewArticle2.aspx?anchorId=247&selectedId=298&contentId=14485. Accessed March 2009.

TLRP and ESRC (Teaching and Learning Research Project and Economic and Social Research Council) (2006) 'A commentary on the Teaching and Research Programme', in Howard-Jones, P. (ed.) Neuroscience in Education' *Issues and Opportunities*. Available online at: http://www.tlrp.org/pub/documents/Neuroscience%20Commentary%20FINAL.pdf. Accessed 11.09.08.

TLRP and ESRC (2007) *Principles into Practice: A Teaching Guide to Research Evidence on Teaching and Learning*, London: Institute of Education, University of London. Available online at: www.tlrp.org. Accessed 26.08.08.

Tod, J. and Powell, S. (2004) *A Systematic Review of How Theories Explain Learning Behaviour in School Contexts*, London: EPPI.

Tomlinson, C. (1999) *The Differentiated Classroom: Responding to the Needs of All Learners*, Alexandria, VA: Association for Supervision and Curriculum.

Tomlinson, S. (2005), 'Race, ethnicity and education under New Labour', *Oxford Review of Education*, 31 (1), 153–171.

Topping, K. J. (1992) 'The effectiveness of paired reading in ethnic minority homes', *Multicultural Teaching to Combat Racism in School and Community*, 10 (2), 19–23.

Torrance, H. (2002) 'Can testing really raise educational standards?' Inaugural professorial lecture, University of Sussex, 11 June.

Travers, C. J. and Cooper, C. L. (1996) *Teachers Under Pressure: Stress in the Teaching Profession*, London: Routledge.

Trend, R., Davis, N. and Loveless, A. (1999) *QTS Information Communication Technology*, London: LETTS Educational.

Triantafillou, E., Pomportsis, A., Demetriadis, S. and Georgiadou, E. (2004) 'The value of adaptivity based on cognitive style: an empirical study', *British Journal of Educational Technology*, 35, 1, 95–106.

Turner, S. (1995) 'Simulations', in J. Frost (ed.) *Teaching Science*, London: Woburn Press.

Twiselton, S. (2000) 'Seeing the wood for the trees: the national literacy strategy and initial teacher education; pedagogical content knowledge and the structures of subjects', *Cambridge Journal of Education*, 30 (3), 391–403.

Underwood, J. (1998) 'Making groups work', in M. Monteith (ed.) *IT for Learning Enhancement*, Exeter: Intellect.

Van Manen, M. (1991) *The Tact of Teaching*, Alberta: The Althouse Press.

Vygotsky, L. S. (1962) *Thought and Language*, Cambridge, MA: MIT Press.

Vygotsky, L. S. (1978) *Mind in Society: The Development of Higher Psychological Processes*, Cambridge, MA: Harvard University Press.

Vygotsky, L.S. (1986) *Thought and Language*, tr. and ed. A. Kozulin, Cambridge, MA: MIT Press.

Vygotsky, L. (1997) *Thought and Language*, tr. and ed. A. Kozulin, London: MIT Press.

Waldon, A. (2000) 'Bridging the Gap', *Journal of Design and Technology Education*, 6, 158–160

Walker, S. (2004) 'Interprofessional work in child and adolescent mental health services', *Emotional and Behavioural Difficulties*, 9 (1), 189–204.

Wallace, B. (2000) *Teaching the Very Able Child*, London: NACE/David Fulton.

Walsh, B. (2006) *Beyond Multiple Choice: e-help seminar*. Available online at: http://educationforum.ipbhost.com/index.php?showtopic=8508. Accessed 01.08.08.

Warburton, N. (2003) *Thinking from A-Z*, 2nd edn, London: RoutledgeFalmer.

Waterhouse, P. (1983) *Managing the Learning Process*, Maidenhead: McGraw-Hill Series for Teachers.

Watkins, C., Carnell, E., and Lodge, C. (2007) *Effective Learning in Classrooms*, London: Paul Chapman.

Watson. G. (2002) 'The National Curriculum', in S. Capel, M. Leask and T. Turner (eds) *Learning to Teach in the Secondary School: A Companion to School Experience*, 3rd edn, London: Routledge.

Weare, K. (2004) *Developing the Emotionally Literate School*. London: Paul Chapman Publishing.

Wegerif, R. and Dawes, L. (2004) *Thinking and Learning with ICT: Raising Achievement in Primary Classrooms*, London: Routledge.

Weiner, B. J. (1972) *Theories of Motivation*, Chicago, IL: Markham.

Wells, J., Barlow, J. and Stewart-Brown, S. (2003) 'A systematic review of universal approaches to mental health promotion in schools', *Health Education*, 103 (4), 197–200.

Welsh Assembly Government (2004) *Learning Pathways Through Statutory Assessment: Key Stages 2 and 3 (Daugherty Report)*. Cardiff, Wales: National Assembly and Department for Training and Education.

Wentzel, K. (1997) 'Student motivation in middle school: the role of perceived pedagogical caring', *Journal of Educational Psychology*, 89, 411–417.

West-Burnham, J. and Coates, M. (2007) *Personalizing Learning: Transforming Learning for Every Child,* London: Network Educational Press Ltd.

Westwood, P (2007) *Commonsense Methods for Children with Special Educational Needs*, 5th edn, London: RoutledgeFalmer.

White, J. (1998) *Do Howard Gardner's Multiple Intelligences Add Up?*, London: Institute of Education, University of London. In the series 'Perspectives on Education Policy'.

White, J. (2005) *Howard Gardner; The Myth of Multiple Intelligences*, London: The Institute of Education, University of London. Viewpoint No. 16 (October).

White, J. (2007) *What Schools Are For and Why*, London: Philosophy of Education Society of Great Britain (IMPACT series No. 14).

White, J. (ed.) (2004) *Rethinking the School Curriculum: Values, Aims and Purposes*, London: Routledge Falmer.

White, R. T. and Gunstone, R. (1992) *Probing Understanding*, London: Falmer Press.

Whylam, H. and Shayer, M. (1978) *CSMS Reasoning Tasks: General Guide*, Windsor: National Foundation for Educational Research Publishing Co.

Wilhelm, K., Dewhurst-Savellis, J. and Parker, G. (2000) 'Teacher stress? An analysis of why teachers leave and why they stay', *Teachers and Teaching: Theory and Practice*, 6, 291–304.

Wiliam, D and Bartholomew, H. (2004) 'It's not which school but which set you're in that matters: the influence on ability-grouping practices on student progress in mathematics', *British Educational Research Journal*, 30 (2), 279–294.

Wiliam, D. (2001) *Level Best? Levels of Attainment in National Curriculum Assessment*, London: ATL (Association of Teachers and Lecturers).

Wiliam, D. (2007) 'Scrap these 19th century GCSEs says expert', *Observer*, 19 August.

Willis, P. (1977) *Learning to Labour: How Working Class Kids Get Working Class Jobs*, Aldershot: Gower.

Wineburg, S. (1997) 'Beyond breadth and depth: subject matter knowledge and assessment', *Theory into Practice*, 36 (4), 255–261.

Wolfe, P. (2001) *Brain Matters: Translating Research into Classroom Practice*, Alexandria, VA: ASCD (Association for Supervision and Curriculum Development).

Wood, D. (1988) *How Children Think and Learn*, Oxford: Blackwell Press.

Woodward, W. (2003) 'Poverty hits exam scores', *The Guardian* newspaper (21.04.03).

Wragg, E. C. (1999) *An Introduction to Classroom Observation*, London: Routledge.

Wragg, E. C. (ed.) (1984) *Classroom Teaching Skills*, London: Croom Helm.

Wragg, E.C. (ed.) (2004) *The RoutledgeFalmer Reader in Teaching and Learning*, London: Routledge Falmer.

Wragg, E. C. and Brown, G. (2001) *Questioning in the Secondary School*, London: RoutledgeFalmer.

Younger, M., Warrington, M. and Williams, J. (1999) 'The gender gap and classroom interactions: reality and rhetoric?', *British Journal of Sociology of Education*, 20 (3), 325–341.

Zwozdiak-Myers, P (2007) 'The reflective practitioner', in P. Zwozdiak-Myers, *Childhood and Youth Studies*. Exeter: Learning Matters: 160–168

Zwozdiak-Myers, P. (2009) 'An enquiry into how reflective practice influenced the professional development of teachers as they engaged in action research to study their own teaching', London: Brunel University. PhD thesis.

AUTHOR INDEX

AUTHOR INDEX ■ ■ ■ ■

SUBJECT INDEX